Practical Stress Analysis in Engineering Design

Third Edition

MECHANICAL ENGINEERING
A Series of Textbooks and Reference Books

Founding Editor

L. L. Faulkner

*Columbus Division, Battelle Memorial Institute
and Department of Mechanical Engineering
The Ohio State University
Columbus, Ohio*

Practical Stress Analysis in Engineering Design

Third Edition

Ronald Huston

Harold Josephs

CRC Press
Taylor & Francis Group
Boca Raton   London   New York

CRC Press is an imprint of the
Taylor & Francis Group, an **informa** business

CRC Press
Taylor & Francis Group
6000 Broken Sound Parkway NW, Suite 300
Boca Raton, FL 33487-2742

© 2009 by Taylor & Francis Group, LLC
CRC Press is an imprint of Taylor & Francis Group, an Informa business

No claim to original U.S. Government works
Printed in the United States of America on acid-free paper
10 9 8 7 6 5 4 3 2 1

International Standard Book Number-13: 978-1-57444-713-2 (Hardcover)

Library of Congress Cataloging-in-Publication Data

Huston, Ronald L., 1937-
    Practical stress analysis in engineering design / Ronald Huston and Harold Josephs. -- 3rd ed.
        p. cm. -- (Dekker mechanical engineering)
    Prev. ed. authored by Alexander Blake.
    Includes bibliographical references and index.
    ISBN 978-1-57444-713-2 (alk. paper)
    1. Strains and stresses. 2. Engineering design. I. Josephs, Harold. II. Blake, Alexander. Practical stress analysis in engineering design. III. Title.

TA648.3.B57 2009
624.1'76--dc22   2008029335

Visit the Taylor & Francis Web site at
http://www.taylorandfrancis.com

and the CRC Press Web site at
http://www.crcpress.com

Contents

PART II Straight and Long Structural Components: Beams, Rods, and Bars

PART III   Special Beam Geometries: Thick Beams, Curved Beams, Stability, and Shear Center

PART IV   Plates, Panels, Flanges, and Brackets

PART V   Dynamic Loadings, Fatigue, and Fracture

PART VI   Piping and Pressure Vessels

PART VII   Advanced and Specialized Problems

Preface

Industrialists, marketing leaders, military planners, and space scientists are continually asking their engineers and designers to produce new designs for all kinds of mechanical systems. Designs that are simultaneously workable, reliable, long-lived, easy to manufacture, safe, and economical are envisioned. Often, system components are required to be concurrently light in weight, strong, and yet fatigue-resistant. At the same time, engineers and designers are being pressed to produce these designs in ever-shortening time intervals. Consequently, they have to quickly produce analyses that are accurate, or if inaccurate, they have to make sure they err on the safe side.

In response to these demands, engineers and designers are increasingly relying upon finite element methods (FEM) and analogous computational procedures for their designs. However, these methods are primarily methods of analysis and are thus most useful for evaluating proposed designs. Moreover, they are often expensive, inaccessible, and sensitive to element selection and assumptions on loadings and support conditions. In short, they are not always free of error. Even with steady improvements in FEM accuracy, accessibility, and ease of use, engineers and designers still need to be able to readily make accurate stress and deformation analyses without undue computation. Recognizing this need, Alexander Blake published his widely used *Practical Stress Analysis* in 1982, just when FEM and related methods were becoming popular.

In this third edition of *Practical Stress Analysis in Engineering Design*, we have completely rewritten and updated the text of the second edition while maintaining Blake's popular style. Our objective is to produce a book to help engineers and designers easily obtain stress and deformation results for the wide class of common mechanical components. In addition, we have attempted to supplement the methodologies with a presentation of theoretical bases. At the end of each chapter, a list of references is provided for a more detailed investigation and also a list of symbols is presented to aid the reader.

This book is divided into seven parts and consists of 40 chapters. In the first part, we review fundamental concepts including basic ideas such as stress, strain, and Hooke's law. We include analysis in two and three dimensions as well as the use of curvilinear coordinates.

In the second part, we review the fundamental concepts of beam bending and twisting of rods. We introduce the use of singularity functions for analysis of complex loadings. These two parts provide the basis for the topics in the remainder of the book. Curvilinear coordinates and singularity functions are two new topics in this edition.

The third part considers special beam geometries focusing upon thick beams, shear stress in beams, curved beams, buckling of beams, and shear centers. In the fourth part, we extend the analysis to plates, panels, flanges, and brackets. We review the fundamentals of plate bending and then apply the theory to special plate configurations with a focus on circular and annular plates, flanges and brackets, panels, and perforated/reinforced plates.

The fifth part is devoted to dynamic effects including the concepts of fracture and fatigue failure. We consider design for seismic loading and impacts and explore stress propagation. We conclude this part with design concepts to control and prevent fatigue and fracture for systems with repeated and periodic loadings.

The sixth part discusses piping and various pressure vessel problems and considers both internal and external pressurized vessels. Bending, buckling, and other vessel responses to high pressure are evaluated. The part concludes with a consideration of some designs for stiffening of cylindrical vessels. The seventh part considers some advanced and specialized topics including stress concentrations, thermal effects, rings, arches, links, eyebars, and springs.

We are grateful for the opportunity provided by CRC Press to revise and update Blake's outstanding writings. We are deeply appreciative of their patience and encouragement. We especially thank Charlotte Better for meticulously typing and preparing the manuscript.

Authors

Ronald L. Huston is a professor emeritus of mechanics and distinguished research professor in the mechanical engineering department at the University of Cincinnati. He is also a Herman Schneider Chair professor. Dr. Huston has been a member of the faculty of the University of Cincinnati since 1962. During that time he was the head of the Department of Engineering Analysis, an interim head of Chemical and Materials Engineering, the director of the Institute for Applied Interdisciplinary Research, and an acting senior vice president and provost. He has also served as a secondary faculty member in the Department of Biomedical Engineering and as an adjunct professor of orthopedic surgery research.

In 1978, Dr. Huston was a visiting professor in applied mechanics at Stanford University. During 1979–1980, he was the division director of civil and mechanical engineering at the National Science Foundation. From 1990 to 1996, he was a director of the Monarch Foundation.

Dr. Huston has authored over 150 journal articles, 150 conference papers, 5 books, and 75 book reviews. He has served as a technical editor of *Applied Mechanics Reviews*, an associate editor of the *Journal of Applied Mechanics*, and a book review editor of the *International Journal of Industrial Engineering*.

Dr. Huston is an active consultant in safety, biomechanics, and accident reconstruction. His research interests are in multibody dynamics, human factors, biomechanics, and sport mechanics.

Harold Josephs has been a professor in the Department of Mechanical Engineering at the Lawrence Technological University in Southfield, Michigan since 1984, after a stint in the industry working for General Electric and Ford Motor Company. Dr. Josephs is the author of numerous publications, holds nine patents, and has presented numerous seminars to industry in the fields of safety, bolting, and joining. He maintains an active consultant practice in safety, ergonomics, and accident reconstruction. His research interests are in fastening and joining, human factors, ergonomics, and safety.

Dr. Josephs received his BS from the University of Pennsylvania, his MS from Villanova University, and his PhD from the Union Institute. He is a licensed professional engineer, a certified safety professional, a certified professional ergonomist, a certified quality engineer, a fellow of the Michigan Society of Engineers, and a fellow of the National Academy of Forensic Engineers.

Part I

Fundamental Relations and Concepts

Our objective in this first and introductory part of the book is to provide a review of elementary force, stress, and strain concepts, which are useful in studying the integrity of structural members. The topics selected are those believed to be most important in design decisions. A clear understanding of these concepts is essential due to the ever increasing safety and economic considerations associated with structural design.

The integrity of a structure, or of a structural component, depends upon its response to loading, that is, to the induced stress. This response, measured as deformation, or strain, and life, depends upon geometric design and material characteristics. For example, the shaft of a machine may be required to sustain twisting and bending loads simultaneously for millions of revolutions while keeping transverse deflections within a preassigned tolerance; or a pipe flange bolt simultaneously subject to axial, transverse, thermal, and dynamic loadings, may be required to maintain a seal under high and varying pressure.

It is obvious that for many structural configurations, there is a complex arrangement of interacting structural components and loading conditions. Under such conditions, the task of obtaining accurate and detailed stress analyses is usually difficult, time consuming, and subject to intense scrutiny. Fortunately, simple and fundamental stress formulas can often provide insight into the validity of complex analyses and thus also the suitability of proposed designs. Therefore, in this first part of the book, we redirect our attention to the fundamental concepts of force, stress, deformation, strain, and stress–strain relations.

1 Forces and Force Systems

1.1 CONCEPT OF A FORCE

Intuitively, a "force" is a "push or a pull." The effect, or consequence, of a force thus depends upon (i) how "hard" or how large the push or pull is (the force "magnitude"); (ii) the place or point of application of the push or pull; and (iii) the direction of the push or pull. The magnitude, point of application, and direction form the "characteristics" or defining aspects of a force. With these characteristics, force is conveniently represented by vectors.

Figure 1.1 depicts a force **F** (written in bold face to designate it as a vector). The figure shows **F** to be acting along a line L which passes through a point P. In this context, L is called the "line of action" of **F**. **F** may be thought of as acting at any place along L. Thus, a force **F** is sometimes thought of as a "sliding vector."

1.2 CONCEPT OF A MOMENT

Intuitively, a "moment" is like a "twisting" or a "turning." The twisting or turning is usually about a point or a line. Alternatively, a moment is often thought of as a product of a force and a distance from a point or a line. A more precise definition may be obtained by referring to Figure 1.2 where **F** is a force acting along a line L and O is a point about which **F** has a moment. Let **p** be a position vector locating a typical point P of L relative to O. Then the moment of **F** about O is defined as

$$\mathbf{M}_O \overset{\mathrm{D}}{=} \mathbf{p} \times \mathbf{F} \tag{1.1}$$

Observe in the definition of Equation 1.1 that the position vector **p** is not necessarily perpendicular to L or **F**. Indeed, **p** is arbitrary in that it can be directed from O to *any* point on L. It is readily seen, however, that the result of the vector product in Equation 1.1 is *independent* of the choice of point P on L. For, if Q is another point on L as in Figure 1.3, and if position vector **q** locates Q relative to O, then $\mathbf{M}_O$ is seen to be

$$\mathbf{M}_O = \mathbf{q} \times \mathbf{F} \tag{1.2}$$

The consistency of Equations 1.1 and 1.2 is arrived at by expressing **q** as

$$\mathbf{q} = \mathbf{p} + \mathbf{PQ} \tag{1.3}$$

where, as suggested by the notation, **PQ** is the position vector locating Q relative to P. By substituting from Equation 1.3 into Equation 1.2 we have

$$\mathbf{M}_O = (\mathbf{p} + \mathbf{PQ}) \times \mathbf{F} = \mathbf{p} \times \mathbf{F} + \mathbf{PQ} \times \mathbf{F} = \mathbf{p} \times \mathbf{F} \tag{1.4}$$

where $\mathbf{PQ} \times \mathbf{F}$ is zero since **PQ** is parallel to **F** [1].

Observe further that if the line of action of the force **F** passes through a point O, then $\mathbf{M}_O$ is zero. Consequently if the line of action of **F** is "close" to O, then the magnitude of $\mathbf{M}_O$ is small.

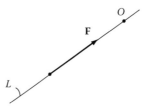

FIGURE 1.1   A force **F**, line of action *L*, and point *P*.

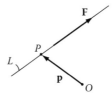

FIGURE 1.2   A force **F**, line of action *L*, point *O*, and a position vector from *O* to a point *P* on *L*.

1.3   MOMENT OF A FORCE ABOUT A LINE

The moment of a force about a point is a vector. The moment of a force about a line is the projection, or component, along the line of the moment of the force about a point on the line. If **F** is a force, *O* is a point, and *L* is a line through *O* as in Figure 1.4, then the moment of **F** about *L*, $\mathbf{M}_L$, is defined as

$$\mathbf{M}_L = (\mathbf{M}_O \cdot \boldsymbol{\lambda})\boldsymbol{\lambda} = [(\mathbf{p} \times \mathbf{F}) \cdot \boldsymbol{\lambda}]\boldsymbol{\lambda} \tag{1.5}$$

where
   $\boldsymbol{\lambda}$ is a unit vector parallel to *L*
   $\mathbf{p}$ is a position vector from *O* to a point on the line of action of **F**

1.4   FORCE SYSTEMS

A force system is simply a collection or set *S* of forces as represented in Figure 1.5. If the system has a large number (say *N*) of forces, it is usually convenient to label the forces by a subscript index as: $\mathbf{F}_1, \mathbf{F}_2, \ldots, \mathbf{F}_N$, or simply $\mathbf{F}_i$ ($i = 1, \ldots, N$) as in Figure 1.6.

A force system is generally categorized by two vectors: (1) the resultant of the system and (2) the moment of the system about some point *O*. The resultant **R** of a force system is simply the sum of the individual forces. That is,

$$\mathbf{R} = \sum_{i=1}^{N} \mathbf{F}_i \tag{1.6}$$

The resultant is a free vector and is not associated with any particular point or line of action.

Correspondingly, the moment of a force system *S* about some point *O* is simply the sum of the moments of the individual forces of *S* about *O*. That is,

$$\mathbf{M}_O^S = \sum_{i=1}^{N} \mathbf{P}_i \times \mathbf{F}_i \tag{1.7}$$

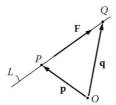

FIGURE 1.3   Points P and Q on the line of action of force **F**.

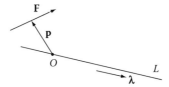

FIGURE 1.4   A force **F** and a line L.

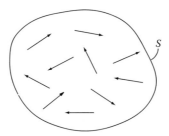

FIGURE 1.5   A force system.

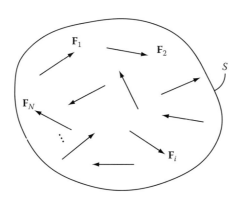

FIGURE 1.6   An indexed set of forces.

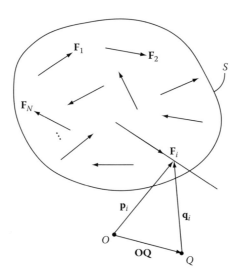

FIGURE 1.7   A force system S and points O and Q.

where $\mathbf{P}_i$ is a position vector from O to a point on the line of action of $\mathbf{F}_i$ ($i = 1, \ldots, N$) as represented in Figure 1.7.

The point O is arbitrary and is usually chosen as a convenient reference point. "Convenient," however, is subjective, and after computing $\mathbf{M}_O^S$ as in Equation 1.7 we may be interested in knowing the moment of S about some other point, say Q. If S contains a large number of forces, the computation in Equation 1.7 could be quite tedious and thus the additional computation for a point Q may not be a welcome task. Fortunately, if $\mathbf{M}_O^S$ and the resultant $\mathbf{R}$ of S are known, we can determine the moment about some point Q *without* doing the potentially tedious computation associated with Equation 1.7. $\mathbf{M}_O^S$ may be expressed in terms of $\mathbf{M}_O^S$ by the simple relation:

$$\mathbf{M}_O^S = \mathbf{M}_Q^S + \mathbf{OQ} \times \mathbf{R} \tag{1.8}$$

The validity of Equation 1.8 is readily established by deriving from Equation 1.7 that $\mathbf{M}_O^S$ and $\mathbf{M}_Q^S$ are

$$\mathbf{M}_O^S = \sum_{i=1}^{N} \mathbf{P}_i \times \mathbf{F}_i \quad \text{and} \quad \mathbf{M}_Q^S = \sum_{i=1}^{N} \mathbf{q}_i \times \mathbf{F}_i \tag{1.9}$$

where, from Figure 1.7, $\mathbf{q}_i$ is the position vector from Q to a point on the line of action of $\mathbf{F}_i$. Also, from Figure 1.7 we see that $\mathbf{p}_i$ and $\mathbf{q}_i$ are related by the connecting position vector $\mathbf{OQ}$. That is,

$$\mathbf{p}_i = \mathbf{OQ} + \mathbf{q}_i \tag{1.10}$$

By substituting from Equation 1.10 in Equation 1.9, we have

$$\mathbf{M}_O^S = \sum_{i=1}^{N} (\mathbf{OQ} + q_i) \times \mathbf{F}_i = \sum_{i=1}^{N} \mathbf{OQ} \times \mathbf{F}_i + \sum_{i=1}^{N} q_i \times \mathbf{F}_i$$

$$= \mathbf{OQ} \times \sum_{i=1}^{N} \mathbf{F}_i + \mathbf{M}_Q^S = \mathbf{OQ} \times \mathbf{R} + \mathbf{M}_Q^S \tag{1.11}$$

1.5   SPECIAL FORCE SYSTEMS

There are several force systems that are useful in stress analyses. These are reviewed in the following sections.

1.5.1   Zero Force Systems

If a force system has a zero resultant and a zero moment about some point, it is called a "zero system." Zero systems form the basis for static analyses.

   Interestingly, if a force system has a zero resultant and a zero moment about *some* point, it then has a zero moment about *all* points. This is an immediate consequence of Equation 1.8. That is, if the resultant $\mathbf{R}$ is zero and if $\mathbf{M}_O$ is zero for some point O, then Equation 1.8 shows that $\mathbf{M}_Q$ is zero for any point Q.

1.5.2   Couples

If a force system has a zero resultant but a nonzero moment about some point O, it is called a "couple." Equation 1.8 shows that a couple has the same moment about *all* points: for, if the resultant $\mathbf{R}$ is zero, then $\mathbf{M}_Q = \mathbf{M}_O$ for any point Q. This moment, which is the same about all points, is called the "torque" of the couple.

   Figure 1.8 depicts an example of a couple. This couple has many forces. If, alternatively, a couple has only two forces, as in Figure 1.9, it is called a "simple couple."

   To satisfy the definition of a couple, the forces of a simple couple must have equal magnitude but opposite directions.

1.5.3   Equivalent Force Systems

Two force systems S_1 and S_2 are said to be "equivalent" if they have (1) equal resultants and (2) equal moments about some point O. Consider two force systems S_1 and S_2 as represented in Figure 1.10 with resultants $\mathbf{R}_1$ and $\mathbf{R}_2$ and moments $\mathbf{M}_O^{S_1}$ and $\mathbf{M}_O^{S_2}$ about some point O. Then, S_1 and S_2 are equivalent if

$$\mathbf{R}_1 = \mathbf{R}_2 \tag{1.12}$$

and

$$\mathbf{M}_O^{S_1} = \mathbf{M}_O^{S_2} \tag{1.13}$$

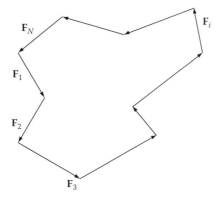

FIGURE 1.8   A couple with many forces.

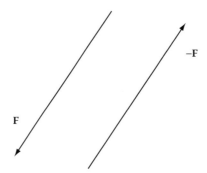

FIGURE 1.9   A simple couple.

It happens that if Equations 1.13 and 1.14 are satisfied, then the moments of S_1 and S_2 about any and all points Q are equal. This is derived by using Equation 1.8 to express the moments of S_1 and S_2 as

$$\mathbf{M}_O^{S_1} = \mathbf{M}_Q^{S_2} + \mathbf{OQ} \times \mathbf{R}_1 \tag{1.14}$$

$$\mathbf{M}_O^{S_2} = \mathbf{M}_Q^{S_2} + \mathbf{OQ} \times \mathbf{R}_2 \tag{1.15}$$

By subtracting these expressions and using Equations 1.12 and 1.13, we have

$$0 = \mathbf{M}_Q^{S_1} - \mathbf{M}_Q^{S_2} \quad \text{or} \quad \mathbf{M}_Q^{S_1} = \mathbf{M}_Q^{S_2} \tag{1.16}$$

For a rigid body, equivalent force systems may be interchanged without affecting either the statics or the dynamics of the body. Thus, if one force system, say S_1, has significantly fewer forces than an equivalent force system S_2, then S_1 will generally call for a simpler analysis.

For a deformable body, however (such as the bodies and structural components considered in this book), equivalent force systems *cannot* be interchanged without changing the stress distribution and deformation of the body.

Consider, for example, two identical bars B_1 and B_2 subjected to equivalent force systems as in Figure 1.11. Each force system is a zero system. The force systems are thus equivalent. Their

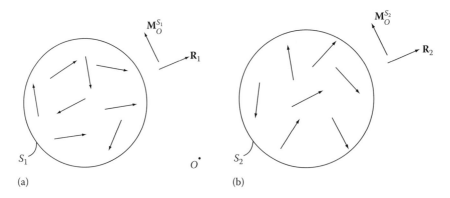

(a)   (b)

FIGURE 1.10   Two force systems.

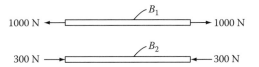

FIGURE 1.11   Identical deformable bars subjected to equivalent but different force systems.

effects on the deformable bars, however, are dramatically different. In the first instance, the bar is in tension and is elongated. In the second, the bar is in compression and is shortened.

This example then raises the question: What is the value, if any, of equivalent force systems for deformable bodies? The answer is provided by Saint Venant's principle [2] as illustrated by the following example: consider two identical cantilever beams subjected to equivalent end loadings as represented in Figure 1.12. St. Venant's principle states that in the region of the beam near the end loading, the stresses and strains are different for the two loadings. However, in regions of the beam far away from the loading, the stresses and strains are the same. This then raises another question: How far from the load is there negligible difference between the stresses and strains for the equivalent loading conditions? Unfortunately, the answer here is not so precise, but what is clear is that the further away a region is from the loading, the more nearly equal are the stresses and strains. For practical purposes, in this example, there will generally be negligible differences in the stresses and strains for the two loadings, when the region is an "order of magnitude" of thickness away from the loading, that is, a distance of $10h$ away where h is the beam thickness.

1.5.4 Equivalent Replacement by a Force and a Couple

Consider any force system S. No matter how large (or small) S is, there exists an equivalent force system S^* consisting of a single force passing through an arbitrary point, together with a couple. To understand this, consider Figure 1.13a, which represents an arbitrary force system S. Let $\mathbf{R}$ be the resultant of S and $\mathbf{M}_O$ be the moment of S about some point O. Let there be a proposed equivalent force system S^* as shown in Figure 1.13b. Let S^* consist of a force $\mathbf{F}$, with line of action passing through O, together with a couple with torque $\mathbf{T}$. Let $\mathbf{F}$ and $\mathbf{T}$ be

$$\mathbf{F} = \mathbf{R} \quad \text{and} \quad \mathbf{T} = \mathbf{M}_O \tag{1.17}$$

We readily, see that S and S^* are equivalent: That is, they have equal resultants and equal moments about O. ($\mathbf{F}$ has no moment about O.)

Observe in Equation 1.17 that the magnitude of $\mathbf{T}$ depends upon the location of O. If O is a point selected within or near S, and if all forces of S have lines of action that are close to O, then the magnitude of $\mathbf{T}$ is small.

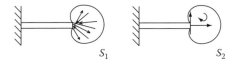

FIGURE 1.12   Identical cantilever beams with equivalent end loadings.

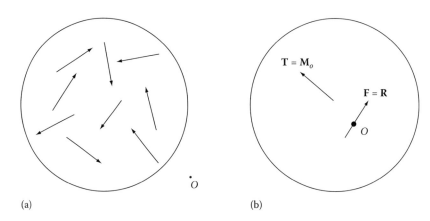

FIGURE 1.13   A given force system S (a) and an equivalent force system S^* (b).

SYMBOLS

B_1, B_2	Bars
$\mathbf{F}$	Force
$\mathbf{F}_i \; (i = 1, \ldots, N)$	Series of N forces
L	Line
$\mathbf{M}_L$	Moment about a line L
$\mathbf{M}_O$	Moment about O
$\mathbf{M}_O^S$	Moment of system S about O
$\mathbf{M}_Q$	Moment about Q
$\mathbf{M}_Q^S$	Moment of system S about Q
O	Point
P	Point
$\mathbf{P}$	Position vector
$\mathbf{PQ}$	Position vector from P to Q
$\mathbf{p}_i \; (i = 1, \ldots, N)$	Position vector from O to force $\mathbf{F}_i$
$\mathbf{q}$	Position vector
Q	Point
$\mathbf{q}_i \; (i = 1, \ldots, N)$	Position vector from Q to force $\mathbf{F}_i$
$\mathbf{R}, \mathbf{R}_1, \mathbf{R}_2$	Resultants
S, S_1, S_2, S^*	Force systems
$\mathbf{T}$	Torque of couple
$\boldsymbol{\lambda}$	Unit vector parallel to L

REFERENCES

1. L. Brand, *Vector and Tensor Analysis*, Wiley, New York, 1947 (chap. 1).
2. I. S. Sokolnikoff, *Mathematical Theory of Elasticity*, McGraw Hill, New York, 1956, p. 89.

2 Simple Stress and Strain: Simple Shear Stress and Strain

2.1 CONCEPT OF STRESS

Conceptually, "stress" is an "area-averaged" or "normalized" force. The averaging is obtained by dividing the force by the area over which the force is regarded to be acting. The concept is illustrated by considering a rod stretched (axially) by a force P as in Figure 2.1. If the rod has a cross-section area A, the "stress" σ in the rod is simply

$$\sigma = P/A \tag{2.1}$$

There are significant simplifications and assumptions made in the development of Equation 2.1: First, recall in Chapter 1, we described a force as a "push" or a "pull" and characterized it mathematically as a "sliding vector" acting through a point. Since points do not have area, there is no "area of application." Suppose that a body B is subjected to a force system S as in Figure 2.2, where S is applied over a relatively small surface region R of B. Specifically, let the forces of S be applied through points of R. Let $\mathbf{F}$ be the resultant of S and let A be the area of R. Then a "stress vector" $\boldsymbol{\sigma}$ may be defined as

$$\boldsymbol{\sigma} = \mathbf{F}/A \tag{2.2}$$

If R is regarded as "small," the area A of R will also be small, as will be the magnitude of $\mathbf{F}$. Nevertheless, the ratio in Equation 2.2 will not necessarily be small. If Q is a point within R, then the stress vector at Q ("point stress vector") $\boldsymbol{\sigma}^Q$ be defined as

$$\boldsymbol{\sigma}^Q = \lim_{A \to 0} \mathbf{F}/A \tag{2.3}$$

The components of the stress vector $\boldsymbol{\sigma}^Q$ are then regarded as stresses at Q, that is, "point stresses." If the resultant force $\mathbf{F}$ in Equation 2.3 is assigned to pass through Q, then the couple torque of the equivalent force system is negligible (see Section 1.5).

Next, referring again to Equation 2.1, the "stress" in the rod is thus an average stress at the points of the cross section of the rod. That is, there is the implied assumption that the stress is the same at all points of the cross section, and that the corresponding stress vectors are directed along the axis of the rod. For a long, slender rod, at cross sections away from the ends, these assumptions are intuitively seen to be reasonable and they can be validated both mathematically and experimentally.

If the rod of Figure 2.1 is deformable, the forces P will tend to elongate the rod. The rod is then regarded as being in "tension" and the corresponding stress is a "tension" or "tensile" stress.

On the contrary, if the rod is being compressed or shortened by forces P as in Figure 2.3 the "stress" in the rod is again P/A, but this time it is called a "compressive stress" or "pressure."

Tensile stress is customarily considered positive while compressive stress is negative.

FIGURE 2.1 A rod subject to a stretching (tensile) force.

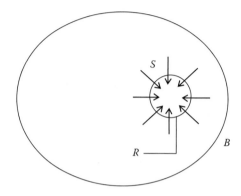

FIGURE 2.2 A body subjected to a force system.

FIGURE 2.3 A rod subjected to a compression force.

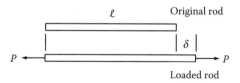

FIGURE 2.4 A rod being elongated by end forces P.

FIGURE 2.5 A rod being shortened by end forces P.

Equation 2.1 shows that the dimensions of stress are force per area (length squared). In the English system, stress is usually measured in pounds per square inch (lb/in.^{2}) or (psi) and in the International System (SI) in Newtons per square meter (N/m^2) or Pascals (Pa). The conversion between these systems is

$$1 \text{ psi} = 6894.76095 \text{ Pa} \qquad (2.4)$$

and

$$1 \text{ Pa} = 0.000145 \text{ psi} \qquad (2.5)$$

2.2 CONCEPT OF STRAIN

Conceptually, "strain" is an average elongation, shortening, deformation, or distortion due to applied forces (or "loading"). The averaging is obtained by dividing the amount of elongation, shortening, deformation, or distortion by an appropriate underlying length. This concept may be illustrated by again considering a rod being stretched, or elongated, by a force P as in Figure 2.4. If ℓ is the length of the unstretched and unloaded rod and if $\ell + \delta$ is the length of the elongated rod, then the average strain ε is defined as the elongation δ divided by the original length ℓ. That is,

$$\varepsilon = \delta/\ell \qquad (2.6)$$

With the rod being elongated, this strain is sometimes called "tensile strain."

On the contrary, if the rod is being compressed or shortened by compressive forces as in Figure 2.5, the average strain is the amount of shortening δ divided by the original length ℓ. When the rod is being shortened, the strain is sometimes called "compressive strain." Compressive strain is customarily considered negative while tensile strain is positive.

Observe from Equation 2.6 that unlike stress, strain is a dimensionless quantity.

2.3 SHEAR STRESS

When the force is directed normal (or perpendicular) to the region (or area) of interest (as in Section 2.1), the stress on the area is called "normal stress" or "simple stress" and the resulting strain is called "normal strain" or "simple strain." If, however, the force is directed tangent (or parallel) to the cross section, it is called a "shear force" and the corresponding stress is called a "shear stress." Figure 2.6 illustrates this concept, where V is a shear (or "shearing") force exerted on a block B.

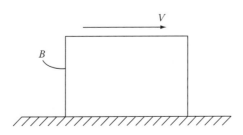

FIGURE 2.6   Block B subjected to a shearing force.

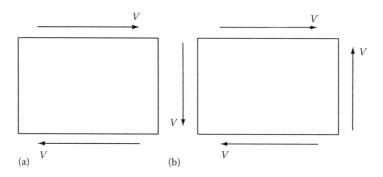

FIGURE 2.7   A block subjected to shearing forces. (a) Block with shearing forces. (b) Block in equilibrium.

The shear stress τ is then defined as

$$\tau = V/A \tag{2.7}$$

where A is the area over which V is acting.

Observe in Figure 2.6 that if we consider a free-body diagram of B, we see that unless there are vertical forces at the support base, the block will not be in equilibrium. That is, if block B is acted upon only by shear forces as in Figure 2.7a, then B is not in equilibrium and will tend to rotate. Thus, to maintain equilibrium, shearing forces with equal magnitudes and opposite directions must be applied, as in Figure 2.7b.

From Figure 2.7b, we note that shearing forces tend to distort the geometry. That is, a square will tend to become diamond in shape. This is discussed in the following section.

Finally, shearing of a block as in Figure 2.7b is called "simple shear" and the resulting stress, "simple shear stress."

2.4   SHEAR STRAIN

Consider a block with height h subjected to a shearing force V as in Figure 2.8. As the block yields to the force and is deformed, the block will have the shape shown (exaggerated) in Figure 2.9, where δ is the displacement of the top edge of the block in the direction of the shearing force. The shear strain γ is then defined as

$$\gamma = \delta/h \tag{2.8}$$

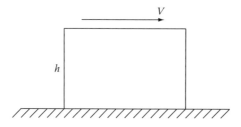

FIGURE 2.8   Block subjected to a shearing force.

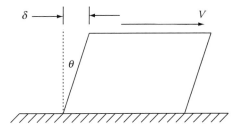

FIGURE 2.9 Block deformed by shearing force.

Observe by comparing Figures 2.8 and 2.9 that if the height h of the block is unchanged during the deformation (a reasonable assumption for small displacement δ), then from Equation 2.8 the shear strain γ may also be expressed as

$$\gamma = \tan\theta \tag{2.9}$$

where θ is the distortion angle shown in Figure 2.9.

Observe further that if δ is small compared with h (as is virtually always the case with elastic structural materials), then $\tan\theta$ is approximately equal to θ and we have the relation:

$$\gamma = \theta = \delta/h \tag{2.10}$$

Finally, observe the similarity in the form of Equations 2.10 and 2.6 for the shear strain γ and the normal strain ε respectively. The shear strain of Equation 2.10 is sometimes called "simple shear strain" or "engineering shear strain."

Referring again to Figure 2.7b, we see that Equation 2.10 may be interpreted as a measure of the distortion of the rectangular block into a parallelogram or diamond shape as illustrated in Figure 2.10. The shear strain is a measure of the distortion of the right angles of the block away from 90°.

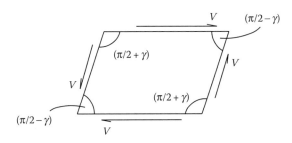

FIGURE 2.10 Distorted block due to shearing forces and shear strain interpretation.

SYMBOLS

A   Area
B   Body
F   Resultant of force system
h   Height
ℓ Length
P   Axial force ("push" or "pull")
Q   Point
R   Surface region
S   Force system
V   Shear force
γ Shear strain
δ Elongation, shortening, displacement
ε Strain, normal strain
θ Distortion angle
σ Stress, normal stress
σ Stress vector
τ Shear stress

3 Hooke's Law and Material Strength

3.1 HOOKE'S LAW IN ONE DIMENSION

A simple statement of Hooke's law is that: "the force is proportional to the displacement" or alternatively (and equivalent) "the stress is proportional to the strain."

As an illustration of this concept, consider a bar or rod being extended by axial loads as in Figure 3.1. If the magnitude of the load is P and the rod length is extended by an amount δ, then Hooke's law may be given as

$$P = k\delta \quad \text{or} \quad \delta = P/k \tag{3.1}$$

where k is a constant.

If the rod of Figure 3.1 has an initial length ℓ and a cross-section area A, then the stress σ in the rod is P/A and the strain ε is δ/ℓ (see Equations 2.1 and 2.6). Thus, P and δ may be expressed in terms of the stress and strain as

$$P = \sigma A \quad \text{and} \quad \delta = \varepsilon\ell \tag{3.2}$$

Then by substituting into Equation 3.1 we have

$$\sigma A = k\varepsilon\ell \quad \text{or} \quad \sigma = (k\ell/A)\varepsilon = E\varepsilon \tag{3.3}$$

and

$$\delta = \sigma A/k = P\ell/AE \tag{3.4}$$

where E is defined as

$$E \overset{D}{=} k\ell/A \tag{3.5}$$

Then

$$k = AE/\ell \tag{3.6}$$

E is commonly referred to as the "modulus of elasticity" or "Young's modulus."

Hooke's law also implies that the rod responds similarly in compression. Consider again the rod of Figure 3.1 subjected to a compressive load P (a "push" instead of a "pull") as in Figure 3.2. If the rod length is shortened by an amount δ, the relation between P and δ is again

$$P = k\delta \tag{3.7}$$

Then, as before, we have the relations

$$\sigma = (k\ell/A)\varepsilon = E\varepsilon \tag{3.8}$$

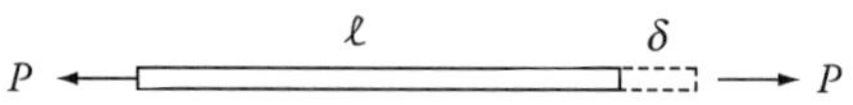

FIGURE 3.1   Rod extended by axial loads.

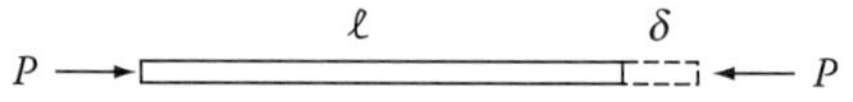

FIGURE 3.2   Rod shortened by axial loads.

and

$$k = AE/\ell \tag{3.9}$$

3.2   LIMITATIONS OF PROPORTIONALITY

It happens that Equations 3.4 and 3.8 are only *approximate* representations of structural material behavior. Nevertheless, for a wide range of forces (or loads), the expressions provide reasonable and useful results. When the loads are very large, however, the linearity of Equations 3.4 and 3.8 is no longer representative of structural material behavior. Unfortunately, a nonlinear analysis is significantly more involved. Indeed, for the range of forces that can be sustained by structural material (such as steel) the stress and strain are typically related as in Figure 3.3.

If the force is large enough to load the material of the rod beyond the proportional limit, the linear relation between the stress and strain is lost. If the material is loaded beyond the yield point, there will be permanent (or "plastic") deformation. That is, when the loading is removed from a rod stressed beyond the yielding point, it does not return to its original length, but instead shows a residual deformation. Alternatively, when the loading is relatively low such that the proportional limit between the stress and strain is not exceeded, the loading is said to be in the "elastic" range.

In many instances, it is difficult to know where precisely the yield point is. In actuality, the apparent linear relation (or line) below the yield point (see Figure 3.3) is a slight curve. In such cases, the limit of proportionality is often arbitrarily defined as the stress where the residual strain is 0.002 (0.2%), as depicted in Figure 3.4. From a design perspective, however, it is recommended that the loads be kept sufficiently small so that the stress remains in the elastic range, well below the yield point. The material is then unlikely to fail and there is the added benefit of a simpler analysis since the relation between the stress and strain is linear, as in Equations 3.4 and 3.8.

The value of the elastic modulus E of Equations 3.4 and 3.8 is dependent upon the material. Table 3.1 provides a tabular listing of approximate elastic modulus values for some commonly used materials [1,2]. But, a note of caution should be added: The values listed are for pure materials (without defects). Actual materials in use may have slightly lower values due to imperfections occurring during manufacture.

3.3   MATERIAL STRENGTH

The "strength" of a material is an ambiguous term in that "strength" can refer to any of the three concepts: (1) yield strength; (2) maximum tensile (or compressive) strength; or (3) breaking (fracture or rupture) strength. These are, however, relatively simple concepts. To illustrate them, consider a bar, or rod, being stretched by axial forces as in Figure 3.5.

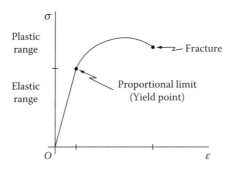

FIGURE 3.3 Stress–strain relation.

If we construct a graph relating the stress and strain, as in Figures 3.3, 3.4, and 3.6, we can identify points on the curve with these three strength concepts. Specifically, the "yield strength" is the stress at which yielding, or alternatively, 0.2% strain occurs (see Figure 3.4). This is also the beginning of plastic deformation. The "maximum strength" is the largest stress attained in the rod. The "breaking strength" is the stress just prior to fracture or rupture. The breaking strength is less than the maximum strength since the sustainable force P decreases rapidly once extensive plastic deformation occurs.

Table 3.2 provides a list of approximate strength values for commonly used materials [1,2].

3.4 HOOKE'S LAW IN SHEAR

Consider again Hooke's law for simple stress and strain of Equation 3.4:

$$\sigma = \mathbf{E}\varepsilon \tag{3.10}$$

We can extend this relation to accommodate simple shear stress and strain. Consider again a block subjected to a shearing force as in Figure 2.6 and as shown again in Figure 3.7. Then, from Equations 2.7 and 2.8 the shear stress τ and the shear strain γ are defined as

$$\tau = V/A \quad \text{and} \quad \gamma = \delta/\ell \tag{3.11}$$

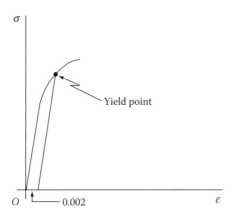

FIGURE 3.4 Yield point definition of a strain of 0.002.

TABLE 3.1

Selected Values of Elastic Constants

Material	10^6 psi (lb/in.^{2})	10^9 Pa (N/m^2)
	E	
Steel	30	207
Aluminum	10	69
Copper	17	117
Concrete	4	28
Wood	1.9	13

FIGURE 3.5   Axial stretching of a rod.

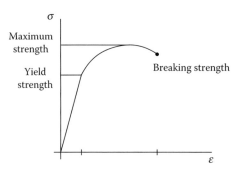

FIGURE 3.6   Stress–strain diagram illustrating yield, maximum, and breaking strength.

TABLE 3.2

Selected Material Strengths

Material	Yield Strength		Maximum Strength	
	10^3 psi (lb/in.^{2})	10^6 Pa (N/m^2)	10^3 psi (lb/in.^{2})	10^6 Pa (N/m^2)
Steel	40–80	275–550	60–120	410–820
Aluminum	35–70	240–480	40–80	275–550
Copper	10–50	70–350	30–60	200–400
Concrete	—	—	4–6	28–40
Wood	—	—	5–10	35–70

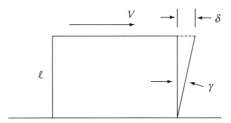

FIGURE 3.7   A block subjected to a shearing force.

TABLE 3.3
Selected Values of the Shear Modulus

Material	10^6 psi (lb/in.^{2})	10^9 Pa (N/m^2)
Steel	11.2	77
Aluminum	3.8	27
Copper	6.4	44

---

where
  V is the shearing force
  A is the area over which V acts
  ℓ is the height of the block
  δ is the horizontal displacement

Then, analogous to Equation 3.10, Hooke's law for simple shear is

$$\tau = G\gamma \qquad\qquad (3.12)$$

where the proportional parameter G is called the "shear modulus," "modulus of elasticity in shear," or the "modulus of rigidity."

  Table 3.3 lists values of the shear modulus for a few commonly used materials [1,2].

SYMBOLS

A Area
E Modulus of elasticity, Young's modulus
G Shear modulus, modulus of rigidity
k Spring constant
ℓ Length
P Axial force
V Shear force
γ Shear strain

δ Elongation, shortening
ε Normal strain
σ Normal stress
τ Shear stress

REFERENCES

1. F. P. Beer and E. R. Johnston, Jr., *Mechanics of Materials*, 2nd ed., McGraw Hill, New York, 1992.
2. T. Baumeister, Ed., *Marks' Standard Handbook for Mechanical Engineers*, 8th ed., McGraw Hill, New York, 1978.

4 Stress in Two and Three Dimensions

4.1 STRESS VECTORS

Consider an elastic body B subjected to surface loads as in Figure 4.1. Consider a cutting plane N dividing B into two parts as shown in edge view in Figure 4.2.

Consider the equilibrium of one of the parts of B, say the left part B_L, as in Figure 4.3. The figure depicts the forces exerted across the dividing plane by the right portion of $B(B_R)$ on the left portion (B_L). Correspondingly, B_L exerts equal and opposite forces on B_R.

Consider next a view of the dividing surface of B_L and a small region R on this surface, as in Figure 4.4 where forces exerted by B_R on B_L across R are depicted. Consider now a force system S, which is equivalent to the system of forces exerted by B_R on B_L across R. Specifically, let S consist of a single force $\mathbf{P}$ passing through a point P of R together with a couple with torque $\mathbf{M}$ (see Section 1.5.3) as represented in Figure 4.5.

Let A be the area of R. Next, imagine that R is decreased in size, or shrunk, around point P. As this happens, consider the ratio: $\mathbf{P}/A$. As R shrinks, A diminishes, but the magnitude of $\mathbf{P}$ also diminishes. In the limit, as A becomes infinitesimally small the ratio $\mathbf{P}/A$ will approach a vector $\mathbf{S}$ given by

$$\mathbf{S} = \lim_{A \to 0} \mathbf{P}/A \tag{4.1}$$

This vector is called the "stress vector on R at P."

From Section 1.5.3, it is apparent that as R gets small the magnitude of the couple torque $\mathbf{M}$ becomes increasingly small. That is,

$$\lim_{A \to 0} \mathbf{M} = 0 \tag{4.2}$$

Observe that, in general, $\mathbf{S}$ is parallel neither to R nor to the normal of R. Observe further that for a different dividing plane, say $\hat{N}$, passing through P, the corresponding stress vector $\hat{\mathbf{S}}$ will be different than $\mathbf{S}$.

Finally, consider a set of mutually perpendicular unit vectors $\mathbf{n}_x$, $\mathbf{n}_y$, and $\mathbf{n}_z$ with $\mathbf{n}_x$ being normal to the plane of R, directed outward from B_L as in Figure 4.6. Let $\mathbf{S}$ be expressed in terms of $\mathbf{n}_x$, $\mathbf{n}_y$, and $\mathbf{n}_y$ as

$$\mathbf{S} = S_x \mathbf{n}_x + S_y \mathbf{n}_y + S_z \mathbf{n}_z \tag{4.3}$$

Then S_x, S_y, and S_z are stresses at P with S_x being a normal (tension or compression) stress and S_y and S_z being tangential (or shear) stresses.

4.2 STRESSES WITHIN A LOADED ELASTIC BODY—NOTATION AND SIGN CONVENTION

Consider again the loaded elastic body B of Figure 4.1 and consider a small rectangular element E in the interior of B as represented in Figure 4.7. Let X, Y, and Z be coordinate axes parallel to the edges

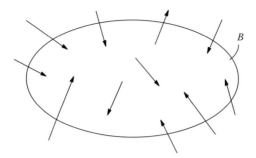

FIGURE 4.1   An elastic body subjected to surface loads.

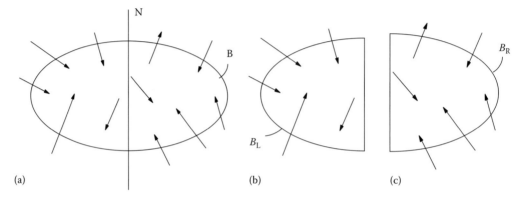

(a)   (b)   (c)

FIGURE 4.2   Edge view of a cutting plane devising the elastic body of Figure 4.1.

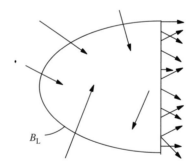

FIGURE 4.3   Equilibrium of the left portion of the elastic body with forces exerted across the dividing plane.

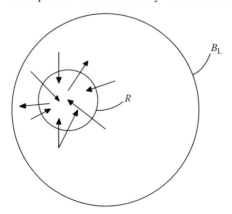

FIGURE 4.4   A small region of R of the dividing plane.

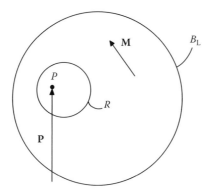

FIGURE 4.5 Equivalent force system exerted across *R*.

of *E* and with origin *O* within *E* as shown. Next, let *E* be shrunk to an infinitesimal element about *O* (as *R* was shrunk about *P* in Section 4.1). As in Section 4.1, imagine the coordinate planes to be cutting planes of *E*, separating *E* into six different parts (two for each cutting plane). Then in the context of the foregoing analysis, each of the six sides (or "cut faces" of *E*) will have an associated stress vector with stress components as in Equation 4.3. Thus, with six faces and three stress components per face, there are 18 stress components (or stresses) associated with element *E*.

For *E* to be in equilibrium, while being infinitesimal, the corresponding stress components on opposite, parallel faces of *E* must be equal and oppositely directed. Thus, we need to consider only nine of the 18 stress components. To make an account (or list) of these components, it is convenient to identify the components first with the face on which they are acting and then with their direction. We can identify the faces of *E* with their normals since each face is normal to one of the *X*, *Y*, or *Z* axes. Since there are two faces normal to each axis, we can think of these faces as being "positive" or "negative" depending upon which side of the origin *O* they occur. Specifically, let the vertices of *E* be numbered and labeled as in Figure 4.8. Then a face is said to be "positive" if when going from the interior of *E* to the exterior across a face, the movement is in the positive axis direction. Correspondingly, a face is "negative" if the movement is in the negative axis direction when crossing the face. Table 4.1 lists the positive and negative faces of *E*.

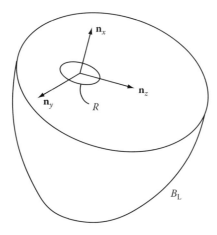

FIGURE 4.6 Unit vectors parallel to normal to region *R* of B_L.

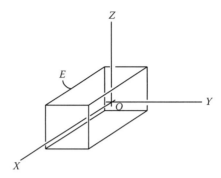

FIGURE 4.7   A small rectangular element E of loaded elastic body B (see Figure 4.1).

To account for the nine stress components, it is convenient to use subscript notation such as σ_{ij} where the subscripts i and j have the values x, y, and z with the first subscript (i) referring to the face upon which the stress is applied and the second subscript (j) referring to the direction of the stress component. We can then arrange the stress components into an array σ as

$$\sigma = \begin{bmatrix} \sigma_{xx} & \sigma_{xy} & \sigma_{xz} \\ \sigma_{yx} & \sigma_{yy} & \sigma_{yz} \\ \sigma_{zx} & \sigma_{zy} & \sigma_{zz} \end{bmatrix} \tag{4.4}$$

The diagonal elements of this array are seen to be the normal stresses (tension/compression) while the off-diagonal elements are shear stresses. The shear stresses are sometimes designated by the Greek letter τ as in Section 2.3.

A stress component is said to be "positive" if the component is exerted on a positive face in a positive direction or on a negative face in a negative direction. On the contrary, a stress component is said to be "negative" if it is exerted on a negative face in the positive direction or a positive face in the negative direction. (With this sign convention, tension is positive and compression is negative.)

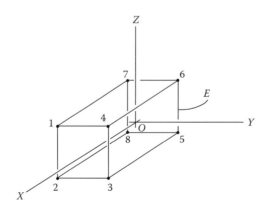

FIGURE 4.8   Numbering the vertices of element E.

TABLE 4.1

Positive and Negative Faces of E

Face	Normal Axis	Face Sign
1234	$+X$	Positive
4356	$+Y$	Positive
6714	$+Z$	Positive
7658	$-X$	Negative
2178	$-Y$	Negative
2853	$-Z$	Negative

4.3 EQUILIBRIUM CONSIDERATIONS—INDEX NOTATION

Consider the small rectangular element of Figure 4.8 as drawn again in Figure 4.9. Let the lengths of the edges be Δx, Δy, and Δz. Consider an "overhead" or Z-direction view of the element as in Figure 4.10 where the shear stresses on the X and Y faces in the X- and Y-directions are shown. Next, imagine a free-body diagram of the element. If the element is sufficiently small, the forces on the element may be represented by force components acting through the centers of the faces with magnitudes equal to the product of the stresses and the areas of the faces as in Figure 4.11. By setting moments about the Z-axis equal to zero, we have

$$\sigma_{xy}\Delta y\Delta z(\Delta x/2) - \sigma_{yx}\Delta x\Delta z(\Delta y/2) + \sigma_{xy}\Delta_y\Delta_z(\Delta x/2) - \sigma_{yx}\Delta x\Delta z(\Delta y/2) = 0 \qquad (4.5)$$

By dividing by the element volume, $\Delta x\Delta y\Delta z$, we obtain

$$\sigma_{xy} = \sigma_{yx} \qquad (4.6)$$

Similarly, by considering moment equilibrium about the Y- and Z-axes, we obtain the expressions

$$\sigma_{xz} = \sigma_{zx} \quad \text{and} \quad \sigma_{zy} = \sigma_{yz} \qquad (4.7)$$

These results show that the stress array σ of Equation 4.4 is symmetric. That is,

$$\sigma = \begin{bmatrix} \sigma_{xx} & \sigma_{xy} & \sigma_{xz} \\ \sigma_{yx} & \sigma_{yy} & \sigma_{yz} \\ \sigma_{zx} & \sigma_{zy} & \sigma_{zz} \end{bmatrix} = \begin{bmatrix} \sigma_{xx} & \sigma_{xy} & \sigma_{xz} \\ \sigma_{xy} & \sigma_{yy} & \sigma_{yz} \\ \sigma_{xz} & \sigma_{yz} & \sigma_{zz} \end{bmatrix} \qquad (4.8)$$

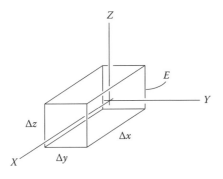

FIGURE 4.9 Small rectangular element with dimensions Δx, Δy, and Δz.

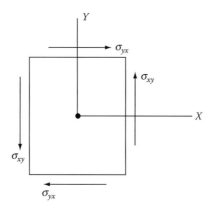

FIGURE 4.10 *X–Y* shear stresses on the element of Figure 4.9.

In short,

$$\sigma_{ij} = \sigma_{ji} \quad i, j = x, y, z \tag{4.9}$$

Next, consider the equilibrium of a small tetrahedron *T* as in Figure 4.12, where three of the sides are normal to coordinate axes. Let **n** be a unit vector normal to the inclined face *ABC* of *T* and let **S**$_n$ be the stress vector exerted on *ABC*. As before, since *T* is small, let the forces on *T* be represented by individual forces passing through the centroids of the faces of *T*. Let these forces be equal to the stress vectors, on the faces of *T*, multiplied by the areas of the respective faces.

Let *A* be the area of face *ABC*, and let A_x, A_y, and A_z be the area of the faces normal to the coordinate axes (*OBC*, *OCA*, and *OAB*). Let **n** be expressed in terms of the coordinate line unit vectors as

$$\mathbf{n} = n_x\mathbf{n}_x + n_y\mathbf{n}_y + n_z\mathbf{n}_z \tag{4.10}$$

Then, it is evident that A_x, A_y, and A_z are

$$A_x = An_x, \quad A_y = An_y, \quad A_z = An_z \tag{4.11}$$

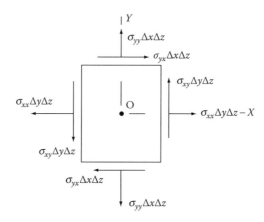

FIGURE 4.11 *X*- and *Y*-direction forces on element *E*.

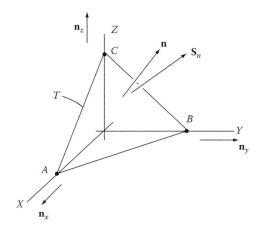

FIGURE 4.12 Small tetrahedron within a loaded elastic body.

Imagine a free-body diagram of T. The forces on T may be represented by the four forces: $\mathbf{S}_x A_x$, $\mathbf{S}_y A_y$, $\mathbf{S}_z A_z$, and $\mathbf{S}_n A$ acting through the centroids of the respective faces, where $\mathbf{S}_x$, $\mathbf{S}_y$, and $\mathbf{S}_z$ are the stress vectors on faces OBC, OCA, and OAB, respectively. The equilibrium of T then leads to the expression:

$$\mathbf{S}_x A_x + \mathbf{S}_y A_y + \mathbf{S}_z A_z + \mathbf{S}_n A = 0 \tag{4.12}$$

Using the notation in Section 4.2, let the stress vectors be expressed in terms of $\mathbf{n}_x$, $\mathbf{n}_y$, and $\mathbf{n}_z$ as

$$\mathbf{S}_x = -\sigma_{xx}\mathbf{n}_x - \sigma_{xy}\mathbf{n}_y - \sigma_{xz}\mathbf{n}_z \tag{4.13}$$

$$\mathbf{S}_y = -\sigma_{yx}\mathbf{n}_x - \sigma_{yy}\mathbf{n}_y - \sigma_{yz}\mathbf{n}_z \tag{4.14}$$

$$\mathbf{S}_z = -\sigma_{zx}\mathbf{n}_x - \sigma_{zy}\mathbf{n}_y - \sigma_{zz}\mathbf{n}_z \tag{4.15}$$

$$\mathbf{S}_n = S_{nx}\mathbf{n}_x + S_{ny}\mathbf{n}_y + S_{nz}\mathbf{n}_z \tag{4.16}$$

where the negative signs in Equations 4.13, 4.14, and 4.15 occur since OBC, OCA, and OAB are "negative" faces (see Section 4.2).

By substituting from Equation 4.11 into Equation 4.12 we obtain

$$n_x\mathbf{S}_x + n_y\mathbf{S}_y + n_z\mathbf{S}_z + \mathbf{S}_n = 0 \tag{4.17}$$

Then, by substituting from Equation 4.13 through 4.16 and setting $\mathbf{n}_x$, $\mathbf{n}_y$, and $\mathbf{n}_z$ components equal to zero, we have

$$S_{nx} = \sigma_{xx}n_x + \sigma_{xy}n_y + \sigma_{xz}n_z \tag{4.18}$$

$$S_{ny} = \sigma_{yx}n_x + \sigma_{yy}n_y + \sigma_{yz}n_z \tag{4.19}$$

$$S_{nz} = \sigma_{zx}n_x + \sigma_{zy}n_y + \sigma_{zz}n_z \tag{4.20}$$

Observe the pattern of the indices of Equations 4.12 through 4.20: repeated indices range through x, y, and z. Otherwise, the terms are the same. Thus, it is often convenient to use numerical indices and summation notation. Let x, y, and z be replaced by 1, 2, and 3. Then Equations 4.17 through 4.20 may be written in a compact form as

$$\sum_{j=1}^{3} n_j \mathbf{S}_j + \mathbf{S}_n = 0 \qquad (4.21)$$

and

$$\mathbf{S}_{ni} = \sum_{i=1}^{3} \sigma_{ij} n_j \quad (i = 1, 2, 3) \qquad (4.22)$$

Since in three dimensional analyses the sums generally range from 1 to 3, it is usually possible to delete the summation sign (Σ) and simply adopt the convention that repeated indices designate a sum over the range of the index. Thus, Equations 4.21 and 4.22 may be written as

$$n_j \mathbf{S}_j + \mathbf{S}_n = 0 \qquad (4.23)$$

and

$$Sn_i = \sigma_{ij} n_j \quad (i = 1, 2, 3) \qquad (4.24)$$

Finally, consider the equilibrium of a small, but yet finite size, rectangular element of a loaded elastic body as in Figure 4.13. Let the lengths of the sides of E be Δx, Δy, and Δz as shown. Let E be sufficiently small so that the forces on the faces of E may be represented by stress vectors acting through the centroids of the faces multiplied by the areas of the respective faces.

Consider the force components in the X-direction. Consider specifically the change in corresponding stresses from one side of E to the other. By using a Taylor series expansion, we can relate these stresses by the expression:

$$\sigma_{xx} \Big|_{\substack{\text{front} \\ \text{face}}} = \sigma_{xx} \Big|_{\substack{\text{rear} \\ \text{face}}} + \frac{\partial \sigma_{xx}}{\partial x} \Big|_{\substack{\text{rear} \\ \text{face}}} \Delta x + \frac{1}{2!} \frac{\partial^2 \sigma_{xx}}{\partial x^2} \Big|_{\substack{\text{rear} \\ \text{face}}} (\Delta x)^2 + \cdots \qquad (4.25)$$

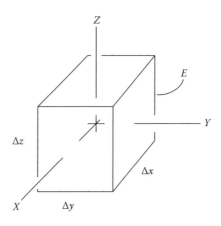

FIGURE 4.13   A small element within a loaded elastic body.

With the element E being small, the terms not shown in the sum of Equation 4.25 are also small. Indeed, these terms as well as the third term on the right-hand side of Equation 4.25 become increasingly small as E gets smaller. Hence, to a reasonable degree of accuracy we have

$$\sigma_{xx} \Big|_{\substack{\text{front} \\ \text{face}}} = [\sigma_{xx} + (\partial \sigma_{xx}/\partial x)\Delta x] \Big|_{\substack{\text{back} \\ \text{face}}} \qquad (4.26)$$

Similar analyses for the shear stresses in the X-direction lead to the expressions:

$$\sigma_{zx} \Big|_{\substack{\text{front} \\ \text{face}}} = [\sigma_{zx} + (\partial \sigma_{zx}/\partial z)\Delta z] \Big|_{\substack{\text{back} \\ \text{face}}} \qquad (4.27)$$

and

$$\sigma_{yx} \Big|_{\substack{\text{front} \\ \text{face}}} = [\sigma_{yx} + (\partial \sigma_{yx}/\partial y)\Delta y] \Big|_{\substack{\text{back} \\ \text{face}}} \qquad (4.28)$$

Consider now the X-direction forces of a free-body diagram of E. As E shrinks to a point, the corresponding stresses on opposite faces become nearly equal in magnitude. Then, a balance of forces leads to the expression:

$$\begin{aligned}
\sigma_{xx}\Delta y \Delta z &- \sigma_{xx}\Delta y \Delta z + [\partial \sigma_{xx}/\partial x]\Delta y \Delta z \\
&+ \sigma_{zx}\Delta y \Delta x - \sigma_{zx}\Delta y \Delta x + [(\partial \sigma_{zx}/\partial z)\Delta z]\Delta y \Delta x \\
&+ \sigma_{yx}\Delta x \Delta z - \sigma_{yx}\Delta x \Delta z + [(\partial \sigma_{yx}/\partial z)\Delta y]\Delta x \Delta z \\
&= (\rho \Delta x \Delta y \Delta z)a_x
\end{aligned} \qquad (4.29)$$

where
  ρ is the mass density of B at the origin O, which could be any typical point P on B
  a_x is the acceleration of P in an inertial reference frame*

By dividing by the element volume, canceling terms, and by the index symmetry for the shear stresses, we see that Equation 4.29 may be written as

$$\partial \sigma_{xx}/\partial x + \partial \sigma_{xy}/\partial y + \partial \sigma_{xz}/\partial z = \rho a_x \qquad (4.30)$$

Similarly, by adding forces in the Y- and Z- directions, we have

$$\partial \sigma_{yx}/\partial x + \partial \sigma_{yy}\partial y + \partial \sigma_{yz}/\partial z = \rho a_y \qquad (4.31)$$

$$\partial \sigma_{zx}/\partial x + \partial \sigma_{zy}\partial y + \partial \sigma_{zz}/\partial z = \rho(a_z - g) \qquad (4.32)$$

where g is the gravity acceleration (9.8 m/s or 32.2 ft/s^2). Except in the case of large structures, the gravity (or weight) is usually inconsequential. Thus, in most cases, Equations 4.30, 4.31, and 4.32 have the same form and by using numerical index notation they may be cast into a compact expression. If we let $x \to x_1$, $y \to x_2$, and $z \to x_3$, that is, letting 1, 2, 3 correspond to x, y, z, then we can write the equations as

$$\partial \sigma_{ij}/\partial x_j = \rho a_i \quad (i = 1, 2, 3) \qquad (4.33)$$

with a sum over the repeated index j.

---

* See Ref. [1]. For static or slowly moving bodies, which comprise the majority of stress analysis problems, a_x will be zero.

Equation 4.33 may be written in a more compact form by using the comma notation for differentiation.* That is,

$$( )_{,i} \equiv \partial( )/\partial x_i \tag{4.34}$$

Then, Equation 4.33 becomes

$$\sigma_{ij,j} = \rho a_i \tag{4.35}$$

A few more comments on notation: whereas repeated indices (such as the j in Equation 4.35) designate a sum (from 1 to 3), nonrepeated (or "free") indices (such as the i in Equation 4.35) can have any of the values: 1, 2, or 3. In this context, in a given equation or expression, indices are either free or repeated. Repeated indices are to be repeated only once, but free indices must occur in each term of an equation. With a repeated index, the letter used for the index is immaterial. That is, any letter can be used for the index that is repeated. Thus, Equation 4.35 may be written as

$$\sigma_{ij,j} = \sigma_{ik,k} = \sigma_{i\ell,\ell} = \cdots = \rho a_i \tag{4.36}$$

4.4 STRESS MATRIX, STRESS DYADIC

As we observed in Section 4.2, it is convenient to assemble the stresses into an array, called the "stress matrix," as

$$\sigma = \begin{bmatrix} \sigma_{xx} & \sigma_{xy} & \sigma_{xz} \\ \sigma_{yx} & \sigma_{yy} & \sigma_{yz} \\ \sigma_{zx} & \sigma_{zy} & \sigma_{zz} \end{bmatrix} \tag{4.37}$$

In numerical index notation, we can express σ as

$$\sigma = [\sigma_{ij}] = \begin{bmatrix} \sigma_{11} & \sigma_{12} & \sigma_{13} \\ \sigma_{21} & \sigma_{22} & \sigma_{23} \\ \sigma_{31} & \sigma_{32} & \sigma_{33} \end{bmatrix} \tag{4.38}$$

Observe that the values of the individual stresses of σ depend upon the orientation of the X-, Y-, Z-axis system and thus upon the direction of the unit vectors $\mathbf{n}_x$, $\mathbf{n}_y$, and $\mathbf{n}_z$, or alternatively upon the direction of unit vectors $\mathbf{n}_1$, $\mathbf{n}_2$, and $\mathbf{n}_3$. A question arising then is: How are the stresses changed if the orientation of the coordinate axes are changed? To answer this question, it is convenient to introduce the concept of a "stress dyadic." A dyadic is simply a product of vectors following the usual rules of elementary analysis (except for communitivity) (see Ref. [3]). As an illustration, consider a pair of vectors $\mathbf{a}$ and $\mathbf{b}$ expressed in terms of mutually perpendicular unit vectors, $\mathbf{n}_i$ $(i = 1, 2, 3)$ as

$$\mathbf{a} = a_1\mathbf{n}_1 + a_2\mathbf{n}_2 + a_3\mathbf{n}_3 = a_i\mathbf{n}_i \tag{4.39}$$

$$\mathbf{b} = b_1\mathbf{n}_1 + b_2\mathbf{n}_2 + b_3\mathbf{n}_3 = b_j\mathbf{n}_j \tag{4.40}$$

where, as before, the repeated indices designate a sum over the range (1 to 3) of the indices. The dyadic product $\mathbf{d}$ of $\mathbf{a}$ and $\mathbf{b}$ may then be expressed as

---

* See Ref. [2], for example.

$$\mathbf{d} = \mathbf{ab} = (a_1\mathbf{n}_1 + a_2\mathbf{n}_2 + a_3\mathbf{n}_3)(b_1\mathbf{n}_1 + b_2\mathbf{n}_2 + b_3\mathbf{n}_3)$$
$$= (a_i\mathbf{n}_i)(b_j\mathbf{n}_i)$$
$$= a_1b_1\mathbf{n}_1\mathbf{n}_1 + a_1b_2\mathbf{n}_1\mathbf{n}_2 + a_1b_3\mathbf{n}_1\mathbf{n}_3$$
$$+ a_2b_1\mathbf{n}_2\mathbf{n}_1 + a_2b_2\mathbf{n}_2\mathbf{n}_2 + a_2b_3\mathbf{n}_2\mathbf{n}_3$$
$$+ a_3b_1\mathbf{n}_3\mathbf{n}_1 + a_3b_2\mathbf{n}_3\mathbf{n}_2 + a_3b_3\mathbf{n}_3\mathbf{n}_3$$
$$= a_ib_j\mathbf{n}_i\mathbf{n}_j$$
$$= d_{ij}\mathbf{n}_i\mathbf{n}_j \tag{4.41}$$

where d_{ij} is defined as the product: a_ib_j. The unit vector products in Equation 4.41 are called "dyads." The order or positioning of the unit vectors in a dyad must be maintained. That is,

$$\mathbf{n}_1\mathbf{n}_2 \neq \mathbf{n}_2\mathbf{n}_1, \quad \mathbf{n}_2\mathbf{n}_3 \neq \mathbf{n}_3\mathbf{n}_2, \quad \mathbf{n}_3\mathbf{n}_1 \neq \mathbf{n}_1\mathbf{n}_3 \tag{4.42}$$

Dyadics are sometimes called "vector-vectors" because they may be viewed as vectors whose components are vectors. The components of a dyadic (as well as those of vectors) are sometimes called "tensors" (of rank 2 and rank 1).

Using these concepts and notation, let the stress dyadic σ be defined as

$$\sigma = \sigma_{ij}\mathbf{n}_i\mathbf{n}_j \tag{4.43}$$

Now suppose we are interested in a different orientation of unit vectors. Let $\hat{\mathbf{n}}_j$ ($j = 1, 2, 3$) be a set of mutually perpendicular unit vectors inclined relative to the $\mathbf{n}_i$ as depicted in Figure 4.14. Then the respective orientations of the $\hat{\mathbf{n}}_j$ relative to the $\mathbf{n}_i$ may be defined in terms of direction cosines T_{ij} given by

$$T_{ij} = \mathbf{n}_i \cdot \hat{\mathbf{n}}_j \tag{4.44}$$

It is then obvious that the $\mathbf{n}_i$ and the $\hat{\mathbf{n}}_j$ are related by the expressions [1]:

$$\mathbf{n}_i = \mathbf{T}_{ij}\hat{\mathbf{n}}_j \quad \text{and} \quad \hat{\mathbf{n}}_j = T_{ij}\mathbf{n}_i \tag{4.45}$$

Observe in Equations 4.44 and 4.45 that the rules regarding free and repeated indices are maintained. That is, the free indices match the terms on either side of the equality and the repeated indices are repeated only once in a given term. Also, in Equation 4.44, the first index (i) of S_{ij} is associated with the $\mathbf{n}_i$ and the second index (j) is associated with the $\mathbf{n}_j$. This association is maintained in Equation 4.45.

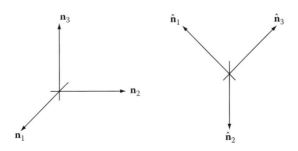

FIGURE 4.14   Unit vector sets.

The stress dyadic σ is expressed in terms of the $\hat{\mathbf{n}}_j$ as

$$\sigma = \hat{\sigma}_{k\ell}\hat{\mathbf{n}}_{\mathbf{k}}\hat{\mathbf{n}}_{\ell} \tag{4.46}$$

then, using Equations 4.45 and 4.43, we get

$$\sigma_{ij} = T_{ik}T_{j\ell}\hat{\sigma}_{k\ell} \quad \text{and} \quad \hat{\sigma}_{k\ell} = T_{ik}T_{j\ell}\sigma_{ij} \tag{4.47}$$

As noted earlier, the σ_{ij} and the $\hat{\sigma}_{k\ell}$ are sometimes called "stress tensors."

4.5 EIGENVECTORS AND PRINCIPAL STRESSES

Equation 4.47 shows that the value of the stress components depends upon the choice of axis system and the corresponding unit vector directions. By using well-established procedures in vector, matrix, and tensor analysis [3], it is seen that the matrix of the stress dyadic can be placed in diagonal form by the appropriate choice of basis unit vectors. When this is done, with the off-diagonal elements being zero, the shear stresses vanish and the normal stresses, occurring on the diagonal, have among them the maximum and minimum normal stresses for all directions. These maximum and minimum stresses are called "principal stresses," or "eigenvalues" of the stress dyadic. The unit vectors producing the diagonal stress matrix are called "eigenvectors" (or "unit eigenvectors"), and they define what are called the "principal directions" of the stress dyadic.

In stress analysis for strength considerations and in mechanical component design, it is of interest to know the values of these principal stresses and the directions of the surfaces over which they act. The following paragraphs outline a procedure for calculating these stresses and directions. (Additional details may be found in Refs. [1] and [3].)

Consider again the stress dyadic σ of Equation 4.46:

$$\sigma = \sigma_{ij}\mathbf{n}_i\mathbf{n}_j \tag{4.48}$$

Let $\mathbf{n}_a$ be a unit vector. $\mathbf{n}_a$ is defined as a unit eigenvector if it satisfies the relation:

$$\sigma \cdot \mathbf{n}_a = \lambda \mathbf{n}_a \tag{4.49}$$

where λ is a scalar. That is, $\mathbf{n}_a$ is an eigenvector if the stress vector associated with $\mathbf{n}_a$ is parallel to $\mathbf{n}_a$.

Let $\mathbf{n}_b$ be a unit vector perpendicular to the unit eigenvector $\mathbf{n}_a$. Then, the shear stress σ_{ab} associated with $\mathbf{n}_a$ and $\mathbf{n}_b$ is zero. That is,

$$\sigma_{ab} = \mathbf{n}_a \cdot \sigma \cdot \mathbf{n}_b = \lambda \mathbf{n}_a \cdot \mathbf{n}_b = 0 \tag{4.50}$$

Recall that σ is symmetric which implies that shear stresses associated with eigenvectors are zero.

The definition of Equation 4.49 may be used to obtain an expression for $\mathbf{n}_a$. Let $\mathbf{n}_a$ be expressed in terms of a convenient set of mutually perpendicular unit vectors $\mathbf{n}_i$ $(i = 1, 2, 3)$ as

$$\mathbf{n}_a = a_1\mathbf{n}_1 + a_2\mathbf{n}_2 + a_3\mathbf{n}_3 = a_i\mathbf{n}_i = a_k\mathbf{n}_k \tag{4.51}$$

Then $\mathbf{n}_a$ is known once the a_i are determined.

By substituting from Equations 4.48 and 4.51 into Equation 4.49, we obtain

$$\sigma \cdot \mathbf{n}_a = \mathbf{n}_i\sigma_{ij}\mathbf{n}_j \cdot a_k\mathbf{n}_k = \mathbf{n}_i\sigma_{ij}a_k\mathbf{n}_j \cdot \mathbf{n}_k$$
$$= n_i\sigma_{ij}a_k\delta_{jk} = \mathbf{n}_i\sigma_{ij}a_j = \lambda a_i\mathbf{n}_i \tag{4.52}$$

where δ_{jk}, called Kronecker's delta function, is defined as

$$\delta_{jk} = \mathbf{n}_j \cdot \mathbf{n}_k = \begin{cases} 0 & j \neq k \\ 1 & j = k \end{cases} \tag{4.53}$$

δ_{jk} has several useful properties. From the definition of Equation 4.53 we see that

$$\delta_{kk} = 3 \tag{4.54}$$

Also, if $\mathbf{v}$ is any vector expressed in component form as $v_j\mathbf{n}_j$, we have

$$\delta_{ij}v_j = v_i \quad (i = 1, 2, 3) \tag{4.55}$$

This property (used in Equation 4.52), has led δ_{ij} at times to be called the "substitution symbol." Finally, the δ_{ij} are the elements of the identity dyadic $\mathbf{I}$ defined as

$$\mathbf{I} = \mathbf{n}_i\mathbf{n}_i = \mathbf{n}_i\delta_{ij}\mathbf{n}_j \tag{4.56}$$

where the matrix of elements δ is defined as

$$\delta = [\delta_{ij}] = \begin{bmatrix} 1 & 0 & 0 \\ 0 & 1 & 0 \\ 0 & 0 & 1 \end{bmatrix} \tag{4.57}$$

The last equality of Equation 4.52 may be written as

$$\sigma_{ij}a_j\mathbf{n}_i = \lambda a_i\mathbf{n}_i \tag{4.58}$$

or in component form as

$$\sigma_{ij}a_j = \lambda a_i \tag{4.59}$$

and in matrix form as

$$\begin{bmatrix} \sigma_{11} & \sigma_{12} & \sigma_{13} \\ \sigma_{21} & \sigma_{22} & \sigma_{23} \\ \sigma_{31} & \sigma_{32} & \sigma_{33} \end{bmatrix} \begin{bmatrix} a_1 \\ a_2 \\ a_3 \end{bmatrix} = \lambda \begin{bmatrix} a_1 \\ a_2 \\ a_3 \end{bmatrix} \tag{4.60}$$

Equations 4.58, 4.59, and 4.60 are equivalent to the scalar equations:

$$\begin{aligned} (\sigma_{11} - \lambda)a_1 + \sigma_{12}a_2 + \sigma_{13}a_3 &= 0 \\ \sigma_{21}a_1 + (\sigma_{22} - \lambda)a_2 + \sigma_{23}a_3 &= 0 \\ \sigma_{31}a_1 + \sigma_{32}a_2 + (\sigma_{33} - \lambda)a_3 &= 0 \end{aligned} \tag{4.61}$$

These equations form a set of three linear algebraic equations for a_1, a_2, and a_3. Thus their solution determines $\mathbf{n}_a$. However, since the equations are "homogeneous" (all right-hand sides are zero), there is a nonzero solution only if the determinant of the coefficients is zero [4]. That is,

$$\begin{vmatrix} (\sigma_{11} - \lambda) & \sigma_{12} & \sigma_{13} \\ \sigma_{21} & (\sigma_{22} - \lambda) & \sigma_{23} \\ \sigma_{31} & \sigma_{32} & (\sigma_{33} - \lambda) \end{vmatrix} = 0 \tag{4.62}$$

By expanding the determinant, we obtain

$$\lambda^3 - \sigma_{\text{I}}\lambda^2 + \sigma_{\text{II}}\lambda - \sigma_{\text{III}} = 0 \tag{4.63}$$

where the coefficients σ_{I}, σ_{II}, and σ_{III} are

$$\sigma_{\text{I}} = \sigma_{11} + \sigma_{22} + \sigma_{33} \tag{4.64}$$

$$\sigma_{\text{II}} = \sigma_{22}\sigma_{33} - \sigma_{32}\sigma_{23} + \sigma_{33}\sigma_{11} - \sigma_{13}\sigma_{31} + \sigma_{11}\sigma_{22} - \sigma_{21}\sigma_{12} \tag{4.65}$$

$$\sigma_{\text{III}} = \sigma_{11}\sigma_{22}\sigma_{33} - \sigma_{11}\sigma_{32}\sigma_{23} + \sigma_{12}\sigma_{31}\sigma_{23} - \sigma_{12}\sigma_{21}\sigma_{33} + \sigma_{21}\sigma_{32}\sigma_{13} - \sigma_{31}\sigma_{13}\sigma_{22} \tag{4.66}$$

It is clear that σ_{I} is the sum of the diagonal elements of the stress matrix, σ_{II} is the sum of the diagonal elements of the matrix of cofactors of the stress matrix, and σ_{III} is the determinant of the stress matrix.

Equation 4.63 is sometimes called the Hamilton–Cayley equation. It is known that with σ being symmetric (that is, $\sigma_{ij} = \sigma_{ji}$), the roots ($\lambda_1$, λ_2, and λ_3) of the equation are real [3]. When the roots are distinct, Equations 4.61 form a set of dependent equations for a_1, a_2, and a_3. That is, at most only two of Equations 4.61 are independent. Thus, there is no solution for a_1, a_2, and a_3 without an additional equation. But since $\mathbf{n}_a$ is a unit vector with magnitude 1, we have

$$a_1^2 + a_2^2 + a_3^2 = 1 \tag{4.67}$$

Observe that the roots, λ_i of Equation 4.63 are themselves the eigenvalues and thus the principal stresses. That is, from Equation 4.49 we have

$$\sigma_{aa} = \mathbf{n}_a \bullet \sigma \bullet \mathbf{n}_a = \mathbf{n}_a \bullet \lambda \mathbf{n}_a = \lambda \tag{4.68}$$

Since there are three eigenvalues, λ_1, λ_2, and λ_3, there are three unit eigenvectors. When the eigenvalues are distinct, these unit eigenvectors can be shown to be mutually perpendicular [1,3].

4.5.1 ILLUSTRATIVE COMPUTATION

To illustrate procedures for calculating values of principal stresses and their corresponding directions (the unit eigenvectors) suppose that the stress matrix relative to a convenient axis system is

$$\sigma = \begin{bmatrix} 5.0625 & 1.1908 & 1.0825 \\ 1.1908 & 3.6875 & 0.6250 \\ 1.0825 & 0.6250 & 5.7500 \end{bmatrix} 10^4 \text{ psi} \tag{4.69}$$

Let $\mathbf{n}_1$, $\mathbf{n}_2$, and $\mathbf{n}_3$ be unit vectors parallel to the axes, and let the unit eigenvectors have the form:

$$\mathbf{n}_a = a_1\mathbf{n}_1 + a_2\mathbf{n}_2 + a_3\mathbf{n}_3 = a_i\mathbf{n}_i \tag{4.70}$$

Then, from Equations 4.61 the equations determining the a_i and the λ are

$$(5.0625 - \lambda)a_1 + 1.1908a_2 + 1.0825a_3 = 0$$
$$1.1908a_1 + (3.6875 - \lambda)a_2 + 0.6250a_3 = 0 \tag{4.71}$$
$$1.0825a_1 + 0.6250a_2 + (5.75 - \lambda)a_3 = 0$$

where the units of the coefficients are 10^4 psi.

The determinantal equation of Equation 4.62 together with Equations 4.64 through 4.66 produces the Hamilton–Cayley equation (Equation 4.63):

$$\lambda^3 - 14.5\lambda^2 + 66\lambda - 94.5 = 0 \tag{4.72}$$

By solving for λ we obtain the results

$$\lambda_1 = 3.0, \quad \lambda_2 = 4.5, \quad \lambda_3 = 7.0 \tag{4.73}$$

Let $\lambda = \lambda_1$ and substitute into Equation 4.71:

$$
\begin{aligned}
2.0625a_1^{(1)} + 1.1908a_2^{(1)} + 1.0825a_3^{(1)} &= 0 \\
1.1908a_1^{(1)} + 0.6875a_2^{(1)} + 0.6250a_3^{(1)} &= 0 \\
1.0825a_1^{(1)} + 0.6250a_2^{(1)} + 3.750a_3^{(1)} &= 0
\end{aligned}
\tag{4.74}
$$

where the superscript (1) refers to λ_1. Since Equations 4.74 are dependent, we can obtain specific values of the $a_i^{(1)}$ by using Equation 4.67:

$$\left(a_1^{(1)}\right)^2 + \left(a_2^{(1)}\right)^2 + \left(a_3^{(1)}\right)^2 = 1 \tag{4.75}$$

By selecting any two Equations 4.74 and using Equation 4.75, we obtain the results

$$a_1^{(1)} = 0.5, \quad a_2^{(1)} = -0.866, \quad a_3^{(1)} = 0.0 \tag{4.76}$$

Similarly, if we let $\lambda = \lambda_2 = 4.5$, we obtain

$$a_1^{(2)} = 0.6124, \quad a_2^{(2)} = 0.3536, \quad a_3^{(2)} = -0.707 \tag{4.77}$$

Finally, if we let $\lambda = \lambda_3 = 7.0$, we obtain

$$a_1^{(3)} = 0.6124, \quad a_2^{(3)} = 0.3536, \quad a_3^{(3)} = 0.7071 \tag{4.78}$$

To summarize these results, the principal stresses are

$$\sigma_1 = 3.0 \times 10^4 \, \text{psi}, \quad \sigma_2 = 4.5 \times 10^4 \, \text{psi}, \quad \sigma_3 = 7.0 \times 10^4 \, \text{psi} \tag{4.79}$$

and the corresponding principal directions are defined by the unit eigenvectors:

$$
\begin{aligned}
\mathbf{n}_a^{(1)} &= 0.5\mathbf{n}_1 - 0.866\mathbf{n}_2 + 0\mathbf{n}_3 \\
\mathbf{n}_a^{(2)} &= 0.6124\mathbf{n}_1 + 0.3536\mathbf{n}_2 - 0.7071\mathbf{n}_3 \\
\mathbf{n}_a^{(3)} &= 0.6124\mathbf{n}_1 + 0.3536\mathbf{n}_2 + 0.7071\mathbf{n}_3
\end{aligned}
\tag{4.80}
$$

Next, suppose that we form a transformation matrix T whose columns are the components of these unit eigenvectors. That is,

$$
T = \begin{bmatrix}
0.5 & 0.6124 & 0.6124 \\
-0.866 & 0.3536 & 0.3536 \\
0 & -0.7071 & 0.7071
\end{bmatrix}
\tag{4.81}
$$

Let $\hat{\sigma}$ be the stress matrix with the principal stresses on the diagonal. That is,

$$\hat{\sigma} = \begin{bmatrix} 3.0 & 0 & 0 \\ 0 & 4.5 & 0 \\ 0 & 0 & 7.0 \end{bmatrix} 10^4 \, \text{psi} \tag{4.82}$$

Then, we have the relation:

$$\hat{\sigma} = T^T \sigma T \tag{4.83}$$

where T^T is the transpose of T. That is,

$$\begin{bmatrix} 3.0 & 0.0 & 0.0 \\ 0.0 & 4.5 & 0.0 \\ 0.0 & 0.0 & 7.0 \end{bmatrix} = \begin{bmatrix} 0.5 & -0.866 & 0 \\ 0.6124 & 0.3536 & -0.7071 \\ 0.6124 & 0.3536 & 0.7071 \end{bmatrix} \begin{bmatrix} 5.0625 & 1.1908 & 1.0825 \\ 1.1908 & 3.6875 & 0.6250 \\ 1.0825 & 0.6250 & 5.75 \end{bmatrix}$$
$$\times \begin{bmatrix} 0.5 & 0.6124 & 0.6124 \\ -0.866 & 0.3536 & 0.3536 \\ 0 & -0.7071 & 0.7071 \end{bmatrix} \tag{4.84}$$

4.5.2 Discussion

Observe in the foregoing analysis that the eigenvalues (the roots of the Hamilton–Cayley equation, Equation 4.72) are real. Observe further that the associated unit eigenvectors of Equations 4.81 are mutually perpendicular, as predicted earlier. It happens that with the stress matrix being symmetric, the roots of the Hamilton–Cayley equation are always real and there always exist three mutually perpendicular unit eigenvectors.

Suppose that instead of there being three distinct eigenvalues, two of them are equal. In this case, it happens that every unit vector, which is perpendicular to the unit eigenvector of the distinct eigenvalue is a unit eigenvector. That is, there are an infinite number of unit eigenvectors parallel to a plane normal to the unit eigenvector of the distinct eigenvalue. If all three of the eigenvalues are equal, every unit vector is a unit eigenvector. That is, all directions are principal directions and we have a state of "hydrostatic pressure."

Finally, observe that the set of eigenvalues, or principal stresses, contain values that are both larger (7×10^4 psi) and smaller (3×10^4 psi) than the normal stresses on the diagonal of the stress matrix of Equation 4.69. We discuss these concepts in more detail in the next section.

4.6 EIGENVALUES AND EIGENVECTORS—THEORETICAL CONSIDERATIONS

In the foregoing discussion, several claims were made about the roots of the Hamilton–Cayley equation and about the associated unit eigenvectors. Specifically, it is claimed that the roots are real and that they contain the values of the maximum and minimum normal stresses. It is also claimed that associated with these roots (or eigenvalues), there exist mutually perpendicular eigenvectors. In this section, we discuss these claims. In subsequent sections, we also show that values of the maximum shear stresses occur on planes inclined at 45° to the planes normal to the unit eigenvectors.

4.6.1 Maximum and Minimum Normal Stresses

Let $\mathbf{n}_a$ be an arbitrary unit vector, and σ be a stress dyadic. Let $\mathbf{n}_i$ ($i = 1, 2, 3$) form a set of mutually perpendicular unit vectors and let $\mathbf{n}_a$ and σ be expressed in terms of the $\mathbf{n}_i$ as

$$\mathbf{n}_a = a_i\mathbf{n}_i \quad \text{and} \quad \sigma = \sigma_{ij}\mathbf{n}_i\mathbf{n}_j \tag{4.85}$$

Then, a review of Equations 4.22 and 4.24 shows that the stress vector $\mathbf{S}_a$ for $\mathbf{n}_a$ and the normal stress σ_{aa} on a plane normal to $\mathbf{n}_a$ are

$$\mathbf{S}_a = \sigma \cdot \mathbf{n}_a = \mathbf{n}_i\sigma_{ij}a_j \quad \text{and} \quad \sigma_{aa} = a_i\sigma_{ij}a_j \tag{4.86}$$

The issue of finding out maximum and minimum values for the normal stress, σ_{aa}, then becomes the problem of finding out the a_i producing the maximum/minimum σ_{aa}. This is a constrained maximum/minimum problem because $\mathbf{n}_a$ is a unit vector, the a_i must satisfy the relation:

$$a_i a_i = 1 \tag{4.87}$$

We can obtain a_i producing the maximum/minimum σ_{aa} subject to the constraint of Equation 4.87 by using the Lagrange multiplier method [5,6]: Let $f(a_i)$ be defined as

$$f(a_i) = \sigma_{aa} + \lambda(1 - a_i a_i) = a_i\sigma_{ij}a_j + \lambda(1 - a_{ii}) \tag{4.88}$$

where λ is a Lagrange multiplier. Then f will have maximum/minimum (extremum) values when

$$\partial f/\partial a_k = 0, \quad k = 1, 2, 3 \tag{4.89}$$

By substituting from Equation 4.88 into 4.89, we have

$$\delta_{ki}\sigma_{ij}a_j + a_i\sigma_{ij}\delta_{jk} - 2\lambda a_i\delta_{ik} = 0 \tag{4.90}$$

where we have used Equation 4.53.* Then by using the properties of δ_{ij} and the symmetry of σ we obtain

$$\sigma_{kj}a_j + a_i\sigma_{ik} - 2\lambda a_k = 0$$

or

$$\sigma_{kj}a_j = \lambda a_k \tag{4.91}$$

By comparing Equations 4.59 and 4.91 we see that the values of the a_i, which produce the eigenvectors are the same a_i, which produce extremal values (maximum/minimum) of the normal stresses. Moreover, the Lagrange multipliers are the eigenvalues.

4.6.2 Real Solutions of the Hamilton–Cayley Equation

To see that the roots of the Hamilton–Cayley equation (Equation 4.63) are real, suppose the contrary, that they are not real. Then, with the equation being a cubic polynomial, there

---

* Note that $\partial a_k/\partial a_k$ is 0 if $i \neq k$ and 1 if $i = k$, that is $\partial a_i/\partial a_k = \delta_{ik}$.

will be one real root and a pair of complex conjugate roots [7]. Let these imaginary roots have the form:

$$\lambda = \mu + iv \quad \text{and} \quad \bar{\lambda} = \mu - iv \tag{4.92}$$

with i being $\sqrt{-1}$.

With imaginary roots, the resulting eigenvectors will also be imaginary. That is, they will have the form:

$$\mathbf{n} = \mathbf{u} + i\mathbf{v} \tag{4.93}$$

where $\mathbf{u}$ and $\mathbf{v}$ are real vectors. Then, from Equation 4.48 we have

$$\sigma \bullet \mathbf{n} = \lambda \mathbf{n} \tag{4.94}$$

or

$$\begin{aligned}
\sigma \bullet (\mathbf{u} + i\mathbf{v}) &= (\mu + iv)(\mathbf{u} + i\mathbf{v}) \\
&= (\mu\mathbf{u} - v\mathbf{v}) + i(v\mathbf{u} + \mu\mathbf{v})
\end{aligned} \tag{4.95}$$

Equating the real and imaginary parts (recalling that σ is real), we have

$$\sigma \bullet \mathbf{u} = \mu\mathbf{u} - v\mathbf{v} \tag{4.96}$$

and

$$\sigma \bullet \mathbf{v} = v\mathbf{u} + \mu\mathbf{v} \tag{4.97}$$

If we multiply the terms of Equation 4.96 by $\mathbf{v}$ and those of Equation 4.97 by $\mathbf{u}$ and subtract, we obtain

$$\mathbf{v} \bullet \sigma \bullet \mathbf{u} - \mathbf{u} \bullet \sigma \bullet \mathbf{v} = -v(\mathbf{v} \bullet \mathbf{v} + \mathbf{u} \bullet \mathbf{u}) \tag{4.98}$$

But, since σ is symmetric, the left-hand side is zero and thus we have

$$0 = v(\mathbf{v}^2 + \mathbf{u}^2) \tag{4.99}$$

Finally, since $(\mathbf{v}^2 + \mathbf{u}^2)$ is positive (otherwise the eigenvector would be zero) we have

$$v = 0 \tag{4.100}$$

Therefore, from Equation 4.92, the roots (or eigenvalues) are found to be real.

4.6.3 Mutually Perpendicular Unit Eigenvectors

Suppose that λ_a and λ_b are distinct roots of the Hamilton–Cayley equation with corresponding unit eigenvectors $\mathbf{n}_a$ and $\mathbf{n}_b$. Then, from Equation 4.48

$$\sigma \bullet \mathbf{n}_a = \lambda_a \mathbf{n}_a \quad \text{and} \quad \sigma \bullet \mathbf{n}_b = \lambda_b \mathbf{n}_b \tag{4.101}$$

If we multiply the first expression by $\mathbf{n}_b$ and the second by $\mathbf{n}_a$ and subtract, we obtain

$$\mathbf{n}_b \bullet \sigma \bullet \mathbf{n}_a - \mathbf{n}_a \bullet \sigma \bullet \mathbf{n}_b = (\lambda_a - \lambda_b)\mathbf{n}_a \bullet \mathbf{n}_b \tag{4.102}$$

Since σ is symmetric, the left-hand side of Equation 4.102 is zero and since λ_a and λ_b are distinct, we have

$$\mathbf{n}_a \bullet \mathbf{n}_b = 0 \tag{4.103}$$

That is, $\mathbf{n}_a$ is perpendicular to $\mathbf{n}_b$. Therefore, if we have three distinct eigenvalues, we will have three mutually perpendicular unit eigenvectors.

Next, let $\mathbf{n}_a$, $\mathbf{n}_b$, and $\mathbf{n}_c$ be a set of mutually perpendicular unit eigenvectors and let them be expressed as

$$\begin{aligned}
\mathbf{n}_a &= a_1\mathbf{n}_1 + a_2\mathbf{n}_2 + a_3\mathbf{n}_3 = a_i\mathbf{n}_i \\
\mathbf{n}_b &= b_1\mathbf{n}_1 + b_2\mathbf{n}_2 + b_3\mathbf{n}_3 = b_i\mathbf{n}_i \\
\mathbf{n}_c &= c_1\mathbf{n}_1 + c_2\mathbf{n}_2 + c_3\mathbf{n}_3 - c_i\mathbf{n}_i
\end{aligned} \tag{4.104}$$

where the $\mathbf{n}_i$ form a convenient set of mutually perpendicular unit vectors. In matrix form, these equations may be written as

$$\begin{bmatrix} \mathbf{n}_a \\ \mathbf{n}_b \\ \mathbf{n}_c \end{bmatrix} = \begin{bmatrix} a_1 & a_2 & a_3 \\ b_1 & b_2 & b_3 \\ c_1 & c_2 & c_3 \end{bmatrix} \begin{bmatrix} \mathbf{n}_1 \\ \mathbf{n}_2 \\ \mathbf{n}_3 \end{bmatrix} = S \begin{bmatrix} \mathbf{n}_1 \\ \mathbf{n}_2 \\ \mathbf{n}_3 \end{bmatrix} \tag{4.105}$$

where T is a matrix defined by inspection. Since $\mathbf{n}_a$, $\mathbf{n}_b$, $\mathbf{n}_c$ and $\mathbf{n}_1$, $\mathbf{n}_2$, $\mathbf{n}_3$ are mutually perpendicular unit vector sets, it is readily seen that S is an orthogonal matrix. That is, the inverse is the transpose. Therefore, we can readily solve Equation 4.105 for the $\mathbf{n}_i$ as

$$\begin{bmatrix} \mathbf{n}_1 \\ \mathbf{n}_2 \\ \mathbf{n}_3 \end{bmatrix} = S^{\mathrm{T}} \begin{bmatrix} \mathbf{n}_a \\ \mathbf{n}_b \\ \mathbf{n}_c \end{bmatrix} = \begin{bmatrix} a_1 & b_1 & c_1 \\ a_2 & b_2 & c_2 \\ a_3 & b_3 & c_3 \end{bmatrix} \begin{bmatrix} \mathbf{n}_a \\ \mathbf{n}_b \\ \mathbf{n}_c \end{bmatrix} \tag{4.106}$$

or

$$\begin{aligned}
\mathbf{n}_1 &= a_1\mathbf{n}_a + b_1\mathbf{n}_b + c_1\mathbf{n}_c \\
\mathbf{n}_2 &= a_2\mathbf{n}_a + b_2\mathbf{n}_b + c_2\mathbf{n}_c \\
\mathbf{n}_3 &= a_3\mathbf{n}_a + b_3\mathbf{n}_b + c_3\mathbf{n}_c
\end{aligned} \tag{4.107}$$

Recall that since λ_a, λ_b, and λ_c are eigenvectors (or principal stresses) and $\mathbf{n}_a$, $\mathbf{n}_b$, and $\mathbf{n}_c$ are unit eigenvectors, the stress dyadic $\boldsymbol{\sigma}$ may be expressed as

$$\boldsymbol{\sigma} = \sigma_{ij}\mathbf{n}_i\mathbf{n}_j = \lambda_a\mathbf{n}_a\mathbf{n}_a + \lambda_b\mathbf{n}_b\mathbf{n}_b + \lambda_c\mathbf{n}_c\mathbf{n}_c \tag{4.108}$$

Then, by substituting from Equations 4.105 and 4.106, we obtain the expression:

$$\begin{bmatrix} a_1 & a_2 & a_3 \\ b_1 & b_2 & b_3 \\ c_1 & c_2 & c_3 \end{bmatrix} \begin{bmatrix} \sigma_{11} & \sigma_{12} & \sigma_{13} \\ \sigma_{21} & \sigma_{22} & \sigma_{23} \\ \sigma_{31} & \sigma_{32} & \sigma_{33} \end{bmatrix} \begin{bmatrix} a_1 & b_l & c_1 \\ a_2 & b_2 & c_2 \\ a_3 & b_3 & c_3 \end{bmatrix} = \begin{bmatrix} \lambda_a & 0 & 0 \\ 0 & \lambda_b & 0 \\ 0 & 0 & \lambda_c \end{bmatrix} \tag{4.109}$$

By comparing Equation 4.109 with Equations 4.83 and 4.84, we see that $S^{\mathrm{T}} = T$ and $S = T^{\mathrm{T}}$ (see Section 4.5.1).

4.6.4 Multiple (Repeated) Roots of the Hamilton–Cayley Equation

If two of the roots of Equation 4.63 are equal, or if all three roots are equal, there still exist sets of mutually perpendicular unit eigenvectors. To see this, recall from algebraic analysis that finding the roots of a polynomial equation is equivalent to factoring the equation [7]. That is, if we know the roots, say λ_1, λ_2 and λ_3, of Equation 4.63, we know the factors. This means that the following equations are equivalent:

$$\lambda^3 - \sigma_I \lambda^2 + \sigma_{II} \lambda - \sigma_{III} = 0 \tag{4.110}$$

and

$$(\lambda - \lambda_1)(\lambda - \lambda_2)(\lambda - \lambda_3) = 0 \tag{4.111}$$

By expanding Equation 4.111 and then comparing the coefficients with those of Equation 4.110 we see that

$$\lambda_1 + \lambda_2 + \lambda_3 = \sigma_I \tag{4.112}$$

$$\lambda_1 + \lambda_2 + \lambda_2\lambda_3 + \lambda_3\lambda_1 = \sigma_{II} \tag{4.113}$$

$$\lambda_1 \lambda_2 \lambda_3 = \sigma_{III} \tag{4.114}$$

Now, suppose that two of the roots, say λ_1 and λ_2, are equal. Let $\mathbf{n}_b$ and $\mathbf{n}_c$ be unit eigenvectors associated with roots λ_2 and λ_3. Then, with λ_2 and λ_3 being distinct, the foregoing analysis (Section 4.6.3) shows that $\mathbf{n}_b$ and $\mathbf{n}_c$ are perpendicular. Let $\mathbf{n}_a$ be $\mathbf{n}_b \times \mathbf{n}_c$. Then, $\mathbf{n}_a$, $\mathbf{n}_b$, and $\mathbf{n}_c$ form a mutually perpendicular set of unit vectors.

Consider the vector $\boldsymbol{\sigma} \cdot \mathbf{n}_a$. Since $\mathbf{n}_a$, $\mathbf{n}_b$, and $\mathbf{n}_c$ form a mutually perpendicular set, let $\boldsymbol{\sigma} \cdot \mathbf{n}_a$ be expressed as

$$\boldsymbol{\sigma} \cdot \mathbf{n}_a = \alpha \mathbf{n}_a + \beta \mathbf{n}_b + \gamma \mathbf{n}_c \tag{4.115}$$

where α, β, and γ are scalars to be determined. Observe that, being a dyadic, σ may be expressed as

$$\boldsymbol{\sigma} = \boldsymbol{\sigma} \cdot \mathbf{I} = \boldsymbol{\sigma} \cdot (\mathbf{n}_a \mathbf{n}_a + \mathbf{n}_b \mathbf{n}_b + \mathbf{n}_c \mathbf{n}_c)$$
$$= (\boldsymbol{\sigma} \cdot \mathbf{n}_a)\mathbf{n}_a + (\boldsymbol{\sigma} \cdot \mathbf{n}_b)\mathbf{n}_b + \boldsymbol{\sigma} \cdot \mathbf{n}_c)\mathbf{n}_c \tag{4.116}$$

where $\mathbf{I}$ is the identity dyadic (see Equation 4.56). Since $\mathbf{n}_a$ and $\mathbf{n}_b$ are unit eigenvectors, we have (see Equation 4.49)

$$\boldsymbol{\sigma} \cdot \mathbf{n}_b = \lambda_2 \mathbf{n}_b \quad \text{and} \quad \boldsymbol{\sigma} \cdot \mathbf{n}_c = \lambda_3 \mathbf{n}_c \tag{4.117}$$

By substituting from Equations 4.115 and 4.117 into 4.116 σ is seen to have the form:

$$\boldsymbol{\sigma} = (\alpha \mathbf{n}_a + \beta \mathbf{n}_b + \gamma \mathbf{n}_c)\mathbf{n}_a + \lambda_2 \mathbf{n}_b \mathbf{n}_b + \lambda_3 \mathbf{n}_c \mathbf{n}_c$$
$$= \alpha \mathbf{n}_a \mathbf{n}_a + \beta \mathbf{n}_b \mathbf{n}_a + \gamma \mathbf{n}_c \mathbf{n}_a + \lambda_2 \mathbf{n}_b \mathbf{n}_b + \lambda_3 \mathbf{n}_c \mathbf{n}_c \tag{4.118}$$

Relative to $\mathbf{n}_a$, $\mathbf{n}_b$, and $\mathbf{n}_c$, the matrix σ of $\boldsymbol{\sigma}$ is then

$$\sigma = \begin{bmatrix} \alpha & 0 & 0 \\ \beta & \lambda_2 & 0 \\ \gamma & 0 & \lambda_3 \end{bmatrix} \tag{4.119}$$

But since σ must be symmetric, we have

$$\beta = \gamma = 0 \qquad (4.120)$$

Hence, $\boldsymbol{\sigma}$ becomes

$$\boldsymbol{\sigma} = \alpha\mathbf{n}_a\mathbf{n}_a + \gamma_2\mathbf{n}_b\mathbf{n}_b + \lambda_3\mathbf{n}_c\mathbf{n}_c \qquad (4.121)$$

and $\boldsymbol{\sigma}\cdot\mathbf{n}_a$ is

$$\boldsymbol{\sigma}\cdot\mathbf{n}_a = \alpha\mathbf{n}_a \qquad (4.122)$$

Therefore $\mathbf{n}_a$ is also a unit eigenvector. Moreover, from Equation 4.122 we see that

$$\alpha + \lambda_2\lambda_3 = \lambda_1 + \lambda_2 + \lambda_3 \qquad (4.123)$$

That is,

$$\alpha = \lambda_1 = \lambda_2 \qquad (4.124)$$

It happens that *any* unit vector parallel to the plane of $\mathbf{n}_a$ and $\mathbf{n}_b$ is a unit eigenvector. To see this, let $\mathbf{n}$ be the unit vector

$$\mathbf{n} = a\mathbf{n}_a + b\mathbf{n}_b \qquad (4.125)$$

with $a^2 + b^2 = 1$. Then,

$$\begin{aligned}
\boldsymbol{\sigma}\cdot\mathbf{n} = \boldsymbol{\sigma}\cdot(a\mathbf{n}_a + b\mathbf{n}_b) &= a\boldsymbol{\sigma}\cdot\mathbf{n}_a + b\boldsymbol{\sigma}\cdot\mathbf{n}_b \\
&= a\lambda_1\mathbf{n}_a + b\lambda_2\mathbf{n}_b = \lambda_1(a\mathbf{n}_a + b\mathbf{n}_b) \\
&= \lambda_1\mathbf{n}
\end{aligned} \qquad (4.126)$$

Thus, $\mathbf{n}$ is a unit eigenvector.

Similarly, suppose that all three roots of the Hamilton–Cayley equation are equal. That is,

$$\lambda_1 = \lambda_2 = \lambda_3 = \lambda \qquad (4.127)$$

Let $\mathbf{n}_a$ be a unit eigenvector associated with the root λ and let $\mathbf{n}_b$ and $\mathbf{n}_c$ be unit vectors perpendicular to $\mathbf{n}_a$ and to each other. Then we have the expressions:

$$\boldsymbol{\sigma}\cdot\mathbf{n}_a = \lambda\mathbf{n}_a \qquad (4.128)$$

and

$$\boldsymbol{\sigma}\cdot\mathbf{n}_b = \alpha\mathbf{n}_a + \beta\mathbf{n}_b + \gamma\mathbf{n}_c, \quad \boldsymbol{\sigma}\cdot\mathbf{n}_c = \hat{\alpha}\mathbf{n}_a + \hat{\beta}\mathbf{n}_b + \hat{\gamma}\mathbf{n}_c \qquad (4.129)$$

The objective is thus to determine α, β, γ, $\hat{\alpha}$, $\hat{\beta}$, and $\hat{\gamma}$.

From Equation 4.116, σ may then be expressed as

$$\begin{aligned}
\boldsymbol{\sigma} = {}& \lambda\mathbf{n}_a\mathbf{n}_a + (\alpha\mathbf{n}_a + \beta\mathbf{n}_b + \gamma\mathbf{n}_c)\mathbf{n}_b + (\hat{\alpha}\mathbf{n}_a + \hat{\beta}\mathbf{n}_b + \hat{\gamma}\mathbf{n}_c)\mathbf{n}_c \\
= {}& \lambda\mathbf{n}_a\mathbf{n}_a + \alpha\mathbf{n}_a\mathbf{n}_b + \hat{\alpha}\mathbf{n}_a\mathbf{n}_c \\
& + 0\mathbf{n}_b\mathbf{n}_a + \beta\mathbf{n}_b\mathbf{n}_b + \hat{\beta}\mathbf{n}_b\mathbf{n}_c \\
& + 0\mathbf{n}_c\mathbf{n}_a + \gamma\mathbf{n}_c\mathbf{n}_b + \hat{\gamma}\mathbf{n}_c\mathbf{n}_c
\end{aligned} \qquad (4.130)$$

Relative to $\mathbf{n}_a$, $\mathbf{n}_b$, and $\mathbf{n}_c$, the matrix σ of σ is then

$$\sigma = \begin{bmatrix} \lambda & \alpha & \hat{\alpha} \\ 0 & \beta & \hat{\beta} \\ 0 & \gamma & \hat{\gamma} \end{bmatrix} \tag{4.131}$$

Since σ is to be symmetric, we have

$$\alpha = \hat{\alpha} = 0 \quad \text{and} \quad \hat{\beta} = \gamma \tag{4.132}$$

By using Equations 4.112 through 4.114 and Equations 4.127 and 4.131 we see that

$$\sigma_{\mathrm{I}} = \lambda + \beta + \hat{\gamma} = 3\lambda \tag{4.133}$$

$$\sigma_{\mathrm{II}} = (\beta\hat{\gamma} - \gamma^2) + \lambda\hat{\gamma} + \lambda\beta = 3\lambda^2 \tag{4.134}$$

$$\sigma_{\mathrm{III}} = \lambda(\beta\hat{\gamma} - \gamma^2) = \lambda^3 \tag{4.135}$$

(Recall that σ_{I}, σ_{II}, and σ_{III} are respectively the sums of the diagonal elements, the sum of the diagonal elements of the matrix of cofactors, and the determinant of the stress matrix.) These equations are found to be redundant,* but a simple solution will be

$$\beta = \hat{\gamma} = \lambda \quad \text{and} \quad \gamma = 0 \tag{4.136}$$

Equations 4.129 then become

$$\sigma \cdot \mathbf{n}_b = \lambda \mathbf{n}_b \quad \text{and} \quad \sigma \cdot \mathbf{n}_c = \lambda \mathbf{n}_c \tag{4.137}$$

Therefore, $\mathbf{n}_b$ and $\mathbf{n}_c$ are unit eigenvectors. In this case, when all three roots of the Hamilton–Cayley equation are equal, *every* unit vector is a unit eigenvector. To see this let $\mathbf{n}$ be the unit vector.

$$\mathbf{n} = a\mathbf{n}_a + b\mathbf{n}_b + c\mathbf{n}_c \tag{4.138}$$

with $a^2 + b^2 + c^2 = 1$. Then,

$$\begin{aligned} \sigma \cdot \mathbf{n} &= \sigma \cdot (a\mathbf{n}_a + b\mathbf{n}_b + c\mathbf{n}_c) \\ &= a\sigma \cdot \mathbf{n}_a + b\sigma \cdot \mathbf{n}_b + c\sigma \cdot \mathbf{n}_c \\ &= a\lambda\mathbf{n}_a + b\lambda\mathbf{n}_b + c\lambda\mathbf{n}_c \\ &= \lambda(a\mathbf{n}_c + b\mathbf{n}_b + c\mathbf{n}_c) \\ &= \lambda\mathbf{n} \end{aligned} \tag{4.139}$$

4.7   STRESS ELLIPSOID

We can obtain a useful geometrical interpretation of the eigenvalue analysis by regarding the product $\sigma \cdot \mathbf{p}$ as an operator on the vector $\mathbf{p}$. That is, as an operator, σ transforms the vector $\mathbf{p}$ into the vector $\mathbf{q}$ as

$$\sigma \cdot \mathbf{p} = \mathbf{q} \tag{4.140}$$

---

* Observe that the issue of redundancy in Equations 4.133 through 4.135 may be addressed by considering an analogous two-dimensional analysis with a stress matrix $\begin{bmatrix} \beta & \hat{\beta} \\ \gamma & \hat{\gamma} \end{bmatrix}$.

Let $\mathbf{p}$ be a position vector from the origin of a Cartesian axis system to the surface of the unit sphere. Specifically, let $\mathbf{p}$ have the form:

$$\mathbf{p} = x\mathbf{n}_a + y\mathbf{n}_b + z\mathbf{n}_c \tag{4.141}$$

with

$$x^2 + y^2 + z^2 = 1 \tag{4.142}$$

and with $\mathbf{n}_a$, $\mathbf{n}_b$, and $\mathbf{n}_c$ being mutually perpendicular unit eigenvectors. (That is, let the X-, Y-, and Z-axes be along the principal directions of the stress of a body at a point.) Then $\boldsymbol{\sigma} \cdot \mathbf{p}$ becomes

$$\boldsymbol{\sigma} \cdot \mathbf{p} = \boldsymbol{\sigma} \cdot (x\mathbf{n}_a + y\mathbf{n}_b + z\mathbf{n}_c) = \lambda_1 x\mathbf{n}_a + \lambda_2 y\mathbf{n}_b + \lambda_3 z\mathbf{n}_c \tag{4.143}$$

From Equation 4.140, if we let $\mathbf{q}$ be $\boldsymbol{\sigma} \cdot \mathbf{p}$ and express $\mathbf{q}$ in the form:

$$\mathbf{q} = X\mathbf{n}_a + Y\mathbf{n}_b + Z\mathbf{n}_c \tag{4.144}$$

then we have

$$X = \lambda_1 x, \quad Y = \lambda_2 y, \quad Z = \lambda_3 z \tag{4.145}$$

Using Equation 4.142 we then have

$$\frac{X^2}{\lambda_1^2} + \frac{Y^2}{\lambda_2^2} + \frac{Z^2}{\lambda_3^2} = 1 \tag{4.146}$$

We can recognize Equation 4.146 as the equation of an ellipsoid with center at the origin and semimajor axes: λ_1, λ_2, and λ_3. This ellipsoid is called the "stress ellipsoid." In Equations 4.140 and 4.142, if we think of $\mathbf{p}$ as a unit vector, then $\mathbf{q}$ is a stress vector and the units of X, Y, and Z are the units of stress.

If $\mathbf{n}$ is an arbitrary unit vector, we see from Equation 4.24 that the stress vector $\mathbf{S}_n$ associated with $\mathbf{n}$ is

$$\mathbf{S}_n = \boldsymbol{\sigma} \cdot \mathbf{n} \tag{4.147}$$

The normal stress $\mathbf{S}_{nn}$ (or σ_{nn}) on a plane normal to $\mathbf{n}$ is then

$$\mathbf{S}_{nn} = \mathbf{n} \cdot \boldsymbol{\sigma} \cdot \mathbf{n} \tag{4.148}$$

From Equations 4.140 and 4.142, $\mathbf{p}$ is a unit vector, we can identify $\mathbf{p}$ with $\mathbf{n}$ and then using Equation 4.144 we have

$$\begin{aligned} \sigma_{nn} &= \mathbf{n} \cdot \boldsymbol{\sigma} \cdot \mathbf{n} = \mathbf{n} \cdot \mathbf{q} \\ &= \mathbf{n} \cdot (X\mathbf{n}_a + Y\mathbf{n}_b + Z\mathbf{n}_c) \end{aligned} \tag{4.149}$$

Therefore, we can interpret σ_{nn} as the distance from the origin of the stress ellipsoid to a point Q on the surface of the ellipsoid, where $\mathbf{n}$ is parallel to $\mathbf{OQ}$. Observe then that the maximum and minimum stresses will occur in the directions of the principal stresses with values among the eigenvalues ($\lambda_1, \lambda_2, \lambda_3$.), or the semimajor and semiminor axes lengths.

Finally, observe from the ellipsoid equation that if two of the eigenvalues, say λ_1 and λ_2 are equal, then the ellipsoid has a circular cross section. If all three eigenvalues are equal, the ellipsoid becomes a sphere and we have a state of "hydrostatic pressure" (see Section 4.5.2).

4.8 MAXIMUM SHEAR STRESS

Consider again the stress dyadic $\boldsymbol{\sigma}$ at a point on a loaded elastic body. Let λ_1, λ_2, and λ_3 be the values of the principal stresses and let $\mathbf{a}_1$, $\mathbf{a}_2$, and $\mathbf{a}_3$ be the corresponding mutually perpendicular unit eigenvectors. Then from Equation 4.49 we have

$$\boldsymbol{\sigma} \boldsymbol{\cdot} \mathbf{a}_1 = \lambda_1 \mathbf{a}_1, \quad \boldsymbol{\sigma} \boldsymbol{\cdot} \mathbf{a}_2 = \lambda_2 \mathbf{a}_2, \quad \boldsymbol{\sigma} \boldsymbol{\cdot} \mathbf{a}_3 = \lambda_3 \mathbf{a}_3 \tag{4.150}$$

Next, let $\mathbf{n}_1$, $\mathbf{n}_2$, and $\mathbf{n}_3$ be any convenient set of mutually perpendicular vectors and let $\mathbf{n}_a$ and $\mathbf{n}_b$ be an arbitrary pair of perpendicular unit vectors with components a_i and b_i relative to the $\mathbf{n}_i$ $(i = 1, 2, 3)$. That is,

$$\mathbf{n}_a = a_i \mathbf{n}_i \quad \text{and} \quad \mathbf{n}_b = b_i \mathbf{n}_i \tag{4.151}$$

Then, the shear stress σ_{ab} for the directions of $\mathbf{n}_a$ and $\mathbf{n}_b$ is

$$\sigma_{ab} = \mathbf{n}_a \boldsymbol{\cdot} \boldsymbol{\sigma} \boldsymbol{\cdot} \mathbf{n}_b = a_j \sigma_{ij} b_j \tag{4.152}$$

Observe that since $\mathbf{n}_a$ and $\mathbf{n}_b$ are perpendicular unit vectors we also have the relations:

$$a_i a_i = 1, \quad b_i b_i = 1, \quad a_i b_i = 0 \tag{4.153}$$

Now, suppose that we are interested in finding the directions of $\mathbf{n}_a$ and $\mathbf{n}_b$ producing the maximum values of the shear stress σ_{ab}. Then, we will be looking for the a_i and b_i, subject to the conditions of Equations 4.153, so that $a_i \sigma_{ij} b_j$ is maximum. Using the Lagrange multiplier method [56], we are looking for the a_i and b_i, which maximize the function: $\phi(a_i, b_i)$ given by

$$\phi = a_i \sigma_{ij} b_j + \alpha(1 - a_i a_i) + \beta(1 - b_i b_i) + \gamma(0 - a_i b_i) \tag{4.154}$$

where α, β, and γ are Lagrange multipliers and the parenthetical expressions are obtained from Equations 4.153. Then if ϕ is to be maximum, we must have

$$\partial \phi / \partial a_i = 0 \quad \text{and} \quad \partial \phi / \partial b_i = 0 \tag{4.155}$$

or from Equation 4.154,

$$\partial \phi / \partial a_i = \sigma_{ij} b_j - 2\alpha a_i - \gamma b_i = 0 \tag{4.156}$$

and

$$\partial \phi / \partial b_i = a_i \sigma_{ji} - 2\beta b_i - \gamma a_i = 0 \tag{4.157}$$

Equations 4.156 and 4.157 may be written in index-free notation as

$$\boldsymbol{\sigma} \boldsymbol{\cdot} \mathbf{n}_b = 2\alpha \mathbf{n}_a + \gamma \mathbf{n}_b \tag{4.158}$$

and

$$\boldsymbol{\sigma} \cdot \mathbf{n}_a = 2\beta \mathbf{n}_b + \gamma \mathbf{n}_a \tag{4.159}$$

By taking the scalar product of these equations with $\mathbf{n}_a$ and $\mathbf{n}_b$, we obtain

$$\mathbf{n}_a \cdot \boldsymbol{\sigma} \cdot \mathbf{n}_b = \sigma_{ab} = 2\alpha, \quad \mathbf{n}_b \cdot \boldsymbol{\sigma} \cdot \mathbf{n}_b = \sigma_{bb} = \gamma \tag{4.160}$$

$$\mathbf{n}_b \cdot \boldsymbol{\sigma} \cdot \mathbf{n}_a = \sigma_{ba} = 2\beta, \quad \mathbf{n}_a \cdot \boldsymbol{\sigma} \cdot \mathbf{n}_a = \sigma_{aa} = \gamma \tag{4.161}$$

Since $\sigma_{ab} = \sigma_{ba}$ we have $\alpha = \beta$. Then, by successively adding and subtracting Equations 4.158 and 4.159 we obtain the expressions:

$$\boldsymbol{\sigma} \cdot (\mathbf{n}_a + \mathbf{n}_b) = (2\alpha + \gamma)(\mathbf{n}_a + \mathbf{n}_b) \tag{4.162}$$

and

$$\boldsymbol{\sigma} \cdot (\mathbf{n}_a - \mathbf{n}_b) = (\gamma - 2\alpha)(\mathbf{n}_a - \mathbf{n}_b) \tag{4.163}$$

Equations 4.162 and 4.163 are identical to Equation 4.49 with $(\mathbf{n}_a + \mathbf{n}_b)$ and $(\mathbf{n}_a - \mathbf{n}_b)$ now being eigenvectors, and $(2\alpha + \gamma)$ and $(\gamma - 2\alpha)$ being the eigenvalues. Therefore, $(\mathbf{n}_a + \mathbf{n}_b)/\sqrt{2}$ and $(\mathbf{n}_a - \mathbf{n}_b)/\sqrt{2}$ are unit eigenvectors along the directions of the principal stresses, and $(2\alpha + \gamma)$ and $(\gamma - 2\alpha)$ are thus values of the principal stresses. Observe further that $(\mathbf{n}_a + \mathbf{n}_b)/\sqrt{2}$ and $(\mathbf{n}_a - \mathbf{n}_b)/\sqrt{2}$ are perpendicular to each other and that $\mathbf{n}_a$ and $\mathbf{n}_b$ are at 45° angles to $(\mathbf{n}_a + \mathbf{n}_b)/\sqrt{2}$ and $(\mathbf{n}_a - \mathbf{n}_b)/\sqrt{2}$ respectively. That is, the maximum values of the shear stress occur on planes bisecting the planes of the principal stresses. Thus, with $(2\alpha + \gamma)$ and $(2\alpha - \gamma)$ being values of the principal stresses, we can make the assignments

$$2\alpha_1 + \gamma_1 = \lambda_1, \quad 2\alpha_2 + \gamma_2 = \lambda_2, \quad 2\alpha_3 + \gamma_3 = \lambda_3 \tag{4.163}$$

$$\gamma_1 - 2\alpha_1 = \lambda_2, \quad \gamma_2 - 2\alpha_2 = \lambda_3, \quad \gamma_3 - 2\alpha_3 = \lambda_1 \tag{4.164}$$

By solving for $2\alpha_1$, $2\alpha_2$, and $2\alpha_3$ we obtain

$$2\alpha_1 = (\lambda_1 - \lambda_2)/2, \quad 2\alpha_2 = (\lambda_2 - \lambda_3)/2, \quad 2\alpha_3 = (\lambda_3 - \lambda_1)/2 \tag{4.165}$$

But, from Equations 4.160 and 4.161, we see that these are the values of the maximum shear stresses.

From these results, we see that large shear stresses occur when there are large differences in the values of the principal stresses and that if a material fails in shear the failure will occur in directions at 45° relative to the direction of the principal stresses.

4.9 TWO-DIMENSIONAL ANALYSIS—MOHR'S CIRCLE

We can obtain additional insight into the concepts of principal stresses and maximum shear stresses by considering a two-dimensional analysis where the stress is primarily planar, or with the shear stresses in a given direction being zero. Consider, for example, the following stress matrix:

$$\sigma = [\sigma_{ij}] = \begin{bmatrix} \sigma_{11} & 0 & 0 \\ 0 & \sigma_{22} & \sigma_{23} \\ 0 & \sigma_{32} & \sigma_{33} \end{bmatrix} \tag{4.166}$$

where the subscripts i and j refer to mutually perpendicular unit vectors $\mathbf{n}_i$ ($i = 1, 2, 3$), parallel to X, Y, Z coordinate axes, with the stress dyadic σ having the usual form:

$$\sigma = \mathbf{n}_i \sigma_{ij} \mathbf{n}_j \tag{4.167}$$

In this context, we see that $\mathbf{n}_1$ is a unit eigenvector and σ_{11} is a principal stress. By following the procedures in Section 4.5, we can easily obtain the other two unit eigenvectors and principal stresses.

To this end, observe that with a stress distribution as in Equation 4.166, the determinantal expression of Equation 4.62 becomes

$$\begin{bmatrix} (\sigma_{11} - \lambda) & 0 & 0 \\ 0 & (\sigma_{32} - \lambda) & \sigma_{23} \\ 0 & \sigma_{32} & (\sigma_{33} - \lambda) \end{bmatrix} = 0 \tag{4.168}$$

By expanding the determinant, the Hamilton–Cayley equation takes the simplified form:

$$(\lambda - \sigma_{11})[\lambda^2 - (\sigma_{22} + \sigma_{33})\lambda + \sigma_{22}\sigma_{33} - \sigma_{23}^2] = 0 \tag{4.169}$$

where, due to the symmetry of the stress matrix, $\sigma_{32} = \sigma_{23}$. By solving Equation 4.169 for λ we obtain

$$\lambda_1 = \sigma_{11} \quad \text{and} \quad \lambda_2, \lambda_3 = \frac{\sigma_{22} + \sigma_{33}}{2} \pm \left[\left(\frac{\sigma_{22} - \sigma_{33}}{2} \right)^2 + \sigma_{23}^2 \right]^{1/2} \tag{4.170}$$

Let $\mathbf{n}_a$ be a unit eigenvector with components a_i relative to the $\mathbf{n}_i$ ($i = 1, 2, 4$). Then, from Equations 4.61 and 4.67, a_i must satisfy the equations:

$$\begin{aligned} (\sigma_{11} - \lambda)a_1 + 0a_2 + 0a_3 &= 0 \\ 0a_1 + (\sigma_{22} - \lambda)a_2 + \sigma_{23}a_3 &= 0 \\ 0a_1 + \sigma_{23}a_2 + (\sigma_{33} - \lambda)a_3 &= 0 \\ a_1^2 + a_2^2 + a_3^2 &= 1 \end{aligned} \tag{4.171}$$

with the first three being dependent. One immediate solution is

$$a_1^{(1)} = 1, \quad a_2^{(1)} = a_3^{(1)} = 0 \tag{4.172}$$

with the corresponding unit eigenvector being

$$\mathbf{n}_a^{(1)} = \mathbf{n}_1 \tag{4.173}$$

To obtain the other two unit eigenvectors, observe that these vectors will be parallel to the Y–Z plane as depicted in Figure 4.15, where the inclination of the unit eigenvectors is determined by the angle θ as shown. Then $\mathbf{n}_a^{(2)}$ and $\mathbf{n}_a^{(3)}$ may be expressed as

$$\mathbf{n}_a^{(2)} = \cos\theta \mathbf{n}_2 + \sin\theta \mathbf{n}_3 \quad \text{and} \quad \mathbf{n}_a^{(3)} = -\sin\theta \mathbf{n}_2 + \cos\theta \mathbf{n}_3 \tag{4.174}$$

and thus the components $a_i^{(2)}$ and $a_i^{(3)}$ are

$$a_1^{(2)} = 0, \quad a_2^{(2)} = \cos\theta, \quad a_3^{(2)} = \sin\theta \tag{4.175}$$

$$a_1^{(3)} = 0, \quad a_2^{(3)} = -\sin\theta, \quad a_3^{(3)} = \cos\theta \tag{4.176}$$

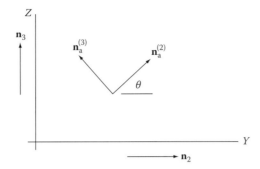

FIGURE 4.15   Unit eigenvectors parallel to the Y–Z plane.

Observing Equations 4.175 and 4.176, the fourth expression of Equation 4.171 is similarly satisfied. The first expression of Equation 4.171 is also satisfied by the first unit eigenvector components. This leaves the second and third equations, which are dependent. For the second unit eigenvector, these are equivalent to the single equation:

$$(\sigma_{22} - \lambda)\cos\theta + \sigma_{23}\sin\theta = 0 \tag{4.177}$$

Solving for $\tan\theta$ we obtain

$$\tan\theta = -\left[\frac{\sigma_{22} - \lambda_2}{\sigma_{23}}\right] = \frac{\lambda_2}{\sigma_{23}} - \frac{\sigma_{22}}{\sigma_{23}} \tag{4.178}$$

By substituting for λ_2 from Equation 4.170, we have

$$\tan\theta = \left[\frac{\sigma_{33} - \sigma_{22}}{2\sigma_{23}}\right] + \left[\left(\frac{\sigma_{22} - \sigma_{33}}{2\sigma_{23}}\right)^2 + 1\right]^{1/2} \tag{4.179}$$

By solving for the radical expression, squaring and simplifying, we obtain

$$\tan^2\theta - \left[\frac{\sigma_{33} - \sigma_{22}}{\sigma_{23}}\right]\tan\theta - 1 = 0$$

or

$$\sin^2\theta + \left[\frac{\sigma_{22} - \sigma_{33}}{\sigma_{23}}\right]\sin\theta\cos\theta - \cos^2\theta = 0$$

or

$$\left[\frac{\sigma_{22} - \sigma_{33}}{2\sigma_{23}}\right]\sin 2\theta = \cos 2\theta$$

thus,

$$\tan 2\theta = \frac{2\sigma_{23}}{\sigma_{22} - \sigma_{33}} \tag{4.180}$$

For the third unit eigenvector and the third eigenvalue, the second expression of Equation 4.171 becomes

$$(\sigma_{22} - \lambda_3)(-\sin\theta) + \sigma_{23}\cos\theta = 0 \tag{4.181}$$

By substituting for λ_3 from Equation 4.170 and simplifying we again obtain

$$\tan 2\theta = \frac{2\sigma_{23}}{\sigma_{22} - \sigma_{33}} \tag{4.182}$$

Although Equations 4.180 and 4.182 are the same, they still produce two values of θ differing by $\pi/2$ radians. That is,

$$\tan 2(\theta + \pi/2) \equiv \tan 2\theta \tag{4.183}$$

so that

$$\theta = \tan^{-1}\left(\frac{\sigma_{23}}{\sigma_{22} - \sigma_{33}}\right) \quad \text{and} \quad \theta = (\pi/2) + \tan^{-1}\frac{\sigma_{23}}{\sigma_{22} - \sigma_{33}} \tag{4.184}$$

Equation 4.184 determines the inclination of the unit eigenvectors $\mathbf{n}_a^{(2)}$ and $\mathbf{n}_a^{(3)}$ in the Y–Z plane. That is, once the value of θ is known from Equation 4.184, Equations 4.175 and 4.176, then the components of $\mathbf{n}_a^{(2)}$ and $\mathbf{n}_a^{(3)}$ relative to the $\mathbf{n}_i$ unit vector system can be determined.

Suppose we select $\mathbf{n}_a^{(1)}$, $\mathbf{n}_a^{(2)}$, and $\mathbf{n}_a^{(3)}$ as a basis system and for simplicity, let us rename these vectors simply as $\mathbf{a}_1$, $\mathbf{a}_2$, and $\mathbf{a}_3$. That is, let

$$\mathbf{n}_a^{(i)} = \mathbf{a}_i \quad (i = 1, 2, 3) \tag{4.185}$$

Suppose that $\hat{\mathbf{a}}_i$ ($i = 1, 2, 3$) form a mutually perpendicular set of unit vectors with $\hat{\mathbf{a}}_1$ parallel to $\mathbf{a}_1$ and $\hat{\mathbf{a}}_2$ and $\hat{\mathbf{a}}_3$ inclined at an angle ϕ relative to $\mathbf{a}_2$ and $\mathbf{a}_3$ as in Figure 4.16. Suppose further that we are interested in determining the stresses in the directions of $\hat{\mathbf{a}}_1$, $\hat{\mathbf{a}}_2$, and $\hat{\mathbf{a}}_3$. To this end, recall that since the $\mathbf{a}_i$ are unit eigenvectors, the stress dyadic σ expressed in terms of $\mathbf{a}_i$ has the relatively simple form:

$$\sigma = \sigma_{ij}\mathbf{a}_i\mathbf{a}_j = \lambda_1\mathbf{a}_1\mathbf{a}_1 + \lambda_2\mathbf{a}_2\mathbf{a}_2 + \lambda_3\mathbf{a}_3\mathbf{a}_3 \tag{4.186}$$

with the stress matrix being

$$\sigma = \begin{bmatrix} \lambda_1 & 0 & 0 \\ 0 & \lambda_2 & 0 \\ 0 & 0 & \lambda_3 \end{bmatrix} \tag{4.187}$$

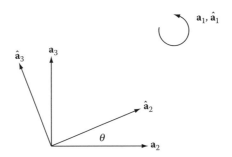

FIGURE 4.16   Unit vector inclinations.

Let S be a transformation matrix between the $\mathbf{a}_i$ and $\hat{\mathbf{a}}_i$ systems with elements of S being S_{ij} defined as

$$S_{ij} = \mathbf{a}_i \cdot \hat{\mathbf{a}}_j \tag{4.188}$$

From Figure 4.16, S is then seen to be

$$S = [S_{ij}] = \begin{bmatrix} 1 & 0 & 0 \\ 0 & \cos\phi & -\sin\phi \\ 0 & \sin\phi & \cos\phi \end{bmatrix} \tag{4.189}$$

Next, let the stress dyadic $\boldsymbol{\sigma}$ be expressed in terms of the $\hat{\mathbf{a}}_i$ system as

$$\boldsymbol{\sigma} = \hat{\sigma}_{ij}\hat{\mathbf{a}}_i\hat{\mathbf{a}}_j \tag{4.190}$$

with the stress matrix $\hat{\sigma}$ then being

$$\hat{\sigma} = \begin{bmatrix} \hat{\sigma}_{11} & \hat{\sigma}_{12} & \hat{\sigma}_{13} \\ \hat{\sigma}_{21} & \hat{\sigma}_{22} & \hat{\sigma}_{23} \\ \hat{\sigma}_{31} & \hat{\sigma}_{32} & \hat{\sigma}_{33} \end{bmatrix} \tag{4.191}$$

From the definition of Equation 4.188, we can relate the $\mathbf{a}_i$ and the $\hat{\mathbf{a}}_i$ unit vectors by the relations:

$$\mathbf{a}_i = S_{ij}\hat{a}_j \quad \text{and} \quad \hat{\mathbf{a}}_i = S_{ji}a_j \tag{4.192}$$

By substituting in Equation 4.190, we obtain the relation:

$$\hat{\sigma}_{ij} = S_{ki}S_{lj}\sigma_{kl} = S_{ki}\sigma_{kl}S_{lj} \tag{4.193}$$

or in the matrix form:

$$\hat{\sigma} = S^{\mathrm{T}}\sigma S \tag{4.194}$$

By substituting from Equations 4.187 and 4.189 we have

$$\hat{\sigma} = \begin{bmatrix} 1 & 0 & 0 \\ 0 & \cos\phi & \sin\phi \\ 0 & -\sin\phi & \cos\phi \end{bmatrix} \begin{bmatrix} \lambda_1 & 0 & 0 \\ 0 & \lambda_2 & 0 \\ 0 & 0 & \lambda_3 \end{bmatrix} \begin{bmatrix} 1 & 0 & 0 \\ 0 & \cos\phi & -\sin\phi \\ 0 & \sin\phi & \cos\phi \end{bmatrix}$$

or

$$\hat{\sigma} = \begin{bmatrix} \lambda_1 & 0 & 0 \\ 0 & (\lambda_2\cos^2\phi + \lambda_3\sin^2\phi) & (\lambda_3 - \lambda_2)\sin\phi\cos\phi \\ 0 & (\lambda_3 - \lambda_2)\sin\phi\cos\phi & (\lambda_2\sin^2\phi + \lambda_3\cos^2\phi) \end{bmatrix} \tag{4.195}$$

Therefore, we have the relation:

$$\hat{\sigma}_{11} = \sigma_{11} = \lambda_1 \tag{4.196}$$

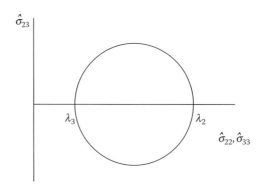

FIGURE 4.17   Stress circle.

$$\hat{\sigma}_{22} = \lambda_2 \cos^2 \phi + \lambda_3 \sin^2 \phi = \frac{\lambda_2 + \lambda_3}{2} + \frac{\lambda_2 - \lambda_3}{2} \cos 2\phi \qquad (4.197)$$

$$\hat{\sigma}_{33} = \lambda_2 \sin^2 \phi + \lambda_3 \cos^2 \phi = \frac{\lambda_2 + \lambda_3}{2} - \frac{\lambda_2 - \lambda_3}{2} \cos 2\phi \qquad (4.198)$$

and

$$\hat{\sigma}_{23} = \hat{\sigma}_{32} = (\lambda_3 - \lambda_2) \sin \phi \cos \phi = -\frac{\lambda_2 - \lambda_3}{2} \sin 2\phi \qquad (4.199)$$

Equations 4.197 through 4.199 may be represented graphically as in Figure 4.17, where we have constructed a "stress circle" with radius $(\lambda_2 - \lambda_3)/2$, positioned on a horizontal axis, which represents the normal stresses: $\hat{\sigma}_{22}$ and $\hat{\sigma}_{33}$. The vertical axis represents the shear stress: $\hat{\sigma}_{23}$. The center of the circle is placed on the horizontal axis at the average stress: $(\lambda_2 + \lambda_3)/2$. With this construction, we can see that the stresses of Equations 4.197 through 4.199 are represented by the ordinates and abscissas of the points A and B on the circle at opposite ends of a diameter inclined at 2ϕ to the horizontal as in Figure 4.18. This construction for planar stress computation is commonly known as "Mohr's circle."

Also, observe that the maximum shear stress occurs on surfaces inclined at 45° relative to the directions of the principal stresses.

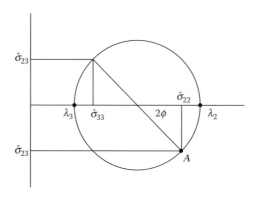

FIGURE 4.18   Mohr's circle for planar stress computations.

SYMBOLS

A	Area
$\mathbf{a}$	Vector
$a_i \, (i = 1, 2, 3)$	Components of $\mathbf{a}$ or $\mathbf{n}_a$ along $\mathbf{n}_i$
A_x, A_y, A_z	Projections of area A normal to X, Y, Z
B	Body
$\mathbf{b}$	Vector
$b_i \, (i = 1, 2, 3)$	Components of $\mathbf{b}$ along $\mathbf{n}_i$
B_L	Left side of B
B_R	Right side of B
$\mathbf{d}$	Dyadic
E	Rectangular element
g	Gravity acceleration
$\mathbf{I}$	Identity dyadic
$\mathbf{n}_i \, (i = 1, 2, 3)$	Mutually perpendicular unit vectors
$\hat{\mathbf{n}}_i \, (i = 1, 2, 3)$	Mutually perpendicular unit vectors
O	Origin
P	Point
$\mathbf{P}$	Force
$\mathbf{M}$	Couple torque
N	Cutting or dividing plane
$\hat{\mathbf{N}}$	Cutting or dividing plane
$\mathbf{n}$	Unit vector
$\mathbf{n}_a$	Unit eigenvectors
$\mathbf{n}_x, \mathbf{n}_y, \mathbf{n}_z$	Mutually perpendicular unit vectors
n_x, n_y, n_z	Components of $\mathbf{n}$ along $\mathbf{n}_x, \mathbf{n}_y, \mathbf{n}_z$
$\mathbf{p}, \mathbf{q}$	Position vectors
R	Surface region
$\mathbf{S}$	Stress vector
$\hat{\mathbf{S}}$	Stress vector
$\mathbf{S}_n$	Stress vector on a plane normal to $\mathbf{n}$
S_{nx}, S_{ny}, S_{nz}	Components of $\mathbf{S}_n$ along $\mathbf{n}_x, \mathbf{n}_y, \mathbf{n}_z$
S_x, S_y, S_z	Components of $\mathbf{S}$ along $\mathbf{n}_x, \mathbf{n}_y, \mathbf{n}_z$
T	Tetrahedron, Transformation matrix
T_{ij}	Direction cosines defined by Equation 4.44
$\mathbf{T}^\mathrm{T}$	Transpose of $\mathbf{T}$
$\mathbf{V}$	Vector
v_i	Components of $\mathbf{V}$ along $\mathbf{n}_i$
$x_i \, (i = 1, 2, 3)$	x, y, z
X, Y, Z	Cartesian (rectangular) coordinate axes
$\Delta x, \Delta y, \Delta z$	Edges of rectangular element
δ	Unit matrix
δ_{ij}	Kronecker's delta symbol, defined by Equation 4.53, Elements of δ
λ	Eigenvalue
$\lambda_1, \lambda_2, \lambda_3$	Eigenvalues
ρ	Mass density
$\boldsymbol{\sigma}$	Stress dyadic
$\boldsymbol{\sigma}$	Stress vectors
$\sigma_{ij} \, (i, j = x, y, z)$	Stress components

$\hat{\sigma}_{ij}$ ($i = 1, 2, 3$)   Stress components

σ_I, σ_{II}, σ_{III} Hamilton–Cayley equation coefficient defined by Equations 4.63 through
                      4.66

τ Shear stress

τ_{ij} ($i, j = x, y, z$)   Shear stress components

REFERENCES

1. R. L. Huston and C. Q. Liu, *Formulas for Dynamic Analysis*, Marcel Dekker, New York, 2001.
2. I. S. Sokolnikoff, *Mathematical Theory of Elasticity*, McGraw Hill, New York, 1956.
3. L. Brand, *Vector and Tensor Analysis*, John Wiley & Sons, New York, 1947.
4. R. A. Usami, *Applied Linear Algebra*, Marcel Dekker, New York, 1987.
5. F. B. Hildebrand, *Advanced Calculus for Applications*, 2nd ed., Prentice Hall, Englewood Cliffs, NJ, 1976, pp. 357–359.
6. M. D. Greenberg, *Foundations of Applied Mathematics*, Prentice Hall, Englewood Cliffs, NJ, 1978, pp. 199–202.
7. H. Sharp, Jr., *Modern Fundamentals of Algebra and Trigonometry*, Prentice Hall, Englewood Cliffs, NJ, 1961, pp. 182–201.

5 Strain in Two and Three Dimensions

5.1 CONCEPT OF SMALL DISPLACEMENT

Just as with stress, we can generalize the concepts of simple strain and simple shear strain from one dimension to two and three dimensions. For most engineering materials, the displacements and deformations are small under usual loadings. (Exceptions might be with polymers and biomaterials.) For small displacements, we can neglect products of displacements and products of displacement derivatives when compared to linear terms. This allows us to make a linear analysis, and thus, a simplified analysis.

Even though the analysis is linear, it may still provide insight into the structures and structural components where the displacements and deformations are not small, as with some polymer and biomaterials. Linear analyses for such materials simply become more valid, as the displacements and deformations become smaller.

5.2 TWO-DIMENSIONAL ANALYSES

Consider a small square element E of an elastic body as it would appear before and after deformation as in Figure 5.1 where the deformation is exaggerated for analysis convenience. Let the vertices of E before and after deformation be A, B, C, D and A', B', C', D' respectively and let the initial sides of E be Δx and Δy (with $\Delta x = \Delta y$).

Let X–Y be a Cartesian axis system parallel to the edges of E before deformation. Let the X–Y components of the displacement of vertex A (to A') be u and v. Then the displacements of the other vertices of E may be approximated by using a truncated Taylor series expansion. For vertices B and D, the X–Y displacements are

$$\begin{array}{ccc} & X & Y \\[2mm] \text{Vertex } B: & u + \dfrac{\partial u}{\partial x}\Delta x + \dfrac{1}{2!}\dfrac{\partial^2 u}{\partial x^2}\Delta x^2 + \cdots & v + \dfrac{\partial v}{\partial x}\Delta x + \dfrac{1}{2!}\dfrac{\partial^2 v}{\partial x^2}\Delta x^2 + \cdots \end{array} \qquad (5.1)$$

$$\begin{array}{ccc} \text{Vertex } D: & u + \dfrac{\partial u}{\partial y}\Delta y + \dfrac{1}{2!}\dfrac{\partial^2 u}{\partial y^2}\Delta y^2 + \cdots & v + \dfrac{\partial v}{\partial y}\Delta y + \dfrac{1}{2!}\dfrac{\partial^2 v}{\partial y^2}\Delta y^2 + \cdots \end{array} \qquad (5.2)$$

To measure the deformation of E, during the displacement, it is convenient to superimpose the before and after representations of Figure 5.1 as in Figure 5.2 where we have neglected the higher order terms of the Taylor series expansion.

As a generalization of the concept of simple strain (see Chapter 2), we can define the strain in the X and Y directions as the normalized elongation (elongation per unit length) of the element E in the X and Y directions. Specifically, the strain in the X direction, written as ε_x, may be approximated as

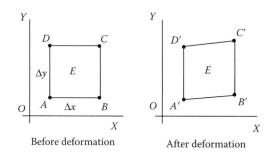

FIGURE 5.1   Square element E before and after deformation.

$$\varepsilon_x = \frac{|A'\,B'| - |AB|}{\Delta x} = \frac{\left[\left(\Delta x + \frac{\partial u}{\partial x}\Delta x\right)^2 + \left(\frac{\partial v}{\partial x}\Delta x\right)^2\right]^{1/2} - \Delta x}{\Delta x}$$

$$= \frac{\left[\Delta x^2 + 2\frac{\partial u}{\partial x}\Delta x^2\right]^{1/2} - \Delta x}{\Delta x} = \frac{\Delta x\left[1 + \frac{1}{2}(2)\frac{\partial u}{\partial x}\right] - \Delta x}{\Delta x}$$

$$= \frac{\partial v}{\partial y} \tag{5.3}$$

where we have used a binomial expansion [1] and neglected quadratic and higher powers of Δy to approximate the square root. Similarly, the strain in the Y-direction, ε_y, is approximately

$$\varepsilon_y = \frac{|A'\,D'| - |AD|}{\Delta y} = \frac{\left[\left(\Delta y + \frac{\partial v}{\partial y}\Delta y\right)^2 + \left(\frac{\partial u}{\partial y}\Delta y\right)^2\right]^{1/2} - \Delta y}{\Delta y}$$

$$= \frac{\left[\Delta y^2 + 2\frac{\partial v}{\partial y}\Delta y^2\right]^{1/2} - \Delta y}{\Delta y} = \frac{\Delta y\left[1 + \frac{1}{2}(2)\frac{\partial v}{\partial y}\right] - \Delta y}{\Delta y}$$

$$= \frac{\partial v}{\partial y} \tag{5.4}$$

Observe again that the approximation used in obtaining Equations 5.3 and 5.4 become increasingly valid the smaller Δx, Δy, $\partial u/\partial x$ and $\partial v/\partial y$ become.

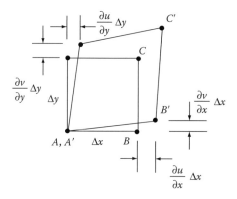

FIGURE 5.2   Superposition of deformed and undeformed element.

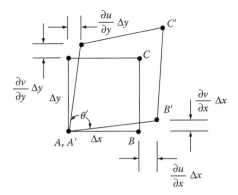

FIGURE 5.3 Superposition of deformed and undeformed element.

5.3 SHEAR STRAIN

Consider again the superposed deformed and undeformed element E of Figure 5.2. Consider the distortion of the element from a square to a rhombic shape. We can quantify the distortion by the shear strain as we did earlier with simple tangential loading (see Figure 2.10). For example, at A, the shear strain, written as γ_{xy}, is the difference between the angle θ' shown in Figure 5.3 and $\pi/2$. That is,

$$\gamma_{xy} = \pi/2 - \theta' \tag{5.5}$$

From Figure 5.3 we see that θ' is

$$\theta' = \pi/2 - \left(\frac{\partial u}{\partial y}\Delta y\right)\Big/\Delta y - \left(\frac{\partial v}{\partial x}\Delta x\right)\Big/\Delta x$$
$$= \pi/2 - \frac{\partial u}{\partial y} - \frac{\partial v}{\partial x} \tag{5.6}$$

Then, by substitution from Equation 5.6 into 5.5, we have

$$\gamma_{xy} = \frac{\partial u}{\partial y} + \frac{\partial v}{\partial x} \tag{5.7}$$

Referring again to Figure 5.3 and Equation 5.7, we can think of the shear strain γ_{xy} as the sum of the angles ψ_x and ψ_y shown in Figure 5.4. By comparing Figures 5.3 and 5.4 we see that these angles are

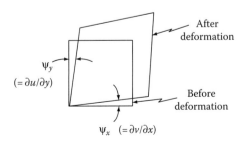

FIGURE 5.4 Distortion angles.

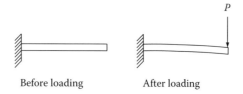

Before loading   After loading

FIGURE 5.5 Cantilever beam before and after end force loading.

simply $\partial v/\partial x$ and $\partial u/\partial y$. Since these angles are not equal in general, it is often convenient to think of the shear strain as the average value of the angles (measured in radians) and designated by ε_{xy}. That is,

$$\varepsilon_{xy} = 1/2\left(\frac{\partial u}{\partial y} + \frac{\partial v}{\partial x}\right) = (1/2)\gamma_{xy} \tag{5.8}$$

To distinguish these two shear strains ε_{xy} is sometimes called the "mathematical shear strain" and γ_{xy} the "engineering shear strain."

5.4 DISPLACEMENT, DEFORMATION, AND ROTATION

Unfortunately the terms "displacement" and "deformation" are occasionally used interchangeably suggesting that they are the same. To be precise, we should think of "displacement" at a point P on an elastic body B as simply the movement of P during the loading of B. Displacement can occur with or without deformation. For example, an elastic body can undergo a "rigid body" movement where the points of the body have relatively large displacements but the body itself has no deformation. On the other hand, "deformation" refers to a distortion, or change in shape, of a body. Whereas displacement can occur without deformation, deformation always involves displacement.

As an illustration, consider a cantilever beam with an end load as in Figure 5.5. Consider a small element E of the beam near the end and at the center (on the neutral axis) of the beam as shown in exaggerated view in Figure 5.6. For all practical purposes, E simply translates and rotates, but it is not deformed.

In general, an element within a loaded elastic body will undergo deformation, translation, and rotation. "Deformation" may be measured and represented by the normal and shear strains defined in Sections 5.2 and 5.3. "Translation" is simply a measure of the change of position of the element. "Rotation," however, prompts further consideration: consider again a small square element E, with vertices A, B, C, D, before and after loading as in Figure 5.7. We can visualize the rotation by

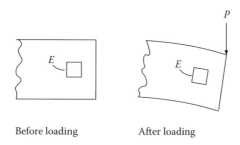

Before loading   After loading

FIGURE 5.6 A small element (in exaggerated view) in the center near the end of the cantilever beam of Figure 5.5.

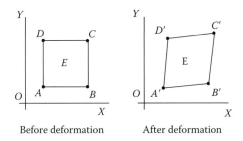

Before deformation   After deformation

FIGURE 5.7   Square element E before and after deformation.

eliminating the translation: that is, by superposing the before and after representations as in Figure 5.8. We can further isolate the rotation by eliminating the strain as in Figure 5.9, resulting in a representation of the element rotation. We can then quantify the rotation of the element in terms of the rotation of its diagonal AC, which in turn is approximately equal to the average of the rotations of the sides AB and AC. With counterclockwise rotation assigned as positive (that is, dextral rotation about the Z-axis), the element rotation ω_z is then approximately

$$\omega_z = 1/2(\partial v/\partial x - \partial u/\partial y) \tag{5.9}$$

5.5   GENERALIZATION TO THREE DIMENSIONS

The foregoing concepts and results are readily generalized to three dimensions: let P be a point on an elastic body B, which is subjected to a general loading as in Figure 5.10. Prior to the loading of B, let P be at the lower rear vertex of a small cubical element E with sides parallel to the coordinate axes and having lengths Δx, Δy, Δz (all equal) as represented in Figure 5.11. Then, as B is loaded and deformed, we can imagine E as being translated and rotated and also deformed as in Figure 5.12. We can visualize normal strains along the elongated (or shortened) edges of E and shear strains, as the faces of E are no longer at right angles to one another, and also visualize the translation and rotation of the element itself.

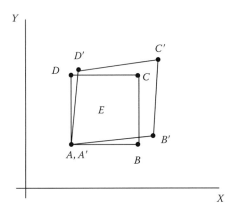

FIGURE 5.8   Superposition of deformed and undeformed element representation.

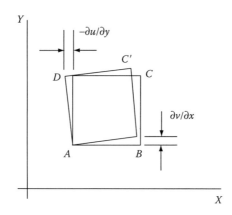

FIGURE 5.9   Element rotation.

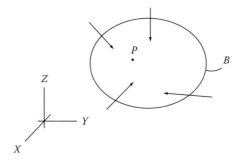

FIGURE 5.10   A loaded elastic body B and a point P in the interior of B.

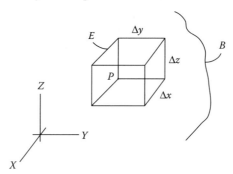

FIGURE 5.11   A small cubical element E of an elastic body.

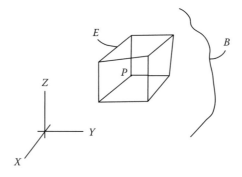

FIGURE 5.12   Translated, rotated, and deformed element E of a loaded elastic body.

Let the displacement of P relative to a convenient set of XYZ axes be u, v, w respectively. Then, as an immediate generalization of Equations 5.3 and 5.4, we obtain the strains along the edges of E intersecting at P as

$$\varepsilon_x = \frac{\partial u}{\partial x}, \quad \varepsilon_y = \frac{\partial v}{\partial y}, \quad \varepsilon_z = \frac{\partial w}{\partial z} \tag{5.10}$$

Similarly, as generalizations of Equations 5.7 and 5.8, the engineering and mathematical shear strains measuring the angle changes of the faces of E intersecting at P are

$$\gamma_{xy} = \frac{\partial u}{\partial y} + \frac{\partial v}{\partial x}, \quad \gamma_{yz} = \frac{\partial v}{\partial z} + \frac{\partial w}{\partial y}, \quad \gamma_{zx} = \frac{\partial w}{\partial x} + \frac{\partial u}{\partial z} \tag{5.11}$$

and

$$\varepsilon_{xy} = \frac{1}{2}\left(\frac{\partial u}{\partial y} + \frac{\partial v}{\partial x}\right), \quad \varepsilon_{yz} = \frac{1}{2}\left(\frac{\partial v}{\partial z} + \frac{\partial w}{\partial y}\right), \quad \varepsilon_{zx} = \frac{1}{2}\left(\frac{\partial w}{\partial x} + \frac{\partial u}{\partial z}\right) \tag{5.12}$$

As a generalization of Equation 5.9, we obtain expressions for the rotation of E about the X, Y, and Z axes as

$$\omega_x = \frac{1}{2}\left(\frac{\partial w}{\partial x} - \frac{\partial u}{\partial y}\right), \quad \omega_y = \frac{1}{2}\left(\frac{\partial v}{\partial z} - \frac{\partial w}{\partial y}\right), \quad \omega_z = \frac{1}{2}\left(\frac{\partial u}{\partial y} - \frac{\partial v}{\partial x}\right) \tag{5.13}$$

Observe the patterns in terms of Equations 5.10 through 5.13: these patterns become evident if we rename variables as

$$
\begin{aligned}
x &\rightarrow x_1, & y &\rightarrow x_2, & z &\rightarrow x_3 \\
u &\rightarrow u_1, & v &\rightarrow u_2, & w &\rightarrow u_3 \\
\varepsilon_x &\rightarrow \varepsilon_{11}, & \varepsilon_y &\rightarrow \varepsilon_{22}, & \varepsilon_z &\rightarrow \varepsilon_{33} \\
\gamma_{xy} &\rightarrow \gamma_{12}, & \gamma_{yz} &\rightarrow \gamma_{23}, & \gamma_{zx} &\rightarrow \gamma_{31} \\
\varepsilon_{xy} &\rightarrow \varepsilon_{12}, & \varepsilon_{yz} &\rightarrow \varepsilon_{23}, & \varepsilon_{zx} &\rightarrow \varepsilon_{31} \\
\omega_z &\rightarrow \omega_{12}, & \omega_y &\rightarrow \omega_{23}, & \omega_x &\rightarrow \omega_{31}
\end{aligned}
\tag{5.14}
$$

Then, Equations 5.10 through 5.13 become

$$\varepsilon_{11} = \frac{\partial u_1}{\partial x_1}, \quad \varepsilon_{22} = \frac{\partial u_2}{\partial x_2}, \quad \varepsilon_{33} = \frac{\partial u_3}{\partial x_3} \tag{5.15}$$

$$\gamma_{12} = \left(\frac{\partial u_1}{\partial x_2} + \frac{\partial u_2}{\partial x_1}\right), \quad \gamma_{23} = \left(\frac{\partial u_2}{\partial x_3} + \frac{\partial u_3}{\partial x_2}\right), \quad \gamma_{31} = \left(\frac{\partial u_3}{\partial x_1} + \frac{\partial u_1}{\partial x_3}\right) \tag{5.16}$$

$$\varepsilon_{12} = \frac{1}{2}\left(\frac{\partial u_1}{\partial x_2} + \frac{\partial u_2}{\partial x_1}\right), \quad \varepsilon_{23} = \frac{1}{2}\left(\frac{\partial u_2}{\partial x_3} + \frac{\partial u_3}{\partial x_2}\right), \quad \varepsilon_{31} = \frac{1}{2}\left(\frac{\partial u_3}{\partial x_1} + \frac{\partial u_1}{\partial x_3}\right) \tag{5.17}$$

$$\omega_{12} = \frac{1}{2}\left(\frac{\partial u_1}{\partial x_2} - \frac{\partial u_2}{\partial x_1}\right), \quad \omega_{23} = \frac{1}{2}\left(\frac{\partial u_2}{\partial x_3} - \frac{\partial u_3}{\partial x_2}\right), \quad \omega_{31} = \frac{1}{2}\left(\frac{\partial u_3}{\partial x_1} - \frac{\partial u_1}{\partial x_3}\right) \tag{5.18}$$

The respective terms of these equations are of the same form and they can be generated from one another by simply permutating the numerical indices (that is, $1 \to 2$, $2 \to 3$, and $3 \to 1$). With this observation, we can simplify the terms in the equations by introducing the notation:

$$\partial(\ )/\partial x_i \overset{D}{=} (\ ),i \quad (i = 1, 2, 3) \tag{5.19}$$

Then, Equations 5.15 and 5.17 may be written in the compact form:

$$\varepsilon_{ij} = \frac{1}{2}(u_{i,j} + u_{j,i}) \quad (i, j = 1, 2, 3) \tag{5.20}$$

Similarly, Equation 5.18 becomes

$$\omega_{ij} = \frac{1}{2}(u_{i,j} - u_{j,i}) \quad (i, j = 1, 2, 3) \tag{5.21}$$

Observe in Equation 5.20 that

$$\varepsilon_{ij} = \varepsilon_{ji} \tag{5.22}$$

Hence, if the ε_{ij} are placed into a strain array, or strain matrix, as

$$\varepsilon = \begin{bmatrix} \varepsilon_{11} & \varepsilon_{12} & \varepsilon_{13} \\ \varepsilon_{21} & \varepsilon_{22} & \varepsilon_{23} \\ \varepsilon_{31} & \varepsilon_{32} & \varepsilon_{33} \end{bmatrix} \tag{5.23}$$

then ε is symmetric (analogous to the stress matrix σ of Chapter 4).

Observe further from Equation 5.21 that

$$\omega_{ij} = -\omega_{ji} \tag{5.24}$$

Hence, if ω_{ij} are placed into a rotation array ω as

$$\omega = \begin{bmatrix} \omega_{11} & \omega_{12} & \omega_{13} \\ \omega_{21} & \omega_{22} & \omega_{23} \\ \omega_{31} & \omega_{32} & \omega_{33} \end{bmatrix} \tag{5.25}$$

we see that the diagonal terms are zero and that the corresponding off-diagonal terms are negative of each other. That is, ω is skew symmetric and it may be written as

$$\omega = \begin{bmatrix} 0 & \omega_{12} & \omega_{13} \\ -\omega_{12} & 0 & \omega_{23} \\ -\omega_{13} & -\omega_{23} & 0 \end{bmatrix} = \begin{bmatrix} 0 & \omega_z & -\omega_x \\ -\omega_z & 0 & \omega_y \\ \omega_z & -\omega_y & 0 \end{bmatrix} \tag{5.26}$$

5.6 STRAIN AND ROTATION DYADICS

Just as the stress matrix elements are scalar components of the stress dyadic, the strain and rotation matrix elements may be regarded as scalar components of strain and rotation dyadics. To this end, let $\mathbf{n}_1$, $\mathbf{n}_2$, and $\mathbf{n}_3$ be mutually perpendicular unit vectors parallel to the X, Y, and Z axes respectively. Then the strain and rotation dyadics are simply

$$\boldsymbol{\varepsilon} = \varepsilon_{ij}\mathbf{n}_i\mathbf{n}_j \quad \text{and} \quad \boldsymbol{\omega} = \omega_{ij}\mathbf{n}_i\mathbf{n}_j \tag{5.27}$$

(As with the stress dyadic, we can regard the components of the strain and rotation dyadics as "tensor components.")

Observe further that with the strain dyadic being symmetric, we can perform an eigenvalue analysis to obtain the values and the directions for the maximum and minimum strains. The procedures of such analyses are exactly the same as the eigenvalue analysis for the stress dyadic in Section 4.5. We can similarly also perform a Mohr circle analysis for two-dimensional (or planar) problems.

5.7 STRAIN AND ROTATION IDENTITIES

Consider again the strain and rotation components expressed in Equations 5.20 and 5.21 as

$$\varepsilon_{ij} = \frac{1}{2}(u_{i,j} + u_{j,i}) \quad \text{and} \quad \omega_{ij} = \frac{1}{2}(u_{i,j} - u_{j,i}) \tag{5.28}$$

We see that both the strain and rotation depend directly upon the rate of change of the displacement from point to point within the body—the so-called "displacement gradients": $u_{i,j}$ (or $\partial u_i / \partial x_j$). Consider the following identity with the $u_{i,j}$:

$$u_{i,j} \equiv \frac{1}{2}(u_{i,j} + u_{j,i}) + \frac{1}{2}(u_{i,j} - u_{j,i}) \tag{5.29}$$

Observe that the terms on the right side of this identity are simply ε_{ij} and ω_{ij}. That is,

$$u_{i,j} \equiv \varepsilon_{ij} + \omega_{ij} \tag{5.30}$$

Consider again the strain components:

$$\varepsilon_{ij} = \frac{1}{2}(u_{i,j} + u_{j,i}) = \frac{1}{2}\left(\frac{\partial u_i}{\partial x_j} + \frac{\partial u_j}{\partial x_i}\right) \tag{5.31}$$

If, during the course of an analysis, we are able to determine the strain components, we can regard Equations 5.31 as a system of partial differential equations for the displacement components. However, since Equation 5.31 is equivalent to six scalar equations, but that there are only three displacement components, the system is overdetermined and thus unique solutions will not be obtained unless there are other conditions or requirements making the equations consistent. These conditions are usually called "compatibility conditions" or "compatibility equations." In theoretical discussion on elasticity and continuum mechanics (see for example, Refs. [2–9]), these compatibility equations are developed in a variety of ways and are found to be [2–10]:

$$\varepsilon_{ij,k\ell} + \varepsilon_{k\ell,ij} - \varepsilon_{ik,j\ell} - \varepsilon_{j\ell,ik} = 0 \tag{5.32}$$

Since each of the indices has integer values 1, 2, and 3, there are a total of 81 of these equations. However, due to symmetry of the strain matrix, and identities of mixed second partial derivatives, only six of the equations are seen to be distinct. These are

$$\varepsilon_{11,23} + \varepsilon_{23,11} - \varepsilon_{31,12} - \varepsilon_{12,13} = 0$$
$$\varepsilon_{22,31} + \varepsilon_{31,22} - \varepsilon_{12,23} - \varepsilon_{23,21} = 0$$
$$\varepsilon_{33,12} + \varepsilon_{12,33} - \varepsilon_{23,31} - \varepsilon_{31,32} = 0$$
$$2\varepsilon_{12,12} - \varepsilon_{11,22} - \varepsilon_{22,11} = 0$$
$$2\varepsilon_{23,23} - \varepsilon_{22,33} - \varepsilon_{33,22} = 0$$
$$2\varepsilon_{31,31} - \varepsilon_{33,11} - \varepsilon_{11,33} = 0$$

(5.33)

It may be convenient to have Equations 5.33 expressed using the usual Cartesian coordinates: x, y, z. In this convention, the compatibility equations are

$$\frac{\partial^2 \varepsilon_{xx}}{\partial y \partial z} + \frac{\partial^2 \varepsilon_{yz}}{\partial x^2} - \frac{\partial^2 \varepsilon_{zx}}{\partial x \partial y} - \frac{\partial^2 \varepsilon_{xy}}{\partial x \partial z} = 0$$
$$\frac{\partial^2 \varepsilon_{yy}}{\partial z \partial x} + \frac{\partial^2 \varepsilon_{zx}}{\partial y^2} - \frac{\partial^2 \varepsilon_{xy}}{\partial y \partial z} - \frac{\partial^2 \varepsilon_{yz}}{\partial y \partial x} = 0$$
$$\frac{\partial^2 \varepsilon_{zz}}{\partial x \partial y} + \frac{\partial^2 \varepsilon_{xy}}{\partial z^2} - \frac{\partial^2 \varepsilon_{yz}}{\partial z \partial x} - \frac{\partial^2 \varepsilon_{zx}}{\partial z \partial y} = 0$$
$$2\frac{\partial^2 \varepsilon_{xy}}{\partial x \partial y} - \frac{\partial^2 \varepsilon_{xx}}{\partial y^2} - \frac{\partial^2 \varepsilon_{yy}}{\partial x^2} = 0$$
$$2\frac{\partial^2 \varepsilon_{yz}}{\partial y \partial z} - \frac{\partial^2 \varepsilon_{yy}}{\partial z^2} - \frac{\partial^2 \varepsilon_{zz}}{\partial y^2} = 0$$
$$2\frac{\partial^2 \varepsilon_{zx}}{\partial z \partial x} - \frac{\partial^2 \varepsilon_{zz}}{\partial x^2} - \frac{\partial^2 \varepsilon_{xx}}{\partial z^2} = 0$$

(5.34)

Observe the pattern of the indices and terms of Equations 5.33 and 5.34.

Finally, in theoretical discussions (see for example Refs. [6,8,9]), it is asserted that only three of the six compatibility equations are independent. This is consistent with the need to constrain the six strain components to obtain a unique set of three displacement components (aside from rigid-body movement). That is, with six equations and three unknowns, only three constraints are needed.

SYMBOLS

$\mathbf{n}_i$ ($i = 1, 2, 3$)	Mutually perpendicular unit vectors
u, v, w	Displacements in X, Y, Z directions
u_i ($i = 1, 2, 3$)	Displacement
x, y, z	Cartesian coordinates
X, Y, Z	Cartesian (rectangular) coordinate axes
γ_{xy}, γ_{yz}, γ_{zx}	Engineering shear strains (Sections 5.3 and 5.5)
γ_{12}, γ_{23}, γ_{31}	Engineering shear strains (Section 5.5)
Δx, Δy, Δz	Element dimensions
ε	Strain matrix
$\boldsymbol{\varepsilon}$	Strain dyadic
ε_{ij} ($i, j = 1, 2, 3$)	Strain components
ε_x, ε_y, ε_z	Normal strains in X, Y, Z direction

$\varepsilon_{xy}, \varepsilon_{yz}, \varepsilon_{zx}$	Mathematical shear strains (Sections 5.3 and 5.5)
ω	Rotation matrix
$\boldsymbol{\omega}$	Rotation dyadic
$\omega_{12}, \omega_{23}, \omega_{31}$	Rotations

REFERENCES

1. H. B. Dwight, *Tables of Integrals and Other Mathematical Data*, 3rd ed., Macmillan, New York, 1957, p. 1.
2. I. S. Sokolnikoff, *Mathematical Theory of Elasticity*, Kreiger Publishing, Malabar, FL, 1983, pp. 25–29.
3. S. P. Timoshenko and J. N. Goodier, *Theory of Elasticity*, 2nd ed., McGraw Hill, New York, 1951, p. 230.
4. Y. C. Fung, *Foundations of Solid Mechanics*, Prentice-Hall, Englewood Cliffs, NJ, 1965, pp. 100–101.
5. C. Truesdell and R. A. Toupin, The classical field theories, in *Encyclopedia of Physics*, S. Flugge, ed., Vol. III/1, Springer-Verlag, Berlin, 1960, pp. 271–273.
6. A. C. Eringen, *Nonlinear Theory of Continuous Media*, McGraw Hill, New York, 1962, pp. 44–47.
7. W. Jaunzemia, *Continuum Mechanics*, Macmillan, New York, 1967, pp. 340–341.
8. H. Reismann and P.S. Pawlik, *Elasticity—Theory and Applications*, John Wiley & Sons, New York, 1980, pp. 106–109.
9. H. Leipolz, *Theory of Elasticity*, Noordhoff, Leyden, the Netherlands, 1974, pp. 92–95.
10. A. E. H. Love, *A Treatise on the Mathematical Theory of Elasticity*, 4th ed., Dover, New York, 1944, p. 49.

6 Curvilinear Coordinates

6.1 USE OF CURVILINEAR COORDINATES

The formulation of the equilibrium equations for stress and the strain–displacement equations are readily developed in Cartesian coordinates, as in Chapters 4 and 5. In practical stress–strain analyses, however, the geometry often is not rectangular but instead cylindrical, spherical, or of some other curved shape. In these cases, the use of curvilinear coordinates can greatly simplify the analysis. But with curvilinear coordinates, the equilibrium equations and the strain–displacement equations have different and somewhat more complicated forms than those with Cartesian coordinates. To determine the equation forms in curvilinear coordinates, it is helpful to review some fundamental concepts of curvilinear coordinate analysis. In the following sections, we present a brief review of these concepts. We then apply the resulting equations using cylindrical and spherical coordinates.

6.2 CURVILINEAR COORDINATE SYSTEMS: CYLINDRICAL AND SPHERICAL COORDINATES

6.2.1 CYLINDRICAL COORDINATES

Probably the most familiar and most widely used curvilinear coordinate system is plane polar coordinates and its extension in three dimensions to cylindrical coordinates: consider an XYZ Cartesian system with a point P having coordinates (x, y, z) as in Figure 6.1.

Next, observe that P may be located relative to the origin O by a position vector $\mathbf{OP}$, or simply $\mathbf{p}$, given by

$$\mathbf{p} = x\mathbf{n}_x + y\mathbf{n}_y + z\mathbf{n}_z \tag{6.1}$$

where $\mathbf{n}_x$, $\mathbf{n}_y$, and $\mathbf{n}_z$ are unit vectors parallel to the X-, Y-, and Z-axes as in Figure 6.2.

Suppose now an image of P, say $\hat{P}$, is projected onto the X–Y plane as in Figure 6.3. Then we can also locate P relative to O by the vector sum $\mathbf{O}\hat{\mathbf{P}} + \hat{\mathbf{P}}\mathbf{P}$ as in Figure 6.4. That is

$$\mathbf{p} = \mathbf{OP} = \mathbf{O}\hat{\mathbf{P}} + \hat{\mathbf{P}}\mathbf{P} \tag{6.2}$$

Let r be the distance from O to $\hat{P}$; z be the distance from $\hat{P}$ to P; and θ be the angle between $\mathbf{O}\hat{\mathbf{P}}$ and the X-axis, as in Figure 6.5. Then we have the expression

$$\mathbf{O}\hat{\mathbf{P}} = r\mathbf{n}_r \quad \text{and} \quad \hat{\mathbf{P}}\mathbf{P} = z\mathbf{n}_z \tag{6.3}$$

where $\mathbf{n}_r$ is a unit vector parallel to $\mathbf{O}\hat{\mathbf{P}}$ as in Figure 6.5.

Here, we observe that $\mathbf{n}_r$ may be expressed in terms of $\mathbf{n}_x$ and $\mathbf{n}_y$ as

$$\mathbf{n}_r = \cos\theta\mathbf{n}_x + \sin\theta\mathbf{n}_y \tag{6.4}$$

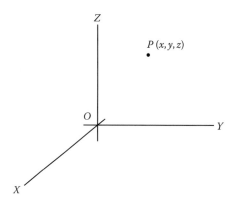

FIGURE 6.1   *XYZ* coordinate system with a point *P* having coordinates (x, y, z).

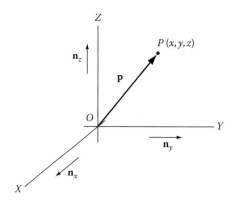

FIGURE 6.2   Position vector **p** locating point *P* relative to origin *O*.

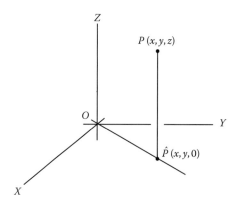

FIGURE 6.3   Projection of *P* onto the *X–Y* plane.

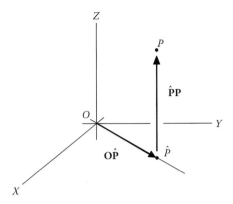

FIGURE 6.4   Position vectors locating P relative to O.

By substituting into Equation 6.3 and by comparing with Equation 6.1 we have

$$x = r \cos \theta, \quad y = r \sin \theta, \quad z = z \tag{6.5}$$

Observe that with the Cartesian coordinates of P being (x, y, z), we see from Equation 6.5 that we can then locate P by specifying the parameters: (r, θ, z), the cylindrical coordinates of P. Observe further that Equation 6.5 may be solved for r, θ, and z in terms of x, y, and z as

$$r = (x^2 + y^2)^{1/2}, \quad \theta = \tan^{-1}(y/x), \quad z = z \tag{6.6}$$

Suppose that in Equation 6.5 we hold θ and z to be constants, but let r be a variable, say t. Then we have

$$x = t \cos \theta, \quad y = t \sin \theta, \quad z = z \tag{6.7}$$

These expressions have the form:

$$x = x(t), \quad y = y(t), \quad z = z(t) \tag{6.8}$$

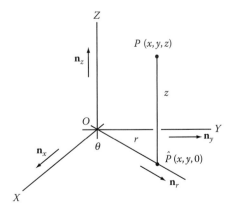

FIGURE 6.5   Locating $\hat{P}$ relative to O by parameters r and θ (cylindrical coordinates).

which may be interpreted as parametric equations [1] with parameter t. Indeed, Equations 6.7 are the parametric equations of radial lines.

Similarly, in Equation 6.5, if we hold r and z constant and let θ be a variable parameter t, we have

$$x = r \cos t, \quad y = r \sin t, \quad z = z \tag{6.9}$$

These are parametric equations of circles.

And, if r and θ are constants and z is varied, we have

$$x = r \cos \theta, \quad y = r \sin \theta, \quad z = t \tag{6.10}$$

These are parametric equations of lines parallel to the Z-axis. The radial lines, the circles, and the axial lines of Equations 6.8 through 6.10 are the coordinate curves of the cylindrical coordinate system.

6.2.2 SPHERICAL COORDINATES

Next to cylindrical coordinates, spherical coordinates appear to be the most widely used of the curvilinear coordinate systems. Figure 6.6 illustrates the parameters ρ, θ, and ϕ commonly used as spherical coordinates where ρ is the distance from the origin O to a typical point P, θ is the angle between line OP and the Z-axis, and ϕ is the angle between line $O\hat{P}$ and the X-axis, where, as before, $\hat{P}$ is the projection of P onto the X–Y plane.

As before, let $\mathbf{p}$ be the position vector $\mathbf{OP}$ locating P relative to O and let $\mathbf{n}_x$, $\mathbf{n}_y$, and $\mathbf{n}_z$ be unit vectors along the X, Y, and Z-axes as in Figure 6.7. Then, we can express $\mathbf{p}$ as

$$\mathbf{p} = \mathbf{OP} = \mathbf{O\hat{P}} + \mathbf{\hat{P}P} = \rho \sin \theta(\cos \phi \mathbf{n}_x + \sin \phi \mathbf{n}_y) + \rho \cos \theta \mathbf{n}_z \tag{6.11}$$

and as

$$\mathbf{p} = x\mathbf{n}_x + y\mathbf{n}_y + z\mathbf{n}_z \tag{6.12}$$

where, as before, x, y, and z are the Cartesian coordinates of P. By comparing Equations 6.11 and 6.12 we have

$$x = \rho \sin \theta \cos \phi, \quad y = \rho \sin \theta \sin \phi, \quad z = \rho \cos \theta \tag{6.13}$$

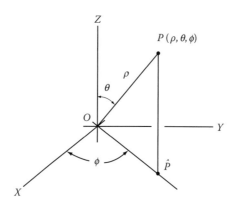

FIGURE 6.6 Spherical coordinate system.

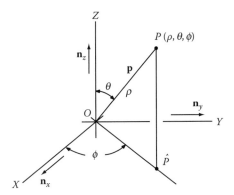

FIGURE 6.7 Position vectors and unit vectors for spherical coordinates.

Equation 6.13 may be solved for ρ, θ, and ϕ in terms of x, y, and z as

$$\rho = (x^2 + y^2 + z^2)^{1/2}, \quad \theta = \cos^{-1}[z/(x^2 + y^2 + z^2)^{1/2}], \quad \phi = \tan^{-1}(y/x) \qquad (6.14)$$

As with cylindrical coordinates, we can use the expressions for x, y, and z (as in Equation 6.13) to obtain parametric equations for the coordinate curves. If we hold θ and ϕ fixed and vary ρ (as parameter t), we have

$$x = (\sin\theta\cos\phi)t, \quad y = (\sin\theta\sin\phi)t, \quad z = (\cos\theta)t \qquad (6.15)$$

There are radial lines projected outward from the origin with inclinations determined by θ and ϕ.
    If we hold ρ and ϕ fixed and vary θ, we have

$$x = \rho\cos\phi\sin t, \quad y = \rho\sin\phi\sin t, \quad z = \rho\cos t \qquad (6.16)$$

These are circles with radius ρ (meridians, great circles of a sphere) with the Z-axis being on a diameter.
    Finally, if we hold ρ and θ fixed and vary ϕ, we have

$$x = \rho\sin\theta\cos t, \quad y = \rho\sin\theta\sin t, \quad z = \rho\cos\theta \qquad (6.17)$$

These are circles, with radius $\rho\sin\theta$ ("parallels" on a sphere), parallel to the X–Y plane.

6.3   OTHER COORDINATE SYSTEMS

Although cylindrical and spherical coordinates are by far the most commonly used curvilinear coordinate systems, there may be occasions when specialized geometry make other coordinate systems useful for simplifying stress analysis. Even with these specialized coordinate systems the geometric complexity will introduce complexity into the analysis.
    Perhaps the simplest of these specialized coordinate systems are parabolic cylindrical coordinates and elliptic cylindrical coordinates. Analogous to Equation 6.5, the parabolic cylindrical coordinates (u, v, z) are defined as [2]

$$x = (u^2 - v^2)/2, \quad y = uv, \quad z = z \qquad (6.18)$$

The corresponding coordinate curves are then families of orthogonally intersecting parabolas in the X–Y plane with axes being the X-axis together with lines parallel to the Z-axis.

Similarly, elliptic cylindrical coordinates (u, v, z) are defined as [2]

$$x = a\cosh u \cos v, \quad y = a \sinh u \sin v, \quad z = z \tag{6.19}$$

where a is a constant. The corresponding coordinate curves are then families of orthogonal, confocal ellipses, and hyperbolas in the X–Y plane with foci at $(a, 0)$ and $(-a, 0)$ together with lines parallel to the Z-axis.

Reference [2] also shows that by rotating the parabolas of the parabolic cylindrical coordinate system about the X-axis, we obtain paraboloidal (u, v, ϕ) coordinates. Similarly, by rotating the ellipses and hyperbolas of the elliptic cylindrical coordinate system about the X- and Y-axes, we obtain prolate spheroidal (ξ, η, ϕ) coordinates and oblate spheroidal coordinates (ξ, η, ϕ), respectively.

Other specialized coordinate systems are bipolar coordinates (u, v, z), toroidal coordinates (u, v, ϕ), and conical coordinates (λ, μ, ν).

6.4 BASE VECTORS

Consider a curve C as represented in Figure 6.10. Let C be defined by parametric equations as

$$x = x(t), \quad y = y(t), \quad z = z(t) \tag{6.20}$$

Let $\mathbf{p}$ be a position vector locating a typical point P on C, as in Figure 6.8. Then $\mathbf{p}$ may be expressed as

$$\mathbf{p} = x(t)\mathbf{n}_x + y(t)\mathbf{n}_y + z(t)\mathbf{n}_z = \mathbf{p}(t) \tag{6.21}$$

where, as before, $\mathbf{n}_x$, $\mathbf{n}_y$, and $\mathbf{n}_z$ are mutually perpendicular unit vectors parallel to X-, Y-, and Z-axes (see Figure 6.8).

This configuration is directly analogous to that encountered in elementary kinematics where $\mathbf{p}(t)$ locates a point P in space as it moves on a curve C. The velocity $\mathbf{v}$ of P is then simply $d\mathbf{p}/dt$. That is

$$v = \lim_{\Delta t \to 0} \frac{\mathbf{p}(t + \Delta t) - \mathbf{p}(t)}{\Delta t} = \lim_{\Delta t \to 0} \frac{\Delta \mathbf{p}}{\Delta t} \tag{6.22}$$

From the last term of Equation 6.22, we see that $\mathbf{v}$ has the direction of $\Delta \mathbf{p}$ as $d\,\Delta t$ becomes small. But $\Delta \mathbf{p}$ is a chord vector of C as represented in Figure 6.9. Thus, as Δt gets small and consequently as $\Delta \mathbf{p}$ gets small, $\Delta \mathbf{p}$ becomes nearly coincident with C at P. Then in the limit, as Δt approaches 0, $\Delta \mathbf{p}$ and thus $\mathbf{v}$ are tangent to C at P.

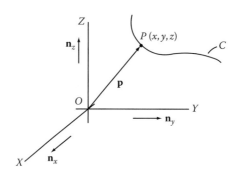

FIGURE 6.8   Curve C defined by parametric equations.

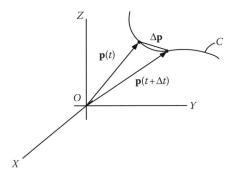

FIGURE 6.9 Chord vector Δp along *C*.

If we regard *C* as a coordinate curve of a curvilinear coordinate system, and imagine *C* as being defined by parametric equations as in Equation 6.20, by as in Equations 6.7, 6.9, and 6.10, and with Equations 6.15 through 6.17 for spherical coordinates, then the derivatives of the coordinate functions $[x(t), y(t), z(t)]$ with respect to the parameter t are the components of vectors tangent to the coordinate curves. These vectors are called: "base vectors."

To illustrate this, consider the cylindrical coordinate system in Section 6.2.1 as defined by Equation. 6.5. The base vectors are simply the derivatives of the position vector **p** of *P* relative to the coordinates which play the role of the parameter t. Specifically, from Equation 6.5, we have

$$\begin{aligned} \mathbf{p} &= x\mathbf{n}_x + y\mathbf{n}_y + z\mathbf{n}_z \\ &= r\cos\,\theta\,\mathbf{n}_x + r\,\sin\theta\mathbf{n}_y + z\mathbf{n}_z \end{aligned} \tag{6.23}$$

Then the base vectors are

$$\partial\mathbf{p}/\partial r \overset{D}{=} \mathbf{g}_r = \cos\theta\mathbf{n}_x + \sin\theta\mathbf{n}_y \tag{6.24}$$

$$\partial\mathbf{p}/\partial\theta \overset{D}{=} \mathbf{g}_\theta = -r\,\sin\theta\,\mathbf{n}_r + r\,\cos\theta\,\mathbf{n}_y \tag{6.25}$$

$$\partial\mathbf{p}/\partial z \overset{D}{=} \mathbf{g}_z = \mathbf{n}_z \tag{6.26}$$

Similarly, for spherical coordinates, the position vector **p** is determined from Equation 6.13 as

$$\mathbf{p} = \rho\,\sin\theta\cos\phi\,\mathbf{n}_x + \rho\,\sin\theta\,\sin\phi\mathbf{n}_y + \rho\cos\theta\mathbf{n}_z \tag{6.27}$$

The base vectors are then

$$\partial\mathbf{p}/\partial\rho = \mathbf{g}_\rho = \sin\theta\,\cos\phi\,\mathbf{n}_x + \sin\theta\sin\phi\,\mathbf{n}_y + \cos\theta\mathbf{n}_z \tag{6.28}$$

$$\partial\mathbf{p}/\partial\theta = \mathbf{g}_\theta = \rho\,\cos\theta\cos\phi\,\mathbf{n}_x + \rho\,\cos\theta\sin\phi\,\mathbf{n}_y - \rho\sin\theta\,\mathbf{n}_z \tag{6.29}$$

$$\partial\mathbf{p}/\partial\phi = \mathbf{g}_\phi = -\rho\sin\theta\sin\phi\,\mathbf{n}_x + \rho\sin\theta\,\cos\phi\,\mathbf{n}_y \tag{6.30}$$

Observe that the cylindrical coordinate base vectors $\mathbf{g}_r$, $\mathbf{g}_\theta$, and $\mathbf{g}_z$ (of Equations 6.24 through 6.26) are mutually perpendicular but they are not all unit vectors (the magnitude of $\mathbf{g}_\theta$ is r). Also, the spherical coordinate base vectors $\mathbf{g}_\rho$, $\mathbf{g}_\theta$, and $\mathbf{g}_\phi$ (of Equations 6.28 through 6.30) are mutually perpendicular, but only $\mathbf{g}_\rho$ is a unit vector. In general, the base vectors are not necessarily even mutually perpendicular, but for three-dimensional systems, they will be noncoplanar.

6.5 METRIC COEFFICIENTS, METRIC TENSORS

The scalar (dot) product of two base vectors, say $\mathbf{g}_i$ and $\mathbf{g}_j$, is called a "metric coefficient" and is written as g_{ij}. That is,

$$g_{ij} \overset{D}{=} \mathbf{g}_i \cdot \mathbf{g}_j \quad (i, j = 1, 2, 3) \tag{6.31}$$

Since there are nine such products, they may be gathered into an array, or matrix, G given by

$$G = [g_{ij}] = \begin{bmatrix} g_{11} & g_{12} & g_{13} \\ g_{21} & g_{22} & g_{23} \\ g_{31} & g_{32} & g_{33} \end{bmatrix} \tag{6.32}$$

Since the scalar product is commutative, G is symmetric. g_{ij} are also called "metric tensor components."

The metric coefficients are directly related to a differential arc length ds of a curve: consider again a curve C as in Figure 6.10. Let $\mathbf{p}$ locate a point P on C and let $\Delta\mathbf{p}$ be an incremental position vector as shown. In the limit, as Δt becomes infinitesimal, $\Delta\mathbf{p}$ becomes the differential tangent vector $d\mathbf{p}$, where t is the parameter as in the foregoing sections. The magnitude of $d\mathbf{p}$ is equal to a differential arc length ds of C.

Let a curvilinear coordinate system have coordinates designated by q^1, q^2, and q^3, or q^i $(i = 1, 2, 3)$, where for convenience in the sequel, we will use superscripts (not to be confused with exponents) to distinguish and label the coordinates. Then with this notation, the base vectors are

$$\mathbf{g}_i = \partial\mathbf{p}/\partial q^i \quad (i = 1, 2, 3) \tag{6.33}$$

where now $\mathbf{p} = \mathbf{p}(q^i)$. Then, using the chain rule for differentiating functions of several variables, the differential vector $d\mathbf{p}$ becomes

$$\begin{aligned} d\mathbf{p} &= \frac{\partial\mathbf{p}}{\partial q^1}dq^1 + \frac{\partial\mathbf{p}}{\partial q^2}dq^2 + \frac{\partial\mathbf{p}}{\partial q^3}dq^3 \\ &= \sum_{i=1}^{3}\frac{\partial\mathbf{p}}{\partial q^i} = \frac{\partial\mathbf{p}}{\partial q^i}dq^i = \mathbf{g}_i dq^i \end{aligned} \tag{6.34}$$

where, as before, we are employing the repeated-index summation convention.

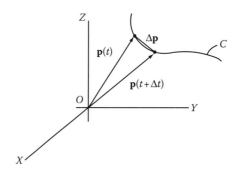

FIGURE 6.10   Curve C with incremental position vector.

Hence, the differential arc length ds is given by

$$(ds)^2 = d\mathbf{p} \cdot d\mathbf{p} = |d\mathbf{p}|^2 = \mathbf{g}_i dq^i \cdot \mathbf{g}_j dq^j = g_{ij} dq^i dq^i \tag{6.35}$$

As an illustration of these concepts, consider again the cylindrical and spherical coordinate systems discussed in Section 6.2. First, for cylindrical coordinates, we have

$$q^1 = r, \quad q^2 = \theta, \quad q^3 = z \tag{6.36}$$

From Equations 6.24 through 6.26 the base vectors are

$$\mathbf{g}_1 = \mathbf{g}_r = \cos\theta \; \mathbf{n}_x + \sin\theta \; \mathbf{n}_y \tag{6.37}$$

$$\mathbf{g}_2 = \mathbf{g}_\theta = -r \; \sin\theta \; \mathbf{n}_x + r \; \cos\theta \; \mathbf{n}_y \tag{6.38}$$

$$\mathbf{g}_3 = \mathbf{g}_z = \mathbf{n}_z \tag{6.39}$$

Then, from Equation 6.31 we have

$$g_{11} = g_{rr} = 1, \quad g_{22} = g_{\theta\theta} = r^2, \quad g_{33} = g_{zz} = 1 \tag{6.40}$$

and

$$g_{ij} = 0, \quad i \neq j \tag{6.41}$$

In matrix form, G is (see Equation 6.32)

$$G = [g_{ii}] = \begin{bmatrix} 1 & 0 & 0 \\ 0 & r^2 & 0 \\ 0 & 0 & 1 \end{bmatrix} \tag{6.42}$$

Then, from Equation 6.35 the arc length ds is given by

$$ds^2 = dr^2 + r^2 d\theta^2 + dz^2 \tag{6.43}$$

Similarly, for spherical coordinates, we have

$$q^1 = \rho, \quad q^2 = \theta, \quad q^3 = \phi \tag{6.44}$$

From Equations 6.28 through 6.30, the base vectors are

$$\mathbf{g}_1 = \mathbf{g}_\rho = \sin\theta \; \cos\phi \; \mathbf{n}_x + \sin\theta \; \sin\phi \; \mathbf{n}_y + \cos\theta \; \mathbf{n}_z \tag{6.45}$$

$$\mathbf{g}_2 = \mathbf{g}_\theta = \rho \; \cos\theta \; \cos\phi \; \mathbf{n}_x + \rho \; \cos\theta \; \sin\phi \; \mathbf{n}_y - \rho \sin \; \theta \; \mathbf{n}_z \tag{6.46}$$

$$\mathbf{g}_3 = \mathbf{g}_\psi = -\rho \sin \; \theta \sin \; \phi \; \mathbf{n}_x + \rho \; \sin\theta \; \cos\phi \; \mathbf{n}_y \tag{6.47}$$

Then, from Equation 6.31 we have

$$g_{11} = g_{\rho\rho} = 1, \quad g_{22} = g_{\theta\theta} = \rho^2, \quad g_{33} = \rho^2 \sin^2\theta \tag{6.48}$$

and

$$g_{ij} = 0 \quad i \neq j \tag{6.49}$$

In matrix form, G is

$$G = [g_{ij}] = \begin{bmatrix} 1 & 0 & 0 \\ 0 & \rho^2 & 0 \\ 0 & 0 & \rho^2 \sin^2 \theta \end{bmatrix} \tag{6.50}$$

Then, from Equation 6.35, the arc length ds is given by

$$ds^2 = d\rho^2 + \rho^2 d\theta^2 + \rho^2 \sin^2 \theta \, d\phi^2 \tag{6.51}$$

6.6 RECIPROCAL BASE VECTORS

The base vectors defined in Section 6.4 are tangent to the coordinate curves and are therefore useful for the expression of stress, strain, and displacement along these directions. The base vectors, however, are not in general unit vectors nor are they necessarily mutually perpendicular. The question arising then is: How are physical quantities to be expressed in terms of these vectors? Or specifically, given a vector $\mathbf{v}$ and noncoplanar base vectors $\mathbf{g}_1$, $\mathbf{g}_2$, and $\mathbf{g}_3$, if $\mathbf{v}$ is to be expressed as

$$\mathbf{v} = (\ )\mathbf{g}_1 + (\ )\mathbf{g}_2 + (\ )\mathbf{g}_3 \tag{6.52}$$

what are the values of the parenthetical quantities (or components) and how are they to be determined?

To answer these questions, recall that if $\mathbf{n}_1$, $\mathbf{n}_2$, and $\mathbf{n}_3$ are mutually perpendicular unit vectors, then a vector $\mathbf{v}$ may be expressed in terms of these vectors as

$$\mathbf{v} = v_1 \mathbf{n}_1 + v_2 \mathbf{n}_2 + v_3 \mathbf{n}_3 = v_i \mathbf{n}_i \tag{6.53}$$

where the scalar components v_i $(i = 1, 2, 3)$ are

$$v_1 = \mathbf{v} \cdot \mathbf{n}_1, \quad v_2 = \mathbf{v} \cdot \mathbf{n}_2, \quad v_3 = \mathbf{v} \cdot \mathbf{n}_3 \quad \text{or} \quad v_i = \mathbf{v} \cdot \mathbf{n}_i \quad (i = 1, 2, 3) \tag{6.54}$$

Thus, $\mathbf{v}$ may be expressed as

$$\mathbf{v} = (\mathbf{v} \cdot \mathbf{n}_1)\mathbf{n}_1 + (\mathbf{v} \cdot \mathbf{n}_2)\mathbf{n}_2 + (\mathbf{v} \cdot \mathbf{n}_3)\mathbf{n}_3 = (\mathbf{v} \cdot \mathbf{n}_i)\mathbf{n}_i \tag{6.55}$$

Consider Equation 6.52 now. Let the parenthetical quantities be designated as v^1, v^2, and v^3 where, again, the superscripts are indices, not to be confused with exponents. Then $\mathbf{v}$ has the form

$$\mathbf{v} = v^1 \mathbf{g}_1 + v^2 \mathbf{g}_2 + v^3 \mathbf{g}_3 = v^i \mathbf{g}_i \tag{6.56}$$

Let the base vectors $\mathbf{g}_i$ $(i = 1, 2, 3)$ be noncoplanar, but not necessarily mutually perpendicular. Then the vector product: $\mathbf{g}_2 \times \mathbf{g}_3$ is nonzero and perpendicular to both $\mathbf{g}_2$ and $\mathbf{g}_3$, and thus normal to the coordinate surface determined by g^2 and g^3. Consider the scalar (dot) product of $\mathbf{g}_2 \times \mathbf{g}_3$ with $\mathbf{v}$:

$$\mathbf{v} \cdot \mathbf{g}_2 \times \mathbf{g}_3 = v^1(\mathbf{g}_1 \cdot \mathbf{g}_2 \times \mathbf{g}_3) + v^2(\mathbf{g}_2 \cdot \mathbf{g}_2 \times \mathbf{g}_3) + v^3(\mathbf{g}_3 \cdot \mathbf{g}_2 \times \mathbf{g}_3)$$
$$= v^1(\mathbf{g}_1 \cdot \mathbf{g}_2 \times \mathbf{g}_3) \tag{6.57}$$

where the last two terms are zero because triple scalar products with duplicate vectors are zero (see Section 3.5.1). Hence v^1 is

$$v^1 = \mathbf{v} \cdot \mathbf{g}_2 \times \mathbf{g}_3 / (\mathbf{g}_1 \times \mathbf{g}_2 \cdot \mathbf{g}_3) \tag{6.58}$$

Similarly, by multiplying by $\mathbf{g}_3 \times \mathbf{g}_1$ times $\mathbf{g}_1 \times \mathbf{g}_2$ we find v^2 and v^3 to be

$$v^2 = \mathbf{v} \cdot \mathbf{g}_3 \times \mathbf{g}_1 / (\mathbf{g}_1 \times \mathbf{g}_2 \cdot \mathbf{g}_3) \tag{6.59}$$

and

$$v^3 = \mathbf{v} \cdot \mathbf{g}_1 \times \mathbf{g}_2 / (\mathbf{g}_1 \times \mathbf{g}_2 \cdot \mathbf{g}_3) \tag{6.60}$$

Considering Equation 6.54 and the results of Equations 6.58 through 6.60, we can simplify the expressions for v^i by introducing "reciprocal base vectors" defined as

$$\mathbf{g}^1 \overset{D}{=} \mathbf{g}_2 \times \mathbf{g}_3 / \mathbf{g}, \quad \mathbf{g}^2 \overset{D}{=} \mathbf{g}_3 \times \mathbf{g}_1 / \mathbf{g}, \quad \mathbf{g}^3 \overset{D}{=} \mathbf{g}_1 \times \mathbf{g}_2 / \mathbf{g} \tag{6.61}$$

where g is defined as

$$\mathbf{g} = \mathbf{g}_1 \times \mathbf{g}_2 \cdot \mathbf{g}_3 = \det G \tag{6.62}$$

Then, Equations 6.58 through 6.60 become

$$v^1 = \mathbf{v} \cdot \mathbf{g}^1, \quad v^2 = \mathbf{v} \cdot \mathbf{g}^2, \quad v^3 = \mathbf{v} \cdot \mathbf{g}^3 \tag{6.63}$$

or simply

$$v^i = \mathbf{v} \cdot \mathbf{g}^i \quad (i = 1, 2, 3) \tag{6.64}$$

Then, Equation 6.56 becomes

$$\mathbf{v} = (\mathbf{v} \cdot \mathbf{g}^1)\mathbf{g}_1 + (\mathbf{v} \cdot \mathbf{g}^2)\mathbf{g}_2 + (\mathbf{v} \cdot \mathbf{g}^3)\mathbf{g}_3 = (\mathbf{v} \cdot \mathbf{g}^i)\mathbf{g}_i \tag{6.65}$$

(Compare these results with Equations 6.54 and 6.55.)

Regarding notation, from Equations 6.54 and 6.65, it is convenient for curvilinear coordinates to adopt the summation convention that there is a sum over repeated indices between subscripts and superscripts, and that free indices must be consistently subscripts or superscripts in each term of an equation.

Next, from Equation 6.61 we have

$$\begin{aligned} \mathbf{g}^1 \cdot \mathbf{g}_1 = 1, \quad \mathbf{g}^2 \cdot \mathbf{g}_2 = 1, \quad \mathbf{g}^3 \cdot \mathbf{g}_3 = 1 \\ \mathbf{g}^1 \cdot \mathbf{g}_2 = \mathbf{g}^2 \cdot \mathbf{g}_1 = 0, \quad \mathbf{g}^2 \cdot \mathbf{g}_3 = \mathbf{g}^3 \cdot \mathbf{g}_2 = 0, \quad \mathbf{g}^3 \cdot \mathbf{g}_1 = \mathbf{g}^1 \cdot \mathbf{g}_3 = 0 \end{aligned} \tag{6.66}$$

or more succinctly

$$\mathbf{g}^i \cdot \mathbf{g}_j = \mathbf{g}_i \cdot \mathbf{g}^j = \delta^i_j = \begin{cases} 1 \ i = j \\ 0 \ i \neq j \end{cases} \tag{6.67}$$

Where δ^i_j is called Kronecker's delta function (see Equation 4.53).

Let reciprocal metric tensor coefficients g^{ij} be defined as

$$g^{ij} = \mathbf{g}^i \cdot \mathbf{g}^j \tag{6.68}$$

Then, it is apparent from Equations 6.31, 6.62, and 6.67 that

$$g^{ik} g_{kj} = \delta^i_j, \quad \mathbf{g}^i = g^{ij} \mathbf{g}_j, \quad \mathbf{g}_i = g_{ij} \mathbf{g}^j, \quad \mathbf{g}^1 \times \mathbf{g}^2 \cdot \mathbf{g}^3 = 1/g \tag{6.69}$$

Also, if we write $\mathbf{v}$ in the form

$$\mathbf{v} = v^i \mathbf{g}_i = v_j \mathbf{g}^i \tag{6.70}$$

then the components are related by the expressions

$$v^i = g^{ij} v_j \quad \text{and} \quad v_i = g_{ij} v^j \tag{6.71}$$

The base vectors $\mathbf{g}_i$ are sometimes called "covariant base vectors" and $\mathbf{g}^i$ are called "contravariant base vectors." Correspondingly in Equations 6.70 and 6.71, the v_i are called "covariant components" and the v^i "contravariant components" [3–5].

6.7 DIFFERENTIATION OF BASE VECTORS

Recall from Equation 4.35 that the equilibrium equations are

$$\sigma_{ij,j} = \rho a_i \quad \text{or} \quad \partial \sigma_{ij}/\partial x_j = \rho a_i \tag{6.72}$$

Recall from Equation 5.32 that the strain–displacement equations are

$$\varepsilon_{ij} = \mathbf{u}_{i,j} + \mathbf{u}_{j,i} \quad \text{or} \quad \varepsilon_{ij} = \frac{1}{2}(\partial \mathbf{u}_i/\partial x_j + \partial \mathbf{u}_j/\partial x_i) \tag{6.73}$$

In these fundamental equations of stress analysis, we see that spatial derivatives occur in most of the terms.

Also, in spatial/vector differentiations, the vector differential operator $\nabla(\ )$ is frequently used as basis of the gradient, divergence, and curl operations [4,6]. In Cartesian coordinates $\nabla(\ )$ is defined as

$$\nabla(\ ) = \mathbf{n}_1 \partial(\ )/\partial x_1 + \mathbf{n}_2 \partial(\ )/\partial x_2 + \mathbf{n}_3 \partial(\ )/\partial x_3 = \mathbf{n}_i \partial(\ )/\partial x_i \tag{6.74}$$

where the $\mathbf{n}_i$ are mutually perpendicular unit vectors parallel to the coordinate axes.

In curvilinear coordinates, Equation 6.74 has the form

$$\nabla(\ ) = \mathbf{g}^1 \partial(\ )/\partial q^1 + \mathbf{g}^2 \partial(\ )/\partial q^2 + \mathbf{g}^3 \partial(\ )/\partial q^3 = \mathbf{g}^i \partial(\ )/\partial q^i \tag{6.75}$$

where $\mathbf{g}^i$ are reciprocal base vectors ("contravariant base vectors") as developed in the previous section, and the $q^i(i = 1, 2, 3)$ are curvilinear coordinates. Equation 6.75 is easily obtained from Equation 6.74 by routine coordinate transformation [4,5,7].

If $\mathbf{v}$ is a vector function of the spatial variables, the gradient, divergence, and curl of $\mathbf{v}$ are

$$\text{Gradient: } \nabla \mathbf{v} ; \quad \text{divergence: } \nabla \cdot \mathbf{v} ; \quad \text{curl: } \nabla \times \mathbf{v} \tag{6.76}$$

In Cartesian coordinates where the unit vectors are constants (not spatially dependent), the gradient, divergence, and curl have relatively simple forms. Specifically, if a vector $\mathbf{v}$ is expressed as

$$\mathbf{v} = v_1 \mathbf{n}_1 + v_2 \mathbf{n}_2 + v_3 \mathbf{n}_3 = v_i \mathbf{n}_i \tag{6.77}$$

where $v_i (i = 1, 2, 3)$ are dependent upon the spatial coordinates, then $\nabla \mathbf{v}$, $\nabla \bullet \mathbf{v}$, and $\nabla \times \mathbf{v}$ become

Gradient

$$\begin{aligned}
\Delta \mathbf{v} = {} & \partial v_1 / \partial x_1 \mathbf{n}_1 \mathbf{n}_1 + \partial v_1 / \partial x_2 \mathbf{n}_1 \mathbf{n}_2 + \partial v_1 / \partial x_3 \mathbf{n}_1 \mathbf{n}_3 \\
& + \partial v_2 / \partial x_1 \mathbf{n}_2 \mathbf{n}_1 + \partial v_2 / \partial x_2 \mathbf{n}_2 \mathbf{n}_2 + \partial v_2 / \partial x_3 \mathbf{n}_2 \mathbf{n}_3 \\
& + \partial v_3 / \partial x_1 \mathbf{n}_3 \mathbf{n}_1 + \partial v_3 / \partial x_2 \mathbf{n}_3 \mathbf{n}_2 + \partial v_3 / \partial x_3 \mathbf{n}_3 \mathbf{n}_3
\end{aligned} \tag{6.78}$$

or alternatively

$$\nabla \mathbf{v} = \partial v_i / \partial x_j \mathbf{n}_i \mathbf{n}_j = v_{i,j} \, \mathbf{n}_i \mathbf{n}_j \tag{6.79}$$

where (as before) the comma designates spatial differentiation.

Divergence

$$\nabla \bullet \mathbf{v} = \partial v_1 / \partial x_1 + \partial v_2 / \partial x_2 + \partial v_3 / \partial x_3 = \partial v_i / \partial x_i = v_{i,i} \tag{6.80}$$

Curl

$$\nabla \times \mathbf{v} = (\partial v_2 / \partial x_3 - \partial v_3 / \partial x_2)\mathbf{n}_1 + (\partial v_2 / \partial x_1 - \partial v_1 / \partial x_3)\mathbf{n}_2 + (\partial v_1 / \partial x_2 - \partial v_2 / \partial x_1)\mathbf{n}_3 \tag{6.81}$$

Suppose that a vector $\mathbf{v}$ is expressed in curvilinear coordinates with either covariant or contravariant base vectors as

$$\mathbf{v} = v^i \mathbf{g}_i = v_j \mathbf{g}^j \tag{6.82}$$

Unlike the unit vectors of Cartesian coordinates, the base vectors are not generally constant. Hence, when we apply the gradient, divergence, and curl operators to $\mathbf{v}$, we need to be able to evaluate the spatial derivatives of the base vectors. To this end, consider first the derivative: $\partial \mathbf{g}_i / \partial q^j$. As before, to simplify the notation, it is convenient to use a comma to designate partial differentiation, where the derivatives are taken with respect to the curvilinear coordinates. That is,

$$(\,),i = \partial(\,)/\partial q^i \tag{6.83}$$

Next, consider that $\partial \mathbf{g}_i / \partial q^j$ or $\mathbf{g}_{i,j}$ is a vector with free indices i and j. As such $\mathbf{g}_{i,j}$ can be represented in the vector form as

$$\mathbf{g}_{i,j} = \Gamma_{ij}^k \mathbf{g}_k \tag{6.84}$$

where the components Γ_{ij}^k are sometimes called Christoffel symbols [3–5,7]. Since the Γ_{ij}^k have three indices, there are 27 values for all the possible combinations of i, j, and k. Fortunately, for the curvilinear coordinate systems of common interest and importance, most of the Γ_{ij}^k are zero. Also, since $\mathbf{g}_i$ is $\partial \mathbf{p}/\partial q^i$, $\mathbf{g}_{i,j}$ may be expressed as

$$\mathbf{g}_{i,j} = \partial^2 \mathbf{p}/\partial q^i \partial q^j = \partial^2 \mathbf{p}/\partial q^j \partial q^i = \mathbf{g}_{j,i} \tag{6.85}$$

Thus Γ_{ij}^k is symmetric in i and j. That is,

$$\Gamma_{ij}^k = \Gamma_{ji}^k \tag{6.86}$$

To evaluate Γ_{ij}^k, it is convenient to introduce the covariant form Γ_{ijk} defined as

$$\Gamma_{ijk} \overset{D}{=} g_{k\ell}\Gamma_{ij}^\ell \tag{6.87}$$

The utility of the Γ_{ijk} is readily seen by differentiating the metric tensor elements. Specifically,

$$\begin{aligned}
g_{ij,k} &= (g_i \bullet g_j)_{,k} = g_{i,k} \bullet g_j + g_i \bullet g_{j,k} \\
&= \Gamma_{ik}^\ell g_\ell \bullet g_j + \Gamma_{jk}^\ell g_\ell \bullet g_i = g_{\ell j}\Gamma_{ik}^\ell + g_{\ell i}\Gamma_{jk}^\ell \\
&= \Gamma_{ikj} + \Gamma_{jki}
\end{aligned} \tag{6.88}$$

By permutating the indices in Equation 6.88 and then by adding and subtracting equations, we obtain

$$\Gamma_{ijk} = (1/2)(g_{ki,j} + g_{kj,i} - g_{ij,k}) \tag{6.89}$$

Then, from the definition of Equation 6.87, Γ_{ij}^k are

$$\Gamma_{ij}^k = g^{k\ell}\Gamma_{ij\ell} \tag{6.90}$$

As an illustration of the values of these expressions, we find that for cylindrical coordinates, Γ_{ijk} and Γ_{ij}^k are

$$\Gamma_{122} = \Gamma_{212} = \Gamma_{r\theta\theta} = \Gamma_{\theta r\theta} = r, \quad \Gamma_{221} = \Gamma_{\theta\theta r} = -1 \tag{6.91}$$

and all other Γ_{ijk} are zero. Also,

$$\Gamma_{12}^2 = \Gamma_{21}^2 = \Gamma_{r\theta}^\theta = \Gamma_{\theta r}^\theta = 1/r, \quad \Gamma_{22}^1 = \Gamma_{\theta\theta}^r = -r \tag{6.92}$$

and all other Γ_{ij}^k are zero.

(Recall from Equation 6.40 that the metric coefficients are: $g_{rr} = 1$, $g_{\theta\theta} = \gamma^2$, $g_{zz} = 1$, and $g_{ij} = 0$, $i \neq j$.)

For spherical coordinates, the Γ_{ijk} and Γ_{ij}^k are

$$\begin{aligned}
\Gamma_{122} &= \Gamma_{212} = \Gamma_{\rho\theta\theta} = \Gamma_{\theta\rho\theta} = r, & \Gamma_{133} &= \Gamma_{313} = \Gamma_{\rho\phi\phi} = \Gamma_{\phi\rho\phi} = r\,\sin^2\theta \\
\Gamma_{233} &= \Gamma_{323} = \Gamma_{\theta\phi\phi} = \Gamma_{\phi\theta\phi} = \rho^2\,\sin\theta\,\cos\theta, & \Gamma_{221} &= \Gamma_{\theta\theta\rho} = -\rho \\
\Gamma_{331} &= \Gamma_{\phi\phi\rho} = -\rho\,\sin^2\theta, & \Gamma_{332} &= \Gamma_{\phi\phi\theta} = -\rho^2\sin\theta\,\cos\theta
\end{aligned} \tag{6.93}$$

and all other Γ_{ijk} are zero. Also,

$$\begin{aligned}
\Gamma_{12}^2 &= \Gamma_{21}^2 = \Gamma_{\rho\theta}^\theta = \Gamma_{\theta\rho}^\theta = -\rho, & \Gamma_{13}^3 &= \Gamma_{31}^3 = \Gamma_{\rho\phi}^\phi = \Gamma_{\phi\rho}^\phi = 1/\rho \\
\Gamma_{23}^3 &= \Gamma_{32}^3 = \Gamma_{\theta\phi}^\phi = \Gamma_{\phi\theta}^\phi = \cos\theta, & \Gamma_{22}^1 &= \Gamma_{\theta\theta}^\rho = -\rho \\
\Gamma_{33}^1 &= \Gamma_{\phi\phi}^\rho = -\rho\sin^2\theta, & \Gamma_{33}^2 &= \Gamma_{\phi\phi}^\theta = -\sin\theta\cos\theta
\end{aligned} \tag{6.94}$$

and all other Γ_{ij}^k are zero.

(Recall from Equation 6.48 that the metric coefficients are $g_{\rho\rho} = 1$, $g_{\theta\theta} = \rho^2$, $g_{\phi\phi} = \rho^2 \sin^2 \theta$, and $g_{ij} = 0$, $i \neq j$.)

Finally, consider the differentiation of the reciprocal (contravariant) base vectors $\mathbf{g}^i$. From Equation 6.67, we have

$$\mathbf{g}_i \cdot \mathbf{g}^j = \delta_i^j \tag{6.95}$$

By differentiating with respect to q^k, we obtain

$$\mathbf{g}_{i,k} \cdot \mathbf{g}^j + \mathbf{g}_i \cdot \mathbf{g}^j_{,k} = 0 \tag{6.96}$$

But, from Equation 6.84, the $\mathbf{g}_{i,k}$ are

$$\mathbf{g}_{i,k} = \Gamma_{ik}^\ell \mathbf{g}_\ell \tag{6.97}$$

Then, by substituting into Equation 6.96, we have

$$\mathbf{g}_i \cdot \mathbf{g}^j_{,k} = -\Gamma_{ik}^j \tag{6.98}$$

Thus, $\mathbf{g}^j_{,k}$ are

$$\mathbf{g}^j_{,k} = -\Gamma_{ij}^j \mathbf{g}^i \tag{6.99}$$

6.8 COVARIANT DIFFERENTIATION

Recall that in our formulation for stress and strain, we found that the stress components satisfy equilibrium equations, where there are derivatives with respect to the space coordinates, and the strain involves derivatives of the displacement vector. In both cases, there are spatial derivatives of vector functions. If we formulate the equations using Cartesian coordinates, the unit vectors along the coordinate axes are constants and the derivatives may be obtained by simply differentiating the components. If, however, we formulate the equations using curvilinear coordinates, the vector functions will be expressed in terms of base vectors, which will generally vary from point-to-point in space. Therefore, in differentiating vector functions expressed in curvilinear coordinates, we need to differentiate the base vectors. We can use the formulation developed in the previous section to obtain these derivatives.

Let $\mathbf{u}$ be a vector, say, a displacement vector, and let $\mathbf{u}$ be expressed in terms of base vectors as

$$\mathbf{u} = u^k \mathbf{g}_k \tag{6.100}$$

Then, using Equation 6.84, the derivative of $\mathbf{u}$ with respect to q^i is

$$
\begin{aligned}
\partial \mathbf{u}/\partial q^i = \partial\left(u^k \mathbf{g}_k\right)/\partial q^i &= u^k_{,i} \mathbf{g}_k + u^k \mathbf{g}_{k,i} \\
&= u^k_{,i} \mathbf{g}_k + u^k \Gamma_{ki}^\ell \mathbf{g}_\ell \\
&= \left(u^k_{,i} + u^\ell \Gamma_{\ell i}^k\right) \mathbf{g}_k
\end{aligned} \tag{6.101}
$$

More succinctly, we can write these expressions as

$$\partial \mathbf{u}/\partial q^i = \mathbf{u}_{,i} = u^k_{;i} \mathbf{g}_k \tag{6.102}$$

where $u_{;i}^k$ is defined as

$$u_{;i}^k = u_{,i}^k + \Gamma_{\ell i}^k u^\ell \qquad (6.103)$$

and is called the "covariant derivative" of u^k.

Next, let $\mathbf{u}$ be expressed in the form

$$\mathbf{u} = u_k \mathbf{g}^k \qquad (6.104)$$

Then, the derivative with respect to q^i is

$$\begin{aligned}
\partial\mathbf{u}/\partial q^i = \partial(u_k \mathbf{g}^k)/q^i &= u_{k,i}\mathbf{g}^k + u_k \mathbf{g}_{,i}^k \\
&= u_{k,i}\mathbf{g}^k - u_k \Gamma_{\ell i}^k \mathbf{g}^\ell \\
&= (u_{k,i} - u_\ell \Gamma_{ki}^\ell)\mathbf{g}^k
\end{aligned} \qquad (6.105)$$

or

$$\partial\mathbf{u}/\partial q^i = \mathbf{u}_{,i} = u_{k;j}\mathbf{g}^k \qquad (6.106)$$

where $u_{k;i}$ is defined as

$$u_{k;i} = u_{k,i} - \Gamma_{ki}^\ell u_\ell \qquad (6.107)$$

and is called the covariant derivative of u_k.

Observe in Equations 6.102 and 6.106 that the derivative of $\mathbf{u}$ may be obtained by simply evaluating the covariant derivative of the components (as defined in Equations 6.103 and 6.107) and leaving the base vectors unchanged. From another perspective, we can view the covariant derivative of a scalar or a nonindexed quantity as the same as a partial derivative. That is, if ϕ is a scalar then

$$\partial\phi/\partial q^i = \phi_{,i} = \phi_{;i} \qquad (6.108)$$

Also, for a vector $\mathbf{u}$ we have

$$\partial\mathbf{u}/\partial q^i = \mathbf{u}_{,i} = \mathbf{u}_{;i} \qquad (6.109)$$

Then, by formally applying the product rule for differentiation, we have

$$\begin{aligned}
\partial\mathbf{u}/\partial q^i = \mathbf{u}_{;i} &= (u^k \mathbf{g}_k)_{;i} \\
&= \mathbf{u}_{;i}^k \mathbf{g}_k + u^k \mathbf{g}_{k;i}
\end{aligned} \qquad (6.110)$$

For the result to be consistent with that of Equation 6.102 the covariant derivative of the base vector $\mathbf{g}_k$ must be zero. That is

$$\mathbf{g}_{k;i} = 0 \qquad (6.111)$$

The validity of Equation 6.111 is evident from Equations 6.107 and 6.97. That is,

$$\mathbf{g}_{k;i} = \mathbf{g}_{k,i} - \Gamma^{\ell}_{ki}\mathbf{g}_{\ell}$$
$$= \Gamma^{\ell}_{ki}\mathbf{g}_{\ell} - \Gamma^{\ell}_{ki}\mathbf{g}_{\ell} = 0 \tag{6.112}$$

Similarly, we see that the covariant derivatives of the base vectors $\mathbf{g}^k$ as well as the metric tensors are zero.

Finally, consider the dyadic $\mathbf{D}$ expressed as

$$\mathbf{D} = d_{ij}\mathbf{g}^i\mathbf{g}^j = d^j_i\mathbf{g}^i\mathbf{g}_j = d^i_j\mathbf{g}_i\mathbf{g}^j = d^{ij}\mathbf{g}_i\mathbf{g}_j \tag{6.113}$$

Then, by following the same procedure as in the differentiation of vectors, we see that the derivative of $\mathbf{D}$ relative to q^k may be expressed as

$$\partial\mathbf{D}/\partial q^k = \mathbf{D}_{,k} = d_{ij;k}\mathbf{g}^i\mathbf{g}_j = d^j_{i;k}\mathbf{g}^i\mathbf{g}_j = d^i_{j;k}\mathbf{g}_i\mathbf{g}_j = d^{ij}\mathbf{g}_i\mathbf{g}_j \tag{6.114}$$

where covariant derivatives $d_{ij;k}$, $d_{ij;k}$, $d^j_{i;k}$, and $d^i_{j;k}$ are

$$d_{ij;k} = d_{ij,k} - \Gamma^{\ell}_{ik}d_{\ell j} - \Gamma^{\ell}_{jk}d_{i\ell} \tag{6.115}$$

$$d^j_{i;k} = d^j_{i,k} - \Gamma^{\ell}_{ik}d^j_{\ell} + \Gamma^{\ell}_{\ell k}d^{\ell}_i \tag{6.116}$$

$$d^i_{j;k} = d^i_{j,k} + \Gamma^i_{\ell k}d^{\ell}_j - \Gamma^j_{jk}d^i_{\ell} \tag{6.117}$$

$$d^{ij}_{;k} = d^{ij}_{,k} + \Gamma^i_{\ell k}d^{\ell j} + \Gamma^j_{\ell k}d^{i\ell} \tag{6.118}$$

6.9 EQUILIBRIUM EQUATIONS AND STRAIN–DISPLACEMENT RELATIONS IN CURVILINEAR COORDINATES

Consider the stress equilibrium equation (Equation 4.35) and the strain–displacement equations (Equation 5.20) again

$$\sigma_{ij,j} = \rho a_i \tag{6.119}$$

and

$$\varepsilon_{ij} = \frac{1}{2}(u_{i,j} + u_{j,i}) \tag{6.120}$$

Although these equations have been developed in Cartesian coordinates, they may be expressed in a vector form, which in turn may be used to express them in curvilinear coordinates. To this end, by comparing Equation 6.119 with Equation 6.80, we see that we can express Equation 6.119 as

$$\nabla \cdot \boldsymbol{\sigma} = \rho\mathbf{a} \tag{6.121}$$

here, $\boldsymbol{\sigma}$ is the stress dyadic (see Section 4.4) and $\mathbf{a}$ is the acceleration vector at a point of the body where the equilibrium equation holds, and where, as before, ∇ is the vector differential operator. Recall from Equation 6.75 that in curvilinear coordinates ∇ has the form

$$\nabla() = \mathbf{g}^1\partial()/\partial q_1 + \mathbf{g}^2\partial()/\partial q_2 + \mathbf{g}^3\partial()/\partial q_3$$
$$= \mathbf{g}^k\partial()/\partial q^k \tag{6.122}$$

Then the left-hand side of Equation 6.121 becomes

$$\nabla \cdot \boldsymbol{\sigma} = \mathbf{g}^k \cdot \partial \boldsymbol{\sigma} / \partial q^k = \mathbf{g}^k \cdot \partial \left(\mathbf{g}^i \sigma_{ij} \mathbf{g}^j \right) / \partial q^k$$
$$= \mathbf{g}^k \cdot \sigma_{ij;k} \mathbf{g}^i \mathbf{g}^j = g^{ik} \sigma_{ij;k} \mathbf{g}^j$$
$$= \sigma_{j;k}^k \mathbf{g}^j = \sigma_{j;k}^k \mathbf{g}^j$$

Then Equation 6.121 becomes

$$\sigma_{j;k}^k \mathbf{g}^j = \rho \mathbf{a} = \rho a_j \mathbf{g}^j \tag{6.123}$$

or in component form

$$\sigma_{j;k}^k = \rho a_j \quad \text{or} \quad \sigma_{i;j}^j = \rho a_i \tag{6.124}$$

Next, recall from Equation 6.79 that if $\mathbf{u}$ is a vector, say, the displacement vector, then in Cartesian coordinates, $\nabla \mathbf{u}$ is

$$\nabla \mathbf{u} = u_{i,j} \mathbf{n}_i \mathbf{n}_j \tag{6.125}$$

where, as before, the $\mathbf{n}_i$ ($i = 1, 2, 3$) are mutually perpendicular unit vectors. In curvilinear coordinates $\nabla \mathbf{u}$ is

$$\nabla \mathbf{u} = u_{i;j} \mathbf{g}^i \mathbf{g}^j \tag{6.126}$$

Thus, we see that in curvilinear coordinates the strain tensor may be expressed as

$$\varepsilon_{ij} = \frac{1}{2} (u_{i;j} + u_{j;i}) \tag{6.127}$$

Observe that in comparing Equations 6.119 and 6.120 with Equations 6.124 and 6.127, the difference is simply that the partial differentiation of Cartesian coordinates is replaced with the covariant differentiation of curvilinear coordinates.

Using the results of Equations 6.115 and 6.107, we see that $\sigma_{i;j}^j$ and $u_{i;j}$ may be expressed as

$$\sigma_{i;j}^j = \sigma_{i,j}^j - \Gamma_{ij}^\ell \sigma_\ell^j + \Gamma_{\ell j}^j \sigma_i^\ell \tag{6.128}$$

and

$$u_{i;j} = u_{i,j} - \Gamma_{ij}^k u_k \tag{6.129}$$

To illustrate the forms of these equations for cylindrical coordinates, we can use the results in Equation 6.92 for the Christoffel symbol components. The equilibrium equations (Equation 6.124) become [8]

$$\frac{\partial \sigma_\pi}{\partial r} + (1/r) \frac{\partial \sigma_{r\theta}}{\partial \theta} + \frac{\partial \sigma_{rz}}{\partial z} + (1/r)(\sigma_\pi - \sigma_{\theta\theta}) = \rho a_r \tag{6.130}$$

$$\frac{\partial \sigma_{r\theta}}{\partial r} + (1/r) \frac{\partial \sigma_{\theta\theta}}{\partial \theta} + \frac{\partial \sigma_{\theta z}}{\partial z} + (2/r) \sigma_{r\theta} = \rho a_\theta \tag{6.131}$$

$$\frac{\partial \sigma_{rz}}{\partial r} + (1/r)\frac{\partial \sigma_{\theta z}}{\partial \theta} + \frac{\partial \sigma_{zz}}{\partial z} + (1/r)\sigma_{rz} = \rho a_z \tag{6.132}$$

and the strain–displacement equations (Equation 6.127) become

$$\varepsilon_{rr} = \frac{\partial u_r}{\partial r} \tag{6.133}$$

$$\varepsilon_{r\theta} = (1/2)\left[(1/r)\frac{\partial u_r}{\partial r} + \frac{\partial u_\theta}{\partial r} - (1/r)u_\theta\right] \tag{6.134}$$

$$\varepsilon_{rz} = (1/2)\left(\frac{\partial u_z}{\partial r} + \frac{\partial u_r}{\partial z}\right) \tag{6.135}$$

$$\varepsilon_{\theta\theta} = (1/r)\left(\frac{\partial u_\theta}{\partial \theta} + u_r\right) \tag{6.136}$$

$$\varepsilon_{\theta z} = (1/2)\left[\frac{\partial u_\theta}{\partial z} + (1/r)\frac{\partial u_z}{\partial \theta}\right] \tag{6.137}$$

$$\varepsilon_{zz} = \frac{\partial u_z}{\partial z} \tag{6.138}$$

Similarly, for spherical coordinates we can use the results of Equation 6.94 for the Christoffel symbol components. The equilibrium equations become [8]

$$\frac{\partial \sigma_{\rho\rho}}{\partial \rho} + (1/\rho \sin\theta)\frac{\partial \sigma_{\rho\phi}}{\partial \phi} + (1/\rho)\frac{\partial \sigma_{\rho\theta}}{\partial \theta} + (1/\rho)(2\sigma_{\rho\rho} - \sigma_{\phi\phi} - \sigma_{\theta\theta} + \sigma_{\rho\theta}\cot\theta) = \hat{\rho}a_\rho \tag{6.139}$$

$$\frac{\partial \sigma_{\rho\phi}}{\partial \rho} + (1/\rho \sin\theta)\frac{\partial \sigma_{\phi\phi}}{\partial \phi} + (1/\rho)\frac{\partial \sigma_{\theta\theta}}{\partial \theta} + (1/\rho)(3\sigma_{\rho\theta} + \sigma_{\theta\theta}\cot\theta - \sigma_{\phi\phi}\cot\theta) = \hat{\rho}a_\theta \tag{6.140}$$

$$\frac{\partial \sigma_{\rho\phi}}{\partial \rho} + (1/\rho \sin\theta)\frac{\partial \sigma_{\phi\phi}}{\partial \phi} + (1/\rho)\frac{\partial \sigma_{\phi\phi}}{\partial \theta} + (1/\rho)(3\sigma_{\rho\phi} + 2\sigma_{\phi\theta}\cot\theta) = \hat{\rho}a_\phi \tag{6.141}$$

where $\hat{\rho}$ is now the mass density. The strain–displacement equations become

$$\varepsilon_{\rho\rho} = \frac{\partial u_\rho}{\partial \rho} \tag{6.142}$$

$$\varepsilon_{\rho\theta} = (1/2)\left[(1/\rho)\frac{\partial u_\rho}{\partial \theta} - (u_\theta/\rho) + \frac{\partial u_\theta}{\partial \rho}\right] \tag{6.143}$$

$$\varepsilon_{\rho\phi} = (1/2)\left[(1/\rho \sin\theta)\frac{\partial u_\rho}{\partial \phi} - (u_\phi/\rho) + \frac{\partial u_\phi}{\partial \rho}\right] \tag{6.144}$$

$$\varepsilon_{\theta\theta} = (1/\rho)\left(\frac{\partial u_\theta}{\partial \theta} + u_\rho\right) \tag{6.145}$$

$$\varepsilon_{\theta\phi} = (1/2)\left[(1/\rho)\frac{\partial u_\phi}{\partial \theta} - (1/\rho)u_\phi\cot\theta + (1/\rho\sin\theta)\frac{\partial u_\theta}{\partial \phi}\right] \tag{6.146}$$

$$\varepsilon_{\phi\phi} = (1/\rho)\left[(1/\sin\theta)\frac{\partial u_\phi}{\partial \rho} + u_\rho + u_\theta\cot\theta\right] \tag{6.147}$$

SYMBOLS

a	Acceleration vector
C	Curve
$\mathbf{D}$	Dyadic
d_{ij}, d_i^j, d^{ij} $(i, j = 1, 2, 3)$	Elements of dyadic matrices
$\mathbf{g}_i$ $(i = 1, 2, 3)$	Base vectors
$\mathbf{g}^i$ $(i = 1, 2, 3)$	Reciprocal base vectors
$\mathbf{g}_{ij}$ $(i, j = 1, 2, 3)$	Metric coefficients, metric tensor components
G	Metric coefficient array
$\mathbf{n}_i$ $(i = 1, 2, 3)$	Mutually perpendicular unit vectors
n_x, n_y, n_z	Mutually perpendicular unit vectors
O	Origin
P	Point
$\mathbf{p}$	Position vector
q^i $(i = 1, 2, 3)$	Curvilinear coordinates
r, θ, z	Cylindrical coordinates
s	Arc length
t	Parameter
$\mathbf{u}$	Displacement vector
u_i $(i = 1, 2, 3)$	Displacements
$\mathbf{V}$	Vector
v_i $(i = 1, 2, 3)$ n_i	$\mathbf{n}_i$ components of vector $\mathbf{v}$
$v_{i,j}$ $(i, j = 1, 2, 3)$	Partial derivative of v_i with respect to q^j
$v_{i;j}$ $(i, j = 1, 2, 3)$	Covariant derivative of v_i with respect to q^j
X, Y, Z	Cartesian (rectangular) coordinate axes
x, y, z	Cartesian coordinates
x_i $(i = 1, 2, 3)$	Cartesian coordinates
Γ_{ijk} $(i, j, k = 1, 2, 3)$	Christoffel symbols (Section 6.7)
Γ_{ij}^k $(i, j, k = 1, 2, 3)$	Christoffel symbols (Section 6.7)
$\Delta\mathbf{p}$	Chord vector
ε_{ij} $(i, j = 1, 2, 3)$	Strain tensor components
ρ, ϕ, θ	Spherical coordinates
$\boldsymbol{\sigma}$	Stress dyadic
σ_{ij} $(i, j = 1, 2, 3)$	Stress tensor components

REFERENCES

1. B. Rodin, *Calculus with Analytic Geometry*, Prentice Hall, Englewood Cliffs, NJ, 1970, pp. 472–476.
2. M. Fogiel (Ed.), *Handbook of Mathematical Formulas, Tables, Functions, Graphs, and Transforms*, Research and Educative Association, New York, 1980, pp. 297–303.
3. A. C. Eringen, *Nonlinear Theory of Continuous Media*, McGraw Hill, New York, 1962, pp. 444–445.
4. P. M. Morse and H. Feshbach, *Methods of Theoretical Physics*, Part I, McGraw Hill, New York, 1953 (chap. 1).
5. I. S. Sokolnikoff, *Tensor Analysis, Theory and Applications to Geometry and Mechanics of Continua*, 2nd ed., John Wiley, New York, 1964 (chap. 2).
6. F. B. Hildebrand, *Advanced Calculus for Applications*, Prentice Hall, Englewood Cliffs, NJ, 1976 (chap. 6).
7. J. G. Papastavrides, *Tensor Calculus and Analytical Dynamics*, CRC Press, Boca Raton, FL, 1999 (chaps. 2, 3).
8. I. S. Sokolnikoff, *Mathematical Theory of Elasticity*, Krieger Publishing, Malabar, FL, 1983, pp. 183–184.

7 Hooke's Law in Two and Three Dimensions

7.1 INTRODUCTION

In Chapter 3, we saw that Hooke's law predicts a linear relationship between simple stress and simple strain. In this chapter, we extend this elementary concept to two and three dimensions. As before, we will restrict our attention to small strain. The resulting relations then continue to be linear between the stresses and strains.

We begin our analysis with a discussion of Poisson's ratio, or "transverse contraction ratio," which quantifies induced strain in directions perpendicular to an applied strain, such as a rod shrinking laterally as it is elongated.

7.2 POISSON'S RATIO

Consider a rod with a square cross section subjected to an axial load as in Figure 7.1. Intuitively as the rod is stretched or elongated, the cross-section area will become smaller. Poisson's ratio is a measure or quantification of this effect. Specifically, for the rod of Figure 7.1, Poisson's ratio ν is defined as

$$\nu = -\varepsilon_{yy}/\varepsilon_{xx} \tag{7.1}$$

where the X-axis is along the rod and the Y-axis is perpendicular to the rod, as shown. From the simple geometry of the rod we see that ε_{xx} and ε_{yy} are

$$\varepsilon_{xx} = \delta_x/\ell \quad \text{and} \quad \varepsilon_{yy} = -\delta_y/a \tag{7.2}$$

where δ_x is the elongation of the rod, δ_y is the shrinking of the cross section, ℓ is the rod length, and a is the cross section side length. Poisson's ratio is then a measure of the contraction of the rod as it is stretched. Consequently, Poisson's ratio is occasionally called the "transverse contraction ratio."

As a further illustration of this concept, consider a circular cross-section rod being elongated as in Figure 7.2. As the rod is stretched, the circular cross section will become smaller as represented (in exaggerated form) in Figure 7.3 where a is the undeformed cross-section radius and δ_r is the radius decrease. During stretching, the radial displacement u_r of a point Q in the cross section is proportional to the distance of Q from the axis as in Figure 7.4. That is,

$$u_r = -\delta_r(r/a) \tag{7.3}$$

where the minus sign indicates that Q is displaced toward O as the cross section shrinks. The radial strain ε_{rr} is then (see Equation 6.133)

$$\varepsilon_{rr} = \partial u_r/\partial r = -\delta_r/a \tag{7.4}$$

If the rod is elongated with a length change (or stretching) δ_x (see Figure 7.2), the axial strain ε_{xx} is

$$\varepsilon_{xx} = \delta_x/\ell \tag{7.5}$$

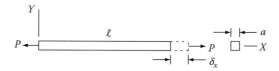

FIGURE 7.1   An elongated rod with a square cross section.

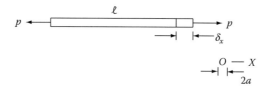

FIGURE 7.2   An elongated rod with a circular cross section.

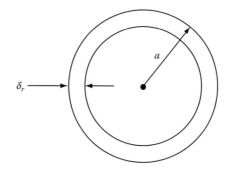

FIGURE 7.3   Shrinkage of rod cross section.

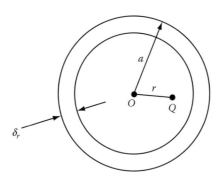

FIGURE 7.4   Rod cross section with typical point Q.

TABLE 7.1
Typical Values of Poisson's Ratio

Upper theoretical limit (perfectly deformable material)	0.50
Lead	0.43
Gold	0.42
Platinum	0.39
Silver	0.37
Aluminum (pure)	0.36
Phosphor bronze	0.35
Tantalum	0.35
Copper	0.34
Titanium (pure)	0.34
Aluminum (wrought)	0.33
Titanium (alloy)	0.33
Brass	0.33
Molybdenum (wrought)	0.32
Stainless steel	0.31
Structural steel	0.30
Magnesium alloy	0.28
Tungsten	0.28
Granite	0.28
Sandstone	0.28
Thorium (induction-melted)	0.27
Cast iron (gray)	0.26
Marble	0.26
Glass	0.24
Limestone	0.21
Uranium (D-38)	0.21
Plutonium (alpha phase)	0.18
Concrete (average water content)	0.12
Beryllium (vacuum-pressed powder)	0.027
Lower theoretical limit (perfectly brittle material)	0.000

Poisson's ratio ν is then simply the negative ratio of the radial and axial strains. Thus ν is

$$\nu = -\varepsilon_{rr}/\varepsilon_{xx} = (\delta_r/\delta_x)(\ell/a) \tag{7.6}$$

Observe that ν is a material property. That is, the values of ν depend upon the character of the material being deformed. Table 7.1 provides a list of typical values of ν for common materials.

Since ν is a measure of the shrinkage of a loaded rod, as in the foregoing examples, it is also a measure of the volume change of a loaded body. Consider again the elongated round bar of Figure 7.2. The undeformed volume V of the rod is simply

$$V = \pi a^2 \ell \tag{7.7}$$

From Figures 7.2 and 7.3 we see that the deformed volume $\hat{V}$ is

$$\hat{V} = \pi(a - \delta_r)^2(\ell + \delta_x) \tag{7.8}$$

Since δ_r and δ_x are small, $\hat{V}$ may reasonably be expressed as

$$\hat{V} = \pi(a^2 - 2a\delta_r)(\ell + \delta_x) = \pi(a^2\ell - 2a\ell\delta_r + a^2\delta_x) \tag{7.9}$$

The volume change ΔV is then

$$\Delta V = \hat{V} - V = \pi(-2a\ell\delta_r + a^2\delta_x) \tag{7.10}$$

But from Equation 7.6, δ_r is

$$\delta_r = \nu\delta_x(a/\ell) \tag{7.11}$$

Hence, ΔV becomes

$$\Delta V = \pi(-2\nu + 1)a^2\delta_x \tag{7.12}$$

Finally, the volumetric strain defined as $\Delta V/V$ is

$$\Delta V/V = (1 - 2\nu)(\delta_x/\ell) = (1 - 2\nu)\varepsilon_{xx} \tag{7.13}$$

As a final example, consider the small rectangular parallelepiped block depicted in Figure 7.5. Let the block be loaded with a uniform, outward (tension) pressure on the face perpendicular to the X-axis. Let the edges of the block before loading be a, b, and c. After being loaded, the block

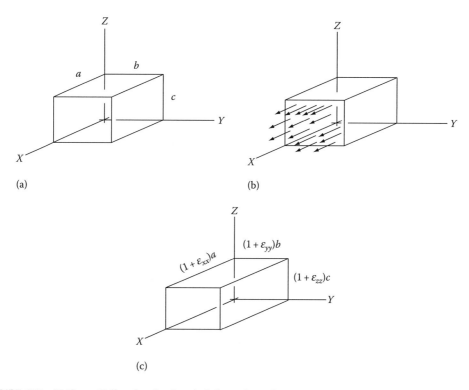

(a)

(b)

(c)

FIGURE 7.5 Uniform X-direction load and deformation of a rectangular block. (a) Rectangular block, (b) uniform X-directional load, and (c) deformed block.

edges will be (for small deformation): $(1 + \varepsilon_{xx})a$, $(1 + \varepsilon_{yy})b$, and $(1 + \varepsilon_{zz})c$. Then the block volumes V and $\hat{V}$, before and after deformation respectively, are

$$V = abc \tag{7.14}$$

and

$$\begin{aligned}
\hat{V} &= (1 + \varepsilon_{xx})a(1 + \varepsilon_{yy})b(1 + \varepsilon_{zz})c \\
&= abc(1 + \varepsilon_{xx} + \varepsilon_{yy} + \varepsilon_{zz})
\end{aligned} \tag{7.15}$$

The volumetric strain $\Delta V/V$ is then

$$\Delta V/V = (\hat{V} - V)/V = \varepsilon_{xx} + \varepsilon_{yy} + \varepsilon_{zz} \tag{7.16}$$

But from Equation 7.1, ε_{yy} and ε_{zz} may be expressed in terms of ε_{xx} as

$$\varepsilon_{yy} = -\nu\varepsilon_{xx} \quad \text{and} \quad \varepsilon_{zz} = -\nu\varepsilon_{xx} \tag{7.17}$$

The volumetric strain is then

$$\Delta V/V = (1 - 2\nu)\varepsilon_{xx} \tag{7.18}$$

7.3 BRITTLE AND COMPLIANT MATERIALS

If a material does not contract or shrink transversely when loaded, that is, if the Poisson's ratio ν is zero, then the material is said to be "brittle." Conversely, if a brittle material has no transverse contraction, then the Poisson's ratio is zero.

Alternatively, if a material shrinks so that the volume of a loaded body remains constant, the material is said to be "incompressible" or "fully compliant." Conversely, for a fully compliant material, the volume change, ΔV, is zero during loading. Then from Equation 7.18, we have

$$1 - 2\nu = 0 \quad \text{or} \quad \nu = 1/2 \tag{7.19}$$

Therefore, Poisson's ratio ranges from 0 to $1/2$.

7.4 PRINCIPLE OF SUPERPOSITION OF LOADING

The principle of superposition states that multiple loadings on an elastic body may be considered individually and in any order, for evaluating the stresses and strains due to the loadings. That is, the state of stress or strain of an elastic body subjected to multiple loads is simply the addition (or "superposition") of the respective stresses or strains obtained from the individual loads.

In other words, individual loads on a body do not affect each other and therefore in stress and strain analyses, they may be considered separately (or independently) in any order.

The principle of superposition is very useful in analysis, but unfortunately, it is not always applicable, especially in heavily loaded bodies with large deformation. However, if the deformation is small and if linear stress–strain equations are applicable, the principle holds.

7.5 HOOKE'S LAW IN TWO AND THREE DIMENSIONS

In Chapter 3, we discussed the fundamental version of Hooke's law, which simply states that for uniaxial (one-dimensional) stress and strain, the stress σ is proportional to the strain ε, that is

$$\sigma = E\varepsilon \quad \text{or} \quad \varepsilon = \sigma/E \tag{7.20}$$

(see Equation 3.4) where E is usually called the "modulus of elasticity," the "elastic modulus," or "Young's modulus." In Chapter 3, we also saw that this fundamental version of Hooke's law is readily extended to shear stresses and shear strains. That is, for simple shearing of a block, the shear stress τ and the shear strain γ are related by

$$\tau = G\gamma \quad \text{or} \quad \gamma = \tau/G \tag{7.21}$$

(see Equation 3.12) where the proportional constant G is sometimes called the "shear modulus," the "modulus of elasticity in shear," or the "modulus of rigidity."

We can use Poisson's ratio and the principle of superposition to extend these fundamental relations to two and three dimensions where we have combined stresses and strains in two or more directions as well as shear stresses and strains in various directions. To this end, consider again a small rectangular elastic block or element subjected to tension as in Figure 7.6. Let the resulting tensile stresses be σ_{xx}, σ_{yy}, and σ_{zz}. Then by the use of Poisson's ratio and the principle of superposition, the strains on the elemental black are

$$\varepsilon_{xx} = (1/E)[\sigma_{xx} - \nu(\sigma_{yy} + \sigma_{zz})] \tag{7.22}$$

$$\varepsilon_{yy} = (1/E)[\sigma_{yy} - \nu(\sigma_{zz} + \sigma_{xx})] \tag{7.23}$$

$$\varepsilon_{zz} = (1/E)[\sigma_{zz} - \nu(\sigma_{xx} + \sigma_{yy})] \tag{7.24}$$

These results are obtained assuming that all the stresses are positive (tension). If, however, some or all of the stresses are negative (compression), the expressions are still valid. We then simply have negative values inserted for those negative stresses in the right side of Equations 7.22, through 7.24.

Observe further that Equations 7.22, through 7.24 are linear in both the stresses and the strains. Therefore, we can readily solve the equations for the stresses in terms of the strains, which give the expressions

$$\sigma_{xx} = E\frac{\nu(\varepsilon_{yy} + \varepsilon_{zz}) + (1 - \nu)\varepsilon_{xx}}{(1 + \nu)(1 - 2\nu)} \tag{7.25}$$

$$\sigma_{yy} = E\frac{\nu(\varepsilon_{zz} + \varepsilon_{xx}) + (1 - \nu)\varepsilon_{yy}}{(1 + \nu)(1 - 2\nu)} \tag{7.26}$$

$$\sigma_{zz} = E\frac{\nu(\varepsilon_{xx} + \varepsilon_{yy}) + (1 - \nu)\varepsilon_{zz}}{(1 + \nu)(1 - 2\nu)} \tag{7.27}$$

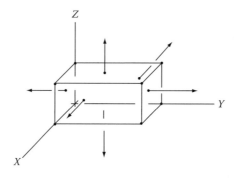

FIGURE 7.6 A rectangular elastic element subjected to tension loading.

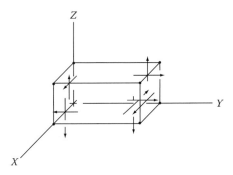

FIGURE 7.7   A rectangular elastic element subjected to shear loading.

Consider the shear stresses and strains: imagine a series of shearing forces applied to an elastic elemental block as in Figure 7.7. As a result of the shear stress–strain relations of Equations 7.21 and the principle of superposition the resulting shear stresses and strains on the block are related by the equations

$$\tau_{xy} = G\gamma_{xy}, \quad \gamma_{xy} = (1/G)\tau_{xy} \tag{7.28}$$

$$\tau_{yz} = G\gamma_{yz}, \quad \gamma_{yz} = (1/G)\tau_{yz} \tag{7.29}$$

$$\tau_{zx} = G\gamma_{zx}, \quad \gamma_{zx} = (1/G)\tau_{zx} \tag{7.30}$$

Alternatively, using the tensor notation of Chapters 4 and 5 (see for example, Equation 5.8), we have

$$\sigma_{xy} = 2G\varepsilon_{xy}, \quad \varepsilon_{xy} = (1/2G)\sigma_{xy} \tag{7.31}$$

$$\sigma_{yz} = 2G\varepsilon_{yz}, \quad \varepsilon_{yz} = (1/2G)\sigma_{yz} \tag{7.32}$$

$$\sigma_{zx} = 2G\varepsilon_{zz}, \quad \varepsilon_{zx} = (1/2G)\sigma_{zx} \tag{7.33}$$

7.6   RELATIONS BETWEEN THE ELASTIC CONSTANTS

The elastic constants E, ν, and G are not independent. Instead only two of these are needed to fully characterize the behavior of linear elastic materials. Consider a square plate with side length placed in tension as in Figure 7.8. As a result of the tensile forces and the resulting tensile stress, the plate will be elongated and narrowed as represented in Figure 7.9, where we have labeled the plate

FIGURE 7.8   A square plate under tension.

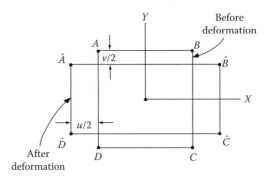

FIGURE 7.9 Plate in tension (X-axis) before and after deformation.

corners as A, B, C, and D before deformation, and as $\hat{A}$, $\hat{B}$, $\hat{C}$, and $\hat{D}$ after deformation. Let the amount of elongation be u and that of narrowing be v. Then the strains in the X- and Y-directions are

$$\varepsilon_{xx} = u/a \quad \text{and} \quad \varepsilon_{yy} = -v/a \tag{7.34}$$

But from the definition of Poisson's ratio (see Equation 7.1), we have

$$\varepsilon_{yy} = -\nu\varepsilon_{xx} \quad \text{or} \quad v = \nu u \tag{7.35}$$

Consider now the diamond $PQRS$ within the plate before and after deformation ($\hat{P}\hat{Q}\hat{R}\hat{S}$) as in Figure 7.10. The difference between the angle ϕ and $90°$ is a measure of the shear strain γ of the diamond. That is,

$$\gamma = \pi/2 - \phi \tag{7.36}$$

To quantify γ in terms of the loading, consider a force analysis or free-body diagram of the triangular plate PQR as in Figure 7.11, where F is the resultant tensile load on the original square plate, σ_{xx} is the uniform tensile stress, t is the plate thickness, and a is the side length of the original square plate. From the symmetry of the loadings, we have equivalent force systems as in Figure 7.12. Consider the force system in the right sketch of the figure: the force components parallel to the edges are shearing forces. They produce shear stresses τ on those inclined edges as

$$\tau = (\sqrt{2}/4)F/(a\sqrt{2}/2)t = F/2at = \sigma_{xx}/2 \tag{7.37}$$

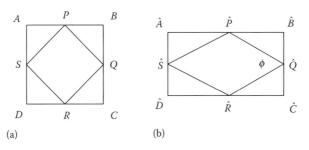

FIGURE 7.10 A diamond $PQRS$ within the undeformed and deformed plate. (a) Before deformation and (b) after deformation.

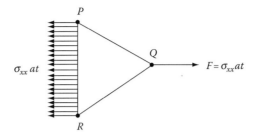

FIGURE 7.11 Force balance on triangular plate PQR of the square plate of Figures 7.8 through 7.10.

where $(a\sqrt{2}/2)t$ is the area of an inclined edge and the final equality is seen in Figure 7.11.

Next, consider the shear strain of Equation 7.36: specifically, consider the deformation of the triangular plate PQR into $\hat{P}\hat{Q}\hat{R}$ as in Figure 7.10 and as shown again with exaggerated deformation in Figure 7.13. Let θ be the half angle at $\hat{Q}$. That is,

$$\theta = \phi/2 \tag{7.38}$$

From Figure 7.13, we see that

$$\begin{aligned}
\tan\theta &= [(a-v)/2]/[(a+u)/2] = (a-v)(a-u) \\
&= [1-(v/a)]/[1+(u/a)] = [1-(v/a)][1+(u/a)]^{-1} \\
&\cong [1-(v/a)][1-(u/z)] = 1 - \frac{v}{a} - \frac{u}{a} \\
&= 1 - \frac{u+v}{a} \tag{7.39}
\end{aligned}$$

where we have used a binomial expansion [1] to approximate $[1+(u/a)]^{-1}$ and where we have neglected higher order terms in the displacement u and v. Then by substituting from Equations 7.34 and 7.35 into 7.39, we have

$$\tan\theta = 1 - (1+v)(u/a) = 1 - (1+v)\varepsilon_{xx} \tag{7.40}$$

From Equation 7.63, the shear strain γ is

$$\gamma = (\pi/2) - \phi = (\pi/2) - 2\theta \tag{7.41}$$

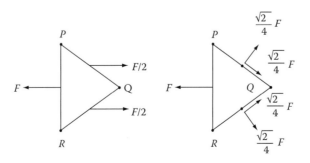

FIGURE 7.12 Equivalent force systems on triangular plate PQR.

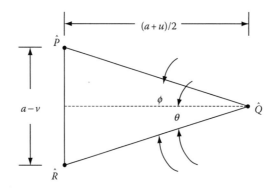

FIGURE 7.13 Deformed triangle PQR into $\hat{P}\hat{Q}\hat{R}$.

Then θ is

$$\theta = (\pi/4) - (\gamma/2) \tag{7.42}$$

Then $\tan \theta$ is

$$
\begin{aligned}
\tan \theta = \tan\left(\frac{\pi}{4} - \frac{\gamma}{2}\right) &\equiv \frac{\tan \pi/4 - \tan \gamma/2}{1 + (\tan \pi/4)(\tan \gamma/2)} \\
&\cong \frac{1 - \gamma/2}{1 + \gamma/2} = \left(1 - \frac{\gamma}{2}\right)\left(1 + \frac{\gamma}{2}\right)^{-1} \\
&\cong (1 - \gamma/2)(1 - \gamma/2) \approx 1 - \gamma
\end{aligned}
\tag{7.43}
$$

where we have used the trigonometric identity [1]:

$$\tan (A - B) \equiv \frac{\tan A - \tan B}{1 + \tan A \tan B} \tag{7.44}$$

and where again we have assumed that γ is small in the binomial expansion of $(1 + \gamma/2)^{-1}$, in the product $(1 - \gamma/2)(1 - \gamma/2)$, and in the approximation of $\tan \gamma/2$ by $\gamma/2$. (That is, we have neglected all but linear terms in γ.)

Next, recall from the fundamental shear stress–strain equation, we have

$$\gamma = \tau/G = \sigma_{xx}/2G \tag{7.45}$$

where the last equality follows from Equation 7.37. Then from Equation 7.43, $\tan \theta$ is

$$\tan \theta = 1 - (1 + v)\varepsilon_{xx} = 1 - \frac{\sigma_{xx}}{2G} \tag{7.46}$$

Finally, by comparing Equations 7.40 and 7.46, we have

$$\tan \theta = 1 - (1 + v)\varepsilon_{xx} = 1 - \frac{\sigma_{xx}}{2G}$$

or

$$(1 + v)\varepsilon_{xx} = \sigma_{xx}/2G \tag{7.47}$$

But ε_{xx} is σ_{xx}/E. Therefore G is

$$G = E/2(1 + v) \tag{7.48}$$

7.7 OTHER FORMS OF HOOKE'S LAW

In Section 7.5, we saw that Hooke's law may be written for the strains in terms of the stresses as (see Equations 7.22 through 7.24, and 7.31 through 7.33):

$$\varepsilon_{xx} = (1/E)[\sigma_{xx} - v(\sigma_{yy} + \sigma_{zz})] \tag{7.49}$$

$$\varepsilon_{yy} = (1/E)[\sigma_{yy} - v(\sigma_{zz} + \sigma_{xx})] \tag{7.50}$$

$$\varepsilon_{zz} = (1/E)[\sigma_{zz} - v(\sigma_{xx} + \sigma_{yy})] \tag{7.51}$$

$$\varepsilon_{xy} = (1/2G)\sigma_{xy} \tag{7.52}$$

$$\varepsilon_{yz} = (1/2G)\sigma_{yz} \tag{7.53}$$

$$\varepsilon_{zx} = (1/2G)\sigma_{zx} \tag{7.54}$$

Also from Equations 7.25 through 7.27, and 7.31 through 7.33, the stresses may be expressed in terms of the strain as

$$\sigma_{xx} = E\frac{v(\varepsilon_{yy} + \varepsilon_{zz}) + (1 - v)\varepsilon_{xx}}{(1 + v)(1 - 2v)} \tag{7.55}$$

$$\sigma_{yy} = E\frac{v1(\varepsilon_{zz} + \varepsilon_{xx}) + (1 - v)\varepsilon_{yy}}{(1 + v)(1 - 2v)} \tag{7.56}$$

$$\sigma_{zz} = E\frac{v(\varepsilon_{xx} + \varepsilon_{yy}) + (1 - v)\varepsilon_{zz}}{(1 + v)(1 - 2v)} \tag{7.57}$$

$$\sigma_{xy} = 2G\varepsilon_{xy} \tag{7.58}$$

$$\sigma_{yz} = 2G\varepsilon_{yz} \tag{7.59}$$

$$\sigma_{zx} = 2G\varepsilon_{zx} \tag{7.60}$$

We can express these equations in more compact form by using index notation and by reintroducing and redefining the expressions

$$\Delta = \varepsilon_{xx} + \varepsilon_{yy} + \varepsilon_{zz}, \quad \Theta = \sigma_{xx} + \sigma_{yy} + \sigma_{zz} \tag{7.61}$$

where, as before, Δ is the dilatation and Θ will be recognized as the sum of the diagonal elements of the stress matrix (see Equation 4.64). To use index notation, let x, y, and z become 1, 2, and 3 respectively. Then Δ and Θ may be expressed as

$$\Delta = \varepsilon_{11} + \varepsilon_{22} + \varepsilon_{33} = \varepsilon_{kk} \quad \text{and} \quad \Theta = \sigma_{11} + \sigma_{22} + \sigma_{33} = \sigma_{kk} \tag{7.62}$$

where, repeated indices designate a sum from 1 to 3. Using this notation, it is readily seen that Equations 7.49 through 7.54 may be combined into a single expression as

$$\varepsilon_{ij} = -(v/E)\Theta\delta_{ij} + \sigma_{ij}/2G \tag{7.63}$$

where, δ_{ij} is Kronecker's delta symbol defined as

$$\delta_{ij} = \begin{cases} 0 & i \neq j \\ 1 & i = j \end{cases} \tag{7.64}$$

Equation 7.61 may be validated by simply writing the individual terms. For example, ε_{11} is

$$\varepsilon_{11} = -(v/E)(\sigma_{11} + \sigma_{22} + \sigma_{33}) + \sigma_{11}/2G \tag{7.65}$$

But from Equation 7.47, $1/2G$ is $(1 + v)/E$. Thus ε_{11} is

$$\varepsilon_{11} = -(v/E)(\sigma_{11} + \sigma_{22} + \sigma_{33}) + (1 + v)\sigma_{11}/E$$
or $\tag{7.66}$
$$\varepsilon_{11} = (1/E)[\sigma_{11} - v(\sigma_{22} + \sigma_{33})]$$

This is similar to Equation 7.49. The other five elements of Equation 7.61 are similarly validated. Also, Equations 7.55 through 7.60 may be written in the compact form

$$\sigma_{ij} = \lambda\delta_{ij}\Delta + 2G\varepsilon_{ij} \tag{7.67}$$

where λ is defined as

$$\lambda = \frac{vE}{(1 + v)(1 - 2v)} \tag{7.68}$$

λ and G are sometimes called Lamé constants.

7.8 HYDROSTATIC PRESSURE AND DILATATION

Equations 7.61 and 7.67 can be used to obtain a relation between the first stress invariant Θ and the dilatation Δ (the first strain invariant). Specifically, in Equation 7.67, by replacing i with j and adding, we obtain

$$\sigma_{jj} = \Theta = \lambda\delta_{jj}\Delta + 2G\varepsilon_{jj} = (3\lambda + 2G)\Delta \tag{7.69}$$

By substituting for λ and G from Equations 7.48 and 7.68, we have

$$3\lambda + 2G = E/(1 - 2v) \tag{7.70}$$

Therefore, Equation 7.69 becomes

$$\Theta = [E/(1 - 2v)]\Delta \quad \text{or} \quad \Delta[(1 - 2v)/E]\Theta \tag{7.71}$$

If each of the normal stresses are equal, we have a state of "hydrostatic pressure." In particular, let the stresses be

$$\sigma_{11} = \sigma_{22} = \sigma_{33} = -p \tag{7.72}$$

where p is the pressure and the negative sign is used since pressure is compressive. Then Θ is

$$\Theta = -3p \tag{7.73}$$

Then Equation 7.71 becomes

$$-p = \left[E/3(1 - 2v)\right]\Delta \overset{D}{=} k\Delta \tag{7.74}$$

where k is called the "bulk modulus" defined by inspection as

$$k = E/3(1 - 2v) = \lambda + (2/3)G \tag{7.75}$$

SYMBOLS

a	Cross section side length, radius
a, b, c	Lengths
E	Modulus of elasticity
F	Force
G	Shear modulus, Lame constant
k	Bulk modulus
p	Hydrostatic pressure
r	Radius
u, v, w	Displacements
u_i $(i = 1, 2, 3; x, y, z)$	Displacement components
V	Volume
$\widehat{V}$	Volume of a deformed body or element
γ	Shear strain
γ_{ij} $(i, j = 1, 2, 3; x, y, z)$	Engineering strain components
δ	Elongation
Δ	Sum of diagonal elements of strain matrix
ΔV	Volume change
ε	Strain
ε $(i, j = 1, 2, 3; x, y, z)$	Strains, Strain matrix elements
Θ	Sum of diagonal elements of stress matrix
θ, ϕ	Angle measure
λ	Lamé constant
v	Poisson's ratio
σ	Stress
σ_{ij} $(i, j = 1, 2, 3; x, y, z)$	Stresses, stress matrix elements
τ	Shear stress
τ_{ij} $(i, j = 1, 2, 3; x, y, z)$	Shear stress components

REFERENCE

1. H. B. Dwight, *Tables of Integrals and Other Mathematical Data*, Macmillan, 3rd ed., New York, 1957.

Part II

Straight and Long Structural Components: Beams, Rods, and Bars

In this second part, we apply the concepts documented in the first part. We start with a discussion of stress, strain, and displacement of beams, rods, and bars. These are the most commonly used structural elements and components in the design of structures and machines.

We focus on thin straight members, looking primarily at the concepts of bending and torsion. In the first part (Chapters 2 and 3), we have already considered simple extension and compression of rods. In this part, we will also look at the consequences of bending and torsion, that is, the resulting stresses, strains, and displacements. In the next part, we will look at thick and curved beams and buckling of beams. In subsequent parts, we will look at assemblages of beams in the form of trusses and frames.

Finally, from an analytical perspective, there is no major difference between a rod, a bar, or a beam. Generally, the distinction refers to the shape of the cross section with beams being rectangular, rods being round, and bars being square or hexagonal. But these are rather arbitrary classifications.

8 Beams: Bending Stresses (Flexure)

8.1 BEAMS

A beam is simply a long, slender member as represented in Figure 8.1, where ℓ is the length of the beam, h is its height, and b is its thickness. An immediate question is: what is meant by "long and slender?" That is, how long is "long" or equivalently, how slender is "slender?" Unfortunately, these questions have no precise answers. We can certainly say that whatever approximations are made, by assuming a beam to be long and slender, become more appropriate the longer (or more slender) the beam becomes. While this is reassuring, and potentially useful, it is still not very specific. A general rule is that a beam is long or slender if its length ℓ is an order of magnitude (i.e., 10 times) larger than the cross section dimensions. That is, in Figure 8.1 the beam may be regarded as long as

$$\ell > 10h \quad \text{and} \quad \ell > 10b \tag{8.1}$$

8.2 LOADINGS

Beams may be loaded in three principal ways: (1) axially (producing longitudinal extension or compression); (2) transversely (producing bending); and (3) in torsion (producing twisting). Figure 8.2 illustrates these loading methods.

Beams may, of course, have combinations of these loadings. When the deformations from combined loadings are small, the resulting displacements and stresses from these loadings may be obtained by superposition.

In Chapters 8 and 9, we will discuss bending, which results from transverse loading. We will consider torsion in Chapter 10 and axial loading in Chapter 11 in connection with buckling. We will also consider axial loading in trusses as a means for developing the finite element method (FEM). Recall that axial loading and deformation are also discussed in Part I (see Chapters 2 and 3).

8.3 COORDINATE SYSTEMS AND SIGN CONVENTIONS

The coordinate system and sign conventions establishing positive and negative directions are essential features of any stress analysis. For beam bending, the sign conventions are especially important, particularly because there is disagreement among analysts as to which convention to use. While each of the various conventions has advantages (and disadvantages), a key to a successful analysis is to stay consistent throughout the analysis.

We will generally follow the sign convention of three-dimensional stress analysis established in Chapter 4. That is, stresses and displacements at points of positive element faces are positive if they are in positive directions, and negative if they are in negative directions. Conversely, stresses and displacements at points of negative element faces are positive if they are in negative directions and negative if they are in positive directions. Recall that a "positive face" of an element is a face where one goes in the positive direction in crossing the face by going from inside to outside of the element. Correspondingly a "negative face" of the element has one going in a negative direction in crossing the face while going from inside to outside of the element.

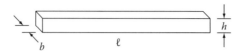

FIGURE 8.1   A rectangular beam.

In our convention, we place the X-axis along the axis of the beam. Since many beams are weight-bearing structures, we choose the Y-axis to be downward, producing positive beam displacements for weight (or gravity) forces. The Z-axis is then inward when viewing the X–Y plane, as in Figure. 8.3, where the origin O is placed at the left end of the beam as shown.

In beam structural analyses, we are principally interested in loadings, shear forces, bending moments, stresses, and displacements. In the following paragraphs and figures, we describe and illustrate the positive direction for these quantities.

First, for loading, since our focus in this chapter is on transverse loadings, the positive direction for the applied forces is in the positive Y-axis as illustrated in Figure. 8.4. (Note that if the beam displacement is small, we can also have transverse loading in the Z-direction and then superpose the analyses results.)

Next, transverse beam loading, as in Figure 8.4, produces transverse shearing forces and bending moments on the beam. Figure 8.5 shows the positive directions for the shearing forces. Observe that the positive shear force V acts on the positive face (cross section) in the positive direction and on the negative cross section in the negative direction.

Figure 8.6 shows the conventional positive directions for the bending moments produced by transverse loadings. Unfortunately, these directions are opposite to those suggested by elasticity theory. In this case, the positive moment on the positive face is directed in the negative Z-direction. The advantage of this departure from elasticity theory is that the resulting stresses are positive in the lower portion of the beam cross section where the bending moment is positive. That is, adopting the convention of Figure. 8.6 leads to the familiar expression

$$\sigma = My/I \tag{8.2}$$

where I is the second moment of area of the beam cross section (that is, $I = \int y^2 dA$). Finally, Figure 8.7 illustrates the positive directions for beam displacement and cross section rotation.

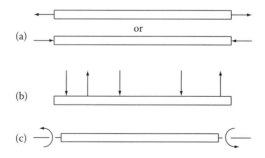

FIGURE 8.2   Methods of beam loading. (a) Axial loading, (b) transverse loading, and (c) torsional loading.

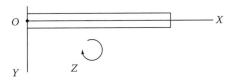

FIGURE 8.3   Beam coordinate axes.

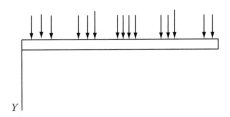

FIGURE 8.4   Positive-directed transverse forces on a beam.

FIGURE 8.5   Positive directions for shearing forces.

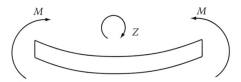

FIGURE 8.6   Positive directions for bending moments.

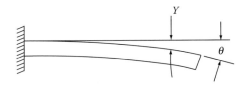

FIGURE 8.7   Positive directions for displacement and rotation of a beam.

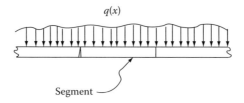

FIGURE 8.8   A segment of a loaded beam.

8.4   EQUILIBRIUM AND GOVERNING EQUATIONS

Consider a short segment of a transversely loaded beam as in Figure 8.8. Let $q(x)$ represent the loading on the beam per unit length. Let Δx be the segment length and V and M be the shear and bending moment on the left end of the segment respectively as in Figure 8.9. With the segment length Δx being small, we can conveniently use the beginning term of a Taylor series expansion to represent the shear and bending moment on the right side of the segment as shown in the figure. Consider a free-body diagram of the segment. As Δx becomes vanishingly small, we can safely neglect the higher order terms in the shear and bending moment expressions on the right side of the beam. Correspondingly, the resultant force on the segment due to the loading function $q(x)$ is then approximately $q\Delta x$, where q is simply an average value of $q(x)$ across the short segment.

Using these approximations, we may envision a free-body diagram of the segment as in Figure 8.10. Then by adding forces vertically, we obtain

$$q\Delta x + \frac{\mathrm{d}V}{\mathrm{d}x}\Delta x = 0 \tag{8.3}$$

or

$$\frac{\mathrm{d}V}{\mathrm{d}x} = -q \tag{8.4}$$

Similarly, by setting moments about the left end equal to zero, we have

$$\frac{\mathrm{d}M}{\mathrm{d}x}\Delta x - V\Delta x - \left(\frac{\mathrm{d}V}{\mathrm{d}x}\Delta x\right)\Delta x - (q\Delta x)\Delta x/2 = 0 \tag{8.5}$$

By again neglecting higher powers in Δx, we obtain

$$\frac{\mathrm{d}M}{\mathrm{d}x} = V \tag{8.6}$$

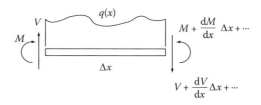

FIGURE 8.9   Beam segment, with loading $q(x)$, shear V, and bending moment M.

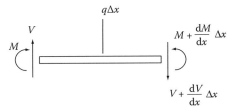

FIGURE 8.10 Free-body diagram of the beam segment.

Finally, by substituting for V from Equation 8.6 in Equation 8.4, we obtain

$$\frac{d^2M}{dx^2} = -q \tag{8.7}$$

8.5 BEAM DEFLECTION DUE TO BENDING

Consider again a portion of a beam being bent due to transverse loads as in Figure 8.11. Consider a segment (or "element") (e) of the beam and let Δx be the length of (e) as shown in Figure 8.11. Let the transverse loading produce a bending moment M on the beam as indicated in the figure. Finally, let an axis system be introduced with origin O at the left end of (e), with the X-axis along (e) and the Y-axis below the plane of (e) as shown in Figure 8.11.

As the beam is bent by the bending moments, it will of course no longer be straight but slightly curved as represented in exaggerated form in Figure 8.12, where Q is the center of the curvature of the arc formed at O by the beam centerline and ρ is the corresponding radius of curvature.

Let N be a centerline axis of the beam which is straight before bending but then curved after bending as shown in Figure 8.12. The principal tenet of elementary beam bending theory is that during bending plane cross sections normal to the beam axis N before bending remain plane and normal to N during and after bending. As a consequence, as the beam is bent upwards (positive bending) as in Figure 8.12, the upper longitudinal fibers of the beam are shortened and correspondingly, the lower longitudinal fibers are lengthened.

Figure 8.13 shows an enlarged view of element (e) of the bent beam. With the upper fibers of the beam, and hence also of (e), being shortened, with the lower fibers being lengthened, and with the cross section normal to the beam axis remaining plane during bending, there will exist at some mid elevation of (e), a fiber that is neither shortened nor lengthened due to the bending. Indeed, if we consider the thickness of the beam in the Z-direction, there will be a strip or surface of the beam, which is neither shortened nor lengthened by the bending. This surface is sometimes called a "neutral surface" of the beam.

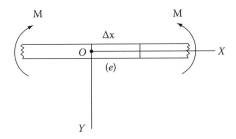

FIGURE 8.11 A beam segment, or element, in bending.

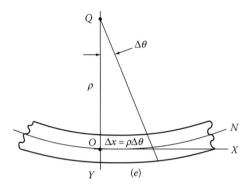

FIGURE 8.12   Exaggerated bending of the beam segment.

Let the X–Y plane be placed at the mid section of the beam in the Z-direction and let the origin O be placed on the neutral surface. Then the intersection of the neutral surface with the X–Y plane is a curve called the "neutral axis" of the beam. We now identify N in Figures 8.12 and 8.13 with this neutral axis. Before bending, N and the X-axis are coincident, but N is a "material" line and the X-axis is a "spatial" line.

Consider a fiber of (e) along the neutral axis N. Since this fiber is neither shortened nor lengthened during bending, its length will remain as Δx. However, after bending, this fiber will be curved forming an arc with radius ρ and subtended angle $\Delta \theta$ as represented in Figure 8.12. Thus, the fiber length is also $\rho \Delta \theta$. That is

$$\Delta x = \rho \Delta \theta \tag{8.8}$$

Consider a fiber of distance y beneath the neutral axis N, shown shaded in Figure 8.13. This fiber will also be curved into an arc. But, although its original length is Δx, its deformed length is $(\rho + y)\Delta \theta z$. The strain ε of this fiber is simply

$$\varepsilon = \frac{(\rho + y)\Delta \theta - \Delta x}{\Delta x} = \frac{(\rho + y)\Delta \theta - \rho \Delta \theta}{\rho \Delta \theta} = \frac{y}{\rho} \tag{8.9}$$

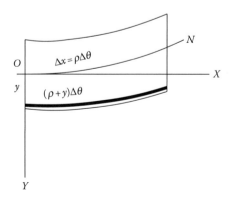

FIGURE 8.13   Enlarged, view of element (e).

8.6 BEAM STRESSES DUE TO BENDING

Equation 8.9 shows that the axial strain of the beam varies linearly across the beam cross section. (This is a direct consequence of the requirement that beam cross sections normal to the beam axis before bending remain plane and normal to the beam axis during and after bending.) From Hooke's law, the axial stress in the beam will also vary linearly across the cross section. Specifically, the axial stress σ is

$$\sigma = E\varepsilon = (E/\rho)y \tag{8.10}$$

Consider the relation between the axial stress and the beam bending moment. Consider particularly the stresses and bending moment at a typical cross section of a beam, as represented in Figure 8.14. For the purpose of simplifying the analysis, let the beam cross section be rectangular, and consider an end view as in Figure 8.15. Let the cross section dimensions be b and h as shown. Then for equilibrium, the stress produced by the applied bending moment must have the same moment about the Z-axis as the bending moment M itself. That is,

$$\int_{-h/2}^{h/2} \sigma y b \, dy = M \tag{8.11}$$

By substituting for σ from Equation 8.10, we have

$$M = \int_{-h/2}^{h/2} (E/\rho)y^2 b \, dy = (E/\rho)(bh^3/12)$$

or

$$M = EI/\rho \tag{8.12}$$

where I is defined as $bh^3/12$ and is generally called "the second moment of area" or the "moment of inertia" of the cross section.

Equation 8.13 holds for other rectangular cross sections, such as that of I-beams as well.

Observe further in the development of Equation 8.13 that for a given cross section E/ρ is a constant across the cross section. That is, E/ρ is independent of y. However ρ will, in general, vary from point to point along the beam axis.

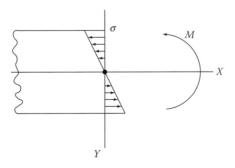

FIGURE 8.14 Bending moment and axial stress in a typical beam cross section.

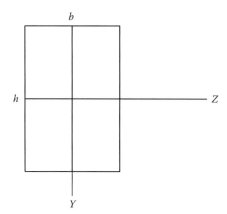

FIGURE 8.15   Beam cross section.

Finally, by eliminating E/ρ between Equations 8.10 and 8.12, we obtain the familiar relation:

$$\sigma = My/I \tag{8.13}$$

For a rectangular cross section, as in Figure 8.15, the maximum value of y is $h/2$. Thus, the maximum bending stress at the top (compression) and bottom (tension) of the beam with values

$$\sigma_{\substack{max \\ min}} = \pm 6M/bh^2 \tag{8.14}$$

For a beam with a nonrectangular cross section, if the maximum distance from a material point of the cross section to the neutral axis is c, we have the widely used expression

$$\sigma_{\substack{max \\ min}} = \pm Mc/I \tag{8.15}$$

SYMBOLS

b	Beam width
c	Half beam height
E	Modulus of elasticity
(e)	Element
h	Beam height
I	Second moment of area
ℓ	Beam length
M	Bending moment
N	Neutral axis
O	Origin
Q	Center of curvature
$q(x)$	Loading
V	Shear force
X, Y, Z	Cartesian (rectangular) coordinate axes
x, y, z	Coordinates relative to X, Y, and Z
ε	Normal strain
ρ	Radius of curvature
σ	Normal stress

9 Beams: Displacement from Bending

9.1 BEAM DISPLACEMENT AND BENDING MOMENT

Equation 8.12 provides the fundamental relationship between the bending moment applied to a beam and the resulting induced curvature of the beam's centerline:

$$M = EI/\rho \tag{9.1}$$

where

M is the bending moment
ρ is the radius of curvature
I is the second moment of area of the beam cross section about the neutral axis
E is the modulus of elasticity

In general, the bending moment is a function of the axial position x along the beam. Thus, in view of Equation 9.1, the radius of curvature is also a function of x.

Consider a planar curve C represented in the X–Y plane by the function: $f(x)$, as in Figure 9.1. It is known [1] that the radius of curvature ρ of C can be expressed in terms of f and its derivatives as

$$\rho = [1 + (dy/dx)^2]^{3/2}/d^2y/dx^2 \tag{9.2}$$

We can readily apply Equation 9.2 with the curved centerline (or neutral axis) of a beam since the induced curvature due to bending is small. Since dy/dx is the beam slope, it will be small and thus the product $(dy/dx)^2$ is negligible compared to 1. That is,

$$(dy/dx)^2 \ll 1 \quad \text{and} \quad \rho \approx 1/d^2y/dx^2 \tag{9.3}$$

Recall that with our sign convention, the Y-axis pointing downward, opposite to that of Figure 9.1. Therefore, to maintain our convention for positive bending, as in Figure 8.6, with the Y-axis pointing downward, Equation 9.3 becomes

$$\rho = -d^2y/dx^2 \tag{9.4}$$

Equation 9.4 provides a differential equation determining the beam axis displacement in terms of the axis curvature, and thus in terms of the bending moment, via Equation 9.1. That is

$$\frac{d^2y}{dx^2} = -M/EI \tag{9.5}$$

9.2 BEAM DISPLACEMENT IN TERMS OF TRANSVERSE SHEAR AND THE LOADING ON THE BEAM

By using Equation 9.3, we can relate the displacement of the transverse shear V and the applied loading function $q(x)$. Recall from Equations 8.4 and 8.6 that the bending moment M, shear V, and load q are related by the simple expressions

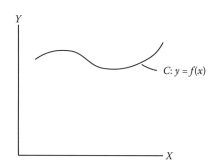

FIGURE 9.1   A planar curve C.

$$dM/dx = V \quad \text{and} \quad dV/dx = -q \tag{9.6}$$

Therefore, by substituting from Equation 9.3 we have

$$\frac{d^3y}{dx^3} = -V/EI \quad \text{and} \quad \frac{d^4y}{dx^4} = q/EI \tag{9.7}$$

The second expression of Equation 9.7 is the governing ordinary differential equation for the displacement of the neutral axis due to the beam loading. Once this equation is solved and $y(x)$ is known, we can immediately determine the transverse shear V and bending moment M along the beam axis using the expressions

$$V = -EId^3y/dx^3 \quad \text{and} \quad M = -EId^2y/dx^2 \tag{9.8}$$

9.3 BEAM SUPPORTS, SUPPORT REACTIONS, AND BOUNDARY CONDITIONS

Equation 9.7 provides a fourth-order, ordinary differential equation for beam displacement once the loading function $q(x)$ is known. Upon solving (or integrating) the equations, there will be four constants of integration to be evaluated. We can evaluate these constants using the auxiliary conditions (or "boundary conditions") required by the beam supports. We discuss these concepts in the following paragraphs.

There are four principal types of supports: (1) built-in ("clamped" or "cantilever"); (2) simple ("pin" or "roller"); (3) free ("unconstrained"); and (4) elastic.

9.3.1 BUILT-IN (CLAMPED OR CANTILEVER) SUPPORT

In this case, the beam end is completely supported or fixed, that is, the beam end is restrained from moving in both translation and rotation, as represented in Figure 9.2. This means that at the support,

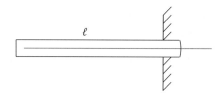

FIGURE 9.2   A beam with a built-in (clamped or cantilever) support at the right end.

FIGURE 9.3   A simple (pin) support.

the beam displacement y and the beam rotation dy/dx are zero. Thus, if the origin of the beam axis is at the left end and the support is at $x = \ell$, we have

$$y(\ell) = 0 \quad \text{and} \quad \frac{dy}{dx}(\ell) = 0 \tag{9.9}$$

Observe that for the conditions of Equation 9.9 to be satisfied, the support will exert a force and a moment on the beam. The magnitude of this force and moment can be determined from the loading conditions using a free-body diagram.

9.3.2   SIMPLE (PIN OR ROLLER) SUPPORT

Here the support provides a vertical constraint for the beam, but it allows for beam rotation, as represented in Figure 9.3. Thus if the support is at say $x = a$ (with the origin being at the left end of the beam), we have

$$y(a) = 0 \quad \text{and} \quad M(a) = 0 \tag{9.10}$$

where, as before, $M(x)$ is the bending moment along the beam axis. Since from Equation 9.8 M is $-EI d^2 y/dx^2$, we have the simple support condition:

$$\frac{d^2 y}{dx^2}(a) = 0 \tag{9.11}$$

Finally, the magnitude of the reaction force, exerted by the support to restrain the vertical movement of the beam, may be obtained using a free-body diagram from the given loading conditions.

9.3.3   FREE (UNCONSTRAINED) SUPPORT

In this case, the beam has no restraint at the support, that is, the shear V and the bending moment M at the support are zero as in Figure 9.4. Thus, if the free end is at $x = \ell$, we have the conditions:

$$V(a) = 0 \quad \text{and} \quad M(a) = 0 \tag{9.12}$$

or in view of Equation 9.8 we have the conditions:

$$\frac{d^3 y}{dx^3}(a) = 0 \quad \text{and} \quad \frac{d^2 y}{dx^2}(a) = 0 \tag{9.13}$$

$$x = a$$

FIGURE 9.4   A free end at $x = a$.

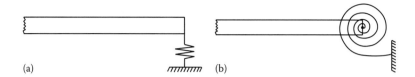

FIGURE 9.5 Elastic force and moment supports.

9.3.4 ELASTIC SUPPORT

In this case, the beam displacement (or rotation) is not fully constrained, but instead it is resisted by a force (or moment) proportional to the displacement (or rotation) as suggested by spring models of Figure 9.5. The shear force V and bending moment M are

$$V(a) = k_V y(a) \quad \text{and} \quad M(a) = k_M \frac{dy}{dx}(a) \tag{9.14}$$

where, as before, $x = 1$ is the location of the support.

9.4 SUMMARY OF GOVERNING EQUATIONS

For convenience, we briefly summarize the pertinent equations:

Bending moment

$$\text{Equation 9.3: } M = -EI d^2y/dx^2 \tag{9.15}$$

Shear

$$\text{Equation 9.7: } V = -EI d^3y/dx^3 \tag{9.16}$$

Load

$$\text{Equation 9.7: } q = EI d^4y/dx^4 \tag{9.17}$$

The support conditions are

Built-in (clamped)

$$\text{Equation 9.9: } y = dy/dx = 0 \tag{9.18}$$

Simple (pin)

$$\text{Equation 9.10: } y = d^2y/dx^2 = 0 \tag{9.19}$$

Free

$$\text{Equation 9.13: } d^2y/dx^2 = d^3y/dx^3 = 0 \tag{9.20}$$

Elastic

$$\text{Equation 9.14: } V = k_V y, \ M = k_M \ dy/dx \tag{9.21}$$

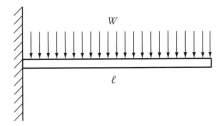

FIGURE 9.6 Uniformly loaded cantilever beam.

9.5 ILLUSTRATIONS

We can illustrate the application of the governing equations of the foregoing sections with a few elementary examples: specifically, we will consider cantilever and simply supported beams under uniform and concentrated loadings. The objective in each case is to determine the displacement, shear, and bending moment along the beam axis.

9.5.1 UNIFORMLY LOADED CANTILEVER BEAM

Consider first a cantilever beam supported (that is, built-in or clamped) at its left end and loaded with a uniform load along its span as in Figure 9.6, where ℓ is the beam length and w is the load per unit length along the beam.

We can determine the reactions at the support by using a free-body diagram as in Figure 9.7 where V_O and M_O are the shear and bending moment applied to the beam by the support at $x = 0$. To find V_O and M_O, it is useful to consider an equivalent free-body diagram as in Figure 9.8. From this figure, it is obvious that V_O and M_O are

$$V_O = w\ell \quad \text{and} \quad M_O = -w\ell^2/2 \tag{9.22}$$

We can now readily determine the bending moment, shear, and displacement along the beam axis using Equations 9.15, 9.16, and 9.17, respectively. Specifically, in Equation 9.17, the load $q(x)$ along the beam is

$$q(x) = w \tag{9.23}$$

Thus, Equation 9.17 becomes

$$EI \, d^4y/dx^4 = w \tag{9.24}$$

By integrating Equation 9.24, we have

$$EI \, d^3y/dx^3 = wx + c_1 = -V \tag{9.25}$$

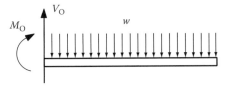

FIGURE 9.7 Free-body diagram of the uniformly loaded cantilever beam.

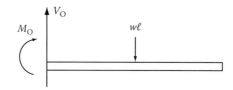

FIGURE 9.8   Equivalent free-body diagram of the uniformly loaded cantilever beam.

where the last equality follows from Equation 9.16. Since the shear V is V_O ($= w\ell$) when $x = 0$, we find the integration constant c_1 to be

$$c_1 = -V_O = -w\ell \tag{9.26}$$

By substituting from Equation 9.26 in Equation 9.25 and integrating again, we have

$$EI\, d^2y/dx^2 = wx^2/2 - w\ell x + c_2 = -M \tag{9.27}$$

where the last equality follows from Equation 9.15. Since the bending moment M is M_O ($= -w\ell^2/2$) where $x = 0$, we find the integration constant c_2 to be

$$c_2 = -M_O = w\ell^2/2 \tag{9.28}$$

By substituting from Equation 9.28 in Equation 9.27 and integrating again, we have

$$EI\, dy/dx = wx^3/6 - w\ell x^2/2 + w\ell^2 x/2 + c_3 \tag{9.29}$$

But from Equation 9.18, dy/dx is zero when $x = 0$, the integration constant c_3 is

$$c_3 = 0 \tag{9.30}$$

Finally, by substituting from Equation 9.30 in Equation 9.29 and integrating again, we find the displacement y to be

$$EIy = wx^4/24 - w\ell x^3/6 + w\ell^2 x^2/4 + c_4 \tag{9.31}$$

But from Equation 9.18, y is zero when $x = 0$, the integration constant c_4 is

$$c_4 = 0 \tag{9.32}$$

In summary, from Equations 9.25, 9.26, and 9.27 the shear, bending moment, and displacement are

$$V(x) = (-wx + w\ell)/EI \tag{9.33}$$

$$M(x) = (-wx^2/2 + w\ell x - w\ell^2/2)/EI \tag{9.34}$$

$$y(x) = (wx^4/24 - w\ell x^3/6 + w\ell^2 x^2/4)/EI \tag{9.35}$$

The maximum bending moment M_{max} occurs at $x = 0$ as

$$M_{max} = -w\ell^2/2EI \tag{9.36}$$

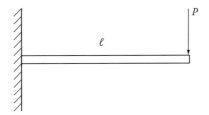

FIGURE 9.9 Cantilever beam with a concentrated end load.

The maximum displacement y_{max} occurs at $x = \ell$ as

$$y_{max} = w\ell^4/8EI \tag{9.37}$$

9.5.2 CANTILEVER BEAM WITH A CONCENTRATED END LOAD

Consider next a cantilever beam, built-in at its left end and loaded with a single vertical force on its right end as in Figure 9.9. Let the beam length be ℓ and the load magnitude be P as indicated in the figure. As before, our objective is to determine the displacement, bending moment, and shear along the length of the beam.

To begin the analysis, consider a free-body diagram of the beam as in Figure 9.10, where V_O and M_O are the shear and bending moment applied to the beam by the support.

Then for equilibrium, V_O and M_O are

$$V_O = 0 \quad \text{and} \quad M_O = -P\ell \tag{9.38}$$

The beam loading may thus be represented as in Figure 9.11.

Next, consider a free-body diagram of a segment of the beam to the left of a cross section which is a distance x from the left end support as in Figure 9.12. Then by considering the equilibrium of the segment, we immediately see that the shear V and bending moment M on the cross section at x are

$$V = P \quad \text{and} \quad M = -P(\ell - x) \tag{9.39}$$

The beam displacement y may now be determined using Equation 9.15:

$$EI \, d^2y/dx^2 = -M = P(\ell - x) \tag{9.40}$$

By integrating, we have

$$EI \, dy/dx = P\ell x - Px^2/2 + c_1 \tag{9.41}$$

FIGURE 9.10 Free-body diagram of the beam of Figure 9.9.

FIGURE 9.11 Beam loading.

But since the beam is clamped at its left end, we have (at $x = 0$):

$$\frac{dy}{dx}(0) = 0 \quad \text{so that } c_1 = 0 \tag{9.42}$$

Then by integrating again, we have

$$EIy = P\ell x^2/2 - Px^3/6 + c_2 \tag{9.43}$$

But since the beam is supported at its left end, we have (at $x = 0$):

$$y(0) = 0 \quad \text{so that } c_2 = 0 \tag{9.44}$$

Therefore, the displacement $y(x)$ is

$$y = (P/EI)\left(\frac{\ell x^2}{2} - \frac{x^3}{6}\right) \tag{9.45}$$

These results show that the maximum beam displacement, y_{max}, occurs at the right end as

$$y_{max} = P\ell^3/3EI \tag{9.46}$$

Also, from Equation 9.39, the maximum bending moment M_{max} is seen to occur at the left end of the beam as

$$M_{max} = -P\ell \tag{9.47}$$

Finally, from Equation 9.39, the shear V is constant along the beam as

$$V = P$$

FIGURE 9.12 Free-body diagram of a left side segment of the beam.

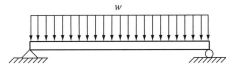

FIGURE 9.13 A uniformly loaded, simply supported beam.

9.5.3 SIMPLY SUPPORTED BEAM WITH A UNIFORM LOAD

Consider now a simply supported beam with a uniform load as in Figure 9.13. As before, let w be the loading per unit beam length and ℓ be the length of the beam. Consider a free-body diagram of the beam as in Figure 9.14 and a free-body diagram with equivalent loading as in Figure 9.15, where V_L and V_R are the shear loadings on the beam at the supports, by the supports. From Figure 9.15, these shear loadings are

$$V_L = w\ell/2 \quad \text{and} \quad V_R = -w\ell/2 \tag{9.48}$$

(Observe that in Figures 9.14 and 9.15 the shear forces are shown in their positive direction using our convention of Section 8.3.)

Consider next a free-body diagram of a segment, say a left segment, of the beam as in Figure 9.16, where V and M are the shear and bending moment respectively on a cross section at a distance x from the left support. Consider also a free-body diagram of the segment with equivalent loading as in Figure 9.17. Then by enforcing equilibrium, by setting the sum of the vertical forces equal to zero and also the sum of the moment of the forces about the left end equal to zero, we obtain

$$-\frac{w\ell}{2} + wx + V = 0 \tag{9.49}$$

and

$$M - wx(x/2) - Vx = 0 \tag{9.50}$$

Solving for V and M, we obtain

$$V = w(\ell/2 - x) \tag{9.51}$$

and

$$M = (wx/2)(\ell - x) \tag{9.52}$$

In knowing the moment distribution along the beam, as in Equation 9.52, we may use Equation 9.15 to determine the displacements:

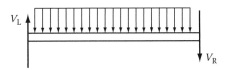

FIGURE 9.14 Free-body diagram of the simply supported, uniformly loaded beam.

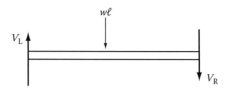

FIGURE 9.15   Free-body diagram with equivalent beam loading.

$$EId^2y/dx^2 = -M = -(wx/2)(\ell - x) \tag{9.53}$$

Then by integrating, we have

$$EIdy/dx = -w\ell x^2/4 + wx^3/6 + c_1 \tag{9.54}$$

and

$$EIy = -w\ell x^3/12 + wx^4/24 + c_1 x + c_2 \tag{9.55}$$

We can determine the integration constants by recalling that y is zero at the supports. That is,

$$y(0) = 0 = c_2 \quad \text{and} \quad y(\ell) = 0 = -w\ell^4/c_2 + w\ell^4/24 + c_1\ell \tag{9.56}$$

Then c_1 and c_2 are

$$c_1 = w\ell^3/24 \quad \text{and} \quad c_2 = 0 \tag{9.57}$$

Therefore, the displacement y of the beam centerline is

$$y = (w/12EI)[x^4/2 - x^3\ell + x\ell^3/2] \tag{9.58}$$

From Equations 9.52 and 9.58, the maximum bending moment and maximum displacement are seen to occur at midspan as

$$M_{max} = w\ell^2/4 \quad \text{and} \quad y_{max} = 5w\ell^4/384EI \tag{9.59}$$

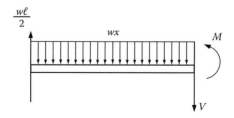

FIGURE 9.16   Free-body diagram of a left-side segment of the beam.

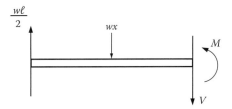

FIGURE 9.17 Free-body diagram of the segment with equivalent loading.

9.5.4 SIMPLY SUPPORTED BEAMS WITH A CONCENTRATED INTERIOR LOAD

As a final illustration of the procedure, consider a simply supported beam with an interior concentrated load as in Figure 9.18, where, ℓ is the beam length, and a and b are the distances of the point of loading from the left and right end supports as shown.

Consider a free-body diagram of the beam as in Figure 9.19, where the support reactions are represented by shear forces V_L and V_R as shown. By setting the sum of the forces equal to zero and by setting the sum of the moment of the forces about the right end equal to zero, we have

$$-V_L + P + V_R = 0 \quad \text{and} \quad -V_L\ell + Pb = 0 \tag{9.60}$$

or

$$V_L = Pb/\ell \quad \text{and} \quad V_R = -Pa/\ell \tag{9.61}$$

Consider next a free-body diagram of a segment of the beam to the left of the load as in Figure 9.20 where x is the segment length and where, as before, V and M are the shear and bending moment respectively on the right end of the segment. By setting the sum of the forces equal to zero and by setting the sum of the moment of the forces about the left end equal to zero, we have

$$V - Pb/\ell = 0 \quad \text{and} \quad M - Vx = 0 \tag{9.62}$$

Thus, the bending moment M is

$$M = Vx = Pbx/\ell \tag{9.63}$$

From Equation 9.15, the displacement of the beam segment may be determined from the expression

$$EI\mathrm{d}^2y/\mathrm{d}x^2 = -M = -Pbx/\ell \tag{9.64}$$

By integrating, we have

$$EI\mathrm{d}y/\mathrm{d}x = -Pbx^2/2\ell + c_1 \tag{9.65}$$

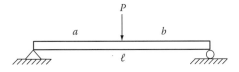

FIGURE 9.18 Simply supported beam with an internal concentrated load.

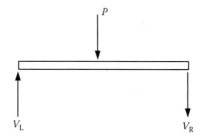

FIGURE 9.19   Free-body diagram of internally loaded, simply supported beam.

and

$$EIy = -Pbx^3/6\ell + c_1 x + c_2 \tag{9.66}$$

With the beam having a simple support at $x = 0$, we have $y(0) = 0$, and thus c_2 is zero. Therefore, for $0 \leq x \leq a$, the beam displacement is

$$EIy = -Pbx^3/6\ell + c_1 x \quad (0 \leq x \leq a) \tag{9.67}$$

Next, consider in a similar manner a free-body diagram of a segment of the beam to the right of the load as in Figure 9.21, where ξ is the segment length and where now V and M are the (positively directed) shear and bending moment on the left end of the segment. By enforcing the equilibrium conditions, we have

$$V + Pa/\ell = 0 \quad \text{and} \quad M + V\xi = 0 \tag{9.68}$$

Thus the bending moment M is

$$M = Pa\xi/\ell \tag{9.69}$$

We can again use Equation 9.15 to determine the beam displacement. To do this, however, it is convenient to consider x as the distance of the left end of the segment from the left support. That is, let ξ be

$$\xi = \ell - x \quad x \geq a \tag{9.70}$$

Then the bending moment of Equation 9.69 is

$$M = Pa(\ell - x)/\ell \tag{9.71}$$

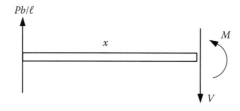

FIGURE 9.20   Free-body diagram of the left segment of the beam.

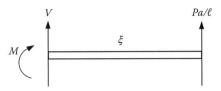

FIGURE 9.21   Free-body diagram of the right segment of the beam.

Equation 9.15 then becomes

$$EId^2y/dx^2 = -M = -Pa(\ell - x)/\ell \qquad (9.72)$$

By integrating, we have

$$EIdy/dx = -(Pa/\ell)\ell x + (Pa/\ell)x^2/2 + c_3 \qquad (9.73)$$

and

$$EIy = -(Pa/\ell)(\ell x^2/2) + (Pa/\ell)(x^3/6) + c_3 x + c_4 \qquad (9.74)$$

With the beam having a simple support at $x = \ell$, we have $y(\ell) = 0$, or

$$0 = -Pa\ell^2/2 + Pa\ell^2/6 + c_3\ell + c_4 \qquad (9.75)$$

or

$$c_4 = -c_3\ell + Pa\ell^2/3 \qquad (9.76)$$

Thus by substituting for c_4 in Equation 9.72, the beam displacement is given by

$$EIy = -(Pa/\ell)(\ell x^2/2) + (Pa/\ell)(x^3/6) + c_3(x - \ell) + Pa\ell^2/3 \qquad (9.77)$$

Equations 9.66 and 9.77 provide expressions for the beam displacement for $0 \leq x \leq a$ (Equation 9.66) and for $a \leq x \leq \ell$ (Equation 9.77). Observe that both equations have undetermined constants: c_1 in Equation 9.66 and c_3 in Equation 9.77. These constants can now be evaluated by requiring that the beam displacement and the beam slope have the same values at $x = a$ as determined from each equation. (That is, the beam displacement must be continuous and smooth at $x = a$.) Therefore, by equating the results for the displacement at $x = a$ from Equations 9.66 and 9.77, we have

$$-Pba^3/6\ell + c_1a = -Pa^3/2 + Pa^4/6\ell + c_3(a - \ell) + Pa\ell^2/3 \qquad (9.78)$$

or since b is $\ell - a$, we have

$$c_1a + c_3(\ell - a) = Pa\ell^2/3 - Pa^3/3 \qquad (9.79)$$

Similarly, by equating the expressions for the displacement derivatives at $x = a$ from Equations 9.65 and 9.73, we have

$$-Pba^2/2\ell + c_1 = -Pa^2 + Pa^3/2\ell + c_3 \qquad (9.80)$$

or since b is $\ell - a$, we have

$$c_1 - c_3 = -Pa^2/2 \tag{9.81}$$

By solving Equations 9.79 and 9.81 for c_1 and c_3, we have

$$c_1 = Pa(\ell/3 + a^2/6\ell - a/2) \quad \text{and} \quad c_3 = pa(\ell/3 + a^2/6\ell) \tag{9.82}$$

Finally, by substituting these results into Equations 9.66 and 9.77, we obtain the beam displacement as

$$EIy = -Pbx^3/6\ell + Pax(\ell/3 + a^2/6\ell - a/2) \quad 0 \le x \le a \tag{9.83}$$

and

$$EIy = (Pa/\ell)(x^3/6 - \ell x^2/2 + \ell^2 x/3 + a^2 x/6 - \ell a^2/6) \quad (a \le x \le \ell) \tag{9.84}$$

In these results, if $a = b = \ell/2$ (that is, a centrally loaded beam), then the displacement under the load is

$$y = P\ell^3/48EI \tag{9.85}$$

From equations 9.63 and 9.69, we can deduce for a centrally loaded beam (that is, $a = b = \ell/2$) the maximum bending moment occurs under the load as

$$M_{\max} = P\ell/4 \tag{9.86}$$

9.6 COMMENT

The illustrations in the above discussion show that by using free-body diagrams to determine beam loading, bending, and shear and by integrating the governing equations of Section 9.4, we can determine beam displacement along the centerline. The last illustration shows that this procedure can be cumbersome with even relatively simple configurations. The difficulty arises primarily with concentrated loading, which leads to singularities and discontinuities. In Chapter 10, we discuss singularity functions, which enable us to overcome the difficulty with concentrated loads and to greatly simplify the procedure.

SYMBOLS

A	Length, value of x
B	Beam depth (in Z-axis direction)
C	Plane curve
c_i ($i = 1, 2, 3, 4$)	Integration constants
E	Elastic modulus
I	Second moment of area
k_M	Spring constant for bending
k_V	Spring constant for shear
ℓ	Beam length

L, R	Subscripts designating left and right
M	Bending moment
O	Origin of X- and Y-axes
P	Load
q	Loading
V	Shear force
w	Uniform load per unit length
x	X-axis coordinate
X, Y, Z	Cartesian (rectangular) coordinate axes
y	Y-axis coordinate, displacement
ξ	Segment length
ρ	Radius of curvature

REFERENCE

1. G. B. Thomas, *Calculus and Analytic Geometry*, 3rd ed., Addison-Wesley, New York, 1960, p. 589.

10 Beam Analysis Using Singularity Functions

10.1 USE OF SINGULARITY FUNCTIONS

The final example of Chapter 9 (a simply supported beam with an interior-concentrated load) illustrates a difficulty in traditional beam analysis with concentrated loads. The solution procedure requires separate analyses on both sides of the load, and the solution itself requires two expressions depending upon the position of the independent variable relative to the load. We can avoid these difficulties by using singularity functions.

In this chapter, we introduce these functions and illustrate their use with some examples as in Chapter 9. We then discuss some less-trivial configurations. Singularity functions are developed using the properties of the Heaviside unit step function and the Dirac delta function [1]. Typically, these functions are defined as follows.

10.1.1 HEAVYSIDE UNIT STEP FUNCTION

The unit step function $\phi(x - a)$ is defined as

$$\phi(x - a) = \begin{cases} 0 & x < a \\ 1 & x \geq a \end{cases} \tag{10.1}$$

Graphically, $\phi(x - a)$ may be represented as in Figure 10.1. Observe that $\phi(x - a)$ appears as a "step" at $x = a$. However, $\phi(x - a)$ is not continuous at $x = a$ and therefore the derivative of $\phi(x - a)$ does not exist at $x = 1$, in the context of elementary analysis. Nevertheless, if we regard $\phi(x - a)$ as a generalized function, we can define its derivative as the Dirac delta function: $\delta(x - a)$.

10.1.2 DIRAC DELTA FUNCTION

This function, which is sometimes called the "impulse" function, is defined as

$$\delta(x - a) = \begin{cases} 0 & x \neq a \\ \infty & x = a \end{cases} \tag{10.2}$$

Graphically, $\delta(x - a)$ may be represented as in Figure 10.2. If $\delta(x - a)$ is to be the derivative of $\phi(x - a)$, we have the relation

$$\phi(x - a) = \int_{6}^{x} \delta(\xi - a)\,d\xi \tag{10.3}$$

where b is a constant. If $b < a < c$, we have

$$\int_{b}^{c} \delta(\xi - a)\,d\xi = 1 \tag{10.4}$$

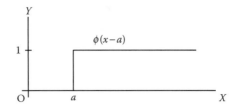

FIGURE 10.1 Unit step function.

Finally, if $f(x)$ is a continuous function, we have the relation

$$f(a) = \int_b^c f(x)\delta(x - a)\mathrm{d}x \tag{10.5}$$

The Dirac delta function is a convenient mathematical model of a concentrated load. Indeed, if we think of a concentrated load as a force exerted over a vanishingly small area, we have a stress concentration or infinite stress. Since our objective here is beam displacement, as opposed to local surface effects, we can eliminate the singularity by integrating as in Equation 10.3. This integration is conveniently performed using the methodology and formalism of singularity functions as developed in the following paragraphs.

10.2 SINGULARITY FUNCTION DEFINITION

Singularity functions are designated using angular brackets: $<\cdot>$ with the following properties [2]:

$$<x - a>^n = \begin{cases} 0 & x < a \text{ for all } n \\ 0 & x > a \text{ for } n < 0 \\ \infty & x = a \text{ for } n = -1 \\ \pm\infty & x = a \text{ for } n = -2 \\ 1 & x \geq a \text{ for } n = 0 \\ (x - a)^n & x \geq a \text{ for } n > 0 \end{cases}$$

10.3 SINGULARITY FUNCTION DESCRIPTION AND ADDITIONAL PROPERTIES

Singularity functions are especially useful for modeling concentrated and discontinuous loadings on structures (particularly beams). Specifically, for a concentrated load with magnitude P at $x = 1$, we

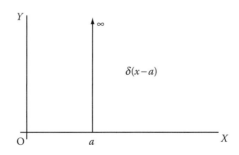

FIGURE 10.2 Unit impulse function (Dirac).

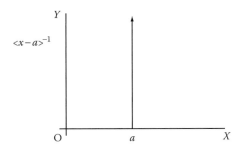

FIGURE 10.3 Representation of $<x - a>^{-1}$ (Dirac delta function).

can use the function $P<x - a>^{-1}$. Figure 10.3 provides a pictorial representation of $<x - a>^{-1}$. Recall that the positive direction for beam loading is downward.

For a uniform load with intensity w_O beginning at $x = 1$, we can use the function $w_o <x - a>^0$. Figure 10.4 provides a pictorial representation of $<x - a>^0$.

For a concentrated moment with intensity M_O at $x = a$, we can use the function $M_O <x - a>^{-2}$. Figure 10.5 provides a pictorial representation of $<x - a>^{-2}$. (Recall again that the positive direction for bending moment is in the negative Z-direction for a moment applied to a positive beam face and in the positive Z-direction for a moment applied to a negative beam face.)

The derivatives and antiderivatives of $<x - a>^n$ are defined by the expressions [2]

$$\frac{d}{dx} <x - a>^{n+1} = <x - a>^n \quad n < 0$$

$$\frac{d}{dx} <x - a>^n = n <x - a>^{n-1} \quad n > 0$$

$$\int_b^x <\xi - a>^n \, d\xi = <x - a>^{n+1} \quad n < 0 \qquad (10.6)$$

$$\int_b^x <\xi - a>^n \, d\xi = <x - a>^{n+1} / n + 1 \quad n \geq 0$$

where $b < a$.

Finally, upon integration, we often encounter the "ramp" and "parabola" functions $<x - a>^1$ and $<x - a>^2$. Figures 10.6 and 10.7 provide a pictorial representation of these functions.

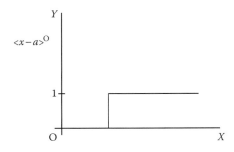

FIGURE 10.4 Representation of $<x - a>^{-1}$ (heavyside unit step function).

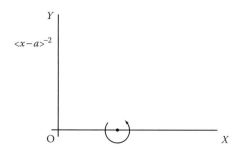

FIGURE 10.5 Representation of the function $<x-a>^{-2}$.

10.4 ILLUSTRATION OF SINGULARITY FUNCTION USE

10.4.1 UNIFORMLY LOADED CANTILEVER BEAM

See Section 9.5.1 and consider again the cantilever beam supported at its left end and loaded with a uniform load along its span as in Figure 10.8, where ℓ is the beam length and w is the load per unit length.

Consider a free-body diagram of the beam to determine the support reactions as in Figure 10.9 where V_O and M_O are the shear and bending moment applied to the beam by the support. From the diagram, V_O and M_O are readily seen to be (see Equation 9.22)

$$V_O = w\ell \quad \text{and} \quad M_O = -w\ell^2/2 \tag{10.7}$$

From the results of Equation 10.7, the loading on the beam including that from the support reaction is that shown in Figure 10.10. Using the singularity functions, the loading function $q(x)$ may be expressed as

$$q(x) = -(w\ell^2/2) <x-0>^{-2} -w\ell <x-0>^{-1} +w <x-0>^0 \tag{10.8}$$

where the origin of the X-axis is at the left end of the beam. Recall also that the positive direction is down for loads and on the left end of the beam, the positive direction is clockwise for bending moments.

Referring to Equation 9.17, the governing equation for the displacement is

$$EI d^4y/dx^4 = q(x) = -(w\ell^2/2) <x-0>^{-2} -w\ell <x-0>^{-1} +w <x-0>^0 \tag{10.9}$$

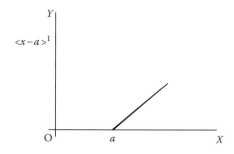

FIGURE 10.6 Ramp singularity function.

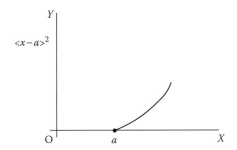

FIGURE 10.7 Parabolic singularity function.

Then by integrating, we have

$$EI d^3y/dx^3 = -V = -(w\ell^2/2) <x-0>^{-1} - w\ell <x-0>^0 + w <x-0>^1 + c_1 \quad (10.10)$$

where V is the shear on the beam cross section. Since V is zero when $x = \ell$, we have

$$0 = 0 - w\ell + w\ell + c_1 \quad (10.11)$$

or

$$c_1 = 0 \quad (10.12)$$

By integrating again, we have

$$EI d^2y/dx^2 = -M = -(w\ell^2/2) <x-0>^0 - w\ell <x-0>^1 + w <x-0>^2/2 + c_2 \quad (10.13)$$

where M is the bending moment on the cross section. Since M is zero when $x = \ell$, we have

$$0 = -(w\ell^2/2) - w\ell^2 + (w\ell^2/2) + c_2 \quad (10.14)$$

or

$$c_2 = w\ell^2 \quad (10.15)$$

By integrating again, we have

$$EI dy/dx = -(w\ell^2/2) <x-0>^1 - w\ell <x-0>^2/2 + w <x-0>^3/6 + w\ell^2 x + c_3 \quad (10.16)$$

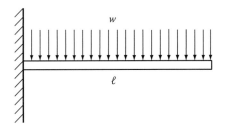

FIGURE 10.8 Uniformly loaded cantilever beam.

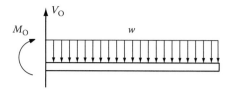

FIGURE 10.9 Free-body diagram of the uniformly loaded cantilever beam.

Since $dy/dx = 0$ when $x = 0$, we have

$$c_3 = 0 \tag{10.17}$$

Finally, by integrating fourth time, we have

$$EIy = -(w\ell^2/2) <x - 0>^2/2 - w\ell <x - 0>^3/6 + w <x - 0>^4/24 + w\ell^2 x^2/2 + c_4 \tag{10.18}$$

Since $y = 0$ when $x = 0$, we have

$$c_4 = 0 \tag{10.19}$$

Thus, the beam displacement is seen to be

$$y = (w/EI)\left[-(\ell^2/4) <x - 0>^2 - (\ell/6) <x - 0>^3 + <x - 0>^4/24 + \ell^2 x^2/2\right] \tag{10.20}$$

The maximum displacement y_{max} will occur at $x = \ell$ as

$$y_{max} = (w\ell^4/EI)[-(1/4) - (1/6) + (1/24) + (1/2)] = w\ell^4/8EI \tag{10.21}$$

Observe that the results of Equations 10.20 and 10.21 match those of Equations 9.35 and 9.38.

10.4.2 CANTILEVER BEAM WITH A CONCENTRATED END LOAD

Consider again the example of Section 9.5.2, the cantilever beam supported at its left end and loaded with a concentrated force at its right end as in Figure 10.11. As before, let the beam length be ℓ and the load magnitude be P. Figure 10.12 presents a free-body diagram of the beam and the support reactions are seen to be

$$V_O = P \quad \text{and} \quad M_O = -P\ell \tag{10.22}$$

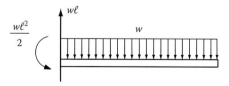

FIGURE 10.10 Loading on the beam.

FIGURE 10.11 Cantilever beam with a concentrated end load.

The loading experienced by the beam is shown in Figure 10.13. The loading function $q(x)$, expressed in terms of singularity functions, is then

$$q(x) = -P <x-0>^{-1} - P\ell <x-0>^{-2} + P <x-\ell>^{-1} \qquad (10.23)$$

From Equation 9.17 the governing equation for the displacement is

$$EId^4y/dx^4 = q(x) = -P <x-0>^{-1} - P\ell <x-0>^{-2} + P <x-\ell>^{-1} \qquad (10.24)$$

By integrating, we have

$$EId^3y/dx^3 = -V = -P <x-0>^0 - P\ell <x-0>^{-1} + P <x-\ell>^0 + c_1 \qquad (10.25)$$

Since the shear is zero at $x=\ell$, we have

$$0 = -P - 0 + P + c_1 \qquad (10.26)$$

or

$$c_1 = 0 \qquad (10.27)$$

By integrating again, we have

$$EId^2y/dx^2 = -M = -P <x-0>^1 - P\ell <x-0>^0 + P <x-\ell>^1 + c_2 \qquad (10.28)$$

Since the bending moment is zero at $x=\ell$, we have

$$0 = -P\ell - P\ell + c_2 \qquad (10.29)$$

or

$$c_2 = 2P\ell \qquad (10.30)$$

FIGURE 10.12 Free-body diagram of the beam of Figure 10.11.

FIGURE 10.13   Beam loading.

By integrating the equation a third time, we have

$$EI dy/dx = -P <x-0>^2/2 - P\ell <x-0>^1 + P <x-\ell>^2/2 + 2P\ell x + c_3 \qquad (10.31)$$

But with the fixed (cantilever) support at $x=0$, we have $dy/dx=0$ at $x=0$ and thus

$$0 = -0 - 0 + 0 + 0 + c_3 \qquad (10.32)$$

or

$$c_3 = 0 \qquad (10.32)$$

Finally, by integrating the equation a fourth time, we have

$$EI y = -P <x-0>^3/6 - P\ell <x-0>^2/2 + P <x-\ell>^3/6 + P\ell x^2 + c_4 \qquad (10.33)$$

But with the fixed support at $x=0$, we have $y=0$ at $x=0$ and thus

$$0 = -0 - 0 + 0 + 0 + c_4 \qquad (10.34)$$

or

$$c_4 = 0 \qquad (10.35)$$

Therefore, the beam displacement is

$$y = (P/EI)[-<x-0>^3/6 - \ell <x-0>^2/2 + <x-\ell>^3/6 + \ell x^2] \qquad (10.36)$$

The maximum beam displacement y_{max} occurs at $x=\ell$ with the value

$$y_{max} = P\ell^3/3EI \qquad (10.37)$$

Observe that the results of Equations 10.36 and 10.37 match those of Equations 9.45 and 9.46.

10.4.3 SIMPLY SUPPORTED BEAM WITH A UNIFORM LOAD

Next consider the example of Section 9.5.3 of a simply supported beam with a uniform load as in Figure 10.14, where again w is the load per unit length and ℓ is the beam length. Consider a free-body diagram of the beam as in Figure 10.15. From the figure, the reaction shearing forces V_L and V_R are (see Equation 9.48)

$$V_L = w\ell/2 \quad \text{and} \quad V_R = -w\ell/2 \qquad (10.38)$$

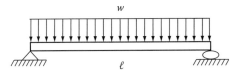

FIGURE 10.14 Uniformly loaded simply supported beam.

Figure 10.16 then illustrates the loading on the beam.

Using the singularity functions, the loading $q(x)$ on the beam may be expressed as

$$q(x) = -(w\ell/2) <x-0>^{-1} + w <x-0>^0 - (w\ell/2) <x-\ell>^{-1} \qquad (10.39)$$

From Equation 9.17, the governing equation for the beam displacement is then

$$EI d^4y/dx^4 = q(x) = -(w\ell/2) <x-0>^{-1} + w <x-0>^0 - (w\ell/2) <x-\ell>^{-1} \qquad (10.40)$$

By integrating, we obtain

$$EI d^3y/dx^3 = -V = -(w\ell/2) <x-0>^0 + w <x-0>^1 - (w\ell/2) <x-\ell>^0 + c_1 \qquad (10.41)$$

By integrating again, we have

$$EI d^2y/dx^2 = -M = -(w\ell/2) <x-0>^1 + w <x-0>^2/2 - (w\ell/2) <x-\ell>^1 + c_1 x + c_2 \qquad (10.42)$$

In Equations 10.41 and 10.42, c_1 and c_2 are integration constants to be determined by the support conditions. Recall that with simple supports the moment exerted by the support is zero. Therefore, $M = 0$ at $x = 0$ and $x = \ell$. Thus from Equation 10.42, we have

$$0 = -0 + 0 - 0 + 0 + c_2 \qquad (10.43)$$

and

$$0 = -w\ell^2/2 + w\ell^2/2 - 0 + c_1\ell \qquad (10.44)$$

or

$$c_1 = 0 \quad \text{and} \quad c_2 = 0 \qquad (10.45)$$

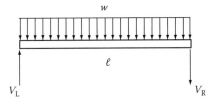

FIGURE 10.15 Free-body diagram of the simply supported, uniformly loaded beam.

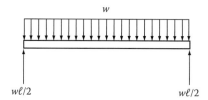

FIGURE 10.16 Loading on the beam.

By integrating Equation 10.42 again, we obtain

$$EI dy/dx = -(w\ell/2) <x - 0>^2/2 + w <x - 0>^3/6 - (w\ell/2) <x - \ell>^2/2 + c_3 \qquad (10.46)$$

And again

$$EIy = -(w\ell/2) <x - 0>^3/6 + w <x - 0>^4/24 - (w\ell/2) <x - \ell>^3/6 + c_3 x + c_4 \qquad (10.47)$$

where the integration constants c_3 and c_4 may be evaluated by recalling that with simple supports the displacements are zero at the supports. Thus, we have at $x = 0$

$$0 = 0 + 0 - 0 + 0 + c_4 \qquad (10.48)$$

and at $x = \ell$

$$0 = -(w\ell/2)(\ell^3/6) + w\ell^4/24 - 0 + c_3\ell \qquad (10.49)$$

Hence

$$c_4 = 0 \quad \text{and} \quad c_3 = w\ell^4/24 \qquad (10.50)$$

Therefore, the displacement $y(x)$ becomes

$$y = (w/24EI)[-2\ell <x - 0>^3 + <x - 0>^4 - 2\ell <x - \ell>^3 + \ell^3 x] \qquad (10.51)$$

The maximum displacement y_{max} will occur at the midspan $(x = \ell/2)$ as

$$y_{max} = (w/24EI)\ell^4[(-2/8) + (1/16) + (1/2)] = \frac{5w\ell^4}{384EI} \qquad (10.52)$$

Observe that the results of Equations 10.51 and 10.52 are the same as those of Equations 9.58 and 9.59.

10.4.4 SIMPLY SUPPORTED BEAM WITH A CONCENTRATED INTERIOR LOAD

Finally, consider the example of Section 9.5.4 of a simply supported beam with a concentrated interior load as in Figure 10.17, where, as before, ℓ is the beam length and a and b are the distances from the load to the left and right ends of the beam, as shown.

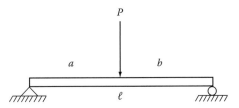

FIGURE 10.17 Simply supported beam with an interior-concentrated load.

Figure 10.18 shows a free-body diagram of the beam with the support reactions represented by the shear forces V_L and V_R. For beam equilibrium, we observe that (see Equation 9.61)

$$V_L = Pb/\ell \quad \text{and} \quad V_R = -Pa/\ell \tag{10.53}$$

Using the singularity functions, the loading $q(x)$ on the beam may then be expressed as

$$q(x) = -(Pb/\ell) <x - 0>^{-1} + P <x - a>^{-1} - (Pa/\ell) <x - \ell>^{-1} \tag{10.54}$$

From Equation 9.17, the governing equation for the beam displacement is then

$$EI d^4 y/dx^4 = q(x) = -(Pb/\ell) <x - 0>^{-1} + P <x - a>^{-1} - (Pa/\ell) <x - \ell>^{-1} \tag{10.55}$$

Then by integrating, we have

$$EI d^3 y/dx^3 = -V = -(Pb/\ell) <x - 0>^0 + P <x - 0>^0 - (Pa/\ell) <x - \ell>^0 + c_1 \tag{10.56}$$

and

$$EI d^2 y/dx^2 = -M = (Pb/\ell) <x - 0>^1 + P <x - a>^1 - (Pa/\ell) <x - \ell>^1 + c_1 x + c_2 \tag{10.57}$$

For a simply supported beam, M is zero at the supports, that is, $M = 0$ at $x = 0$ and $x = \ell$. Then at $x = 0$, we have

$$0 = 0 + 0 - 0 + 0 + c_2 \quad \text{or} \quad c_2 = 0 \tag{10.58}$$

and at $x = \ell$, we have

$$0 = -(Pb/\ell)\ell + P(\ell - a) - 0 + c_1 \ell$$

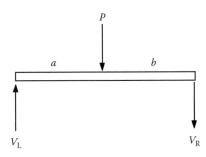

FIGURE 10.18 Free-body diagram of internally loaded, simply supported beam.

or

$$0 = -Pb + Pb + c_1\ell \quad \text{or} \quad c_1 = 0 \tag{10.59}$$

By integrating the equation a third and fourth time, we obtain

$$EI\,dy/dx = -(Pb/\ell) <x-0>^2/2 + P <x-a>^2\, 2\,/ - (Pa/\ell) <x-\ell>^2/2 + c_3 \tag{10.60}$$

and

$$EIy = -(Pb/\ell) <x-0>^3/6 + P <x-a>^3/6 - (Pa/\ell) <x-\ell>^3/6 + c_3x + c_4 \tag{10.61}$$

But with the simple supports, y is zero at the beam ends. Therefore, at $x = 0$, we have

$$0 = -0 + 0 - 0 + 0 + c_4 \quad \text{or} \quad c_4 = 0 \tag{10.62}$$

and at $x = \ell$, we have

$$0 = -(Pb/\ell)(\ell^3/6) + P(\ell-a)^3/6 - 0 + c_3\ell \quad \text{or} \quad c_3 = (Pb/6\ell)(\ell^2 - b^2) \tag{10.63}$$

By substituting from Equations 10.62 and 10.63 in 10.61 the displacement y becomes

$$y = (1/6EI)\left[(-Pb/\ell) <x-0>^3 + P <x-a>^3 - (Pa/\ell) <x-\ell>^3 + (Pb/\ell)(\ell^2 - b^2)x\right] \tag{10.64}$$

Observing the result of Equation 10.64 if $0 \le x \le a$, we have

$$\begin{aligned} EIy &= (-Pb/\ell)(x^3/6) + (Pb/\ell)(\ell^2 - b^2)(x/6) \\ &= -Pbx^3/6\ell + Pax(\ell/3 + a^2/6\ell - a/2) \end{aligned} \tag{10.65}$$

which is identical to the result of Equation 9.83. Similarly, in Equation 10.64 if $a \le x \le \ell$, we have

$$\begin{aligned} EIy &= -(Pb/6\ell)x^3 + (P/6)(x-a)^3 + (Pb/6\ell)(\ell^2 - b^2)x \\ &= (Pa/\ell)(x^3/6 - \ell x^2/2 + \ell^2 x/3 + a^2 x/6 - \ell a^2/6) \end{aligned} \tag{10.66}$$

which is identical to the result of Equation 9.84. (In Equations 9.69 and 9.70, the validation is obtained by letting $b = \ell - a$ and by performing routine analysis.)

Finally, in Equation 10.64 if $x = a = b = \ell/2$, we obtain the maximum displacement as

$$y_{\max} = P\ell^3/48EI \tag{10.67}$$

This result matches that of Equation 9.85.

10.5 DISCUSSION AND RECOMMENDED PROCEDURE

The principal advantages of singularity functions are their simplicity in use and their broad range of application. They are particularly useful in modeling concentrated loads and discontinuous loadings. This utility is immediately seen in comparing the two analyses of the simply supported beam with the interior-concentrated load (in Sections 9.5.4 and 10.4.4). With the traditional method in

Section 9.5.4, we needed to use multiple free-body diagrams and separate equations for locations such as to the left and right of the load. With singularity functions, however, we simply model the load with the function $P <x - a>^{-1}$ and then integrate, as we solve the governing equation.

Specifically, the steps in using singularity functions for beam bending analyses are as follows:

1. For a given beam loading $q(x)$ and support conditions, construct a free-body diagram of the beam to determine the support reactions.
2. Model the loading function $a(x)$ and the support reactions by using the singularity functions of Section 10.2. ($M <x - x_0>^{-2}$ is a concentrated moment M at x_0; $P <x - x_0>^{-1}$ is a concentrated force P at x_0; $q <x - x_0>^{0}$ is a uniform load q beginning at x_0; etc.)
3. Form the governing differential equation:

$$EId^4y/dx^4 = q(x)$$

(see Equation 9.17.)
4. Determine the boundary conditions (auxiliary conditions) from the support conditions.
5. Integrate the governing equation and evaluate the integration constants by using the auxiliary conditions.

10.6   COMMENTS ON THE EVALUATION OF INTEGRATION CONSTANTS

Observe that in the process of integrating the governing differential equation for the beam displacement, we first obtain an expression for the transverse shear V in the beam, and then by integrating again, an expression for the bending moment M in the beam. These expressions, together with the support conditions, may be used to evaluate constants of integration. Thus if displacement conditions are also used to evaluate the constants, we have a means of checking the values obtained.

To illustrate these ideas, consider again the simply supported beam with a concentrated interior load as shown in Figures 10.17 and 10.19. Recall from Sections 9.5.3 and 10.4.4 that the left and right support reactions are Pb/ℓ and Pa/ℓ, and that the beam loading including the support reactions may be modeled as in Figure 10.20.

Next, recall from Equation 10.54 that the loading function $q(x)$ on the beam is

$$q(x) = -(Pb/\ell) <x - 0>^{-1} + P <x - a>^{-1} - (Pa/\ell) <x - \ell>^{-1} \qquad (10.68)$$

and from Equation 10.55 that the governing equation for the beam displacement is

$$EId^4y/dx^4 = q(x) = -(Pb/\ell) <x - 0>^{-1} + P <x - a>^{-1} - (Pa/\ell) <x - \ell>^{-1} \qquad (10.69)$$

Finally, by integrating we have an expression for the shear V in the beam as (see Equation 10.56)

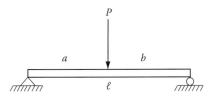

FIGURE 10.19   Simply supported beam with an interior-concentrated load.

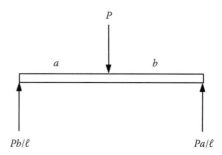

FIGURE 10.20 Equivalent loading on the beam of Figure 10.19.

$$EId^3y/dx^3 = -V = -(Pb/\ell)<x-0>^0 + P<x-a>^0 - (Pa/\ell)<x-\ell>^0 + c_1 \qquad (10.70)$$

where c_1 is an integration constant.

Recall that in the solution of this problem in Section 10.4.4, we discovered that the integration constant c_1 is zero as a result of the bending moments at the supports being zero. We can also see that c_1 is zero by examining the shear forces in the beam at the supports. Consider, for example, the support at the left end of the beam as depicted in Figure 10.21. The reaction force of the support is shown in Figure 10.22. Recall that the shear force V is positive when it is exerted on a positive face (cross section) of the beam in the positive direction, or on a negative face in the negative direction. Correspondingly, the shear is negative when exerted on a positive face in the negative direction, or on a negative face in the positive direction.

Consider cross sections just to the left ($x=0^-$) and just to the right ($x=0^+$) of the support as in Figure 10.23a and b. In the first case (a), the shear is zero, whereas in the second case the shear is: $V=+Pb/\ell$ (positive because of a negatively directed force on a negative face). Similarly for beam cross sections just to the left and to the right of the right-end support (where the reaction force magnitude is Pa/ℓ), we have the shear zero on the face $x=\ell^+$ and $-Pa/\ell$ on the face $x=\ell^-$ (negative since the force is negatively directed on a positive face), as represented in Figure 10.24. Table 10.1 lists these results.

Referring now to Equation 10.69 (written again here), we see that each of the four conditions of Table 10.1 leads to $c_1=0$.

$$V = (Pb/\ell)<x-0>^0 - P<x-a>^0 + (Pa/\ell)<x-\ell>^0 - c_1 \qquad (10.71)$$

10.7   SHEAR AND BENDING MOMENT DIAGRAMS

Stresses in beams discussed in most books on strength and mechanics of materials give considerable emphasis usually to the construction of shear and bending moment diagrams. These diagrams are graphical representations of the shear force V and the bending moment M along the beam span. The

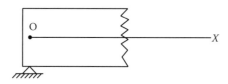

FIGURE 10.21 Left-end support of the beam of Figure 10.19.

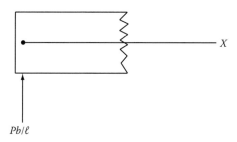

FIGURE 10.22   Left-end support reaction force.

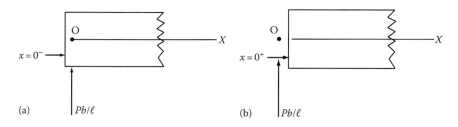

FIGURE 10.23   Beam cross sections just to the left (a) and just to the right (b) of the left-end support.

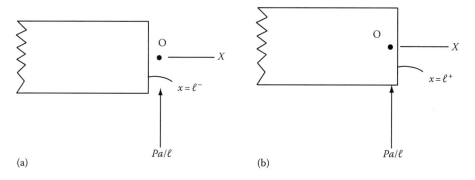

FIGURE 10.24   Beam cross section just to the left and just to the right of the right-end support.

TABLE 10.1

Shear on Cross Section Faces Near the Beam Supports

Face	Shear V
$x = 0^-$	0
a. Cross Section Just Left of the Support	b. Cross Section Just Right of the Support
$x = 0^+$	Pb/ℓ
$x = \ell^-$	$-Pa/\ell$
$x = \ell^+$	0

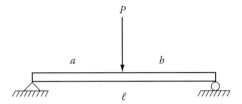

FIGURE 10.25 Simply supported beam with an interior-concentrated load.

shear and bending moment are needed to determine the shear stress (in relatively thick beams) and the flexural, or bending, stress as in the formula $\sigma = Mc/I$. Shear and bending moment diagrams are thus convenient not only for determining the shear and bending stresses, but also for finding the positions along the beam where the maximum values of these stresses occur.

Singularity functions are especially useful for constructing shear and bending moment diagrams. To illustrate this, consider again the simply supported beam with the interior-concentrated load as in the foregoing sections and as shown again in Figure 10.25. From Equations 10.69 and 10.70, the shear V is

$$V = (Pb/\ell) <x - 0>^0 - P <x - a>^0 + (Pa/\ell) <x - \ell>^0 \qquad (10.72)$$

Figure 10.26 shows a graph of this function, where the ordinate V is positive upward.

From Equation 10.57, the bending moment M is (note that c_1 and c_2 are zero)

$$M = (Pb/\ell) <x - 0>^1 - P <x - a>^1 + (Pa/\ell) <x - \ell>^1 \qquad (10.73)$$

Figure 10.27 shows a graph of this function.

As a further illustration of the use of singularity functions to construct shear and bending moment diagrams, consider the cantilever beam with a uniform load as in Figure 10.28. From Figure 10.10, the loading on the beam is as shown in Figure 10.29. From Equation 10.8, the loading $q(x)$ on the beam is

$$q(x) = -(w\ell^2/2) <x - 0>^{-2} - w\ell <x - 0>^{-1} + w <x - 0>^0 \qquad (10.74)$$

From Equations 10.10 and 10.12, the shear V along the beam axis is

$$V = (w\ell^2/2) <x - 0>^{-1} + w\ell <x - 0>^0 - w <x - 0>^1 \qquad (10.75)$$

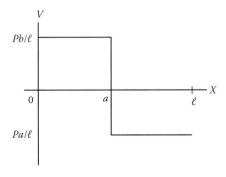

FIGURE 10.26 Transverse shear diagram for a simply supported beam with an interior-concentrated load.

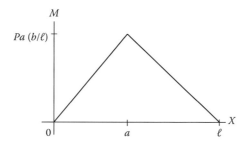

FIGURE 10.27   Bending moment diagram for a simply supported beam with an interior-concentrated load.

Figure 10.30 shows a graph of this function, where the positive direction is upward.
     From Equations 10.13 and 10.15, the bending moment M along the beam axis is

$$M = (w\ell^2/2) <x-0>^0 + w\ell <x-0>^1 - w <x-0>^2/2 - w\ell^2 \qquad (10.76)$$

Figure 10.31 shows a graph of this function.

10.8   ADDITIONAL ILLUSTRATION: CANTILEVER BEAM WITH UNIFORM LOAD OVER HALF THE SPAN

To illustrate the use of singularity functions with a somewhat less simple example, consider the cantilever beam with a right-end support but loaded with a uniform load over the first half of the beam, starting at the free end as in Figure 10.32, where, as before, ℓ is the length of the beam and w is the uniform load per unit length. Also, as before, let the objective of the analysis be to determine the beam displacement g, together with expressions for the shear loading V (transverse or perpendicular to the beam axis), and the bending moment M.
     We can readily determine y, V, and M by following the procedure of Section 10.5: first, we can determine the support reaction from the free-body diagram of Figure 10.33, where we represent the built-in support reaction by the shear force V_ℓ and bending moment M_ℓ, as shown. From this figure, we immediately find V_ℓ and M_ℓ to be

$$V_\ell = -w\ell/2 \quad \text{and} \quad M_\ell = -3W\ell^2/8 \qquad (10.77)$$

Next, from Figure 10.33, the loading function $q(x)$ along the beam is

$$q(x) = w <x-0>^0 - w <x-\ell/2>^0 - (w\ell/2) <x-\ell>^{-1} - (3w\ell^2/8) <x-\ell>^{-2} \quad (10.78)$$

Third, from Equation 9.17, the governing differential equation is

$$EId^4y/dx^4 = q(x) = w <x-0>^0 - w <x-\ell/2>^0 - (w\ell/2) <x-\ell>^{-1} - (3w\ell^2/8) <x-\ell>^{-2}$$
$$(10.79)$$

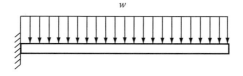

FIGURE 10.28   Uniformly loaded cantilever beam.

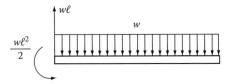

FIGURE 10.29   Loading on the beam.

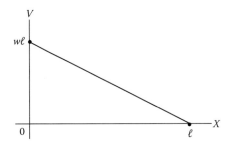

FIGURE 10.30   Transverse shear diagram for the uniformly loaded cantilever beam.

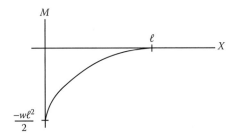

FIGURE 10.31   Bending moment diagram for the uniformly loaded cantilever beam.

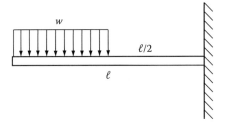

FIGURE 10.32   Cantilever beam with a half span uniform load.

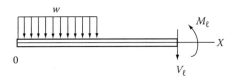

FIGURE 10.33   Free-body diagram for the beam of Figure 10.32.

Finally, since the beam has a free end at $x=0$ and a fixed end at $x=\ell$, we have the auxiliary condition:

$$\text{at } x = 0: \quad V = M = 0 \quad \text{or} \quad \mathrm{d}^3y/\mathrm{d}x^3 = \mathrm{d}^2y/\mathrm{d}x^2 = 0 \tag{10.80}$$

$$\text{at } x = \ell: \quad y = \mathrm{d}y/\mathrm{d}x = 0 \tag{10.81}$$

and also at $x = \ell$: $\quad V = -EI\mathrm{d}^3y/\mathrm{d}x^3 = -w\ell/2 \quad \text{and} \quad M = -EI\mathrm{d}^2y/\mathrm{d}x^2 = -3w\ell^2/8$ (10.82)

By integrating Equation 10.79, we have

$$EI\mathrm{d}^3y/\mathrm{d}x^3 = -V = w <x-0>^1 - w <x-\ell/2>^1 - (w\ell/2) <x-\ell>^0 - (3w\ell^2/8) <x-\ell>^{-1} + c_1 \tag{10.83}$$

Since $V=0$ when $x=0$, we have

$$c_1 = 0 \tag{10.84}$$

Note also at $x = \ell^-$, $V = -w\ell/2$, so that

$$w\ell/2 = w\ell - w(\ell/2) - 0 + c_1$$

or

$$c_1 = 0 \tag{10.85}$$

By integrating again, we have

$$EI\mathrm{d}^2y/\mathrm{d}x^2 = -M = w <x-0>^2/2 - w <x-\ell/2>^2/2 - (w\ell/2) <x-\ell>^1$$
$$- (3w\ell^2/8) <x-\ell>^0 + c_2 \tag{10.86}$$

Since $M=0$ when $x=0$, we have

$$c_2 = 0 \tag{10.87}$$

Note also at $x = \ell^{-1}$, $M = -3w\ell^2/8$, so that

$$3w\ell^2/8 = w\ell^2/2 - w(\ell/2)^2/2 - 0 - 0 + c_2$$

or

$$c_2 = 0 \tag{10.88}$$

By integrating the equation a third time, we have

$$EI\mathrm{d}y/\mathrm{d}x = w <x-0>^3/6 - w <x-\ell/2>^3/6 - (w\ell/2) <x-\ell>^2/2 - (3w\ell^2/8) <x-\ell>^1 + c_3 \tag{10.89}$$

But $\mathrm{d}y/\mathrm{d}x = 0$ when $x = \ell$, so that

$$0 = w\ell^3/6 - w(\ell/2)^3/6 - 0 - 0 + c_3$$

or

$$c_3 = -7w\ell^3/48 \qquad (10.90)$$

Finally, by integrating a fourth time, we have

$$EIy = w <x-0>^4/24 - w <x-\ell/2>^4/24 - (w\ell/2) <x-\ell>^3/6$$
$$- (3w\ell^2/8) <x-\ell>^2/2 - 7w\ell^3x/48 + c_4 \qquad (10.91)$$

But $y = 0$ when $x = \ell$, so that

$$0 = w\ell^4/24 - w(\ell/2)^4/24 - 0 - 0 - 7w\ell^4/48 + c_4$$

or

$$c_4 = (41/384)w\ell^4 \qquad (10.92)$$

To summarize, from Equations 10.83, 10.84, 10.86, 10.87, 10.91, and 10.92, the shear V, bending moment M, and displacement y are

$$V = (w/EI)\left[-<x-0>^1 + <x-\ell/2>^1 + (\ell/2) <x-\ell>^0 + (3\ell^2/8) <x-\ell>^{-1} \right] \quad (10.93)$$

$$M = (w/EI)\left[-<x-0>^2/2 + <x-\ell/2>^2/2 + (\ell/2) <x-\ell>^1 + (3\ell^2/8) <x-\ell>^0 \right]$$
$$(10.94)$$

and

$$y = (w/EI)\left[<x-0>^4/24 - <x-\ell/2>^4/24 - \ell <x-\ell>^3/12 - 3\ell^2 <x-\ell>^2/16 \right.$$
$$\left. - 7\ell^3x/48 + 41\ell^4/384 \right] \qquad (10.95)$$

SYMBOLS

a, b, c	Values of x
$c_i (i = 1, 2, 3, 4)$	Integration constants
E	Elastic modulus
I	Second moment of area
ℓ	Beam length
ℓ^-, ℓ^+	Values of x just to the left and just to the right of $x = \ell$
L, R	Subscripts designating left and right
M	Bending moment
M_O	Concentrated moment
n	Integer
O	Origin at X, Y, and Z-axes
P	Concentrated load
$q(x)$	Loading function
V	Shear
w	Uniform load per unit length
w_O	Uniform load per unit length

x	X-axis coordinate
$<x-a>^n$	Singularity function (see Section 10.2)
X, Y, Z	Cartesian (rectangular) coordinate axes
y	Displacement, Y-axis coordinate
$0^-, 0^+$	Values of x just to the left and just to the right of the origin O
$\delta(x-a)$	Dirac delta function, impulse function
$\phi(x-a)$	Heavyside unit step function

REFERENCES

1. F. B. Hildebrand, *Advanced Calculus for Applications*, 2nd ed., Prentice Hall, Englewood Cliffs, NJ, 1977, p. 66.
2. B. J. Hamrock, B. Jacobson, and S.R. Schmid, *Fundamentals of Machine Elements*, McGraw Hill, New York, 1999, p. 41.

11 Beam Bending Formulas for Common Configurations

11.1 PROSPECTUS

Recall from Equation 9.17 that the governing differential equation for the beam displacement y is

$$EID\mathrm{d}^4y/\mathrm{d}x^4 = q(x) \tag{11.1}$$

where
- E is the elastic modulus
- I is the second moment of area of the beam cross section
- x is the axial dimension
- $q(x)$ is the loading function

This equation has been integrated and solved for a large number of diverse loading and support conditions. References [1–3] provide a comprehensive list of results of these integrations.

In the following sections, we provide lists of a few of the more common and presumably, most useful of these results. We tabulate these results in Section 11.6.

11.2 CANTILEVER BEAMS

11.2.1 LEFT-END SUPPORTED CANTILEVER BEAM

Figure 11.1a through c shows the positive directions for the loading, the support reactions, and the displacements for a cantilever beam supported at its left end. Recall that at the built-in support, the displacement y and the slope $\mathrm{d}y/\mathrm{d}x$ of the beam are zero.

11.2.2 CANTILEVER BEAM, LEFT-END SUPPORT, AND CONCENTRATED END LOAD

Figure 11.1b and c shows loading, support reaction, and displacement results for the left-end supported cantilever beam with a concentrated right-end load.

Analytically, the shear V, bending moment M, and displacement y, may be expressed as

$$V = P <x - 0>^0 + P\ell <x - 0>^{-1} - P <x - \ell>^0 \tag{11.2}$$

(See Equations 10.25 and 10.26.)

$$M = P <x - 0>^1 + P\ell <x - 0>^0 - P <x - \ell>^1 - 2P\ell \tag{11.3}$$

(See Equations 10.28 and 10.30.)
and

$$y = (P/EI)[-<x - 0>^3/6 - \ell <x - 0>^2/2 + <x - \ell>^3/6 + \ell x^2] \tag{11.4}$$

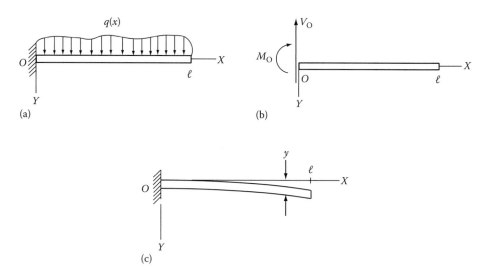

FIGURE 11.1 Positive loading, reaction, and displacement directions for a left-end supported cantilever beam. (a) Positive loading direction (Y-direction), (b) positive direction for left-end support reactions, and (c) positive transverse displacement direction (Y-direction).

with

$$y_{\max} = P\ell^3/3EI \quad \text{at} \quad x = \ell \qquad (11.5)$$

Figure 11.3a through c provides graphical representations of these shear, bending moment, and displacement results. (Note that the positive ordinate direction is downward.)

11.2.3 Cantilever Beam, Left-End Support, and Uniform Load

Figure 11.4a through c shows loading, support reactions, and displacement results for the left-end supported cantilever beam with a uniformly distributed load.

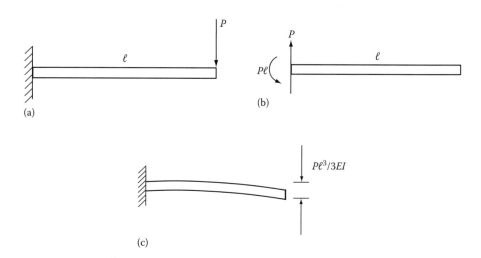

FIGURE 11.2 Concentrated end loading, support reactions, and end displacement for a left-end supported cantilever beam. (a) Concentrated end loading (beam length ℓ, load magnitude P), (b) support reactions ($V_O = P$, $M_O = -P\ell$), and (c) end displacement (elastic modulus E, second area moment I).

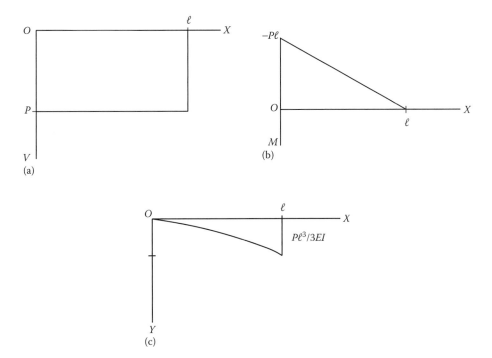

FIGURE 11.3 Shear, bending moment, and displacement of left-end supported cantilever beam with a concentrated right-end load. (a) Transverse shear, (b) bending moment, and (c) displacement.

Analytically, the shear V, bending moment M, and displacement y, may be expressed as

$$V = (w\ell^2/2) <x - 0>^{-1} + w\ell <x - 0>^0 - w <x - 0>^1 \qquad (11.6)$$

(See Equations 10.10 and 10.12.)

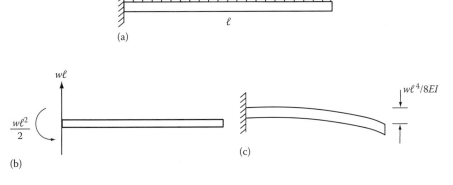

FIGURE 11.4 Uniform loading, support reaction, and end displacement for a left-end supported cantilever beam. (a) Uniform load (beam length ℓ, load intensity w per unit length), (b) support reaction ($V_O = w\ell$, $M_O = -w\ell^2/2$), and (c) end displacement (elastic modulus E, second area moment I).

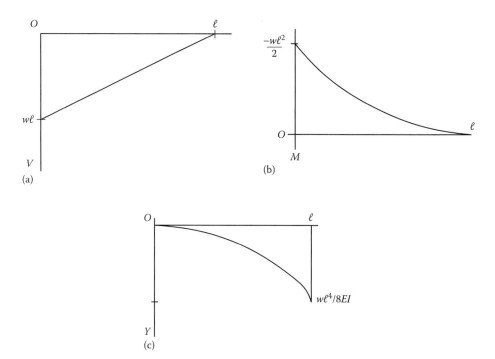

FIGURE 11.5 Shear, bending moment, and displacement of left-end supported cantilever beam with a uniformly distributed load. (a) Transverse shear, (b) bending moment, and (c) displacement.

$$M = (w\ell^2/2) <x - 0>^0 + w\ell <x - 0>^1 - w<x - 0>^2/2 - w\ell^2 \tag{11.7}$$

(See Equations 10.13 and 10.15.)
and

$$y = (w/EI)[-(\ell^2/4) <x - 0>^2 - (\ell/6) <x - 0>^3 + <x - 0>^4/24 + \ell^2 x^2/2] \tag{11.8}$$

with

$$y_{max} = w\ell^4/8EI \quad \text{at} \quad x = \ell \tag{11.9}$$

Figure 11.5a through c provides graphical representations of these results. (Note that the positive ordinate direction is downward.)

11.2.4 RIGHT-END SUPPORTED CANTILEVER BEAM

Figure 11.6a through c shows the positive directions for loading, support reactions, and displacements for a cantilever beam supported at its right end. Recall that at the built-in support, the displacement y, and the slope dy/dx, of the beam are zero.

11.2.5 CANTILEVER BEAM, RIGHT-END SUPPORT, AND CONCENTRATED END LOAD

Figure 11.7a through c shows loading, support reaction, and displacement results for the right-end supported cantilever beam with a concentrated right-end load.

Analytically, the shear V, bending moment M, and displacement y may be expressed as

$$V = -P <x - 0>^0 + P <x - \ell>^0 + P\ell <x - \ell>^{-1} \tag{11.10}$$

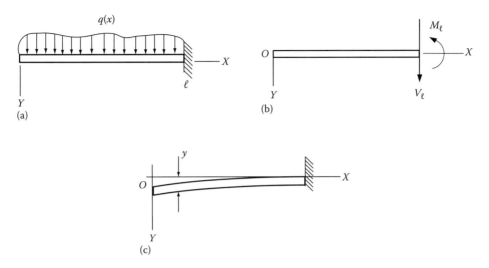

FIGURE 11.6 Positive loading, reaction, and displacement directions for a right-end supported cantilever beam. (a) Positive loading (*Y*-direction), (b) positive direction for right-end support reactions, and (c) positive transverse displacement direction (*Y*-direction).

$$M = -P <x - 0>^1 + P <x -\ell>^1 + P\ell <x -\ell>^0 \qquad (11.11)$$

$$y = (P/EI)[<x - 0>^3/6 - <x -\ell>^3/6 - P\ell <x -\ell>^2/2 - P\ell^2 x/2 + P\ell^3/3] \qquad (11.12)$$

with

$$y_{\text{max}} = P\ell^3/3EI \quad \text{at} \quad x = \ell \qquad (11.13)$$

Figure 11.8a through c provides graphical representations of these shear, bending moment, and displacement results. (Note that the positive ordinate direction is downward as before.)

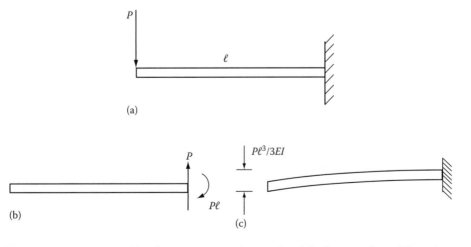

FIGURE 11.7 Concentrated end loading, support reactions, and end displacement for a right-end supported cantilever beam. (a) Concentrated end loading (beam length ℓ, load magnitude P), (b) support reaction ($V_\ell = -P$, $M_\ell = -P\ell$), and (c) end displacement (elastic modulus E, second area moment I).

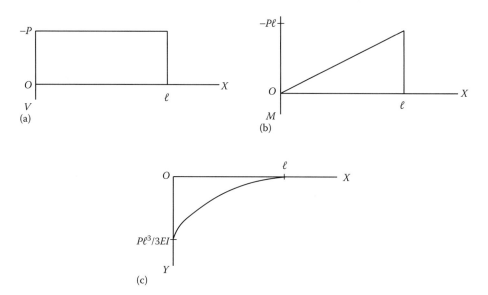

FIGURE 11.8 Shear, bending moment, and displacement of a right-end supported cantilever beam with a concentrated left-end load. (a) Transverse shear, (b) bending moment, and (c) displacement.

11.2.6 CANTILEVER BEAM, RIGHT-END SUPPORT, AND UNIFORM LOAD

Figure 11.9a through c shows loading, support reactions, and displacement results for the right-end supported cantilever beam with a uniformly distributed load.

Analytically, the shear V, bending moment M, and displacement y may be expressed as

$$V = -w <x - 0>^1 + w\ell <x - \ell>^0 + (w\ell^2/2) <x - \ell>^{-1} \tag{11.14}$$

$$M = -(w/2) <x - 0>^2 + w\ell <x - \ell>^{-1} + (w\ell^2/2) <x - \ell>^0 \tag{11.15}$$

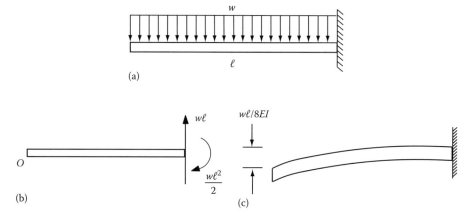

FIGURE 11.9 Uniform loading, support reactions, and end displacement for a right-end supported cantilever beam. (a) Uniform load (beam length ℓ, load intensity w per unit length), (b) support reaction ($V_\ell = -2\ell$, $M_\ell = -w\ell^2/2$), and (c) end displacement (elastic modulus E, second area moment I).

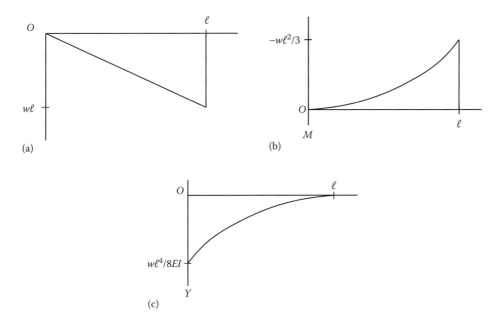

FIGURE 11.10 Shear, bending moment, and displacement of right-end supported cantilever beam with a uniformly distributed load. (a) Transverse shear, (b) bending moment, and (c) displacement.

and

$$y = (w/EI)[<x-0>^4/24 - (\ell/6)<x-\ell>^3 - (\ell^2/4)<x-\ell>^2 - \ell^3 x/6 + \ell^4/8] \qquad (11.16)$$

Figure 11.10a through c provides graphical representations of these results. (Note that the position ordinate direction is downward.)

11.3 SIMPLY SUPPORTED BEAMS

11.3.1 POSITIVE DIRECTIONS

Figure 11.11a through c shows the positive direction for loading, support reactions, and displacement for simply supported beams. Recall that at the supports the displacement y, and the moment M of beam are zero.

11.3.2 SIMPLY SUPPORTED BEAM AND CONCENTRATED CENTER LOAD

Figure 11.12a through c shows loading, support reactions, and displacement results for a simply supported beam with a centrally placed concentrated load.

Analytically, the shear V, bending moment M, and displacement y may be expressed as

$$V = (P/2)<x-0>^0 - P<x-\ell/2>^0 + (P/2)<x-\ell>^0 \qquad (11.17)$$

(See Equation 10.56.)

$$M = (P/2)<x-0>^1 - P<x-\ell/2>^1 + (P/2)<x-\ell>^1 \qquad (11.18)$$

(See Equation 10.57.)

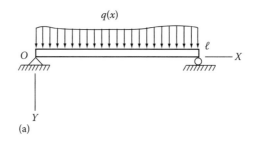

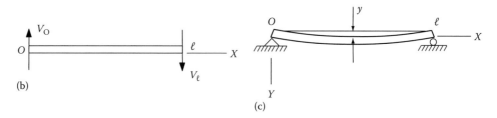

FIGURE 11.11 Positive loading, reaction, and displacement direction for a simply supported beam. (a) Positive loading direction (Y-direction), (b) positive direction for support reactions, and (c) positive transverse displacement direction (Y-direction).

and

$$y = (P/12EI)[(3/4)\ell^2 x - 2 <x - 0>^3 + <x - \ell/2>^3 - <x - \ell>^3] \tag{11.19}$$

with

$$y_{max} = P\ell^3/48EI \quad \text{at} \quad x = \ell/2 \tag{11.20}$$

(See Equation 10.59.)

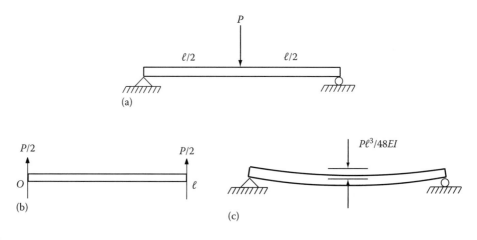

FIGURE 11.12 Loading, support reactions, and displacement of a simply supported beam with a concentrated center load. (a) Concentrated center load (beam length ℓ, load magnitude P), (b) support reactions ($V_O = P/2$, $V_\ell = -P/2$, $M_O = M_\ell = 0$), and (c) center beam displacement (elastic modulus E, second area moment I).

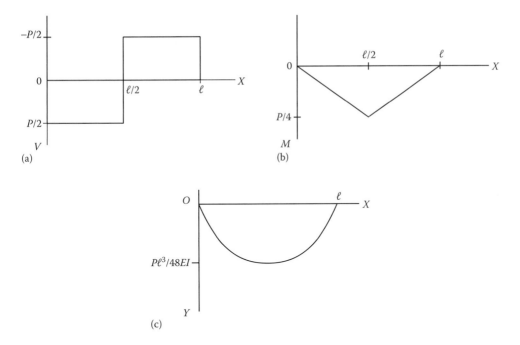

FIGURE 11.13 Shear, bending moment, and displacement of a simply supported beam with a concentrated center load. (a) Transverse shear, (b) bending moment, and (c) displacement.

Figure 11.13a through c provides graphical representations of these shear, bending moment, and displacement results. (Note that the positive ordinate direction is downward.)

11.3.3 SIMPLY SUPPORTED BEAM AND CONCENTRATED OFF-CENTER LOAD

Figure 11.14a through c shows loading, support reactions, and displacement results for a simply supported beam with an off-center concentrated load.

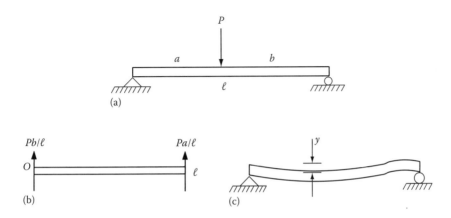

FIGURE 11.14 Loading, support reactions, and displacement of a simply supported beam with an off-center concentrated load. (a) Off-center, concentrated load (beam length ℓ, load magnitude P), (b) support reactions ($V_O = Pb/\ell$, $V_\ell = -Pa/\ell$, $M_O = M_\ell = 0$), and (c) beam displacement.

Analytically, the shear V, bending moment M, and displacement y may be expressed as

$$V = (Pb/\ell) <x - 0>^0 - P <x - a>^0 + (Pa/\ell) <x - \ell>^0 \tag{11.21}$$

$$M = (Pb/\ell) <x - 0>^{-1} - P <x - a>^1 + (Pa/\ell) <x - \ell>^1 \tag{11.22}$$

and

$$y = (P/6EI)[(b/\ell)(\ell^2 - b^2)x - (b/\ell) <x - 0>^3 + <x - a>^3 - (a/\ell) <x - \ell>^3] \tag{11.23}$$

In this case, the maximum displacement is not under the load but instead is between the load and the center of the beam. Specifically, if the load is to the left of center, that is, if $a < \ell/2$, y_{max} is [1]:

$$y_{max} = (Pa/3EI\ell)[(\ell^2 - a^2)/3]^{3/2} \quad (a < \ell/2) \tag{11.24}$$

occurring at

$$x = x_m = \ell - [(\ell^2 - a^2)/3]^{1/2} \quad (a < \ell/2) \tag{11.25}$$

Similarly, if the load is to the right of center, that is, if $a > \ell/2$, y_{max} is [2,3]:

$$y_{max} = (Pb/3EI\ell)[(\ell^2 - b^2)/3]^{3/2} \quad (a > \ell/2) \tag{11.26}$$

occurring at

$$x = x_m = [(\ell^2 - b^2)/3]^{3/2} \quad (a > \ell/2) \tag{11.27}$$

Observe that Equations 11.24 through 11.27 are also valid if $a = b = \ell/2$. (See also Equation 11.20.)

Figure 11.15a through c provides graphical representation of these shear, bending moment, and displacement results. (Note that the positive ordinate direction is downward.)

11.3.4 Simply Supported Beam and Uniform Load

Figure 11.16a through c shows loading, support reactions, and displacement results for a simply supported beam with a uniform load.

Analytically, the shear V, the bending moment M, and the displacement y may be expressed as

$$V = (w\ell/2) <x - 0>^0 - w <x - 0>^1 + (w\ell/2) <x - \ell>^0 \tag{11.28}$$

(See Equation 10.41.)

$$M = (w\ell/2) <x - 0>^1 - w <x - 0>^2/2 + (w\ell/2) <x - \ell>^1 \tag{11.29}$$

(See Equation 10.42.)
and

$$y = (w/24EI)[-2\ell <x - 0>^3 + <x - 0>^4 - 2\ell <x - \ell>^3 + \ell^3 x] \tag{11.30}$$

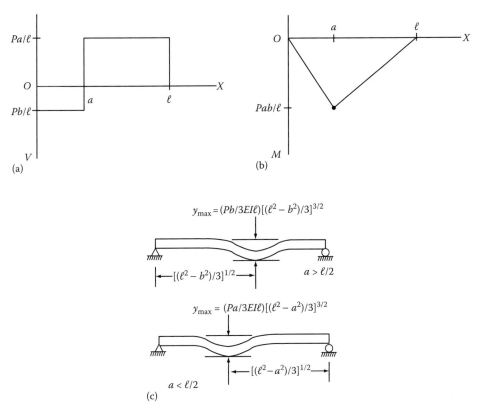

FIGURE 11.15 Transverse shear, bending moment, and displacement results for a simply supported beam with an off-center concentrated load. (a) Transverse shear, (b) bending moment, and (c) maximum displacement (elastic moment E, second area moment I).

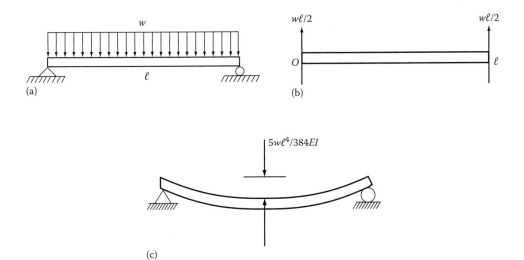

FIGURE 11.16 Loading, support reactions, and displacement of a simply supported beam with a uniform load. (a) Uniform load (beam length ℓ, load intensity w per unit length), (b) support reaction ($V_O = w\ell/2$, $V_\ell = -w\ell/2$, $M\ell = O$), and (c) center beam displacement (elastic modulus E, second area moment I).

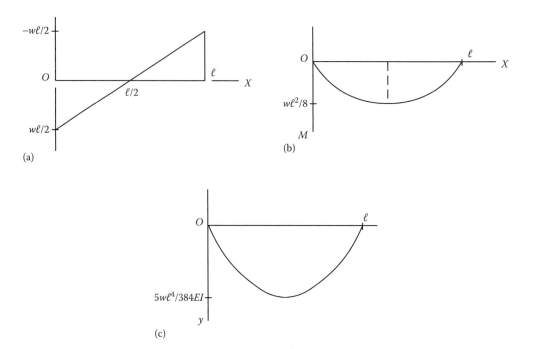

FIGURE 11.17 Transverse shear, bending moment, and displacement of a simply supported beam with a uniform load. (a) Transverse shear, (b) bending moment, and (c) displacement.

with

$$y_{\max} = 5w\ell^4/38EI \quad \text{at} \quad x = \ell/2 \tag{11.31}$$

(see Equations 10.51 and 10.52.)

Figure 11.17a through c provides graphical representations of these shear, bending moment, and displacement results. (Note that the positive ordinate direction is downward.)

11.4 DOUBLE BUILT-IN BEAMS

11.4.1 POSITIVE DIRECTIONS

Figure 11.18a through c shows the positive direction for loading, support reactions, and displacement for double built-in beams. Recall that at the supports the displacement y and the displacement slope dy/dx are zero.

11.4.2 DOUBLE BUILT-IN SUPPORTED BEAM AND CONCENTRATED CENTER LOAD

Figure 11.19a through c shows loading, support reactions, and displacement results for a doubly built-in supported beam with a centrally placed concentrated load.

Analytically, the shear V, bending moment M, and displacement y may be expressed as

$$V = (P/2) <x - 0>^0 - (P\ell/8) <x - 0>^{-1} - P <x - \ell/2>^2 + (P/2) <x - \ell>^0$$
$$+ (P\ell/8) <x - \ell>^{-1} \tag{11.32}$$

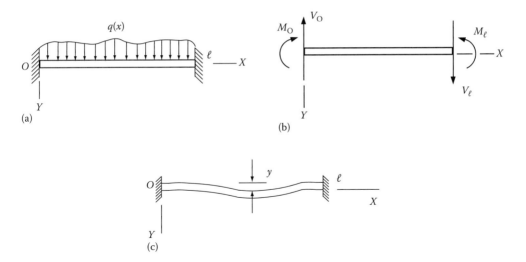

FIGURE 11.18 Positive loading, reaction, and displacement directions for double built-in beams. (a) Positive loading direction (Y-direction), (b) positive direction for support reactions, and (c) positive transverse displacement direction (Y-direction).

$$M = -(P\ell/8) <x - 0>^0 + (P/2) <x - 0>^1 - P <x - \ell/2>^1 + (P/2) <x - \ell>^1$$
$$+ (P\ell/2) <x - \ell>^0 \tag{11.33}$$

and

$$y = (P/2EI)[\ell <x - 0>^2/8 - <x - 0>^3/6 + <x - \ell/2>^3/3$$
$$- <x - \ell>^3/6 - \ell <x - \ell>^2/8] \tag{11.34}$$

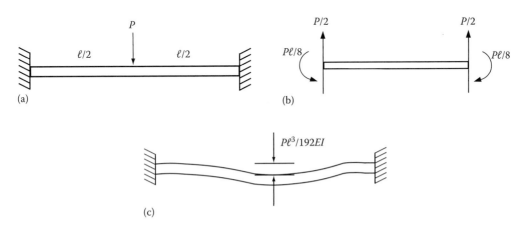

FIGURE 11.19 Loading, support reactions, and displacement of a doubly built-in beam with a concentrated center load. (a) Concentrated center load (beam length ℓ, load magnitude P), (b) support reactions ($V_O = P/2$, $V_\ell = -P/2$, $M_O = -P\ell/8$, $M_\ell = -P\ell/8$), and (c) center beam displacement (elastic modulus E, second area moment I).

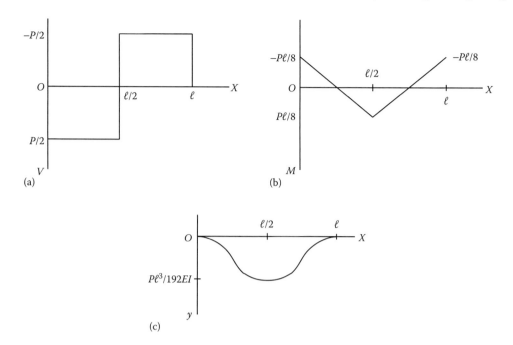

FIGURE 11.20 Shear, bending moment, and displacement of a doubly built-in beam with a concentrated center load. (a) Transverse shear, (b) bending moment, and (c) displacement.

with

$$y_{max} = P\ell^3/192EI \quad \text{at} \quad x = \ell/2 \tag{11.35}$$

Figure 11.20a through c provides positive graphical representations of these shear, bending moment, and displacement results. (Note the positive ordinate direction is downward.)

11.4.3 DOUBLE BUILT-IN SUPPORTED BEAM AND CONCENTRATED OFF-CENTER LOAD

Figure 11.21a through c shows loading, support reactions, and displacement results for a doubly built-in supported beam with an off-center concentrated load.

Analytically, the shear V, bending moment M, and displacement y may be expressed as

$$V = P(b^2/\ell^3)(3a + b) <x - 0>^0 - P(ab^2/\ell^2) <x - 0>^{-1} - P <x - a>^0$$
$$+ P(a^2/\ell^3)(a + 3b) <x - \ell>^0 + P(a^2b/\ell^2) <x - \ell>^{-1} \tag{11.36}$$

$$M = P(b^2/\ell^3)(3a + b) <x - 0>^1 - P(ab^2/\ell 2) <x - 0>^0 - P <x - a>^1$$
$$+ P(a^2/\ell^3)(a + 3b) <x - \ell>^1 + Pa^2b/\ell^2 <x - \ell>^0 \tag{11.37}$$

and

$$y = (P/EI\ell^3)[-b^2(3a + b) <x - 0>^3/6 + ab^2\ell <x - 0>^2/2 + \ell^3 <x - a>^3/6$$
$$- a^2(a + 3b) <x - \ell>^3/6 - a^2b\ell <x - \ell>^2/2] \tag{11.38}$$

In this case, the maximum displacement is not under the load but instead between the load and the center of the beam (see Figure 11.21c). Specifically, if the load is to the left of center, that is, if $a < \ell/2$, y_{max} is

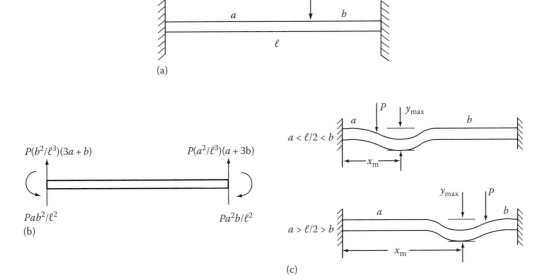

FIGURE 11.21 Loading, support reactions, and displacement of a double built-in supported beam with an off-center concentrated load. (a) Off-center, concentrated load (beam length ℓ, load magnitude P), (b) support reaction ($V_O = P(b^2/\ell^3)(3a + b)$, $M_O = -Pab^2/\ell^2$, $V_\ell = -P(a^2/\ell^3)(a + 3b)$, $M_\ell = -Pa^2b/\ell^2$), and (c) beam displacement (see Equations 11.39 through 11.42 for values of y_{max} and x_m.).

$$y_{max} = 2Pa^2b^3/3(a + 3b)^2 EI \quad (a < \ell/2) \tag{11.39}$$

occurring at

$$x = x_m = \ell - \frac{2b\ell}{a + 3b} = \ell^2/(a + 3b) \tag{11.40}$$

Similarly, if the load is to the right of center, that is, if $a > \ell/2$, y_{max} is

$$y_{max} = 2Pa^3b^2/3(3a + b)^2 EI \quad (a > \ell/2) \tag{11.41}$$

occurring at

$$x = x_m = 2a\ell/(3a + b) \tag{11.42}$$

Observe that Equations 11.39 through 11.42 are also valid if $a = b = \ell/2$. (see also Equation 11.35.)

Figure 11.22a through c provides graphical representations of these shear, bending moment, and displacement results. (Note that the positive ordinate direction is downward.)

11.4.4 Double Built-In Supported Beam and Uniform Load

Figure 11.23a through c shows loading, support reactions, and displacement results for a double built-in supported beam with a uniform load.

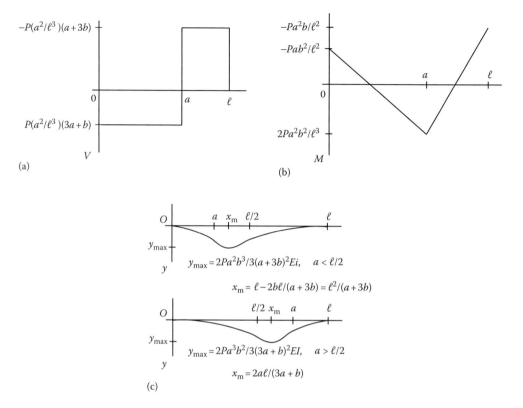

FIGURE 11.22 Shear, bending moment, and displacement results for a doubly built-in beam with an off-center concentrated load. (a) Transverse shear, (b) bending moment, and (c) displacement.

Analytically, the shear V, the bending moment M, and the displacement y may be expressed as

$$V = -w <x - 0>^1 + (w\ell/2) <x - 0>^0 - (w\ell^2/12) <x - 0>^{-1}$$
$$+ (w\ell^2/12) <x - \ell>^{-1} + (w\ell/2) <x - \ell>^0 \tag{11.43}$$

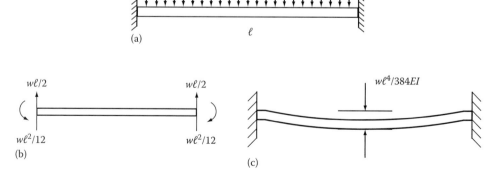

FIGURE 11.23 Loading, support reactions, and displacement of a doubly built-in beam with a uniform load. (a) Uniform load (beam length ℓ, load intensity w per unit length), (b) support reactions ($V_O = w\ell / 2$, $V_\ell = -w\ell/2$, $M_O = M_\ell = w\ell^2/2$), and (c) center beam displacement (elastic modulus E, second area moment I).

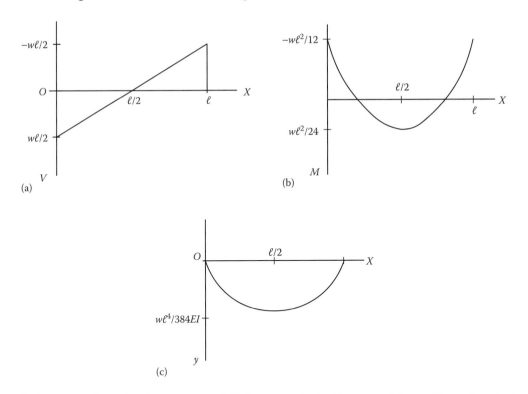

FIGURE 11.24　Shear, bending moment, and displacement of a double supported beam with a uniform load. (a) Transverse shear, (b) bending moment, and (c) displacement.

$$M = -w <x - 0>^2/2 + (w\ell/2) <x - 0>^1 - (w\ell^2/12) <x - 0>^0$$
$$+ (w\ell^2/12) <x - \ell>^0 + (w\ell/2) <x - \ell>^1 \tag{11.44}$$

$$y = (w/24EI)[<x - 0>^4 - 2\ell) <x - 0>^3 + \ell^2 <x - 0>^2$$
$$- \ell^2 <x - \ell>^2 - \ell <x - \ell>^3] \tag{11.45}$$

with

$$y_{max} = w\ell^4/38EI \quad \text{at} \quad x = \ell/2 \tag{11.46}$$

Figure 11.24a through c provides graphical representations of these shear, bending moment, and displacement results. (Note that the positive ordinate direction is downward.)

11.5 PRINCIPLE OF SUPERPOSITION

In most beam problems of practical importance, the loading is not as simple as those in the previous sections or even as those in more comprehensive lists, as in Refs. [1–3]. By using the principle of superposition, however, we can use the results listed for the simple loading cases to solve problems with much more complex loadings. The procedure is to decompose the given complex loading into simpler loadings of the kind listed above, or of those in the references. The principle of superposition then states that the shear, bending moment, and displacement for the beam with the complex loading may be obtained by simply combining (that is, "superposing") the respective results of the simpler cases making up the complex loading.

The principle is a direct result of the linearity of the governing differential equation. To observe this, suppose a loading function $q(x)$ is expressed as

$$q(x) = q_1(x) + q_2(x) \tag{11.47}$$

The governing differential equation is then

$$EId^4y/dx^4 = q(x) = q_1(x) + q_2(x) \tag{11.48}$$

The general solution of this equation may be expressed as [4]:

$$y = y_h + y_p \tag{11.49}$$

where y_h is the solution of the homogeneous equation

$$d^4y/dx^4 = 0 \tag{11.50}$$

and y_p is any ("particular") solution of Equation 11.48. Suppose that y_{p1} and y_{p2} are solutions of the equations

$$EId^4y/dx^4 = q_1(x) \quad \text{and} \quad EId^4y/dx^4 = q_2(x) \tag{11.51}$$

That is,

$$EId^4y_{p1}/dx^4 = q_1(x) \quad \text{and} \quad EId^4y_{p2}/dx^4 = q_2(x) \tag{11.52}$$

Then by adding the respective sides of Equation 11.52, we have

$$EId^4y_{p1}/dx^4 + EId^4y_{p2}/dx^4 = q_1(x) + q_2(x) \tag{11.53}$$

or

$$EId^4(y_{p1} + y_{p2})/dx^4 = q_1(x) + q_2(x) = q(x) \tag{11.54}$$

or

$$y_p = y_{p1} + y_{p2} \tag{11.55}$$

Equation 11.55 shows that the linearity of the governing equation allows individual solutions to equations with individual parts of the loading function to be added to obtain the solution to the equation with the complete loading function. This establishes the superposition principle.

To illustrate the procedure, suppose a simply supported beam has a uniform load w and a concentrated center load P as in Figure 11.25. Then by using the principle of superposition, we can

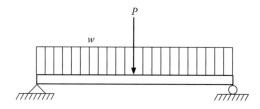

FIGURE 11.25 A simply supported beam with a uniform load w and a concentrated center load P.

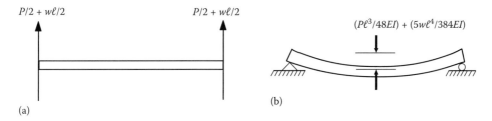

FIGURE 11.26 Support reactions and displacement of a simply supported beam with a uniform load and a concentrated center load. (a) Support reactions ($V_O = P/2 = w\ell/2$, $V_\ell = -P/2 - w\ell/2$, $M_O = M_\ell = 0$) and (b) center beam displacement (elastic modulus E, second area moment I).

combine the results of Figures 11.16 and 11.17 to obtain representations of the support reactions and displacement. Figure 11.26 shows the results.

Similarly, by combining Equations 11.17 through 11.20 with Equations 11.28 through 11.31, respectively, we obtain analytical representations of the shear (V), bending moment (M), and displacement (y) results. That is,

$$V = [(P/2) + (w\ell/2)] <x-0>^0 - P <x - \ell/2>^0 - w <x-0>^1$$
$$+ [(P/2) + (w\ell/2)]<x - \ell>^0 \tag{11.56}$$

$$M = [(P/2) + (w\ell/2)] <x-0>^1 - P <x - \ell/2>^1 - w <x - 0>^2/2$$
$$+ [(P/2) + (w\ell/2)] <x - \ell>^1 \tag{11.57}$$

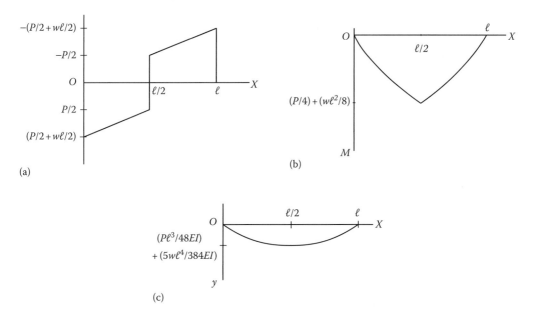

FIGURE 11.27 Shear, bending moment, and displacement of a simply supported beam with a uniform load and a concentrated center load. (a) Transverse shear, (b) bending moment, and (c) displacement.

and

$$y = (1/12EI)[P(3/4)\ell^2 x - 2P <x-0>^3 + P <x-\ell/2>^3 - P <x-\ell>^3$$
$$- w\ell <x-0>^3 + w <x-0>^4/2 - w\ell <x-\ell>^3 + w\ell^3 x/2]$$
(11.58)

with

$$y_{max} = (P\ell^3/48EI) + (5w\ell^4/384EI)$$
(11.59)

Finally by combining Figures 11.13 and 11.17, we have a graphical representation of these results as shown in Figure 11.27a through c. (Note that, as before, the positive ordinate direction is downward.)

Finally, observe that in the superposition process the location of the position of the maximum displacement and maximum moment can shift away from the position with the elementary component loading.

11.6 SUMMARY AND FORMULAS FOR DESIGN

Tables 11.1 through 11.3 provide a concise summary of the foregoing results together with additional results for (1) cantilever, (2) simple support, and (3) double built-in support beams.*

TABLE 11.1

Cantilever Beams: Maximum Bending Moment and Maximum Displacement for Various Loading Conditions

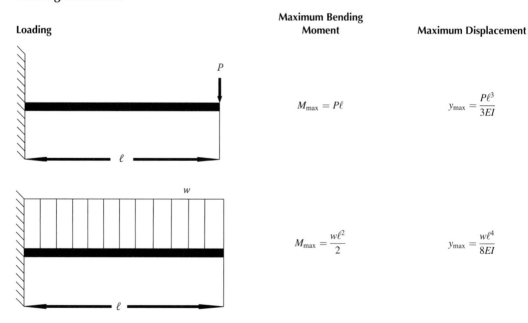

Loading	Maximum Bending Moment	Maximum Displacement
	$M_{max} = P\ell$	$y_{max} = \dfrac{P\ell^3}{3EI}$
	$M_{max} = \dfrac{w\ell^2}{2}$	$y_{max} = \dfrac{w\ell^4}{8EI}$

---

* These tables were part of Alexander Blake's second edition of *Practical Stress Analysis in Engineering Design*, Marcel Dekker, New York, 1990.

TABLE 11.1 (continued)
Cantilever Beams: Maximum Bending Moment and Maximum Displacement for Various Loading Conditions

Loading	Maximum Bending Moment	Maximum Displacement
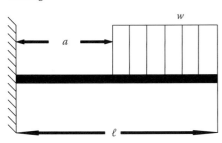	$M_{max} = \dfrac{w(\ell^2 - a^2)}{2}$	$y_{max} = \dfrac{w(\ell - a)\left[6\ell(\ell + a)^2 - 3a(a^2 + 2\ell^2)\right]}{48EI}$
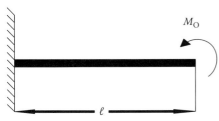	$M_{max} = M_O$	$y_{max} = -\dfrac{M_O \ell^2}{2EI}$
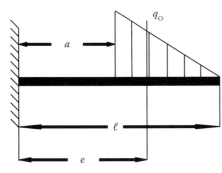	$M_{max} = \dfrac{q_O(\ell - a)}{2}$	$y_{max} = \dfrac{q_O(\ell - a)[14(\ell - a)^3 + 405\ell e^2 - 135 e^3}{1620EI}$ $e = \dfrac{2a + \ell}{3}$
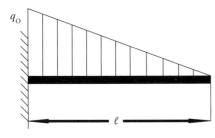	$M_{max} = \dfrac{q_O \ell^2}{6}$	$y_{max} = \dfrac{q_O \ell^4}{30EI}$
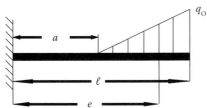	$M_{max} = \dfrac{q_O(\ell - a)}{2}$	$y_{max} = \dfrac{q_O(\ell - a)[17(\ell - a)^3 + 90 e^2(7\ell - a)]}{3240EI}$ $e = \dfrac{a + 2\ell}{3}$

(continued)

TABLE 11.1 (continued)

Cantilever Beams: Maximum Bending Moment and Maximum Displacement for Various Loading Conditions

Loading	Maximum Bending Moment	Maximum Displacement
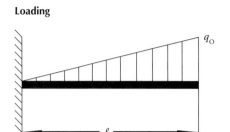	$M_{max} = \dfrac{q_0 \ell^2}{3}$	$y_{max} = \dfrac{11 q_0 \ell^4}{120 EI}$

TABLE 11.2

Simply Supported Beams: Maximum Bending Moment and Maximum Displacement for Various Loading Conditions

Loading	Maximum Bending Moment	Maximum Displacement
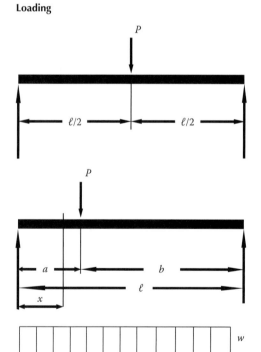	$M_{max} = \dfrac{P\ell}{4}$	$y_{max} = \dfrac{P\ell^3}{48EI}$
	$M_{max} = \dfrac{Pab}{\ell}$	$y_{max} = \dfrac{Pab(a+2b)[3a(a+2b)]^{1/2}}{27EI\ell}$ at $x = 0.58(a^2 + 2ab)^{1/2}$ when $a > b$
	$M_{max} = \dfrac{w\ell^2}{8}$	$y_{max} = \dfrac{5w\ell^4}{384EI}$

TABLE 11.2 (continued)
Simply Supported Beams: Maximum Bending Moment and Maximum Displacement for Various Loading Conditions

Loading	Maximum Bending Moment	Maximum Displacement
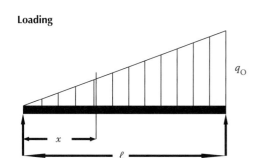	$M_{max} = 0.064w\ell^2$ at $x = 0.58\ell$	$y_{max} = \dfrac{0.0065q_O\ell^4}{EI}$ $x = 0.52\ell$
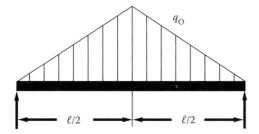	$M_{max} = \dfrac{q_O\ell^2}{12}$	$y_{max} = \dfrac{q_O\ell^4}{120EI}$
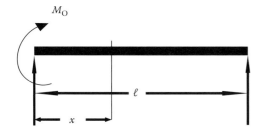	$M_{max} = \dfrac{q_O\ell^2}{24}$	$y_{max} = \dfrac{3q_O\ell^4}{640EI}$
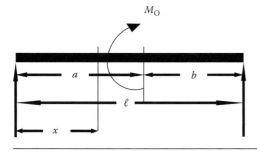	$M = M_O$	$y_{max} = \dfrac{0.064M_O\ell^2}{EI}$ at $x = 0.42\ell$
	$0 < x < a$ $M_{max} = -\dfrac{M_O x}{\ell}$ $a < x < \ell$ $M = M_O - \dfrac{M_O x}{\ell}$	$y_{max} = \dfrac{M_O x(3b^2 - \ell^2 + x^2)}{6EI\ell}$ $y_{max} = -\dfrac{M_O(\ell - x)(3a^2 - 2\ell x^2 + x^2)}{6EI\ell}$

TABLE 11.3

Doubly Built-In Beams: Maximum Pending Moment and Maximum Displacement for Various Loading Conditions

Loading	Maximum Bending Moment	Maximum Displacement
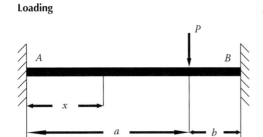	For $a < b$ $$M_A = \frac{Pab^2}{\ell^2}$$ For $a > b$ $$M_B = \frac{Pa^2b}{\ell^2}$$	For $a < b$ and $x = \dfrac{2\ell b}{\ell + 2b}$ $$y_{max} = \frac{2pa^2b^2}{3EI(\ell + 2b)^2}$$ For $a > b$ and $x = \dfrac{2\ell}{\ell + 2a}$ $$y_{max} = \frac{2pa^3b^2}{3EI(\ell + 2a)^2}$$
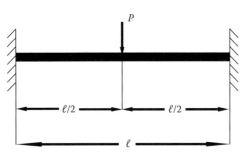	$$M_{max} = \frac{PL}{8}$$	$$y_{max} = \frac{P\ell^3}{192EI}$$
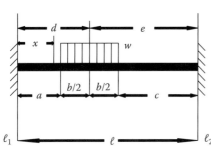	$$R_1 = \left(\frac{wb}{4\ell^3}\right) \times$$ $$[4e^2(\ell + 2d) - b^2(c - a)]$$ $$M_1 = \left(\frac{wb}{24\ell^2}\right) \times$$ $$\{b^2(\ell + 3(c - a)] - 24e^2d\}$$	For $0 < x < a$ $$y_{max} = \frac{3M_1x^2 + R_1x^3}{6EI}$$ For $a < x < (a + b)$ $$y_{max} =$$ $$\frac{12M_1x^2 + 4R_1x^3 - q(x - a)^4}{24EI}$$
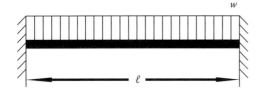	$$M = \frac{w\ell^2}{12}$$	$$y_{max} = \frac{w\ell^4}{384EI}$$

TABLE 11.3 (continued)
Doubly Built-In Beams: Maximum Pending Moment and Maximum Displacement for Various Loading Conditions

Loading	Maximum Bending Moment	Maximum Displacement
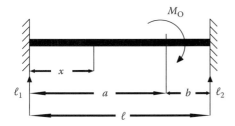	$R_1 = -\dfrac{6M_O ab}{\ell^3}$ $M_1 = -\dfrac{M_O b(\ell - 3a)}{\ell^2}$ $M_2 = -\dfrac{M_O a(2\ell - 3a)}{\ell^2}$ $R_2 = -R_1$	For $x = -\dfrac{2M_1}{R_1}$ and $a > \dfrac{\ell}{3}$ $y_{max} = -\dfrac{2M_1^3}{3EIR_1^2}$ For $x = \ell - \dfrac{2M_2}{R_2}$ and $a < \dfrac{\ell}{3}$ $y_{max} = -\dfrac{2M_2^3}{3EIR_2^2}$
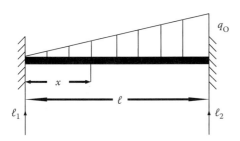	$R_1 = \dfrac{3q_O \ell}{20}$ $R_2 = \dfrac{7q_O \ell}{20}$ $M_B = \dfrac{q_O \ell^2}{20}$ (maximum)	$y_{max} = \dfrac{q_O \ell^4}{764EI}$ (maximum) at $x = 0.525\ell$

SYMBOLS

a, b	Length coordinates
E	Elastic modulus
I	Second moment of area
ℓ	Beam length
M	Bending moment
M_ℓ	Moment at end ℓ
M_O	Moment at end O
O	Coordinate system origin
P	Concentrated load
$q(x)$	Loading
V	Shear force
V_O	Shear force at end O
W	Uniform loading
X, Y, Z	Cartesian (rectangular) coordinate axes
X	X-axis coordinate
$<x - a>$	Singularity function (see Section 10.2)
Y	Y-axis coordinate; displacement
Y_ℓ	Shear force at end ℓ

REFERENCES

1. W. C. Young and R. C. Budynas, *Roark's Formulas for Stress and Strain*, 7th ed., McGraw-Hill, New York, 2002, pp. 189–201.
2. W. D. Pilkey, *Formulas for Stress, Strain, and Structural Matrices*, Wiley, New York, 1994, pp. 520–529.
3. T. Baumeister (Ed.), *Mark"s Standard Handbook for Mechanical Engineers*, McGraw-Hill, New York, 1978, pp. 5-24–5-27.
4. E. D. Rainville and P. E. Bedient, *Elementary Differential Equations*, 5th ed., Macmillan, London, 1974, p. 89.

12 Torsion and Twisting of Rods

12.1 INTRODUCTION

Beams and rods often fail in shear due to excessive twisting or torsion. This frequently occurs with overly tightened bolt/nut systems, with torsion bars, drive shafts, and beams with off-center loads. In this chapter, we review the fundamentals of the torsion of rods with the corresponding stress and displacement analyses. Since most rods subjected to torsion have circular cross sections, we begin our discussion with round bars. After that, we briefly consider bars with noncircular cross sections and hollow tubes.

12.2 BASIC ASSUMPTIONS IN THE TWISTING OF RODS OR ROUND BARS

When a rod (round bar) is subjected to a twisting moment, every cross section is in a state of pure shear. The shear stresses across the cross section then produce a resultant moment over the cross section, which is equal to the applied twisting moment but oppositely directed.

To study this behavior in greater detail, we make the following basic assumptions:

1. The rod material is homogeneous and isotropic
2. Hooke's law is applicable so that the shear stress at a point is proportional to the shear strain at that point
3. Plane circular cross sections remain plane during twisting (for round bars)
4. Radial lines of the cross section remain straight and radial during twisting (for round bars)

The last assumption has special meaning because it implies that the stresses and strains at a point are directly proportional to the radial coordinates of the point. Hence, the maximum shear stress occurs at the perimeter of the cross section. (This behavior is thus different from the case of transverse shear of the beams due to bending, where the maximum shear stress is found at the neutral axis as discussed in Chapter 13.)

12.3 STRESSES, STRAINS, AND DEFORMATION (TWISTING) OF ROUND BARS

Assumptions 3 and 4 (cross sections remain plane and radial lines remain straight during twisting) form the basis for our stress and strain analysis of twisting circular bars. The rationale for these assumptions stems from the circular symmetry of the cross section (see Refs. [1–6]). These assumptions have also been validated experimentally.

To develop the analysis, consider a circular cross sectional rod R subjected to a twisting moment T as in Figure 12.1. Consider a segment AB of R as in Figure 12.2. Let the length of AB be ℓ.

During the twisting of segment AB, let the cross section at A be regarded as fixed. Then consider the rotation of the cross section at B relative to cross section at A. On the surface between A and B, let PQ be a longitudinal line which is initially parallel to the rod axis as represented in Figure 12.3. Next, suppose that section B is rotated relative to A through an angle θ (due to the twisting moment) as represented in Figure 12.4. During the twisting let Q be rotated to Q' and let γ be the angle

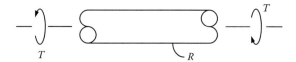

FIGURE 12.1   A circular cross-sectional rod subjected to a twisting moment.

between PQ and PQ', measured at P, as shown. If the twist angle θ is small and consequently γ is also small, we can identify γ with the shear strain on the rod surface. Thus if the segment AB has length ℓ and radius r, we see from Figure 12.4 that θ and γ are related by the simple expressions

$$r\theta = \ell\gamma \quad \text{or} \quad \gamma = (r/\ell)\theta \tag{12.1}$$

Next, imagine an interior cylindrical segment of AB having radius ρ as in Figure 12.5. By similar reasoning we see that the shear strain γ at an interior point $\hat{P}$ at end A is

$$\gamma = (\rho/\ell)\theta \tag{12.2}$$

The shear stress τ at $\hat{P}$ is then (see Equation 3.12):

$$\tau = G\gamma = G\rho\theta/\ell \tag{12.3}$$

where G is the shear modulus.

Observe that Equation 12.3 shows that the shear stress on a cross section varies linearly along a radial line. It is zero on the axis and it reaches its maximum value on the perimeter. The same remarks hold for the shear strain.

Observe further that the radial and axial (normal) stresses are zero. Finally, observe that these characteristics of the stress distribution are direct consequences of the assumptions of Section 12.2.

We can use Equation 12.3 to obtain an expression for the twisting torque T in terms of the twist angle θ. Specifically, equilibrium requires that the twisting torque must be equal to the sum of the moments of the shear stresses on the cross section about the axis of the cross section. To develop this, consider the shear stress on a small element (e) of the cross section as represented in Figure 12.6, where ρ and ϕ are the polar coordinates of (e). The twisting torque T is then

$$T = \int_0^{2\pi}\!\!\int_0^r \rho\tau\rho \, d\rho \, d\phi = \int_0^{2\pi}\!\!\int_0^r \rho(G\rho\theta/\ell)\rho \, d\rho \, d\phi$$

$$= (G\theta/\ell)\int_0^{2\pi}\!\!\int_0^r \rho^3 \, d\rho \, d\phi = (G\theta/\ell)(\pi r^4/2) \tag{12.4}$$

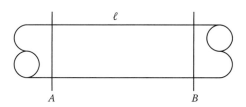

FIGURE 12.2   A segment AB of the twisted rod in Figure 12.1.

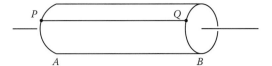

FIGURE 12.3 Rod segment AB with longitudinal line PQ.

FIGURE 12.4 Twisting of segment AB through an angle θ at B.

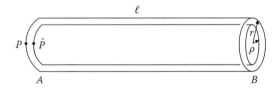

FIGURE 12.5 An interior cylindrical segment of AB.

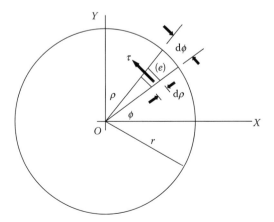

FIGURE 12.6 Shear stress (τ) and element (e) of the cross section of a twisted cylindrical segment.

or

$$T = (G\theta/\ell)J \qquad (12.5)$$

where J, defined as $\pi r^4/2$, is the second polar moment of area (or "polar moment of inertia") of the cross section.

Equation 12.5 may be solved for the twist angle θ as

$$\theta = T\ell/JG \qquad (12.6)$$

Finally, by substituting for θ in Equation 12.3, we have

$$\tau = T\rho/J \qquad (12.7)$$

Equations 12.6 and 12.7 are the fundamental relations for the stress and twist angle for twisted circular rods. Note the similarity between Equations 12.6 and 3.4 and between Equations 12.7 and 8.2.

12.4 TORSION OF NONCIRCULAR CROSS-SECTIONAL BARS

If a twisted bar has a noncircular cross section, the axial symmetry is lost and consequently the simplifying assumptions of Section 12.2 are no longer valid. Indeed, if a twisted bar has a noncircular cross section, as in Figure 12.7, it is unreasonable to expect that before twisting plane cross sections normal to the axis will remain plane during and after twisting. Instead, in the absence of applied axial forces, the asymmetry produces warping of the cross section as the bar is twisted. Correspondingly, radial lines in the cross section, stemming from the axis, no longer remain straight during twisting. The warped cross section and the curving of the radial lines make the stress and deformation analysis significantly more difficult, although it is still mathematically tractable. It simply involves solving a second-order linear partial differential equation.

From a design perspective, however, bars with circular cross sections are suitable for the vast majority of torsion applications. Thus the rather complex analysis of twisted bars with noncircular cross sections is relatively of minor importance from a design perspective. Therefore, we will omit it here, but interested readers may see Ref. [7] and [8] for details of the analysis.

Nevertheless, there are occasions when twisting of a noncircular cross-sectional bar may be of interest. In these cases, the most common cross section shapes are either square, rectangular, or composite combinations of rectangular shapes.

For rectangular cross sections, the absence of symmetry (as opposed to that of circular cross sections) means that the perimeter will be distorted during twisting. This in turn produces warping of the cross section, and as noted earlier, radial lines going outward from the bar axis do not remain straight. Consequently, simple expressions for the stress distribution across the cross section (such as Equation 12.7) do not hold for rectangular cross sections.

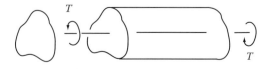

FIGURE 12.7 A twisted bar with a noncircular cross section.

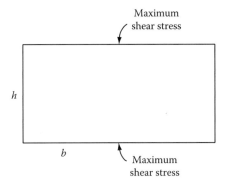

FIGURE 12.8   Points of maximum shear stress in the cross section of a twisted rectangular bar.

From a practical stress analysis and design perspective what is of greatest interest is the magnitude and location of the maximum stress, and the angle of twist of the bar. The theoretical analyses show that the maximum stress occurs at the perimeter or boundary of the cross section at those points which are closest to the bar axis [1,7,8]. Thus, for a rectangular cross-sectional bar the maximum stress (shear stress) occurs at the midpoint of the long side (see Figure 12.8). Interestingly, the corners of the rectangular cross section are found to be without stress.

For rectangular bars the magnitudes of the maximum shear stress τ_{max} and the twist angle θ can conveniently be approximated from the expressions

$$\theta = T\ell/GK \quad \text{and} \quad \tau_{max} = T/K_s \tag{12.8}$$

where the factors K and K_s are listed in Table 12.1, and b and h are the dimensions of the rectangular cross section as in Figure 12.8.

Observe the similarity of Equations 12.8 for a rectangular cross section, to Equations 12.6 and 12.7 for a circular cross section. In this regard, K and K_s of Table 12.1 may be interpreted as "section moduli."

12.5   ILLUSTRATION: TWISTING OF A RECTANGULAR STEEL BAR

Consider a rectangular steel bar B with length 16 in. and cross section dimensions 4.2 in. and 1.4 in. Let B be subjected to a twisting moment of 4000 in. lb. Let the shear modulus G be 11.6×10^6 psi. Suppose we want to determine the angle of twist θ and the maximum shear stress τ_{max}. To do this, in Table 12.1 let $b = 4.2$ in. and $h = 1.4$ in. For $b/h = 3$ K and K_s are given by

$$K/bh^3 = 0.263 \quad \text{and} \quad K_s/bh^2 = 0.267 \tag{12.9}$$

TABLE 12.1
Torsional Parameters for Rectangular Sections

b/h	1.0	1.2	1.5	2.0	2.5	3.0	4.0	5.0	10
K/bh^3	0.141	0.166	0.196	0.229	0.249	0.263	0.281	0.291	0.312
K_s/bh^2	0.208	0.219	0.231	0.246	0.258	0.267	0.282	0.291	0.312

Thus K and K_s are

$$K = 3.031 \quad \text{and} \quad K_s = 2.198 \tag{12.10}$$

Then from Equation 12.8, we have

$$\theta = \frac{(4000)(16)}{(11.6)(10^6)(3.031)} = 0.00182 \text{ rad} = 0.104° \tag{12.11}$$

and

$$\tau_{\max} = 4000/2.198 = 1820 \text{ psi} \tag{12.12}$$

12.6 TORSION OF NONCIRCULAR, NONRECTANGULAR BARS

By solving the governing partial differential equation for the torsion of prismatic bars, it is possible to obtain data for stress distribution and as well as moment/twist relations for a wide variety of cross section shapes [7,8]. References [5] and [9] provide a summary of some of these results.

The solution of the partial differential equation involves the evaluation and the approximation of infinite series or alternative numerical procedures. It happens that there are other approximate procedures which are simpler and perhaps more intuitive but which can provide quite accurate results for torsion problems, particularly for noncircular and nonrectangular cross sections.

Among the most popular of these alternative approximation procedures is the "soap-film" or "membrane" analogy. In this procedure, a tube or a duct is formed whose cross section has the same shape as that of a given bar. The end of the duct is then covered with an elastic membrane. Finally, the interior of the duct is pressurized, causing an outward bulging or deformation of the membrane. Figure 12.9 illustrates the concept.

It happens that the slope of the deformed membrane satisfies the same partial differential equation as that of the shear stress in the cross section of the twisted noncircular bar. That is, the shear stress at any point of the bar cross section is proportional to the slope of the inflated membrane. The direction of the shear stress is perpendicular to the direction for measuring the membrane slope. Finally, the angle of twist of the bar is proportional to the volume created by the deformed membrane.

If we can visualize an inflated membrane covering a duct with a cross section of interest, then by focusing upon the slope of the membrane, we have a qualitative impression of the shear stress across the cross section of the analogous twisted bar.

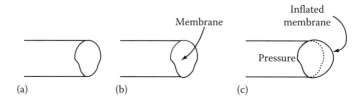

FIGURE 12.9 Illustration of deformed membrane covering a pressurized duct (having a cross section same as that of a given noncircular bar). (a) Cross section of a noncircular bar, (b) duct with the same cross section with a membrane cover, and (c) pressurized duct and outward deformed membrane.

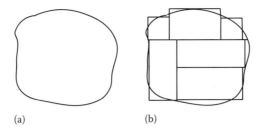

FIGURE 12.10 Approximation of a cross-section shape by rectangular elements. (a) Given cross section and (b) composite rectangular approximation.

Another useful procedure for torsional stress analysis of prismatic bars, with noncircular and nonrectangular cross sections, is the "superposition method." In this procedure, we simply approximate a given cross section shape by a combination of rectangles, as illustrated in Figure 12.10. The torsional parameters K and K_s for the composite section, and hence also for the given original cross section, are approximately equal to the sums of the K and K_s values, respectively, of the individual rectangles as obtained from Table 12.1.

The accuracy of the superposition method depends upon the "goodness-of-fit" of the rectangles approximating the given original cross section. The approximation is improved by increasing the number of rectangles but then the computational effort is also increased.

12.7 TORSION OF THIN-WALLED DUCTS, TUBES, AND CHANNELS

Consider again a circular bar or a rod. This time, let the rod be hollow, having an annular cross section as in Figure 12.11. Let the inner and outer radii be r_i and r_o. Then the polar moment of inertia J is

$$J = (\pi/2)(r_o^4 - r_i^4) \tag{12.13}$$

From Equations 12.6 and 12.7 the shear stress τ and the twist angle θ are

$$\tau = Tr/J \quad \text{and} \quad \theta = T\ell/JG \tag{12.14}$$

where
  T is the applied twisting moment
  ℓ is the rod length
  G is the shear modulus

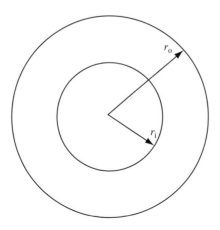

FIGURE 12.11 Circular rod with an annular cross section.

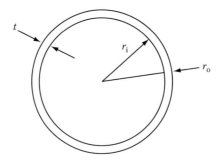

FIGURE 12.12   A thin-walled annular cross section.

Suppose that the annular wall is thin with thickness t as in Figure 12.12. The inner and outer radii are then related by the simple expression

$$r_o = r_i + t \tag{12.15}$$

Then J becomes

$$
\begin{aligned}
J &= (\pi/2)(r_o^4 - r_i^4) \\
&= (\pi/2)\left[(r_i + t)^4 - r_i^4\right] \\
&= (\pi/2)(4r_i^3 t + 6r_i^2 t^2 + 4r_i t^3 + t^4)
\end{aligned} \tag{12.16}
$$

If t is small compared with r_i, J is approximately

$$J \approx 2\pi r_i^3 t \tag{12.17}$$

Consequently the shear stress τ and the twist angle θ are then approximately

$$\tau \approx T/2\pi r_i^2 t \quad \text{and} \quad \theta \approx T\ell/2\pi r_i^3 t G \tag{12.18}$$

or

$$\tau \approx T/2At \quad \text{and} \quad \theta \approx T\ell/2A r_i t G \tag{12.19}$$

where A is the cross section area of the thin-walled cylinder or "tube."

We can generalize these results for application with thin-walled tubes with noncircular cross section. Consider such a tube, with wall thickness t, subjected to a twisting moment T, as represented in Figure 12.13. Consider an element (e) of the tube as in Figure 12.14 and as shown in enlarged views in Figure 12.15.

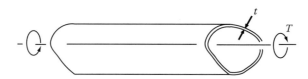

FIGURE 12.13   A noncircular cross-sectional tube subjected to twisting (torsion).

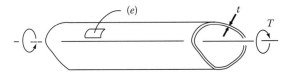

FIGURE 12.14   Element (e) of thin-walled twisted tube.

Finally, consider a free-body diagram of (e) as in Figure 12.16 where q, called the "shear flow," is the sum of the shear stress across the tube thickness. That is,

$$q = \int \tau \, dt \tag{12.20}$$

By considering the equilibrium of the element, we see that the shear flow on the upper edge of the element q_u is equal to the shear flow on the lower edge of the element q_ℓ. That is,

$$q_\mathrm{u} = q_\ell = q \tag{12.21}$$

This result means that the shear flow is constant around the perimeter of the tube.

We can relate the twist moment T to the shear flow around the perimeter by adding the moments of the shear flow q on differential perimeter elements, about the centroid O of the tube cross section, as represented in Figure 12.17. Specifically, T is given by

$$T = v \int |\mathbf{r} \times q \, \mathbf{ds}| = q \int |\mathbf{r} \times \mathbf{ds}| \tag{12.22}$$

where
   ds is a differential length vector
   r is the position vector from the centroid O to the differential length element
   the integral is a line integral along the perimeter of the tube

Observe that the vector product magnitude $|\mathbf{r} \times \mathbf{ds}|$ may be expressed as

$$|\mathbf{r} \times \mathbf{ds}| = |\mathbf{r}||\mathbf{ds}||\sin \phi| \tag{12.23}$$

where ϕ is the angle between **r** and **ds** as represented in Figure 12.18. Observe further that $|\mathbf{ds}|\sin \phi$ may be visualized as the base of a very slender isosceles triangle whose height is $|\mathbf{r}|$ as suggested in Figure 12.19. The area dA of the triangle is then simply

$$dA = (1/2)|\mathbf{r}||\mathbf{ds}| \sin \phi \tag{12.24}$$

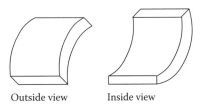

Outside view   Inside view

FIGURE 12.15   Views of tube element.

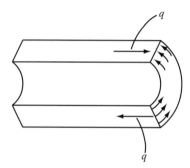

FIGURE 12.16 Free body diagram of tube element.

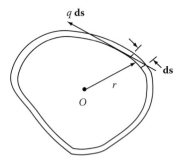

FIGURE 12.17 Shear flow at an element of the twisted tube. (Note that a differential length element does not have a finite length, but it is shown as finite in the figure simply to illustrate the differential shear flow force.)

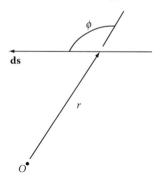

FIGURE 12.18 Angle ϕ between **r** and **ds**.

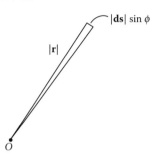

FIGURE 12.19 Differential isosceles triangle forming a differential cross-section area dA.

Then by substituting Equation 12.24 into Equations 12.23 and 12.22, we see that the twist moment T is simply

$$T = 2Aq \tag{12.25}$$

where A is the cross section area of the tube.

Finally, observe that the shear stress τ is approximately equal to q/t, where t is the tube wall thickness. Thus τ is approximately

$$\tau = T/2At \tag{12.26}$$

(See Equation 12.14.)

A review of the analysis shows that the results of Equations 12.25 and 12.26 are valid even if the tube wall thickness t is not uniform. Also, note that these equations are not valid if the tube is split (i.e., without a continuous perimeter).

SYMBOLS

A	Cross-section leader
A, B	Cross sections of rod R
b, h	Dimensions of a rectangular cross-sectional bar
G	Shear modulus
I	Second polar moment of area
K, K_s	Torsion factors (see Table 12.1)
ℓ	Rod length
P, Q	Points on the surface of R
q	Shear flow
R	Rod
r	Radial coordinate; rod radius
s	Arc length
T	Twisting moment
t	Thickness of cylindrical shell
γ	Shear strain
θ	Twisting angle, angular coordinate
ρ	Radial coordinate
τ	Shear stress
ϕ	Angular coordinate

REFERENCES

1. F. P. Beer and E. Russell Johnston, Jr., *Mechanics of Materials*, 2nd ed., McGraw Hill, New York, 1992, pp. 114–178.
2. E. P. Popou, *Mechanics of Materials*, 2nd ed., Prentice Hall, Englewood Cliffs, NJ, 1976, pp. 57–82.
3. J. N. Cernica, *Strength of Materials*, Holt, Rinehart, and Winston, New York, 1966, pp. 102–131.
4. F. L. Singer, *Strength of Materials*, 2nd ed., Harper and Row, New York, 1962, pp. 60–77.
5. W. D. Pilkey, *Formulas for Stress, Strain, and Structural Matrices*, Wiley-Interscience, New York, 1994, pp. 593–599.

6. J. P. Den Hartog, *Advanced Strength of Materials*, McGraw Hill, New York, 1952, pp. 1–48.
7. S. Timoshenko and J. N. Goodier, *Theory of Elasticity*, 2nd ed., McGraw Hill, New York, 1951, pp. 258–313.
8. J. S. Sokolnikoff, *Mathematical Theory of Elasticity*, McGraw Hill, New York, 1956, pp. 107–134.
9. W. C. Young and R. G. Budynas, *Roark's Formulas for Stress and Strain*, 7th ed., McGraw Hill, New York, 2002, pp. 381–426.

Part III

Special Beam Geometries:
Thick Beams, Curved Beams,
Stability, and Shear Center

In this part, we consider deviations from the classical beam and loading conditions of Part I. Even though these deviations take us away from the simple flexural and torsional loading of straight beams, the deviations and changes are not uncommon in their occurrence and use in structural designs. Indeed, most structural members, which resemble beams are in reality, not long straight members with simple and/or fixed end supports.

To determine the effects of geometrical changes on beam stresses and deformations, we initially consider thick (or short) beams where transverse shear stress maybe important. We then look at curved beams and the effects of curvature on beam stresses (Chapter 14). We consider application with hooks and clamps.

In Chapter 15, we examine the conditions where beams and columns can buckle and fail under axial compression loading. This failure usually occurs before yield stress is reached. We consider the effects of various support conditions and the length of the members. We also briefly consider buckling resistance of plates and panels.

Finally, in Chapter 16, we consider the effects of load placement within a cross section and how changes in the line of action of a load can affect the stresses and stability of a beam.

13 Thick Beams: Shear Stress in Beams

13.1 DEVELOPMENT OF SHEAR STRESS IN A BEAM

Consider a layered medium, such as a deck of cards, in the shape of a beam as suggested in Figure 13.1. Let the layered structure be subjected to both pure bending and to transverse loading as represented in Figures 13.2 and 13.3. Let the pure bending be simulated by rigid plates attached to the ends of the beam as represented in Figure 13.4, where an exaggerated representation of the resulting beam deformation is also given. Observe that the plates create tension in the lower layers of the beam and compression in the upper layers so that adjacent layers do not slide relative to one another.

Next, consider the case of bending via transverse loading as in Figure 13.3. If there is no friction between the layers, the layers will slide relative to one another producing a deformation pattern as in Figure 13.5. If, however, there is friction between the layers or if the layers are bonded, tangential forces (shear forces) will prevent the layers from sliding relative to one another. These shear forces will in turn create shear stresses on the surfaces of the layers. We will quantify these stresses in the following paragraphs.

13.2 SHEAR LOADING ANALYSIS

Consider again the beams subjected to pure bending and to concentrated force transverse loading as in Figures 13.2 and 13.3 and as shown again in Figures 13.6 and 13.7. In case of pure bending, the bending moment M is constant along the beam length. Consequently the flexural stress σ is the same for all cross sections along the beam length. In case of the concentrated transverse load, however, the bending moment is not constant along the beam length and thus in this case the flexural stress varies from cross section to cross section.

Recall from Section 11.3.2 that for a simply supported, center-loaded beam, the bending moment M is (see Equations 10.57 and 11.18):

$$M = (P/2) <x - 0>^1 - P <x - \ell/2>^1 + (P/2) <x - \ell>^1 \qquad (13.1)$$

where, as before, the bracket notation $<\cdot>$ designates the singularity function (see Section 10.2), ℓ is the beam length, and x is the coordinate along the beam axis with the origin at the left end. This expression is best represented graphically as in Figure 13.8 (Figure 11.13b).

If the bending moment M varies along the beam length, causing a variation in flexural or axial force from cross section to cross section, then a longitudinal fiber element will not be in equilibrium in the absence of shear stress on the element. That is, with varying bending moment, shear stress is needed to maintain equilibrium of the longitudinal element. To see this, consider a longitudinal element (e) of a narrow rectangular cross-sectional beam as in Figure 13.9, where the cross section dimensions are b and h as shown, and the element dimension are Δx (length), Δy (height), and b (depth), also as shown. (As before, the positive X-axis is to the right along the neutral axis of the beam and the positive Y-axis is downward.)

Suppose that the loading on the beam produces a bending moment M, which varies along the beam span. Then in Figure 13.9, the bending moment at cross section A will be different from that at cross section B. This means that the stress on (e) at end A is different from that at end B.

FIGURE 13.1   Layered media in the shape of a beam.

FIGURE 13.2   Pure bending of a layered beam.

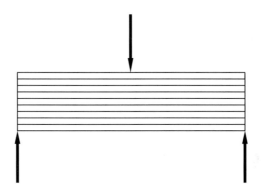

FIGURE 13.3   Transverse loading of a layered beam.

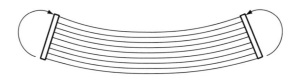

FIGURE 13.4   Simulation of pure bending of a layered beam.

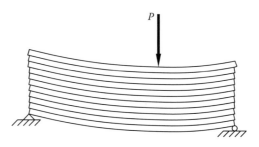

FIGURE 13.5   Simple support simulation of bending via transverse loading.

FIGURE 13.6   Pure bending of a beam.

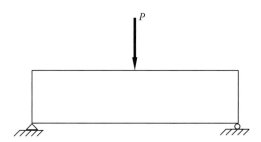

FIGURE 13.7   Bending via transverse load.

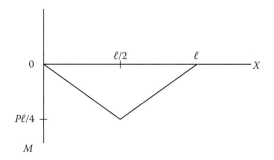

FIGURE 13.8   Bending moment for simply supported center-loaded beam.

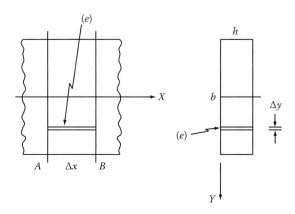

FIGURE 13.9   Longitudinal element (e) of a rectangular cross-sectional beam.

To quantify this difference, recall the fundamental flexural stress expression (see Equations 8.2 and 8.13):

$$\sigma = My/I \tag{13.2}$$

where I is the second moment of area of the cross section ($I = bh^3/12$ for a rectangular cross section). Thus the stresses at sections A and B are

$$\sigma_A = M_A y/I \quad \text{and} \quad \sigma_B = M_B y/I \tag{13.3}$$

We may relate the moments at A and B relative to each other, using a Taylor series as

$$M_B = M_A + \left.\frac{dM}{dx}\right|_A \Delta x + \cdots \tag{13.4}$$

If the element length Δx is small, the unwritten terms of Equation 13.4 can be neglected so that the moments on the beam cross sections at A and B can be represented as in Figure 13.10.

$$M_A + \left.\frac{dM}{dx}\right|_A \Delta x$$

Observe now from Equation 13.1 that with the moments differing from cross section A to B by the amount $(dM/dx)\Delta x$, the stresses at the ends of element (e) differ by $(dM/dx)(\Delta x)(y/I)$. This stress difference in turn will produce a difference in end loadings on (e) by the amount $(dM/dx)(\Delta x)(y/I)b\Delta y$. Then to maintain equilibrium of (e), the shear stresses on the upper and lower surfaces of (e) need to produce a counterbalancing shear force on (e) equal to $(dM/dx)(\Delta x)(y/I)b\Delta y$.

To further quantify the shear forces consider a free-body diagram of that portion of the beam segment between A and B and beneath (e) as in Figure 13.11 where F_A and F_B are the axial force resultants on the beam segment at ends A and B, and S is the resultant of the shearing forces due to the shear stresses on the segment at the interface with (e). Then by setting forces in the axial direction equal to zero (to maintain equilibrium), we have

$$-F_A - S + F_B = 0 \quad \text{or} \quad S = F_B - F_A \tag{13.5}$$

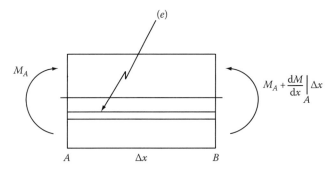

FIGURE 13.10   Moments on beam cross sections A and B.

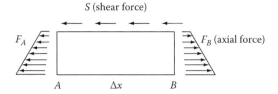

FIGURE 13.11 Free-body diagram of lower portion of beam segment.

For a rectangular cross-sectional beam with width b and depth h, F_A, F_B, and S may be expressed as

$$F_A = \int_y^{h/2} \sigma_A b \ dy = \int_y^{h/2} M_A(y/I)b \ dy \tag{13.6}$$

$$F_B = \int_y^{h/2} \sigma_B b \ dy = \int_y^{h/2} M_B b \ dy = \int_y^{h/2} \left(M_A + \frac{dM_A}{dx}\Delta x\right)(y/I)b \ dy \tag{13.7}$$

and

$$S = \tau_{xy} b \Delta x \tag{13.8}$$

By substituting these expressions into Equation 13.5, we have

$$\tau_{xy} b \Delta x = \int_y^{h/2} \left(M_A + \frac{dM_A}{dx}\Delta x\right)(y/I)b \ dy - \int_y^{h/2} M_A(y/I)b \ dy$$

$$= \int_y^{h/2} \frac{dM_A}{dx}\Delta x (y/I)b \ dy \tag{13.9}$$

$$\tau_{xy} = \int_y^{h/2} (dM_A/dx)(y/I)dy$$

From Equation 8.6 dM_A/dx is the transverse shear V at A. Hence, the shear stress (for the rectangular cross section) is

$$\tau_{xy} = (V/I) \int_y^{h/2} y \ dy = (V/I)(y^2/2) \Big|_y^{h/2} = (V/I)\left(\frac{h^2}{8} - \frac{y^2}{2}\right) \tag{13.10}$$

In general, for a nonrectangular cross section, the shear stress is

$$\tau_{xy} = (V/Ib) \int_y^{h/2} y \ dA \stackrel{D}{=} VQ/Ib \tag{13.11}$$

where by inspection Q is defined as $\int_h^{h/2} y \, dA$, where dA is the cross section area element and b is the cross section width at elevation y. Q is the moment of area of the cross section, beneath y, about the line in the cross section parallel to the Z-axis, and at elevation y.

13.3 MAXIMUM TRANSVERSE SHEAR STRESS

As an illustration of the use of Equation 13.11, consider the special case of a rectangular cross section, governed by Equation 13.10. In this case, Q is

$$Q = \int_y^{h/2} y \, dA = \int_y^{h/2} yb \, dA = by^2/2 \Big|_y^{h/2} = b\left(\frac{h^2}{8} - \frac{y^2}{2}\right) \tag{13.12}$$

By substituting into Equation 13.11 the shear stress has the form

$$\tau_{xy} = VQ/Ib = (V/I)\left(\frac{h^2}{8} - \frac{y^2}{2}\right) \tag{13.13}$$

Recalling now that for a rectangular cross section the second moment of area I is

$$I = bh^3/12 \tag{13.14}$$

Therefore, the shear stress becomes

$$\tau_{xy} = \frac{6V}{6h^3}\left(\frac{h^2}{A} - y^2\right) \tag{13.15}$$

Equation 13.15 shows that the shear stress distribution across the cross section is parabolic, having values zero on the upper and lower surfaces and maximum value at the center, as represented in Figure 13.12.

From Equation 13.15, the maximum shear stress (occurring a $y = 0$) is

$$\tau_{max} = (3/2)(V/A) \tag{13.16}$$

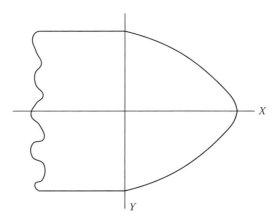

FIGURE 13.12 Shear stress distribution on a rectangular beam cross section.

Equation 13.16 shows that for a rectangular cross section, the maximum shear stress is 1.5 times as large as the average shear stress (V/A) across the cross section or 50% larger than the average shear stress. In Section 13.4, we present analogous results for commonly used nonrectangular cross section shapes.

Finally, observe that the maximum shear stress occurs at the neutral axis where the normal (flexural) stress is minimum (zero), whereas the maximum flexural stress occurs at the upper and lower surfaces where the shear stress is minimum (zero).

13.4 NONRECTANGULAR CROSS SECTIONS

Consider a beam with a circular cross section. An analysis similar to that in Section 13.3 shows that the shear stress distribution across the cross section is parabolic. Here, however, the maximum shear stress is found to be $4/3$ times the average shear stress. That is

$$\tau_{max} = (4/3)(V/A) \quad \text{(circular cross section)} \tag{13.17}$$

Other beam cross sections of interest include hollow circular cross sections, I-beam sections, and open-channel cross sections. Table 13.1 shows the shear stress distribution together with a listing of maximum values for these cross sections.

Beams with open cross sections, such as I-beams or T-type beams transmit the shear loads primarily through the webs, and the maximum stress closely approximates that obtained by dividing the shear load by the area of the web. The effect of a flange on the shear stress distribution is small and can usually be neglected. In rapidly changing cross section geometry, however, some judgment is required to determine which portions of the cross section are likely to behave as flanges and which should be treated as webs.

It should be noted that the foregoing analyses and maximum shear stress values are valid only when the shear loading is equivalent to a single force acting through the centroid of the cross section ("centroidal loading"). When this occurs, no torsional moments are created. The condition of centroidal loading is usually satisfied when the beam cross section is symmetric about the Y-axis.

Finally, consider the case of the tubular cross section (hollow cylinder beam). The result listed in Table 13.1 for the maximum shear stress is obtained by assuming that the wall thickness t is small compared with the radius R. There can be occasions, however, when an annular cross-sectional beam has a thick wall. To address this, R. C. Stephens [1] has developed an expression for the maximum shear stress as a function of the inner and outer radii r and R as

$$\tau_{max} = t_{av} \frac{4(R^2 + R^r + r^2)}{3(R^2 + r^2)} \tag{13.18}$$

where, as before, τ_{av} is the average shear stress across the cross section (V/A).

Table 13.2 provides a tabular listing of computation results using Equation 13.17 for various r/R ratios.

13.5 SIGNIFICANCE OF BEAM SHEAR STRESS

A question arises from a design perspective: how significant is the shear stress in beams? To answer this question, consider a simply supported beam with a concentrated central load as in Figure 13.13.

In this case, at the center of the beam, we have a relatively large shear load with a relatively small bending moment. Specifically, recall from Section 11.3.2 that the shear and bending moment diagrams for the beam are as in Figures 13.14 and 13.15.

TABLE 13.1

Shear Stress Distribution and Maximum Shear Stress for Various Beam Cross Sections

Type of Cross Section	Shear Stress Distribution	Maximum Shear Stress

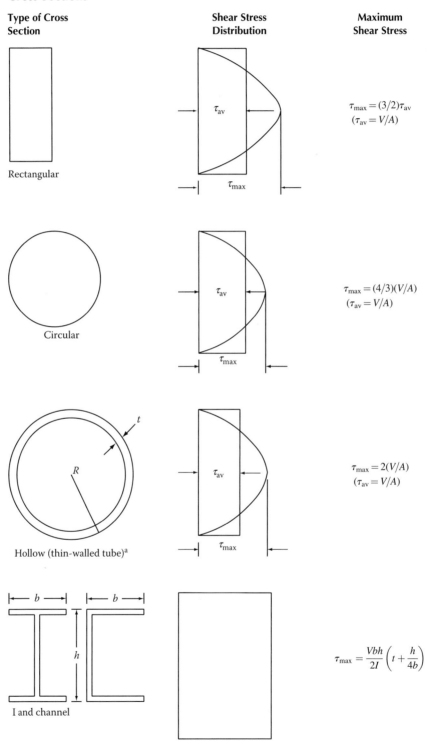

Rectangular — τ_{av}, τ_{max}

$$\tau_{max} = (3/2)\tau_{av}$$
$$(\tau_{av} = V/A)$$

Circular — τ_{av}, τ_{max}

$$\tau_{max} = (4/3)(V/A)$$
$$(\tau_{av} = V/A)$$

Hollow (thin-walled tube)[a] — τ_{av}, τ_{max}

$$\tau_{max} = 2(V/A)$$
$$(\tau_{av} = V/A)$$

I and channel

$$\tau_{max} = \frac{Vbh}{2I}\left(t + \frac{h}{4b}\right)$$

[a] See Table 13.2 for additional data for tubes.

TABLE 13.2
Maximum Shear Stresses for Tubular Cross-Section Beams
with Various Wall Thicknesses

r/R	0	0.2	0.4	0.6	0.8	1.0
τ_{max}/τ_{av}	1.333	1.590	1.793	1.922	1.984	2.000

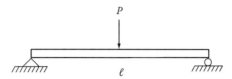

FIGURE 13.13   A simple support beam with a concentrated center load.

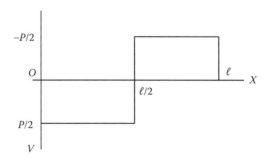

FIGURE 13.14   Shear diagram for simply supported center-loaded beam.

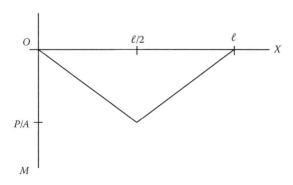

FIGURE 13.15   Bending moment diagram for simply supported center-loaded beam.

From these figures we see that at the center of the beam, the shear and bending moment are

$$V = P/2 \quad \text{and} \quad M = P\ell/4 \tag{13.19}$$

For a rectangular cross section, the maximum shear stress τ_{max} is (see Table 13.1):

$$\tau_{max} = (3/2)(V/A) = (3/2)(P/2bh) \tag{13.20}$$

where, as before, b and h are the cross section width and depth. Correspondingly, the maximum flexural stress σ_{max} is

$$\sigma_{max} = M_{max}c/I = (P\ell/4)(h/2)/bh^3/12)$$

or

$$\sigma_{max} = 3P\ell/2bh^2 \tag{13.21}$$

The ratio of maximum shear to flexural stress is then

$$\tau_{max}/\sigma_{max} = \frac{3P/4bh}{3P\ell/2bh^2} = h/2\ell \tag{13.22}$$

Equation 13.22 shows that in this relatively common loading and support configuration, for long thin beams, the shear stress is small and unimportant. For thick beams (say $h > \ell/5$), the shear stress is as large as one tenth or more of the flexural stress.

SYMBOLS

A	Area
A, B	Beam sections
b	Beam depth
F_A, F_B	Axial force resultants on sections A and B
h	Beam thickness
I	Second moment of area of the beam
ℓ	Beam length
M	Bending moment
P	Concentrated force
Q	Defined by Equation 13.11
r, R	Tube radii
S	Shear force
t	Wall thickness
V	Shear force
X	X-axis coordinate
$<x-a>^n$	Singularity functions (see Section 10.2)
X, Y, Z	Cartesian (rectangular) coordinate axes
y	Y-axis coordinate
$\Delta x, \Delta y$	Element (e) dimensions
σ	Normal stress
τ_{xy}	Shear stress

REFERENCE

1. R. C. Stephens, *Strength of Materials—Theory and Examples*, Edward Arnold, London, 1970.

14 Curved Beams

14.1 HISTORICAL PERSPECTIVE

Curved beams and other relatively thin curved members are commonly found in machines and structures such as hooks, chain links, rings, and coils. The design and analysis of curved members has interested structural engineers for over 150 years. Early developments are attributed to Winkler [1,2] in 1867. However, experimental verification of the theory did not occur until 1906 when tests were conducted on chain links, at the University of Illinois [3]. These tests were later expanded to circular rings providing good agreement with Winkler's work.

Winkler's analysis (later to be known as the Winkler-Bach formula) expressed the maximum flexural stress σ_{max} for a curved beam as

$$\sigma_{max} = \frac{M}{AR}\left[1 + \frac{c}{\lambda(R+c)}\right] \tag{14.1}$$

where
  M is the applied bending moment
  A is the beam cross section area
  R is the distance from the area centroid to the center of curvature of the unstressed beam
  c is the distance from the centroid to the inner perimeter of the beam
  λ is a geometric parameter defined as

$$\lambda = 1/A \int \frac{\eta}{R+\eta}\,dA \tag{14.2}$$

where the integration is over the cross section and η is the distance of a differential area element from the centroid C as in Figure 14.1.

Observe in Equation 14.1 that unlike the flexural stress expression for straight beams ($\sigma = Mc/I$, Equation 8.15), the stress is nonlinearly related to the distance c from the centroid axis to the outer fibers of the cross section. Also observe that the use of Equation 14.1 requires knowledge of the geometric parameter λ of Equation 14.2. Table 14.1 provides series expressions for λ for a variety of cross section shapes.

Even though Winkler's results appeared as early as 1867, English and American practices did not adopt his analysis until 1914 when Morley published a discussion about curved beam design [4], giving support to the Winkler-Bach theory. The adequacy of the theory was later (1926) supported by tests conducted by Winslow and Edmonds [5].

Although the stress represented by Equations 14.1 and 14.2 has been demonstrated to provide reasonable results, it needs to be remembered that it is nevertheless approximate. More exact and more useful expressions may be obtained by accounting for a shift in the neutral axis position, as discussed in the following section.

14.2 NEUTRAL AXIS SHIFT

Consider a segment of a curved beam, subjected to bending as in Figure 14.2. Recall that for straight beams we assumed that plane sections normal to the beam axis, prior to bending, remain plane during and after bending. Interestingly, it happens that the same assumption is reasonable for curved

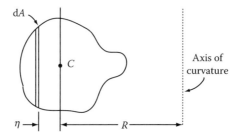

FIGURE 14.1   Cross-section geometry of a curved beam.

beams. Indeed, experiments have shown that for a reasonable range of loading and beam deformation, there is very little distortion of sections normal to the beam axis during bending.

With straight beams the preservation of planeness of the normal cross sections leads to the linear stress distribution across the cross section, that is, $\sigma = Mc/I$ (see Equation 8.13). With curved beams, however, the preservation of planeness of normal cross sections leads to a *nonlinear* stress distribution, and consequently a shaft in the neutral axis toward the center of curvature of the beam.

TABLE 14.1
Values of λ of Equation 14.2

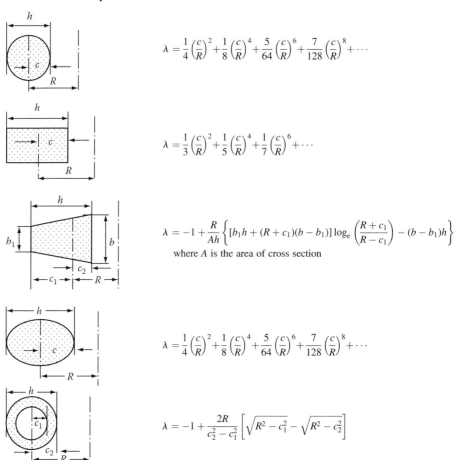

$$\lambda = \frac{1}{4}\left(\frac{c}{R}\right)^2 + \frac{1}{8}\left(\frac{c}{R}\right)^4 + \frac{5}{64}\left(\frac{c}{R}\right)^6 + \frac{7}{128}\left(\frac{c}{R}\right)^8 + \cdots$$

$$\lambda = \frac{1}{3}\left(\frac{c}{R}\right)^2 + \frac{1}{5}\left(\frac{c}{R}\right)^4 + \frac{1}{7}\left(\frac{c}{R}\right)^6 + \cdots$$

$$\lambda = -1 + \frac{R}{Ah}\left\{[b_1 h + (R + c_1)(b - b_1)]\log_e\left(\frac{R + c_1}{R - c_1}\right) - (b - b_1)h\right\}$$

where A is the area of cross section

$$\lambda = \frac{1}{4}\left(\frac{c}{R}\right)^2 + \frac{1}{8}\left(\frac{c}{R}\right)^4 + \frac{5}{64}\left(\frac{c}{R}\right)^6 + \frac{7}{128}\left(\frac{c}{R}\right)^8 + \cdots$$

$$\lambda = -1 + \frac{2R}{c_2^2 - c_1^2}\left[\sqrt{R^2 - c_1^2} - \sqrt{R^2 - c_2^2}\right]$$

TABLE 14.1 (continued)
Values of λ of Equation 14.2

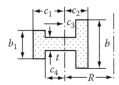

$$\lambda = -1 + \frac{R}{A}[b_1 \log_e (R + c_1) + (t - b_1) \log_e (R + c_4)$$
$$+ (b - t) \log_e (R - c_3) - b \log_e (R - c_2)]$$

where A is the area of cross section

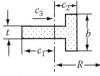

In the expression for the unequal I given above
make $c_4 = c_1$ and $b_1 = t$, so that

$$\lambda = -1 + \frac{R}{A}[t \log_e (R + c_1) + (b - t) \log_e (R - c_3) - b \log_e (R - c_2)]$$

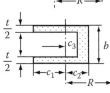

Area $= A = tc_1 - (b - t)c_3 + bc_2$ (applies to U and T sections)

To understand the reason for the nonlinear stress distribution and the neutral axis shaft, consider an enlarged view of the bending of a beam segment as in Figure 14.3. Specifically, consider the rotation of section B_iB_O and the stretching and compression of circular arc fibers PQ and $P'Q'$. If section B_iB_O is to remain plane during bending, fibers near the outside of the segment, such as PQ,

TABLE 14.2
Stress Concentration Factors for Curved Beams

	R/c	ϕ_o Inner Face	ϕ_o Outer Face	δ/R
	1.2	3.41	0.54	0.224
	1.4	2.40	0.60	0.151
	1.6	1.96	0.65	0.108
	1.8	1.75	0.68	0.084
	2.0	1.62	0.71	0.069
	3.0	1.33	0.79	0.030
	4.0	1.23	0.84	0.016
	6.0	1.14	0.89	0.0070
	8.0	1.10	0.91	0.0039
	10.0	1.08	0.93	0.0025
	1.2	2.89	0.57	0.305
	1.4	2.13	0.63	0.204
	1.6	1.79	0.67	0.149
	1.8	1.63	0.70	0.112
	2.0	1.52	0.73	0.090
	3.0	1.30	0.81	0.041
	4.0	1.20	0.85	0.021
	6.0	1.12	0.90	0.0093
	8.0	1.09	0.92	0.0052
	10.0	1.07	0.94	0.0033

(continued)

TABLE 14.2 (continued)

Stress Concentration Factors for Curved Beams

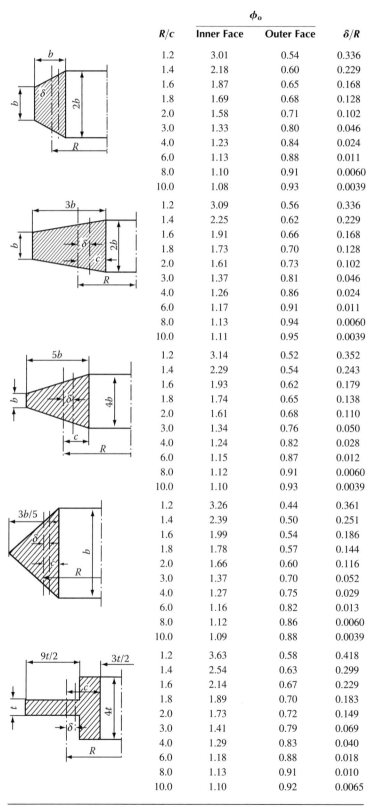

R/c	ϕ_o		δ/R
	Inner Face	Outer Face	
1.2	3.01	0.54	0.336
1.4	2.18	0.60	0.229
1.6	1.87	0.65	0.168
1.8	1.69	0.68	0.128
2.0	1.58	0.71	0.102
3.0	1.33	0.80	0.046
4.0	1.23	0.84	0.024
6.0	1.13	0.88	0.011
8.0	1.10	0.91	0.0060
10.0	1.08	0.93	0.0039
1.2	3.09	0.56	0.336
1.4	2.25	0.62	0.229
1.6	1.91	0.66	0.168
1.8	1.73	0.70	0.128
2.0	1.61	0.73	0.102
3.0	1.37	0.81	0.046
4.0	1.26	0.86	0.024
6.0	1.17	0.91	0.011
8.0	1.13	0.94	0.0060
10.0	1.11	0.95	0.0039
1.2	3.14	0.52	0.352
1.4	2.29	0.54	0.243
1.6	1.93	0.62	0.179
1.8	1.74	0.65	0.138
2.0	1.61	0.68	0.110
3.0	1.34	0.76	0.050
4.0	1.24	0.82	0.028
6.0	1.15	0.87	0.012
8.0	1.12	0.91	0.0060
10.0	1.10	0.93	0.0039
1.2	3.26	0.44	0.361
1.4	2.39	0.50	0.251
1.6	1.99	0.54	0.186
1.8	1.78	0.57	0.144
2.0	1.66	0.60	0.116
3.0	1.37	0.70	0.052
4.0	1.27	0.75	0.029
6.0	1.16	0.82	0.013
8.0	1.12	0.86	0.0060
10.0	1.09	0.88	0.0039
1.2	3.63	0.58	0.418
1.4	2.54	0.63	0.299
1.6	2.14	0.67	0.229
1.8	1.89	0.70	0.183
2.0	1.73	0.72	0.149
3.0	1.41	0.79	0.069
4.0	1.29	0.83	0.040
6.0	1.18	0.88	0.018
8.0	1.13	0.91	0.010
10.0	1.10	0.92	0.0065

TABLE 14.2 (continued)
Stress Concentration Factors for Curved Beams

		ϕ_0		
	R/c	Inner Face	Outer Face	δ/R
	1.2	3.55	0.67	0.409
	1.4	2.48	0.72	0.292
	1.6	2.07	0.76	0.224
	1.8	1.83	0.78	0.178
	2.0	1.69	0.80	0.144
	3.0	1.38	0.86	0.067
	4.0	1.26	0.89	0.038
	6.0	1.15	0.92	0.018
	8.0	1.10	0.94	0.010
	10.0	1.08	0.95	0.065
	1.2	2.52	0.67	0.408
	1.4	1.90	0.71	0.285
	1.6	1.63	0.75	0.208
	1.8	1.50	0.77	0.160
	2.0	1.41	0.79	0.127
	3.0	1.23	0.86	0.058
	4.0	1.16	0.89	0.030
	6.0	1.10	0.92	0.013
	8.0	1.07	0.94	0.0076
	10.0	1.05	0.95	0.0048
	1.2	3.28	0.58	0.269
	1.4	2.31	0.64	0.182
	1.6	1.89	0.68	0.134
	1.8	1.70	0.71	0.104
	2.0	1.57	0.73	0.083
	3.0	1.31	0.81	0.038
	4.0	1.21	0.85	0.020
	6.0	1.13	0.90	0.0087
	8.0	1.10	0.92	0.0049
	10.0	1.07	0.93	0.0031
	1.2	2.63	0.68	0.339
	1.4	2.97	0.73	0.280
	1.6	1.66	0.76	0.205
	1.8	1.51	0.78	0.159
	2.0	1.43	0.80	0.127
	3.0	1.23	0.86	0.058
	4.0	1.15	0.89	0.031
	6.0	1.09	0.92	0.014
	8.0	1.07	0.94	0.0076
	10.0	1.06	0.95	0.0048

Source: Wilson, B. J. and Quereau, J. F., A simple method of determining stress in curved flexural members, Circular 16, Engineering Experiment Station, University of Illinois, 1927.

are lengthened, whereas fibers near the inside of the segment, such as $P'Q'$, are shortened. Then at some point N in the interior of section B_iB_O, the circular arc fibers will neither be lengthened nor shortened during the bending. Indeed an entire surface of such points will occur thus defining a "neutral surface" composed of zero length-change fibers. Suppose that fibers PQ and $P'Q'$ are at

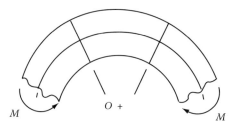

FIGURE 14.2   Bending of a beam segment.

equal distances from the neutral surface on the outer and inner sides of the surface. Then with the cross section remaining plane, the amount δ_{PQ} that PQ is lengthened is equal to the amount $\delta_{P'Q'}$ that $P'Q'$ is shortened during the segment bending, as illustrated in Figure 14.4, where the postbending orientation of section B_iB_O is superposed upon the prebending orientation of B_iB_O. Due to the curvature of the beam, the length of fiber PQ is greater than that of $P'Q'$ (see Figure 14.3). Therefore, the magnitude of the strain in PQ is less than that in $P'Q'$. That is,

$$\left|\delta_{PQ}\right| = \left|\delta_{P'Q'}\right| \quad \text{but} \quad |PQ| > |P'Q'| \tag{14.3}$$

In this notation, the strains are

$$\left|\varepsilon_{PQ}\right| = \left|\delta_{PQ}\right|/|PQ| \quad \text{and} \quad \left|\varepsilon_{P'Q'}\right| = \left|\delta_{P'Q'}\right|/|P'Q'| \tag{14.4}$$

and thus in view of Equation 14.3, we have

$$\left|\varepsilon_{PQ}\right| < \left|\varepsilon_{P'Q'}\right| \tag{14.5}$$

With the stress being proportional to the strain, we then have

$$\left|\sigma_{PQ}\right| < \left|\sigma_{P'Q'}\right| \tag{14.6}$$

Equation 14.6 demonstrates the nonlinear stress distribution along the cross section. That is, stresses at equal distances from the neutral surface do not have equal magnitudes. Instead, the stresses are larger in magnitude at those points closer to the inner portion of the beam, as represented in Figure 14.5.

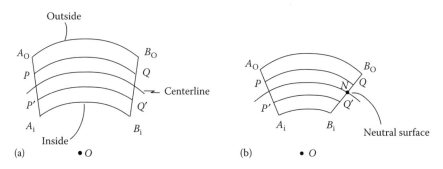

FIGURE 14.3   Deformation of a beam segment. (a) Before bending. (b) After bending.

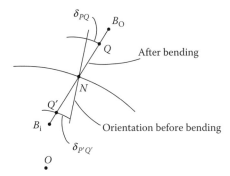

FIGURE 14.4   Rotation of section B_iB_O during bending.

The nonlinear stress distribution over the cross section causes a shift in the neutral axis toward the inner surface of the beam. To see this, observe that in pure bending the resultant normal force on the cross section is zero. Then with larger stresses near the inner surface, the neutral axis must be shifted toward the inner surface to reduce the area where these larger stresses occur so that the resultant normal force on the cross section will be zero.

We can quantify the neutral axis shift by evaluating the normal cross section forces and then setting their resultant equal to zero. To this end, consider further the displacement, rotation, and strain of a cross section. Specifically, consider a circumferential fiber parallel to the central axis of the beam and at a distance y outwardly beyond or "above" the neutral axis as in Figure 14.6. In this figure the inner and outer radii of the beam are R_i and R_O, measured as before, from the center of curvature O of the beam. R is the radius of the centroidal axis and ρ is the radius of the neutral axis. The difference δ between the centroidal and neutral axes radii $(R - \rho)$ is a measure of the shift in the neutral axis (to be determined).

Let r be the distance from O to the circumferential fiber as shown in Figure 14.6. Let the fiber subtend an angle $\Delta\theta$ as shown. The length ℓ of the fiber is then

$$\ell = r\Delta\theta \tag{14.7}$$

Finally, let y be the radial distance from the fiber to the neutral axis.

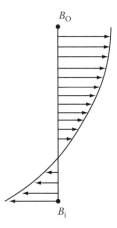

FIGURE 14.5   The form of the stress distribution across the cross section of a curved beam.

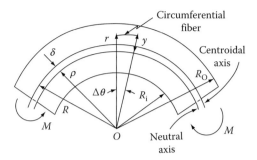

FIGURE 14.6 Curved beam geometry/notation.

Let the beam be subjected to a bending moment M, which tends to increase the curvature (thus decreasing the radius of curvature) of the beam. Then if the circumferential fiber is above or beyond the neutral axis, it will be lengthened by the deforming beam. Let u_θ represent this increase in length. Then the circumferential strain ε_θ (normal strain) at the radial location r of the fiber is

$$\varepsilon_\theta = u_\theta/\ell = u_\theta/r\Delta\theta \tag{14.8}$$

Next, observe that if the cross section remains plane during bending, the circumferential displacement u_θ will be linearly related (proportional) to y. That is,

$$u_\theta = ky \tag{14.9}$$

where k is a constant. Also observe from Figure 14.6 that the radial distance r from the curvature center O to the circumferential fiber may be expressed as

$$r = R - \delta + y \tag{14.10}$$

Hence the normal strain becomes

$$\varepsilon_\theta = ky/(R - \delta + y) \tag{14.11}$$

The corresponding circumferential stress (normal stress) σ_θ is then

$$\sigma_\theta = E\varepsilon_\theta = Eky/(R - \delta + y) \tag{14.12}$$

In this analysis, we have assumed that the beam deformation and consequently the cross section rotation are due to pure bending through an applied bending moment M. That is, there was no applied circumferential loading. Thus the resultant of the circumferential (or normal) loading due to the normal stress must be zero. That is,

$$\int_A \sigma_\theta dA = 0 \tag{14.13}$$

where A is the cross section area. By substituting from Equation 14.12 we have

$$\int_A \frac{Eky\ dA}{(R - \delta + y)} = 0 \quad \text{or} \quad \int_A \frac{y\ dA}{(R - \delta + y)} = 0 \tag{14.14}$$

Again, from Figure 14.6 we see that y may be expressed in terms of the neutral axis radius ρ as

$$y = r - \rho = r - (R - \delta) \quad \text{and} \quad r = y + R - \delta \tag{14.15}$$

Then Equation 14.14 yields

$$\int_A \frac{y \, dA}{(R - \delta + y)} = \int_A \frac{r - R + \delta}{r} dA = \int_A \left[1 + \left(\frac{\delta - R}{r}\right)\right] dA$$

$$= A + (\delta - R) \int_A \frac{dA}{r} = 0 \tag{14.16}$$

Hence, the neutral axis shift δ and the neutral axis radius of curvature ρ are

$$\delta = R - A \bigg/ \int_A (dA/r) \quad \text{and} \quad \rho = A \bigg/ \int_A (dA/r) \tag{14.17}$$

Observe that the neutral axis shift δ away from the centroidal axis is a function of the curved beam geometry and not the loading on the beam.

14.3 STRESSES IN CURVED BEAMS

Consider again a segment of a curved beam subjected to bending moments as in Figure 14.2 and as shown again in Figure 14.7. Recall from Equation 14.12 that the stress σ_θ at a point y above the neutral axis is

$$\sigma_\theta = Eky/(R - \delta + y) \tag{14.18}$$

where
   k is a constant introduced in Equation 14.9, to describe the preservation of the planeness of the cross section during bending
   δ is the amount of inward shift of the neutral axis away from the centroidal axis (given by Equation 14.17)
   E is the modulus of elasticity

As noted earlier and as seen in Equation 14.18, σ_θ is not linear in y but instead σ_θ has a nonlinear distribution across the cross section, as represented in Figure 14.5 and as represented again in Figure 14.7.

   Although Equation 14.18 provides an expression for the stress distribution across the cross section, it is of limited utility without knowledge of k. To determine k, observe that we can express k

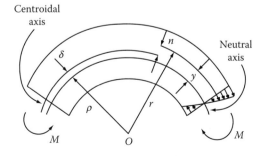

FIGURE 14.7   Stress distribution on a cross section of a beam element.

in terms of the applied bending moment M by evaluating the sum of the moments of the stress-induced elemental forces ($\sigma_\theta dA$) about the neutral axis. That is,

$$M = \int_A y\sigma_\theta dA = \int_A \frac{Eky^2 dA}{R - \delta + y} \tag{14.19}$$

We can simplify the evaluation of this integral by following an analysis used by Singer [6]. Specifically, in the numerator of the integrand let one of the y factors be replaced by the identity

$$y \equiv y + (R - \delta) - (R - \delta) = (R - \delta + y) - (R - \delta) \tag{14.20}$$

Then Equation 14.19 becomes

$$M = Ek\left[\int \frac{y^2 dA}{R - \delta + y}\right] = Ek\int \frac{y[R - \delta + y) - (R - \delta)]dA}{(R - \delta + y)}$$

or

$$M = Ek\left[\int y\, dA - (R - \delta)\int \frac{y\, dA}{R - \delta + y)}\right] \tag{14.21}$$

The first integral of Equation 14.21 is simply the area moment about the neutral axis, or $A\delta$. That is,

$$\int y\, dA = A\delta \tag{14.22}$$

To see this, let η be the distance from the centroidal axis to a typical fiber as in Figure 14.7. Then by the definition of a centroid [7], we have

$$\int \eta\, dA = 0 \tag{14.23}$$

But from Figure 14.7 we see that η is

$$\eta = y - \delta \tag{14.24}$$

By substituting this expression for η into Equation 14.23, we immediately obtain Equation 14.22.
Next, in view of Equation 14.14, we see that the second integral in Equation 14.21 is zero. Therefore, using Equation 14.22 the bending moment M of Equation 14.21 is

$$M = EkA\delta \tag{14.25}$$

Then k is

$$k = M/EA\delta \tag{14.26}$$

Finally, by substituting for k into Equation 14.18, we have

$$\sigma_\theta = My/A\delta(R - \delta + y) = My/A\delta r \tag{14.27}$$

where, from Figure 14.7, r takes the value $R - \delta + y$.

Although Equation 14.27 provides the bending stress distribution across the cross section, it may be practical nor convenient to use due to the need to know the neutral axis shift δ which neither may not be easily evaluated for a given cross section. (See Equation 14.17 for an analytical expression of δ.)

In practical design problems, however, we are generally interested in knowing the maximum bending stress in the beam, that is, the stress at the inner radius. To simplify the procedure for finding this stress, Wilson and Quereau [8] conducted an extensive series of tests on curved beams with various cross sections, measuring the strain and then evaluating the stress. Using the results of these tests, they determined that the maximum bending stress in a curved beam (at the inner radius R_i) may be estimated by the simple formula:

$$\sigma_\theta = \kappa Mc/I \tag{14.28}$$

where
   κ is a stress concentration factor
   c is the distance from the centroid axis to the inner surface (or "face")

The same formula may be used to determine the stress at the outer surface. Table 14.2 provides values of κ as well as neutral axis shift expressions, for a variety of common cross section shapes.

Finally, it should be noted that Equations 14.27 and 14.28 provide the value of normal stress over the cross section due to bending. The total (or resultant) stress on the cross section is the superposition of the bending stress and other normal (axial) and shear stresses on the cross section, arising from the applied loads.

14.4 APPROXIMATION OF STRESS CONCENTRATION FACTORS

When we examine the values of the stress concentration factors (κ) in Table 14.2, we see that in spite of rather large differences in cross section geometry, the factors themselves do not vary much over a wide range of R/c or $R/(R - R_i)$ ratios. Figure 14.8 provides a general graphical description of the variation of κ with $R/(R - R_i)$. [Recall that R is the radius of the centroidal axis and R_i is the inner radius of the beam (see Figure 14.6).]

In curved beam and hook design, it is usually a common practice to incorporate generous factors of safety. Therefore, the use of approximate stress concentration factors based upon Figure 14.8 may be quite acceptable. This approximation becomes increasingly accurate as the cross section becomes more compact and as the beam radii become larger.

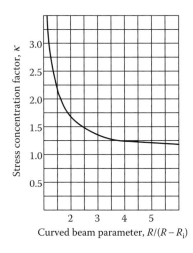

FIGURE 14.8 Approximate stress concentration factor as a function of geometric parameter $R/(R-R_i)$.

14.5 APPLICATION: STRESSES IN HOOKS

The common lifting hook is an example of a curved beam. We can use the foregoing concepts to obtain the insight and data for the stresses in these important structural components. To this end, consider the model of a hook shown in Figure 14.9. From our foregoing analysis, we know that the maximum tension and compression stresses occur at the inner and outer extremes of section A_1A_2. Using the Winkler-Bach analysis (see Equation 14.1), Gough et al. [9] developed expressions for the tensile stress σ_t (at the inner surface) and the compressive stress σ_c (at the outer surface) as

$$\sigma_t = \frac{P}{A}\left[\left(\cos\theta - \frac{H_o}{R}\right)\frac{c}{\lambda(R-c)} + \frac{H_o}{R}\right] \tag{14.29}$$

and

$$\sigma_c = -\frac{P}{A}\left[\left(\cos\theta - \frac{H_o}{R}\right)\frac{d-c}{\lambda(R+d-c)} - \frac{H_o}{R}\right] \tag{14.30}$$

where
  P is the load on the hook
  A is the cross section area
  θ is the angular coordinate (measured relative to the horizontal, see Figure 14.9)
  R is the centroidal axis radius of curvature
  H_o is the horizontal distance between the centers of curvature of the inner and outer surfaces
  d and c are cross section width and distance from the centroidal axis to the inner surface (see
    Figure 14.9)
  λ is the geometric parameter of Equation 14.2 and as listed in Table 14.1

In many cases H_o is small or zero so that Equations 14.29 and 14.30 reduce to

$$\sigma_t = \frac{Pc\cos\theta}{A\lambda(R-c)} \tag{14.31}$$

and

$$\sigma_c = -\frac{P(d-c)\cos\theta}{A\lambda(R+d-c)} \tag{14.32}$$

Finally, if the hook cross section is reasonably uniform, the maximum stresses occur, where θ is zero, as

$$\sigma_t = \frac{Pc}{A\lambda(R-c)} \tag{14.33}$$

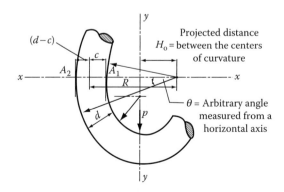

FIGURE 14.9 Notation for a working portion of a machine hook.

and

$$\sigma_c = -\frac{P(d-c)}{A\lambda(R+d-c)} \tag{14.34}$$

Observe that although lifting hooks are examples of curved beam, they have features which complicate the analysis: (1) their cross section is generally not uniform; (2) the cross section shape is usually neither circular nor rectangular; (3) there is a significant axial load creating normal stresses on the cross section, which need to be superimposed upon the bending stresses; and (4) at the point of application of the load, there will be contact stresses which could be as harmful as the bending stresses. Nevertheless, the foregoing analysis is believed to be useful for providing reasonably accurate estimates of the hook stresses as such the analysis is likely to be more convenient than numerical methods (e.g., finite element analyses) and experimental analyses.

14.6 EXAMPLE OF CURVED BEAM COMPUTATIONS

14.6.1 Flexure of a Curved Machine Bracket

Consider a curved machine bracket as in Figure 14.10. Let the bracket form a semicircle and let the cross section have a T-shape with dimensions as shown. For the 10,000 lb (45,480 N) load, the objective is to estimate the maximum stress in the bracket.

From the foregoing analysis, we know that the maximum stress occurs at point B. From the cross section geometry, we have (in the notation of Section 14.3)

$$A = 2.625 \text{ in.}^2, \quad c = 1.0 \text{ in.}, \quad R = 6.0 \text{ in.}, \quad I = 2 \text{ in.}^4 \tag{14.35}$$

From Table 14.2, the stress concentration factor κ is 1.18. Then using Equation 14.28, the stress σ_b due to bending is $\kappa Mc/I$. Also, the load geometry creates axial loading at the support end. The axial stress σ_a from this loading is $\kappa P/A$. Hence, the resultant stress σ_B at B is

$$\sigma_B = \kappa(P/A + Mc/I)$$
$$= (1.18)\left[\frac{10,000}{2.625} + \frac{(10,000)(12.0)(1.0)}{(2.0)}\right]$$

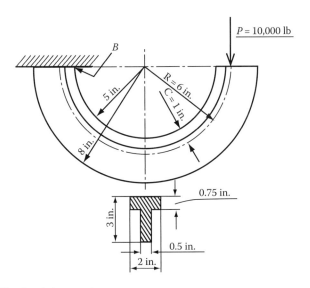

FIGURE 14.10   Machine bracket geometry.

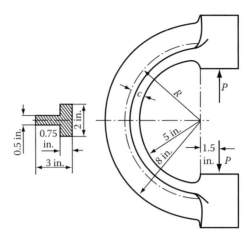

FIGURE 14.11   Half-circle clamp.

or

$$\sigma_B = 75{,}300\,\text{lb/in.}^2 \quad (519\,\text{N/min}^2) \tag{14.36}$$

Observe that the contribution of the axial loading (P/A) is relatively small compared with that of the bending (Mc/I).

14.6.2   EXPANSION OF A MACHINE CLAMP

Consider next a semicircular machine clamp as in Figure 14.11. Let the cross section be the same as that in the previous example. Let a load magnitude P of 12,000 lb be applied to the interior ends of the clamp as shown. The objective, as before, is to determine the maximum stress in the clamp.

From the foregoing analysis, we see that the maximum stress will occur at point B on the inner surface of the bracket.

As before, from the cross section geometry, we have

$$A = 2.625\,\text{in.}^2, \quad c = 1.0\,\text{in.}, \quad R = 6.0\,\text{in.}, \quad I = 2\,\text{in.}^4 \tag{14.37}$$

From Table 14.2, the stress concentration factor κ is again 1.18. Thus from Equation 14.28, by including the axial loading (P/A), the stress σ_B at B is

$$\sigma_B = \kappa(P/A + Mc/I)$$

$$= (1.18)\left[\frac{12{,}000}{2.625} + \frac{(12{,}000)(7.5)(1.0)}{(2.0)}\right]$$

or

$$\sigma_B = 58{,}490\,\text{lb/in.}^2 \quad (403\,\text{N/min}^2) \tag{14.38}$$

If we envision the clamp as a hook we can use Equation 14.33 to compute the stress, where λ may be obtained from Table 14.1 (the next to the last entry) as 0.0163. That is,

$$\sigma_B = \frac{Pc}{A\lambda(R-c)}$$
$$= \frac{(12{,}000)(1.0)}{(2.625)(0.0163)(6.0-1.0)}$$

or

$$\sigma_B = 56{,}090\,\text{lb/in.}^2 \quad (387\,N/\text{min}^2) \tag{14.39}$$

The results of Equation 14.38 and 14.39 differ by approximately 4%. In the majority of practical designs such differences should be well within the customary factors of safety.

14.7 FURTHER COMMENTS ON THE STRESSES IN CURVED BEAMS AND HOOKS

Our focus, in this chapter, has been on the *stress* in curved beams, with application in brackets and hooks. We have not similarly discussed displacements. The reason is that curved members are used primarily for strength. The curvature causes the loads to be supported both axially and in the transverse directions. With straight members, the loading is generally supported either only axially or only in flexure. Thus curved members can support considerably higher loads than their straight counterparts.

With the focus on strength or the ability to support loads, the displacements are usually of less concern. If, for example, a lifting hook is sufficiently strong to support its load, its displacement and deformation is of little or no concern.

Regarding hooks, the stresses on the inner and outer surfaces can be quite different from that on the inner surface being largest. Thus for material and weight efficiency, it is reasonable to design the cross section with greater thickness at the inner surface. From a manufacturing perspective, it is convenient to use a trapezoidal ("bull-head") shaped cross section or an "I" or a "T" cross section.

Once the basic cross section shape is determined, the parameter of greatest interest is the ratio of the depth of the cross section to the inner radius of curvature for a given stress. From a practical design perspective, Gough et al. [9] suggest using the following expressions for the depth D of circular and trapezoidal cross section shapes.

CIRCULAR SECTION

$$D = 0.023\,P^{1/2} + 0.18\,R_{\text{i}} \tag{14.40}$$

TRAPEZOIDAL OR BULL-HEAD SECTION

$$D = 0.026\,P^{1/2} + 0.20\,R_{\text{i}} \tag{14.41}$$

where D is in inches and P is the hook load in pounds.

We may also use finite element methods to obtain insights into the stresses in curved members and hooks. Some caution is encouraged, however, when estimating maximum stresses. If the stresses exceed the yield stress of the material, plastic deformation can occur changing the geometry and redistributing the loading and thus altering the stress values and the stress distribution.

SYMBOLS

A	Beam cross section area
A, B	Beam sections
C	Centroid
c	Distance from neutral axis to beam surface in straight beams; distance from centroid to inner surface
D	Depth of circular and trapezoidal cross section shapes
d	Cross-section width
E	Elastic modulus
H_o	Horizontal distance between centers of curvature
I	Second moment of area
k	Constant
ℓ	Arch length
M	Bending moment
N	Neutral surface
P	Hook load
$PQ, P'Q'$	Curved fibers
R	Distance from area cross section to center of curvature of curved beam
R_i, R_O	Inner and outer radii
u_θ	Tangential displacement
X, Y, Z	Cartesian (rectangular) coordinate system
y	Y-axis coordinate
γ	Radial coordinate
δ	Increment
δ_{PQ}	Lengthening of fiber PQ
$\delta_{P'Q'}$	Shortening of fiber $P'Q'$
ε	Normal strain
ε_θ	Circumferential strain
η	Length measure within a cross section as in Equation 14.2
θ	Angular coordinate
κ	Stress concentration faction
λ	Geometric parameter of Equation 14.2
ρ	Radius of curvature
σ	Normal stress
σ_c	Compressive stress
σ_t	Tensile stress
σ_θ	Tangential stress

REFERENCES

1. S. P. Timoshenko, *History of Strength of Materials*, McGraw Hill, New York, 1953, pp. 152–155.
2. E. Winkler, *Die Lehre von der Elastizität und Festigkeit*, Prague, Polytechnic Institute Press, 1867.
3. G. A. Goodenaugh and L. E. Moore, Strength of chain links, Bulletin 18, Engineering Experiment Station, University of Illinois, Urbana, IL, 1907.
4. A. Morley, Bending stresses in hooks and other curved beams, *Engineering*, London, 1914, p. 98.
5. A. M. Winslow and R. H. G. Edmonds, Tests and theory of curved beams, University of Washington, Engineering Experiment Stations Bulletin, 42, 1927, pp. 1–27.
6. F. L. Singer, *Strength of Materials*, 2nd ed., Harper & Row, New York, 1962, pp. 494–499.

7. S. L. Salas and E. Hille, *Calculus—One and Several Variables*, 3rd ed., Wiley, New York, 1978, p. 803.

8. B. J. Wilson and J. F. Quereau, A simple method of determining stress in curved flexural members, Circular 16, Engineering Experiment Station, University of Illinois, 1927.

9. H. J. Gough, H. L. Cox, and D. C. Sopwith, Design of crane hooks and other components of lifting gear, *Proceedings of the Institute of Mechanical Engineers*, 1935, pp. 1–73.

15 Stability: Buckling of Beams, Rods, Columns, and Panels

15.1 INTRODUCTION

In testing the material strength, the focus is generally directed toward determining material response to tension. Most strength tests are performed by simply measuring the elongation of a rod as a function of an axial tensile load. Homogeneous and isotropic materials (particularly metals) are then assumed to have similar strength properties when compressed. Compression tests are thus often not conducted. Even though Hooke's law (with a linear stress–strain relation) is generally found to be valid for both tension and compression, compression loading often produces changes in structural geometry, which can then lead to buckling even before a yield stress is reached. In this chapter, we look at the phenomenon of buckling and the associated concepts of stability of beams, rods, columns, and panels.

15.2 LONG BARS SUBJECTED TO COMPRESSION LOADING

Consider a long, slender rod or bar subjected to an axial compressive loading as in Figure 15.1. If the geometry is ideal and the loads are centered on the bar axis, the bar will simply shorten due to the loading. If, however, the geometry is not perfect, the bar may bend and buckle as represented in Figure 15.2. As the bar buckles, the greatest deflection will occur at midspan as indicated.

As the bar is buckling, it will experience a bending moment along its length due to the lateral displacement. By inspection, in Figure 15.2 we see that the bending moment M at a cross section at distance x along the bar is simply

$$M = Py \tag{15.1}$$

Recall from Equation 9.3 that the bending moment is related to the displacement by the moment–curvature relation:

$$M = EI\,d^2y/dx^2 \tag{15.2}$$

Then by substituting from Equation 15.1 into Equation 15.2, we have the governing differential equation:

$$d^2y/dx^2 + (P/EI)y = 0 \tag{15.3}$$

The general solution of Equation 15.3 may be written as

$$y = A \cos \sqrt{P/EI}\, x + B \sin \sqrt{P/EI}\, x \tag{15.4}$$

where the integration constants A and B may be determined from the boundary conditions. The bar of Figure 15.1 may be regarded as having simple or pinned supports. This means that the displacements at the ends are zero. That is,

$$y(0) = y(\ell) = 0 \tag{15.5}$$

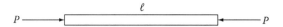

FIGURE 15.1 Long bar subjected to axial compressive load.

By imposing these conditions, we have

$$A = 0 \quad \text{and} \quad B \sin \sqrt{P/EI} \, \ell = 0 \tag{15.6}$$

where the second condition is satisfied if either B is zero or if $\sqrt{P/EI} \, \ell$ is zero. If B is zero, the bar is straight without buckling. If $\sqrt{P/EI} \, \ell$ is zero, we have

$$\sqrt{P/EI} \, \ell = n\pi \tag{15.7}$$

where n is an integer.

Therefore, the smallest load P_{cr} satisfying this expression is

$$P_{cr} = EI\pi^2/\ell^2 \tag{15.8}$$

This load is called "the Euler critical buckling load."

Observe in Equation 15.8 that P_{cr} is independent of the strength of the bar material, instead it depends only upon the elastic modulus (the stiffness) of the bar and upon the bar geometry. Observe further that P_{cr} decreases as the square of the length. Finally, observe that if the bar has a rectangular but not a square cross section, as in Figure 15.3, then the lowest buckling load will occur with bending about the short side axis (or the Y-axis) in the figure.

15.3 BUCKLING WITH VARIOUS END-SUPPORT CONDITIONS

15.3.1 CLAMPED (NONROTATING) ENDS

Consider a bar being compressed axially, whose ends are restrained from rotation as represented in Figure 15.4. Although in reality there are no supports which are completely rigid or without rotation, we know that bolted, welded, and bonded end supports can greatly restrict rotation. With such supports, the compressed bar being kept from rotating at its ends, is less likely to buckle than a similar bar with pinned ends. Nevertheless, as the compression load is increased, the bar will buckle. It will deform into a shape shown with exaggerated displacement as in Figure 15.5.

The symmetry of the bar supports and the loading, and the assumed rigidity of the supports require that the bar have zero slope at its ends ($x = 0$ and $x = \ell$). and at its middle ($x = \ell/2$). Also the symmetry requires that there be inflection points at quarter spans: $x = \ell/4$ and $x = 3\ell/4$. Since an inflection point has no curvature (i.e., $d^2y/dx^2 = 0$), the bending moment at such points is zero.

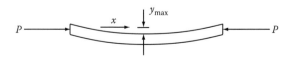

FIGURE 15.2 Buckled bar under axial compressive loading.

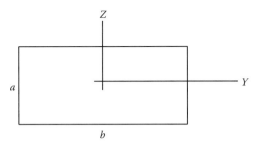

FIGURE 15.3   Rectangular cross section.

From a buckling perspective, these observations indicate that the bar will behave as a pinned-end bar (a simply supported bar) with end supports at $x = \ell/4$ and $x = 3\ell/4$. Thus for buckling, the bar is equivalent to a pinned-end bar with length $\ell/2$. Therefore, from Equation 15.8 we see that the critical buckling load P_{cr} is

$$P_{cr} = EI\pi^2/(\ell/2)^2 \qquad (15.9)$$

15.3.2   A Clamped (Nonrotating) and a Free End

Consider next a bar with one end clamped or fixed (nonrotating) and the other end free, as a cantilever beam. Let the bar be loaded at its free end by an axial load P as represented in Figure 15.6. As P is increased, the bar will buckle as represented (with exaggerated displacement) in Figure 15.7.

The bar displacement as in Figure 15.7 may be viewed as being of the same shape as the right half of the buckled bar with free ends of Figure 15.2 and as shown again in Figure 15.8.

That is, due to symmetry, the center of the bar has no rotation and is thus equivalent to clamped support at that point. Therefore, the clamped-free end support bar with length ℓ behaves as an unsupported end bar with length 2ℓ. Therefore, from Equation 15.8, the buckling load P_{cr} is

$$P_{cr} = EI\pi^2/(2\ell)^2 = EI\pi^2/4\ell^2 \qquad (15.10)$$

15.3.3   A Clamped (Nonrotating) and a Pinned End

Finally, consider an axially compressed bar with clamped and pinned ends as represented in its buckled state in Figure 15.9. As before, let the origin O be at the left end of the bar, which in this case is the pinned (roller supported) end.

As the beam is buckled with the left end constrained from vertical displacement, there will occur a reaction moment at the right end (the clamped end). This in turn means that except at the left end (at the pin support), there will be a bending moment throughout the bar.

FIGURE 15.4   Axially compressed bar with clamped (nonrotating) ends.

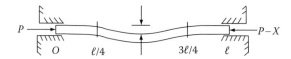

FIGURE 15.5   Buckled bar with clamped (nonrotating) ends.

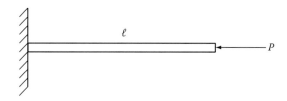

FIGURE 15.6   Axially loaded bar with fixed and free supports.

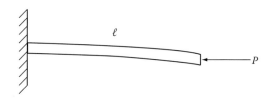

FIGURE 15.7   Buckled bar with clamped and free-end supports.

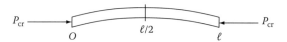

FIGURE 15.8   Buckled bar with axial loading on unsupported ends.

FIGURE 15.9   Pinned/clamped axially loaded compressed rod.

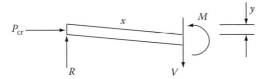

FIGURE 15.10   Free-body diagram of a left end segment of the buckled, pinned/clamped bar.

To qualify the bending moment, consider a free-body diagram of a segment of the left end of the bar as in Figure 15.10 where x is the segment length, R is the reaction force at the pin support and M and V are the bending moment and shear on the right end. By setting moments about the right end equal to zero, we have

$$M = Rx + P_{cr}y \tag{15.11}$$

where, as before, y is the displacement of the bar due to buckling.

Recall again from Equation 9.3 that the bending moment M is related to the curvature d^2y/dx^2 as

$$EId^2y/dx^2 = -M \tag{15.12}$$

Thus the governing differential equation for the bar displacement is

$$EId^2y/dx^2 = -(Rx + P_{cr}y)$$

or

$$d^2y/dx^2 + (P_{cr}/EI)y = -R/EI \tag{15.13}$$

The general solution of this equation is seen to be (homogeneous and particular solutions)

$$y = A \sin kx + B \cos kx - (R/EI)x \tag{15.14}$$

where k is given by

$$k^2 = P_{cr}/EI \quad \text{or} \quad k = \sqrt{P_{cr}/EI} \tag{15.15}$$

The auxiliary (boundary) conditions are

$$y = 0 \text{ at } x = 0 \quad \text{and} \quad y = dy/dx = 0 \text{ at } x = \ell \tag{15.16}$$

The first of these leads to

$$B = 0 \tag{15.17}$$

Then at $x = \ell$, we have

$$0 = A \sin k\ell - (R/EI)\ell \quad \text{and} \quad dy/dx = 0 = kA \cos k\ell - R/EI \tag{15.18}$$

By eliminating R/EI between the last two expressions, we have

$$0 = A \sin k\ell - k\ell A \cos k\ell \quad \text{or} \quad \tan k\ell = k\ell \qquad (15.19)$$

Equation 15.19 is a transcendental equation whose solution (roots) may be obtained numerically and are listed in various mathematical tables (see e.g., [1]). The smallest root is

$$k\ell = 4.49341 \qquad (15.20)$$

Then from Equation 15.15, the critical buckling load P_{cr} is

$$P_{cr} = (4.49341)^2 EI/\ell^2 \qquad (15.21)$$

15.4 SUMMARY OF RESULTS FOR LONG BAR BUCKLING WITH COMMONLY OCCURRING END CONDITIONS

Table 15.1 provides a listing of the foregoing results for buckling loads and Figures 15.11 through 15.14 provide a pictorial representation of the results.

TABLE 15.1
Axial Buckling Load for Long Bars

End-Supports	Buckling Load, P_{cr}
1. Pinned–pinned	$P_{cr} = \pi^2 EI/\ell^2 = 9.87 EI/\ell^2$
2. Clamped–clamped	$P_{cr} = 4\pi^2 EI/\ell^2 = 39.48 EI/\ell^2$
3. Clamped–free	$P_{cr} = \pi^2 EI/4\ell^2 = 2.47 EI/\ell^2$
4. Clamped–pinned	$P_{cr} = (4.4934)^2 EI/\ell^2 = 20.19 EI/\ell^2$

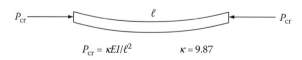

FIGURE 15.11 Pinned–pinned buckled bar.

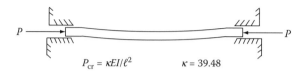

FIGURE 15.12 Clamped–clamped buckled bar.

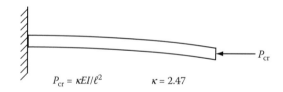

FIGURE 15.13 Clamped–free buckled bar.

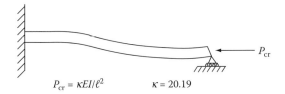

$$P_{cr} = \kappa EI/\ell^2 \qquad\qquad \kappa = 20.19$$

FIGURE 15.14 Clamped–pinned buckled bar.

In the table and figures, ℓ is the distance between the supports. Recall again that these results are valid for long, slender bars, that is, $\ell/\kappa > 0$ where κ is the minimum radius of gyration of the cross section.

If the bars are not so slender, they may fail in compression before buckling. In these cases, minor changes in support conditions are important. We will discuss these concepts in the following sections.

15.5 INTERMEDIATE LENGTH BARS AND COLUMNS—JOHNSON FORMULA

Consider again the foregoing results: Specifically, for a long axially loaded bar, the critical load P_{cr} leading to buckling is (see Equation 15.8)

$$P_{cr} = \pi^2 EI/\ell^2 \tag{15.22}$$

Observe again that although P_{cr} is proportional to the elastic modulus (or stiffness) of the bar material, P_{cr} does not depend upon the strength of the bar material. That is, a long bar could buckle and lead to structural failure before the bar reaches the yield stress.

For short bars buckling is not usually an issue, but as axial loads are increased, high compressive stresses can occur. For intermediate length bars, however, as axial loads become large, the bar may fail either by buckling or by yielding to compressive stress. The transition between failure modes, that is, between buckling and compressive yielding, is of particular interest.

To explore this, consider the compression stress σ_{cr} on a bar as it is about to buckle. Specifically, let σ_{cr} be defined as

$$\sigma_{cr} = P_{cr}/A \tag{15.23}$$

where, as before, A is the cross section area of the bar. Then for the pin–pin supported bar we have

$$\sigma_{cr} = \pi^2 E(I/A)/\ell^2 = \pi^2 E/(\ell/\kappa)^2 \tag{15.24}$$

where we have replaced the second moment of area I by $A\kappa^2$, with κ being the "radius of gyration" of the cross section. The ratio ℓ/κ is called the "slenderness ratio" of the bar.

Figure 15.15 provides a graphical representation of Equation 15.24 where σ_{cr} (the "critical stress") is expressed in terms of the slenderness ratio (ℓ/κ). Observe that for short bars, where the slenderness ratio is small, Equation 15.24 shows that a large load is required to buckle the bars. However, for large loads the unbuckled bar will attain large compressive stresses, ultimately yielding due to the stress.

For design purposes, engineers have suggested that axial loading for nonbuckled bars should be bounded so that the stress is no more than half the yield stress S_y [2–4]. Thus for design, Figure 15.15 is replaced by Figure 15.16.

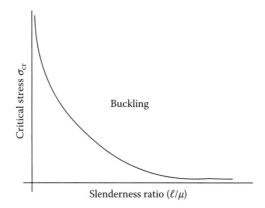

FIGURE 15.15   Critical stress as a function of the slenderness ratio.

For a refinement in practical design considerations, J.B. Johnson [3,5] proposed that the curve of Figure 15.16 be replaced by a smoother curve as in Figure 15.17, where the left end of the curve is based upon Johnson's formula for critical stress, σ_{cr}:

$$\sigma_{cr} = S_y - \left(S_y^2/4\pi^2 E\right)(\ell/\kappa)^2 \tag{15.25}$$

where, as before, S_y is the compressive yield stress.

In Figure 15.17, the transition point between the curves of Equations 15.24 and 15.25 is found by simply equating the stress values. That is,

$$\pi^2 E/(\ell/\kappa)^2 = S_y - \left(S_y^2/4\pi^2 E\right)(\ell/\kappa)^2 \tag{15.26}$$

Solving for ℓ/κ, we obtain the slenderness ratio at the transition point to be

$$\ell/\kappa = (2\pi^2 E/S_y)^{1/2} \tag{15.27}$$

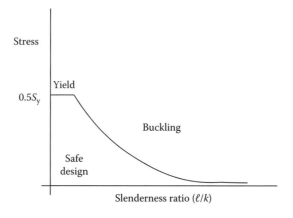

FIGURE 15.16   Design curve for axially loaded bars.

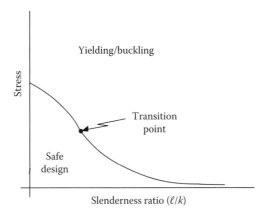

FIGURE 15.17 Johnson curve for axially loaded bars.

15.6 INTERMEDIATE LENGTH BARS AND COLUMNS—ECCENTRIC LOADING AND THE SECANT FORMULA

Consider again an axially loaded bar in compression as in Figure 15.18. Recall that Saint Venant's principle states that equivalent force systems (see Section 1.5.3) exerted on a body produce the same stress state at locations away from the loading site, but different stress states at locations near the loading site [6–8]. Therefore, if an axially loaded bar is long the stress state away from the ends is insignificantly affected by the method of application of the loads at the ends. However, if the bar is short the means of loading can make a measurable difference along the bar in the stress state.

To explore this, consider the end loading of a relatively short axially loaded bar as represented in Figure 15.19. In an actual bar, however, the loading geometry is not perfect and insofar as the loading can be represented by a single axial force P as in Figure 15.19, P will not be precisely on the axis but instead it will be displaced away from the axis by a small distance e as in Figure 15.20. This load displacement or eccentricity gives rise to a bending moment with magnitude Pe in the bar, which in turn can affect the stresses and buckling tendency of the bar.

FIGURE 15.18 An axially loaded bar.

FIGURE 15.19 End loading of a short bar.

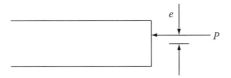

FIGURE 15.20 Off-axis axial load.

To quantify the effect of the eccentric loading, recall again the bending moment/curvature relation of Equation 9.3:

$$EI d^2 y / dx^2 = -M \qquad (15.28)$$

As the loading is increased and the beam begins to deflect and buckle, the bending moment M at a typical cross section will be

$$M = P(y + e) \qquad (15.29)$$

Then by substituting into Equation 15.28, we have

$$d^2 y / dx^2 + (P/EI) y = -Pe/EI \qquad (15.30)$$

If we regard the bar as having pinned–pinned end supports, the auxiliary boundary conditions of Equation 15.30 are

$$y = 0 \text{ at } x = 0 \quad \text{and} \quad x = \ell \qquad (15.31)$$

The general solution of Equation 15.30 is

$$y = y_h + y_p \qquad (15.32)$$

where

y_h is the general solution of the homogeneous equation (right side zero)
y_p is a particular solution

It is readily seen that y_h and y_p may be written as

$$y_h = A \cos \sqrt{P/EI} x + B \sin \sqrt{P/EI} x \qquad (15.33)$$

and

$$Y_p = -e \qquad (15.34)$$

Therefore, the general solution of Equation 15.30 is

$$y = A \cos \sqrt{P/EI} x + B \sin \sqrt{P/EI} x - e \qquad (15.35)$$

By enforcing the end conditions of Equation 15.31 we have at $x = 0$,

$$0 = A - e \quad \text{or} \quad A = e \qquad (15.36)$$

and then at $x = \ell$,

$$0 = e \cos \sqrt{P/EI} \ell + B \sin \sqrt{P/EI} \ell - e$$

or

$$B = e[1 - \cos \sqrt{P/EI}\ell]/ \sin \sqrt{P/EI}\ell \tag{15.37}$$

Therefore, from Equation 15.35 the displacement is

$$y = e\left\{\cos \sqrt{P/EI}x + [(1 - \cos \sqrt{P/EI}\ell) \sin \sqrt{P/EI}x/ \sin \sqrt{P/EI}\ell] - 1\right\} \tag{15.38}$$

By symmetry, we see that the maximum displacement y_{max} occurs at midspan ($x = \ell/2$). Thus y_{max} is

$$y_{max} = e\left\{-1 + \cos \sqrt{P/EI}\frac{\ell}{2} + \left[(1 - \cos \sqrt{P/EI}\ell) \sin \sqrt{P/EI}\frac{\ell}{2}\right] / \sin \sqrt{P/EI}\ell\right\}$$

$$= e\left\{-1 + \left[\cos \sqrt{P/EI}\frac{\ell}{2} \sin \sqrt{P/EI}\ell + \sin \sqrt{P/EI}\frac{\ell}{2} - \cos \sqrt{P/EI}\ell \sin \sqrt{P/EI}\frac{\ell}{2}\right] / \sin \sqrt{P/EI}\ell\right\}$$

$$= e\left\{-1 + \left[\sin \sqrt{P/EI}\frac{\ell}{2} + \sin \sqrt{P/EI}\frac{\ell}{2}\right] / \sin \sqrt{P/EI}\ell\right\}$$

$$= e\left\{-1 + \left[2\sin \sqrt{P/EI}\frac{\ell}{2} \cos \sqrt{P/EI}\frac{\ell}{2} / \sin \sqrt{P/EI}\ell \cos \sqrt{P/EI}\frac{\ell}{2}\right]\right\}$$

$$= e\left\{-1 + \sin \sqrt{P/EI}\ell / \sin \sqrt{P/EI}\ell \cos \sqrt{P/EI}\frac{\ell}{2}\right\}$$

$$= e\left\{-1 + 1/\cos \sqrt{P/EI}\frac{\ell}{2}\right\}$$

or

$$y_{max} = e\left(-1 + \sec \sqrt{P/EI}\frac{\ell}{2}\right) \tag{15.39}$$

From Equation 15.29, the maximum bending moment is then

$$M_{max} = P(y_{max} + e) = Pe \sec\left(\sqrt{P/EI}\frac{\ell}{2}\right) \tag{15.40}$$

The maximum compressive stress σ_{max} due to the combination of axial loading, buckling, and bending is then

$$\sigma_{max} = (P/A) + (M_{max} c/I) = (P/A) + \left(Pce \sec \sqrt{P/EI}\frac{\ell}{2}\right) / I$$

$$= (P/A)\left[1 + (ceA/I) \sec \sqrt{P/EI}\frac{\ell}{2}\right]$$

or

$$\sigma_{max} = (P/A)\left[1 + (ce/\kappa^2) \sec\left(\sqrt{P/EA}\frac{\ell}{2\kappa}\right)\right] \tag{15.41}$$

where as before
   c is the distance from the neutral axis to the most distant perimeter
   κ is the radius of gyration of the cross section area moment

From a design perspective, if we restrict the maximum stress to say S (perhaps a fraction of the compressive yield stress), we then have from Equation 15.41,

$$S = (P/A)\left[1 + (ce/\kappa^2)\ \sec\left(\sqrt{P/EA}\,\frac{\ell}{2\kappa}\right)\right]$$

or

$$P = SA \left/ \left[1 + (ce/\kappa^2)\ \sec\left(\sqrt{P/EA}\,\frac{\ell}{2\kappa}\right)\right]\right. \tag{15.42}$$

Equation 15.42 is the so-called "secant column formula." It is applicable for intermediate length bars ($10 < \ell/\kappa < 100$). It provides a design guide for the applied load P. The formula, however, has the obvious problem of being rather cumbersome as P appears nonlinearly (in the square root of the secant argument). Thus for a given geometry and bar material, an iterative procedure is probably the most practical procedure for finding P.

15.7 BUCKLING OF PLATES

Plate and panel buckling form another class of problems in elastic structure design. Local instabilities can occur during compressive loading which may or may not lead to global buckling. Figure 15.21 illustrates a plate subjected to a typical compressive load where ℓ and b are the plate dimensions, in its plane, t is the thickness, and S is the compressive loading (force per area: bt).

As noted, buckling resistance of a plate is necessarily not lost when local distortion occurs. Indeed, significant residual strength can remain even with local distortions. Therefore, a design analysis may take a twofold approach: (1) we may opt to have no buckling deformation at all or alternatively (2) we may allow local buckling as long as the structural integrity of the overall structural design is not compromised.

The general form of the stress expression for buckling in a plate as in Figure 15.21 is

$$\sigma_{\mathrm{cr}} = \kappa_{\mathrm{p}} E (t/b)^2 \tag{15.43}$$

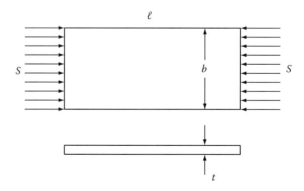

FIGURE 15.21 Plate subjected to compressive in-plane loading.

where

σ_{cr} is the stress where buckling occurs (the "critical stress")
κ_p is a buckling stress coefficient
E is the elastic modulus

The buckling coefficient κ_p is analogous to the column factor κ of Figures 15.11 through 15.14. As with the column factor, the buckling coefficient depends upon the edge constraint. The buckling coefficient is a nondimensional quantity and it is sometimes called the "plate coefficient."

When the critical stress σ_{cr} calculated from Equation 15.43 is less than the yield strength of the material, the buckling is considered to be "elastic." The design value of σ_{cr} obtained from Equation 15.43 should be regarded as an upper limit as stresses that actually occur are usually smaller. The difference is mainly due to geometric irregularities, which inevitably occur in actual designs. Such occurrences and stress differences increase as the plate thickness t decreases.

When the critical stress exceeds the yield strength of the plate material, the buckling is "inelastic." The yield stress is thus a natural limit for the critical stress.

In many practical problems of plate buckling, the ratio of plate length to width, ℓ/b, is greater than 5. In such cases the buckling coefficient, κ_p is virtually independent of the length. For lower length to width ratios, the buckling coefficient increases somewhat but it is a common conservative practice to ignore this change and to consider the edge supports as the primary controlling factor. The choice of the value for κ_p, however, depends to a large extent upon engineering judgment. Table 15.2 presents usual accepted values of κ_p for a variety of common supports. These values are intended for materials characterized by a Poisson's ratio of 0.3. Again, as in the case of structural columns, the fixed supports of condition 5 are practically never realized. Unless the weight requirements are such that the fixed-end condition must be satisfied, the most practical design solution is to assume either simple supports (condition 3) or the simple support-free support (condition 1).

When a long plate of width b is supported along the two long sides and is loaded in compression, condition 3 of Table 15.2 provides a good model for estimating the buckling load.

If the nonloaded edges of the plate are free of support (unlike the conditions of Table 15.2), they are no longer compelled to remain straight. The plate then behaves like a column or an axially compressed bar.

Since all the condition illustrated in Table 15.2 provide some degree of edge restraint, plates with those support conditions will not buckle as a compressed rod. Instead, upon buckling there will

---

TABLE 15.2

Buckling Stress Coefficients for Edge-Loaded Flat Plates

(Poisson's Ratio $\nu = 0.3$)

Simple support	1.	Free	$K_p = 0.38$
Fixed support	2.	Free	$K_p = 1.15$
Simple support	3.	Simple support	$K_p = 3.62$
Simple support	4.	Fixed support	$K_p = 4.90$
Fixed support	5.	Fixed support	$K_p = 6.30$

Note: All loaded edges are simply supported and plates are considered to be relatively long. Loading is perpendicular to the plane of the paper.

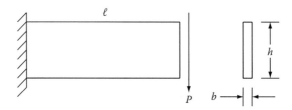

FIGURE 15.22   A cantilever beam with a narrow rectangular cross section.

be interval twisting and bending. This is why the buckling loads of plates and panels are considerably higher than those of bars and columns.

15.8   BUCKLING DUE TO BENDING

When the cross section of a beam is narrow and rectangular, as for example in Figure 15.22, the beam will have a tendency to buckle due to lateral bending and twisting. This bending/twisting failure occurence depends upon the magnitude of the loading, the end support conditions and the cross section geometry. While the mathematical analysis of this problem is somewhat detailed (see [8,9]), it is possible to provide some design guidelines as outlined in the following paragraphs.

Consider a beam with a tall/narrow rectangular cross section as that in Figure 15.22. Let h be the height of the cross section and let b be the base width, and let ℓ be the beam length as shown. Consider four loading and end support cases as represented in Figure 15.23.

Depending upon the supports and load positions, the critical load leading to lateral bending and twisting has the form [9,10]:

$$P_{cr} = \kappa(b^3 h/\ell^2)[1 - 0.63(b/h)EG]^{1/2} \tag{15.44}$$

where as before
   E and G are the moduli of elasticity and rigidity (shear modulus)
   κ is a numerical coefficient as listed in Table 15.3

In the first case (ends-free), P_{cr} in Equation 15.44 is to be replaced by M_{cr}/ℓ. In the third case (simple-support), the ends are held vertical but still allowed to rotate in the vertical plane.

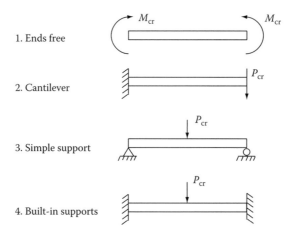

FIGURE 15.23   Loading and end supports for lateral buckling of a narrow rectangular cross-section beam.

TABLE 15.3
Critical Loading Coefficient κ of Equation 15.44
for Loading Cases of Figure 15.23

Loading Case	Coefficient κ
1. Free ends, end moment	0.524
2. Cantilever, end force	0.669
3. Simple support, midspan force	2.82
4. Built-in supports, midspan force	26.6

Several comments may be useful for design considerations: First, observe the product EG in Equation 15.44. This arises from the combined bending and twisting as a beam buckles.

Next, note that I-beams are used in many structural applications, whereas Equation 15.44 and Table 15.3 are applicable only for beams with rectangular cross sections. I-beams, which can have various cross section dimensions, are thus more difficult to study than beams with a rectangular profile. Nevertheless data for lateral bending/buckling are provided as Refs. [9,10].

Further in Equation 15.44 and Table 15.3 we have considered only concentrated point loads. Many applications have uniform loading or combined loadings. Here again Refs. [9,10] provide useful data for critical loading. From a structural design perspective, however, concentrated loading is more harmful than uniform loading. Thus for conservative design the data of Table 15.3 provide a safer design.

Finally, care should be taken when using any such data as in actual design exact ideal geometry will not occur. The actual critical buckling loads may thus be lower than those predicted by using the tabular data. Therefore, generous factors of safety should be used.

15.9   BUCKLING OF COLUMNS LOADED BY THEIR OWN WEIGHT

Tall heavy columns commonly occur as chimneys, towers, and poles. The buckling analysis is similar to the previous analyses although more detailed [9]. Nevertheless we can still obtain estimates of critical weight density γ_{CR}. Most column structures do not have a uniform diameter, but instead they are larger at the base. A conservative (safer) analysis is then to simply consider a structure with uniform diameter as represented in Figure 15.24, where q is the load per unit length.

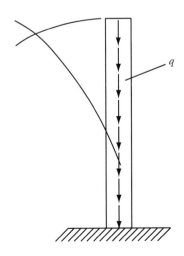

FIGURE 15.24   Heavy uniform cross-section column.

Using a standard flexural analysis [9] and an energy analysis [11], the critical load per unit length q_{CR} is approximately

$$q_{CR} = 7.89EI/\ell^3 \qquad (15.45)$$

We can illustrate the use of this expression with an elementary design problem: suppose we want to determine the maximum length ℓ_{max} of a hollow, thin-walled aluminum column with an average radius r of 2.0 in. and a wall thickness t of 0.2 in. Let the elastic modulus E be 10×10^6 psi and the weight density γ be 0.098 lb/in.^{3}.

For a relatively thin pipe with radius r and wall thickness t, the weight q per unit length is approximately

$$q = 2\pi rt\gamma \qquad (15.46)$$

Also, the second moment of area I is approximately

$$I = \pi r^3 t \qquad (15.47)$$

Then by substituting these expressions into Equation 15.45, we have

$$2\pi rt\gamma = 7.89\pi \; Er^3t/\ell^3$$

or

$$\ell = \ell_{max} = 1.58(Er^2/\gamma)^{1/3} \qquad (15.48)$$

Hence for the given data, ℓ_{max} is

$$\ell_{max} = 97.7 \text{ ft} = 29.77 \text{ m} \qquad (15.49)$$

This is a relatively tall column. Interestingly, the maximum stress is small. Indeed, the stress σ at the base is only

$$\sigma = W/A = 2\pi rt\gamma\ell_{max}/2\pi rt = \gamma\ell_{max} \qquad (15.50)$$

or

$$\sigma = (0.098)(97.7)(12) = 114.9 \text{ psi} \qquad (15.51)$$

Observe that in this example the design is governed by stability rather than stress limitations. Also note that the stability calculations assume ideal geometry. An eccentricity or other geometric irregularity will make the structure less stable. Therefore, factors of safety should be incorporated into stability computations for the design of actual structures.

15.10 OTHER BUCKLING PROBLEMS: RINGS AND ARCHES

The buckling of circular rings and arches is a classical problem, which has been studied extensively. Reference [9] provides an analysis of the phenomenon and Figure 15.25 provides a summary of the more important common cases, where the notation is the same as in the foregoing section. Case 3 for

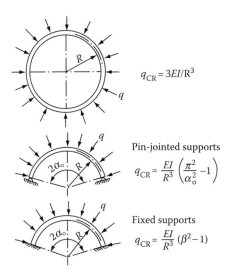

FIGURE 15.25   Buckling loads for rings and arches.

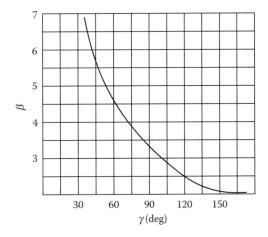

FIGURE 15.26   Arch buckling factor β as a function of the half arch angle α_o.

a fixed ends arch, involves a parameter β (the "arch buckling factor") which satisfies the transcendental equation

$$\beta \tan \alpha_o \cot(\alpha_o \beta) = 1 \qquad\qquad (15.52)$$

Figure 15.26 provides a graphical representation of the relationship between β and α_O.

15.11   SUMMARY REMARKS

In this chapter, we have attempted to simply document and list the most important and most widely used buckling and stability formulas. The references, particularly Refs. [9,10,12], provide additional information for more specialized cases.

   In structural design, when compressive loading occurs, stability (as opposed to material failure) is often the controlling factor in the integrity of the structure. The analyses and resulting formulas of this chapter have, for the most part, assumed ideal geometry and centralized loading condition. In actual design, however, this is a rare case. Thus the critical load values may be too high. Therefore as noted earlier, caution and generous factors of safety need to be employed in the design of those structures and components which may be subjected to high compressive loads.

SYMBOLS

A	Cross-section area
A, B	Integration constants
A, b	Length, width measurements
E	Elastic modulus
E	Load offset (see Figure 15.20)
G	Shear modulus
H	Depth measurement
I	Second moment of area
k	Constant (see Equation 15.15)
K_p	Buckling coefficient; plate coefficient
ℓ	Length
M	Bending moment
M_{cr}	Critical bending moment
P	Applied load
P_{cr}	Euler buckling load
q	Distributed loading
R	Reaction force; radius
r	Radius
S_y	Compressive yield stress
t	Pipe thickness
V	Shear
W	Weight
X, Y, Z	Cartesian (rectangular) coordinate axes
x	X-axis coordinate
y	Y-axis coordinate
y_h	General solution of homogeneous differential equation
y_p	Particular solution of differential equation
α_o	Angle (see Figure 15.25)
β	Buckling arch factor
γ	Weight density
κ	Cross-section "radius of gyration"
σ_{cr}	Stress at buckling

REFERENCES

1. M. Abramowitz and I. A. Stegun (Eds.), *Handbook of Mathematical Functions*, Dover, New York, 1965, p. 224.
2. J. E. Shigley, C. R. Mischke, and R. G. Budynas, *Mechanical Engineering Design*, 7th ed., McGraw Hill, New York, 2004, p. 220.
3. B. J. Hamrack, B.O. Jacobson, and S. R. Schmid, *Fundamentals of Machine Elements*, WCB/McGraw Hill, New York, 1999, p. 368.
4. F. L. Singer, *Strength of Materials*, 2nd ed., Harper & Row, New York, p. 395.
5. R. C. Juvinall and K. M. Marshek, *Fundamentals of Machine Component Design*, Wiley, New York, 1991, pp. 189–193.

6. I. S. Sokolnikoff, *Mathematical Theory of Elasticity*, Robert E. Krieger Publishing, Malabar, FL, 1983, p. 89.

7. S. P. Timoshenko and J. N. Goodier, *Theory of Elasticity*, McGraw Hill, New York, 1951, p. 33.

8. F. P. Beer and E. Russell Johnston, Jr., *Mechanics of Materials*, 2nd ed., McGraw Hill, New York, 1981, pp. 631–688.

9. S. P. Timoshenko and J. M. Gere, *Theory of Elastic Stability*, McGraw Hill, New York, 1961, pp. 251–294.

10. W. C. Young and R. G. Budynas, *Roark's Formulas for Stress and Strain*, 7th ed., McGraw Hill, New York, 2002, pp. 728–729.

11. J. H. Faupel, *Engineering Design*, Wiley, New York, 1964, pp. 566–642.

12. F. Bleich, *Buckling Strength of Metal Structures*, McGraw Hill, New York, 1952.

16 Shear Center

16.1 INTRODUCTORY COMMENTS

Recall in Chapter 15, on stability, we saw that for the buckling of tall, thin cross section beams, the cross section may rotate and warp at loads well below the yield stress loads (see Figures 16.1 and 16.2). If the geometry is ideal and the line of action of the loading is centered in the cross section, the warping is less likely to occur. If a thin-web cross section geometry is less simple, as is usually the case, warping is likely unless the load is carefully placed.

To illustrate this further, consider a cantilever beam with a cross section in the shape of a U-section or channel, as in Figure 16.3. (This is a classic problem discussed in a number of texts [1–3].) Suppose the cross section is oriented that the open side is up as in Figure 16.4. Thus with ideal geometry a carefully placed and centered load will not produce warping. If, however, the cross section is rotated through say 90°, as in Figure 16.5, it is not immediately clear where the load should be placed to avoid warping.

Surprisingly, it happens that if the line of action of the load is placed through the centroid of the cross section, the beam will still tend to warp. But there is a point, called the "shear center," through which the load can be placed where warping will not occur. Our objective in this chapter is to establish the existence and location of the shear center.

16.2 SHEAR FLOW

The warping, twisting, and buckling of beam cross sections is most pronounced when the cross section has webs, flanges, or other thin-walled components. With webs or flanges, the strength is primarily in the plane of the web or flange. That is, a web, a flange, a panel, or a plate has most of its strength in directions parallel to the plane of the member (so-called "membrane strength"). Webs, panels, flanges, and plates have far less resistance to forces directed normal to their planes than to in-plane loading.

For beams with cross sections composed of thin-walled members, such as an I-beam or a channel beam, external loads are then largely supported by in-plane forces in the thin-walled sections. These forces in turn give rise to shear stresses in these thin-walled sections. To evaluate these stresses, it is helpful to reintroduce the concept of "shear flow," which we discussed briefly in Section 12.7.

Consider a web or thin-walled portion of a beam cross section, which is subjected to a shear force as in Figure 16.6. Consider an element e of the web as in Figures 16.7 and 16.8. Let the web thickness be t and the shear stress on the shear-loaded face of the web be τ.

Then from Equation 12.20, we define the shear flow q on the web simply as the integral of the shear stress across the web. That is

$$q \overset{D}{=} \int \tau \, dz \tag{16.1}$$

Specifically from Figure 16.8, q is

$$q = \int_{0}^{t} \tau_{xy} \, dz \tag{16.2}$$

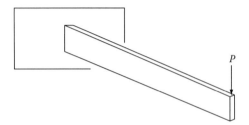

FIGURE 16.1   Tall, thin cross section cantilever beam with end loading.

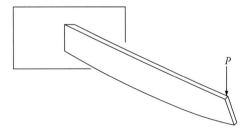

FIGURE 16.2   Warping of tall, thin cross section cantilever beam with end loading.

FIGURE 16.3   Channel shape cross section.

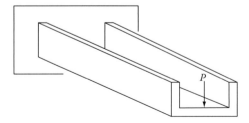

FIGURE 16.4   Cantilever, channel cross section, beam with end side up.

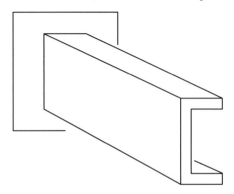

FIGURE 16.5   Cantilever, channel cross section, beam with open side to the right.

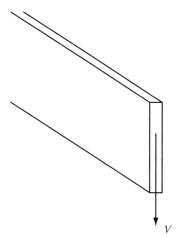

FIGURE 16.6   A thin-walled beam cross section web subjected to a shear force.

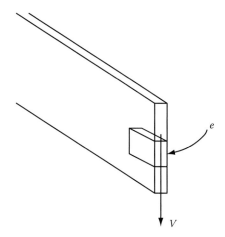

FIGURE 16.7   Element e of a shear-loaded web.

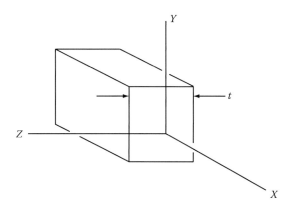

FIGURE 16.8   Element e of the shear-loaded web.

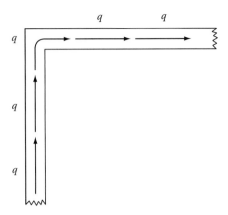

FIGURE 16.9   Shear flow in a narrow web.

Shear flow is often interpreted as being analogous to the flow of a liquid in a narrow channel as represented in Figure 16.9. We will develop this analogy in the following section showing how the "flow" can go around a corner.

Finally, shear flow is useful for determining the shear force on a web. For example, in a right-angle web as in Figure 16.9 and as shown again in Figure 16.10, we can obtain the horizontal and vertical shear forces (H and V) on the web by simply integrating the shear flow in the horizontal and vertical directions. That is,

$$H = \int_{B}^{C} q\ dx \quad \text{and} \quad V = \int_{A}^{B} q\ dy \tag{16.3}$$

16.3   APPLICATION WITH NARROW WEB BEAM CROSS SECTION

Recall from Section 4.3 that in the interior of bodies subjected to loading, equilibrium or rectangular elements requires that shear stresses on abutting perpendicular faces have equal magnitude. Consider for example, the element shown in Figure 16.11 with faces perpendicular to the coordinate axes. Then moment equilibrium requires that the shear stresses shown satisfy the relations

$$\tau_{yx} = \tau_{xy}, \quad \tau_{zy} = \tau_{yz}, \quad \tau_{xz} = \tau_{zx} \tag{16.4}$$

(See Section 4.3.)

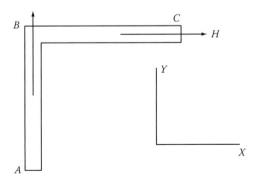

FIGURE 16.10   Horizontal and vertical shear forces on a right-angle web.

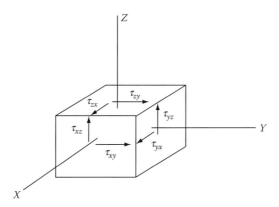

FIGURE 16.11   Element in the interior of a body under loading

In more general index notation, Equations 16.4 are contained within the expression

$$\sigma_{ij} = \sigma_{ji} \tag{16.5}$$

where σ_{ij} represents the stress on the "i-face" in the j direction.

The interpretation and application of Equation 16.4 with shear flow is that the shear flows at perpendicular mating surfaces are of equal magnitude but with opposing directions as illustrated in Figure 16.12.

Remarkably these opposing shear flows on perpendicular abutting surfaces cause the uniformly directed shear flow around corners in a plane, as in Figure 16.9. To see this, consider a right-angle cross section of a web as in Figure 16.13 where there is a vertical shear force V, causing a shear flow q as shown.

Next, consider three portions, or subsections, of the right-angle section as shown in Figure 16.14, where we have named points A, B, C, and D to aid in the identification. Consider first subsection ① as shown in Figure 16.15: from the shear flow pattern shown in Figure 16.12, we obtain the resulting shear on the top face of subsection ①. Observe that this face is in contact with the bottom face of the long square subsection ②.

Consider next subsection ② as shown in enlarged view in Figure 16.16. From the action–reaction principle, the shear on the bottom face of the subsection ② has the same magnitude but opposite direction to that on the top face of subsection ①. From Figures 16.13 and 16.14, we see that the top and left side faces of subsection ② are free surfaces and thus free of shear stresses. Therefore, to maintain equilibrium of the subsection the back face of the subsection must have a balancing shear to that on the bottom face. Figure 16.17 presents a representation of this back face shear.

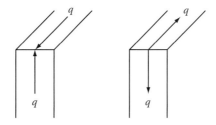

FIGURE 16.12   Shear flow at abutting perpendicular surfaces.

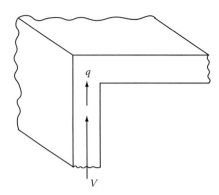

FIGURE 16.13   Shear at a right-angle web.

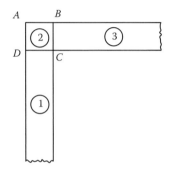

FIGURE 16.14   Subsections of a right-angle web.

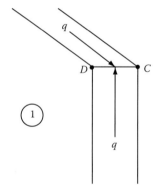

FIGURE 16.15   Shear flow on subsection ①.

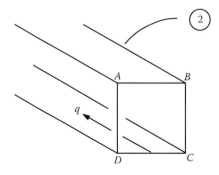

FIGURE 16.16   Shear flow on the bottom face of subsection ②.

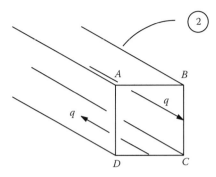

FIGURE 16.17   Equilibrium of shear forces on the bottom and back faces of subsection ②.

Finally, consider the equilibrium of subsection ③: Figure 16.18 shows an enlarged view of the subsection and the shear flow in the interior abutting face with subsection ② and on the web. Again by the principle of action–reaction with the shear directed toward edge *BC* on ② it is directed away from *BC* on ③. Then referring again to Figure 16.12 we see that the shear flow in the web of subsection ③ is directed away from edge *BC* as shown.

Considering the result of Figure 16.18 it is obvious that the shear flow in the right-angle web of Figure 16.13 is directed around the corner as in Figure 16.9 and as shown again in Figure 16.19.

16.4   TWISTING OF BEAMS/SHEAR CENTER

The foregoing results provide a basis for understanding the shear distribution on a webbed beam cross section. To develop this, consider again the channel cross section cantilever beam of Figure 16.4 as shown again in Figures 16.20 and 16.21. Let the beam has an end load *P* as in Figure 16.22.

Consider now a free-body diagram of a beam segment at the loaded end as in Figure 16.23. The figure shows the shear flow distribution over the interior cross section of the segment. Observe that the shear flow creates a counterclockwise axial moment (from the perspective of the figure). The extent of the resulting axial rotation (or twist) is dependent on the lateral placement of the end load *P*.

If there is to be no rotation, the load *P* must create an equal magnitude but oppositely directed moment to that of the shear flow. Specifically, in view of Figure 16.23, if there is to be no twist of the beam, the load must be applied outside the closed end of the cross section as represented in Figure 16.24. The point *C** where the line of action of *P* intersects the neutral axis is called the "shear center." In the following sections, we illustrate the procedure for locating the shear center.

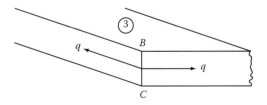

FIGURE 16.18   Equilibrium of shear forces on subsection ③.

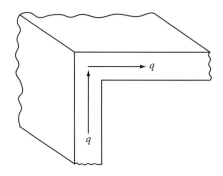

FIGURE 16.19 Shear flow around a corner in a right-angle web.

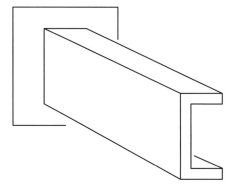

FIGURE 16.20 Cantilever, channel cross section, beam with open side to the right.

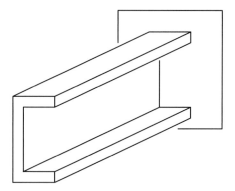

FIGURE 16.21 Right side view of cantilever, channel cross section, beam.

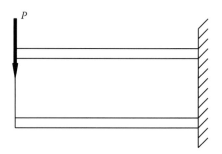

FIGURE 16.22 Right side view of end-loaded cantilever beam with a channel cross section.

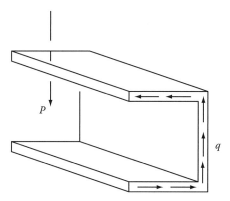

FIGURE 16.23 Rear view of the free-body diagram of the right beam segment.

16.5 EXAMPLE: SHEAR CENTER OF A CHANNEL BEAM

The concepts discussed above can be illustrated and quantified by continuing our consideration of the end-loaded cantilever channel beam example, as represented in Figure 16.25. As noted earlier, this problem is often cited in texts on strength and mechanics of materials (see "References").

From the discussion of Section 16.4, it is clear that the shear center will be located on the left side, or outside of the channel as in Figure 16.24 as well as in Figure 16.26.

To determine the precise location of the shear center C^* (dimension d in Figure 16.26) let the dimensions of the cross section be as shown in Figure 16.27 where b and h are the nominal base and height of the cross section respectively with the web thickness t being small compared to b and h.

Recall from Chapter 13 that for thick beams a varying bending moment along the beam produces a shear force V on the cross sections, which in turn leads to horizontal shear stresses in the beam. Specifically, Equation 13.11 states that the horizontal shear stress τ is

$$\tau = VQ/Ib \tag{16.6}$$

where Q is the moment of the area above (or beneath) the site where the shear stress is to be calculated, with the moment taken about the neutral axis; I is the second moment of area of the cross section about the neutral axis; and b is the width of the cross section at the site where the shear stress is to be calculated.

We can conveniently use Equation 16.6 to obtain an expression for the shear flow q in a web. To deduce this, recall from Section 16.3 that shear flows at perpendicular mating surfaces are of equal magnitude but with opposite directions as illustrated in Figure 16.12. Thus the shear flow in a web has the same magnitude with opposite direction to the horizontal shear flow obtained from the horizontal shear stress. The horizontal shear stress in turn is immediately obtained from Equation 16.1 by integrating through the web thickness.

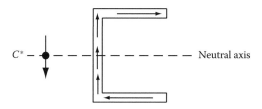

FIGURE 16.24 Load placement to counteract the shear flow moment and to produce zero twist.

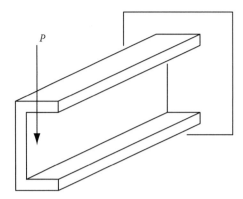

FIGURE 16.25 End-loaded cantilever beam with a channel cross section.

In Equation 16.6, the base width b becomes the web thickness t. Then with the thickness being small the shear flow q is simply

$$q = \int_{O}^{b} \tau \, dt \approx b\tau = VQ/I \tag{16.7}$$

By using Equation 16.7, we can determine the shear flow at all points of a webbed cross section. To illustrate and develop this, let a coordinate axis system (ξ, η) be placed upon the channel cross section with origin A, as shown in Figure 16.28. Let A, B, C, D, and E be selected points in the cross section. Let q_A be the shear flow at the origin A. Then from Equation 16.7, q_A is simply

$$q_A = VQ_A/I \tag{16.8}$$

where Q_A is the moment of the cross section area above A about the neutral axis. That is,

$$Q_A = (h/2)(t)(h/4) + (bt)(h/2) = (th^2/8) + (bth/2) \tag{16.9}$$

where again the web thickness t is assumed to be small compared to the base and height dimensions b and h. Therefore, q_A is

$$q_A = (V/I)[(th^2/8) + (bth/2)] \tag{16.10}$$

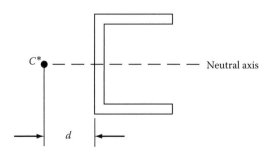

FIGURE 16.26 Shear center location.

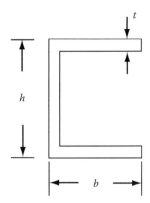

FIGURE 16.27   Channel cross section dimensions.

Next, for point B we have

$$q_B = (V/I)Q_B$$

where from Figure 16.28, Q_B is seen to be

$$Q_B = t[h/2) - \eta]\{\eta + [(h/2) - \eta]/2\} + (bt)(h/2)$$
$$= (t/2)[(h^2/4) - \eta^2] + bth/2 \tag{16.11}$$

where η is the vertical coordinate of B above the origin A. Therefore, q_B is

$$q_B = (V/I)\{(t/2)[(h^2/4) - \eta^2] + (bth/2)\} \tag{16.12}$$

For point C, we have

$$q_C = (V/I)Q_C \tag{16.13}$$

where from Figure 16.28, Q_C is seen to be

$$Q_C = (bt)(h/2) \tag{16.14}$$

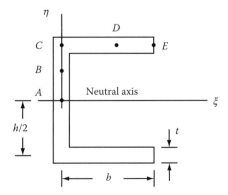

FIGURE 16.28   Coordinate axes and selected points of the channel cross section.

Therefore, q_C is

$$q_C = (V/I)(bt)(h/2) \tag{16.15}$$

For point D, we have

$$q_D = (V/I)Q_D \tag{16.16}$$

where from Figure 16.28, Q_D is seen to be

$$Q_D = (b - \xi)t(h/2) \tag{16.17}$$

where ξ is the horizontal coordinate of D. Therefore, Q_D is

$$Q_D = (V/I)[(bth/2) - (\xi th/2)] \tag{16.18}$$

Finally, for point E, at the end of the flange Q_E is seen to be zero. Therefore, q_E is

$$q_E = 0 \tag{16.19}$$

By summarizing the foregoing results, we can see the pattern of the shear flow throughout the web:

$$
\begin{aligned}
q_A &= (V/I)[(th^2/8) + (bth)/2)] \\
q_B &= (V/I)\{(t/2)[(h^2/4) - \eta^2] + (bth/2)\} \\
q_C &= (V/I)(bt)(h/2) \\
q_D &= (V/I)[(bth/2) - (\xi th/2)] \\
q_E &= 0
\end{aligned}
\tag{16.20}
$$

Observe that the shear flow has a quadratic distribution in the vertical web and a linear distribution in the horizontal flanges.

We can now obtain the resulting shear forces on the flanges and the vertical web by simply integrating the shear flow along the length of the flanges and web: specifically, let the shear forces in the flanges and web be H and $\hat{V}$, as in Figure 16.29, where notationally we use $\hat{V}$ to distinguish from the applied shear force V over the cross section. Then from Equations 16.3 and 16.18 H is seen to be

$$
\begin{aligned}
H &= \int_0^b q_D \mathrm{d}\xi = \int_0^b (V/I)[(bth/2) - (\xi th/2)]\mathrm{d}\xi \\
&= (V/I)[(bth/2)\xi - (th/2)(\xi^2/2)] \Big|_0^b
\end{aligned}
$$

or

$$H = (V/I)b^2 th/4 \tag{16.21}$$

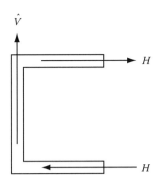

FIGURE 16.29 Resultant shear forces on the flanges and web of the channel beam cross section.

Similarly, from Equations 16.3 and 16.22, the vertical force $\hat{V}$ is

$$\hat{V} = \int_{h/2}^{h/2} q_B d\eta = \int_{-h/2}^{h/2} (V/I)\{(t/2)[h^2/4) - \eta^2] + (bth/2)\}d\eta$$

$$= (V/I)[(t/2)(h^2 4)\eta - (t/2)(\eta^3/3) + (bth/2)\eta] \Big|_{-h/2}^{h/2}$$

or

$$\hat{V} = (V/I)[(th^3/12) + (tbh^2/2)] \tag{16.22}$$

Remembering that I is the second moment of area of the beam cross section about the neutral axis, we see by inspection of Figure 16.28 that for small t, I is approximately

$$I = (th^3/12) + (tbh^2/2) \tag{16.23}$$

Then by comparing Equations 16.22 and 16.23, it is clear that the computed shear force $\hat{V}$ on the vertical web is approximately the same as the applied shear over the cross section. That is, for small t, we have

$$\hat{V} = V \tag{16.24}$$

Finally, to locate the shear center we simply need to place the line of action of the given load P outside (or to the left in the end view) of the beam cross section so that the axial moment created by the shear forces H and $\hat{V}$, on the flanges and web, is counteracted by the moment created by P. Specifically, in Figure 16.30 the line of action of P must be placed a distance d to the left of the cross section so that the system of forces shown is a zero system (see Section 1.5.1), that is, in both force and moment equilibrium.

From Figure 16.30 it is clear that force equilibrium occurs if

$$P = \hat{V} \tag{16.25}$$

Moment equilibrium will occur if

$$Pd = hH \tag{16.26}$$

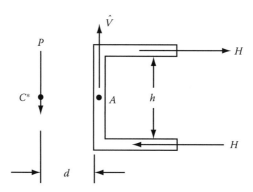

FIGURE 16.30 Placement of beam end load P to counteract the shear forces on the beam cross section.

Hence, from Equation 16.21 and Equations 16.23, 16.24, and 16.25, d is

$$d = hH/P = h(V/I)b^2th/4P$$
$$= b^2th^2/4I = (b^2th^2/4)[(th^3/12) + (tbh^2/2)]$$
$$= b^2/[(h/3) + 2b]$$

or

$$d = 3b/[6 + (h/b)] \tag{16.27}$$

This analysis and the results are of course specific to an end-loaded cantilever beam with an open-channel cross section. The procedures, however, are directly applicable for other web/flange cross section beams.

16.6 A NUMERICAL EXAMPLE

The expressions in the foregoing section are immediately applicable in locating the shear center, in determining the twisting moment if the line of action of a load is not through the shear center, and in determining the resulting distortion, for an end-loaded, open channel, cantilever beam. To illustrate the magnitude of the effects, consider a 25 ft. long beam with cross section dimensions as in Figure 16.31. As before, let C^* be the shear center and G be the centroid as represented in Figure 16.32.

With this configuration, the load will induce a shear flow through the flanges and web of the cross sections as illustrated in Figure 16.33. Figure 16.34 illustrates the resultant shear forces H and V in the flanges and web.

From the given data and Equations 16.21 and 16.24 it is obvious that H and $\hat{V}$ are

$$\hat{H} = (V/I)(thb^2/4) = (3000/126)[(0.25)(12)(5)^2/4] = 446.4\,\text{lb} \tag{16.28}$$

and

$$\hat{V} = V = 3000\,\text{lb} \tag{16.29}$$

where, from Equation 16.23, I is seen to be

$$I = (th^3/12) + (tbh^2/2) = [(0.25)(12)^3/12] + [(0.25)(5)(12)^2/2] = 126\,\text{in.}^4 \tag{16.30}$$

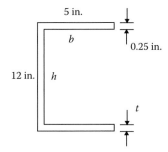

FIGURE 16.31   Channel beam cross section dimensions.

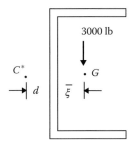

FIGURE 16.32   Shear center and centroid locations.

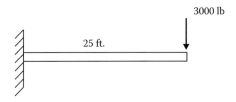

FIGURE 16.33   Beam support and loading.

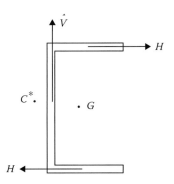

FIGURE 16.34   Resultant shear forces in the flanges and web of the cross section.

Equation 16.27 shows that the shear center C^* is located a distance d outside the web given by

$$d = 3b/[6 + h/b] = (3)(5)/[6 + (12/5)] = 1.786 \text{ in.} \tag{16.31}$$

A question arising is: to what extent is a twisting moment induced if the line of action of the end load is placed through the centroid G of the cross section? To answer this question, consider that G is located inside the channel cross section a distance $\bar{\xi}$ from the web as illustrated in Figure 16.32. An elementary analysis shows that for thin flanges and web $\bar{\xi}$ is

$$\bar{\xi} = b^2/(2b + h) = (5)^2/[(2)(5) + 12] = 1.136 \text{ in.} \tag{16.32}$$

The induced twisting moment T is then

$$T = P(d + \bar{\xi}) = 3000(1.786 + 1.136) = 8766 \text{ in./lb} \tag{16.33}$$

To put this in perspective, the rotation θ, or distorting twist, of the beam due to a misplaced load through the centroid is (see Equation 12.6):

$$\theta = T\ell/JG \tag{16.34}$$

where J is the second polar moment of area of the cross section relative to the shear center, ℓ is the beam length, and G is the shear modulus. For the cross section dimensions of Figure 16.31, J is seen to be

$$J = J_{\text{web}} + J_{\text{flange}} + J_{\text{flange}} \tag{16.35}$$

where J_{web} and J_{flange} are

$$J_{\text{web}} = (th^3/12) + (th)d^2 = [(0.25)(12)^3/12] + [(0.25)(12)(1.786)^2] = 45.57 \text{ in.}^4 \tag{16.36}$$

and

$$
\begin{aligned}
J_{\text{flange}} &= (tb^3/12) + (th)[(h/2)^2 + (d + b/2)^2] \\
&= [(0.25)(5)^3/12] + (0.25)(12)[(12/2)^2 + (1.786 + 5/2)^2] \\
&= 165.7 \text{ in.}^4
\end{aligned} \tag{16.37}
$$

Then J is

$$J = 45.57 + (2)(165.7) = 376.98 \text{ in.}^4 \tag{16.38}$$

If the beam is made of steel with G being approximately 11.5×10^6 psi, and its length ℓ is 25 ft., the twist is

$$
\begin{aligned}
\theta &= (8766)(25)(12)/(376.98)(11.5)(10)^6 = 6.066 \times 10^{-1} \text{ rad} \\
&= 3.476 \times 10^{-2} \text{ degrees}
\end{aligned} \tag{16.39}
$$

For many cases of practical importance, this would seem to be a relatively small and unimportant distortion. Thus for many webbed sections, where the shear center would seem to be important, the resulting distortion from randomly placed loading may not be harmful.

SYMBOLS

B	Beam width
C^*	Shear center
E	Web element
G	Shear modulus
H	Horizontal force
h	Beam height
I	Second moment of area
J	Second polar moment of area
ℓ	Beam length
P	Loading
q	Shear flow (see Equations 16.1 and 16.3)
Q	Moment of the area above (or beneath) a point where shear stress is to be calculated
t	Web thickness
T	Applied torque
V	Shear force on a cross section
$\hat{V}$	Computed shear force from the shear flow
X, Y, Z	Rectangular (Cartesian) coordinate axes
x, y, z	Point coordinates relative to X, Y, Z
θ	Twist angle
ξ, η	Coordinate axes
σ_{ij}	($i, j = 1, 2, 3$) Stress matrix components
τ	Shear stress
τ_{ij}	($i, j = x, y, z$) Shear stress on the I-face in the J-direction
τ_{xz}	Shear stress on the X-face in the Z-direction

REFERENCES

1. F. L. Singer, *Strength of Materials*, 2nd ed., Harper and Row, New York, 1962, pp. 478–486.
2. F. P. Beer and E. Russell Johnston, Jr., *Mechanics of Materials*, McGraw Hill, New York, 1981, pp. 252–281.
3. W. D. Pilkey and O. H. Pilkey, *Mechanics of Solids*, Quantum Publishers, New York, 1974, pp. 314–315.

Part IV

Plates, Panels, Flanges, and Brackets

Second only to beams, plates are the most widely used of all structural components. In buildings, plates and panels are used for floors, walls, roofs, doors, and windows. In vehicles, they also form flooring, windows, and door components. In addition, curved plates and panels make up the external structures of cars, trucks, boats, ships, and aircraft. In machines, these thin members form virtually all the structural components and many of the moving parts.

The design and analysis of plates, panels, flanges, and brackets is considerably more complex than that for beams, rods, or bars. The loading on plate structures, however, is usually simpler, and often consists of only a uniform pressure. Also, the behavior of beams provides insight into the behavior of plates, particularly in response to flexural-type loadings.

In this fourth part, we review the fundamental equations governing the structural behavior of plates and other associated thin-walled members. We consider various modelings and approximation methods that simplify the analysis without compromising the accuracy of the stress and displacement evaluations.

We begin with the flexural response of simple plates and then go on to more complex geometries and loadings in applications with panels, flanges, and brackets.

17 Plates: Bending Theory

17.1 HISTORICAL PERSPECTIVE AND INTRODUCTORY REMARKS

Plate theory and the behavior of plates as structural components, have been fascinating and popular subjects for analysts and structural engineers for hundreds of years. The study of plates dates back to the eighteenth century, long before the development of elasticity theory. Well known theorists associated with plate theory include Euler, Bernoulli, Lagrange, Poisson, Navier, Fourier, Kirchoff, Kelvin, Tait, Boussinesque, Levy, Love, VonKarman, and Reissner. The most important developments occurred during the nineteenth century in France, stimulated in part by Napoleon.

Plates are regarded as two-dimensional, thin, flat structures. Of particular interest is their response to loadings directed normal to their plane. The analytical focus is thus upon flexure (or bending) as opposed to in-plane loading.

At times, plates have been thought of as two-dimensional beams, particularly when they are bent in only one direction. More rigorous analyses require the solution of partial differential equations with various kinds of boundary conditions. As such, the number of simplifying assumptions needed to obtain closed-form solutions is staggering. As a consequence, analysts have been continually searching for approximation methods providing insight into plate behavior, enabling efficient structural design.

Plates may be divided into four general categories:

1. Thick plates or slabs (shear is the predominant consideration)
2. Plates with average thickness (flexure is the predominant consideration)
3. Thin plates (both flexural stress and in-plane tension are important considerations)
4. Membranes (in-plane tension and stretching are the most important considerations)

In this chapter, we focus on the second category, that is, plates sufficiently thin that shear effects can be neglected but also thick enough that in-plane forces are negligible.

17.2 MODELING AND SIMPLIFYING ASSUMPTIONS

A plate is modeled as a thin, initially flat, uniformly thick structural component supported on its edges and loaded in the direction normal to its plane. Figure 17.1 shows a portion of a plate together with coordinate axes directions.

Most modern theories of plate behavior such as those of Vinson et al. [1–4] are developed using the three-dimensional equations of linear elasticity and then reducing them to a two-dimensional form by integrating through the thickness of the plate. A number of simplifying assumptions enable this development. These are

1. The plate is initially flat with uniform thickness.
2. The plate thickness is small compared with the edge dimensions (rectangular plates) or the diameter (circular plates).
3. The plate is composed of a homogeneous, isotropic, and linear elastic material.
4. Loading is applied normal to the plane of the plate.

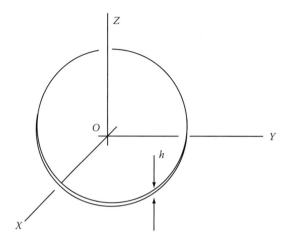

FIGURE 17.1   Coordinate axes for a plate segment.

5. The plate supports the loading by its resistance to flexure (bending). Equivalently in-plane ("membrane") forces are negligible in their support of loadings normal to the plate.
6. The maximum displacement of the plate is less than the thickness of the plate.
7. Line elements normal to the middle surface before loading remain straight and normal to the middle surface during and after loading.
8. Line elements normal to the middle surface undergo neither lengthening nor shortening during loading.
9. Stresses normal to the plane of the plate are small compared with the flexural stresses.
10. Slopes of the plate surfaces, due to bending, are small.

Analytically, these assumptions imply that the displacement components (u, v, w), in Cartesian coordinates as in Figure 17.1, have the following forms:

$$u(x, y, z) = z\alpha\,(x, y) \tag{17.1}$$

$$v(x, y, z) = z\beta\,(x, y) \tag{17.2}$$

$$w(x, y, z) = w\,(x, y) \tag{17.3}$$

17.3   STRESS RESULTANTS

As noted earlier, we can obtain governing equations for plate flexure by integrating the equations of elasticity through the plate thickness. The resulting plate equations are then simpler and fewer in number than the elasticity equations.

In the process, as we integrate the stresses through the thickness, we obtain "stress resultants," and in a similar manner as we integrate the moments of the stresses about coordinate axes, we obtain bending and twisting moments.

To develop this, it is helpful to first recall the sign conventions of elasticity as discussed in Chapter 4 (see Section 4.2). Specifically, positive directions are in the positive (increasing value) coordinate axis directions. A "positive face" of an element is a surface normal to a coordinate axis such that when crossing the surface, from inside the element to the outside, a point moves in the positive axis direction. Negative directions and negative faces are similarly defined. Stresses are

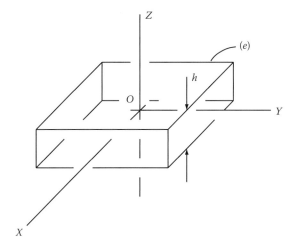

FIGURE 17.2   Plate element with coordinate axes.

then positive or negative as follows: a stress is positive if it is exerted on a positive face in a positive direction or on a negative face in a negative direction. A stress is negative if it is exerted on a positive face in a negative direction or on a negative face in a positive direction.

Consider a rectangular element (e) of a plate as in Figure 17.2. For convenience, let the X-, Y-, Z-axes be oriented and placed relative to the element as shown, with origin O at the center of the element.

As before, let the stresses on the faces of (e) be designated by σ_{ij} where i and j can be x, y, or z, with the first subscript pertaining to the face and the second to the direction.

Consider the stresses on the positive X-face of (e): σ_{xx}, σ_{xy}, and σ_{xz}. First, for σ_{xx}, by integration through the thickness h we have

$$\int_{-h/2}^{h/2} \sigma_{xx} dz \overset{D}{=} N_{xx} \tag{17.4}$$

where N_{xx} is a force per unit edge length and directed normal to the X-face, as represented in Figure 17.3. Let the line of action of N_{xx} be placed through the center of the X-face. N_{xx} is thus along the midplane of (e) and is a "membrane" force.

Next, for σ_{xy}, we have

$$\int_{-h/2}^{h/2} \sigma_{xy} dz \overset{D}{=} S_{xy} \tag{17.5}$$

where S_{xy} is a shear force per unit edge length and, like N_{xx}, let its line of action be placed through the center of the X-face. S_{xy} is directed along the Y-axis and it is also in the midplane of (e). Therefore S_{xy} is also a membrane force.

Finally, for σ_{xz}, we have

$$\int_{-h/2}^{h/2} \sigma_{xz} dz \overset{D}{=} Q_{xz} \tag{17.6}$$

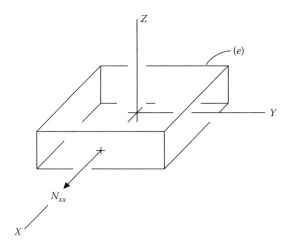

FIGURE 17.3 Stress resultant normal to the X-face.

where Q_{xz} is also a shear force per unit length and, like N_{xx} and S_{xy}, let its line of action be placed through the center of the X-face. Q_{xz} is directed along the Z-axis and is therefore perpendicular to the midplane of (e). Thus, unlike N_{xx} and S_{xy}, Q_{xz} is not a membrane force.

Figure 17.4 provides a representation of N_{xx}, S_{xy}, and Q_{xz}.

Consider now the stresses on the positive Y-face of (e): σ_{yx}, σ_{yy}, and σ_{yz}. By an analysis similar to that on the X-face we have

$$\int_{-h/2}^{h/2} \sigma_{yx}dz \overset{D}{=} S_{yx}, \qquad \int_{-h/2}^{h/2} \sigma_{yy}dz \overset{D}{=} N_{yy}, \qquad \int_{-h/2}^{h/2} \sigma_{yz}dz \overset{D}{=} Q_{yz} \qquad (17.7)$$

where the stress resultants S_{yx}, N_{yy}, and Q_{yz} are forces per unit edge length and are directed parallel to the X-, Y-, and Z-axes, respectively. If we let the lines of action of S_{yx}, N_{yy}, and Q_{yz} pass through the center of the Y-face, we see that S_{yx} and N_{yy} are membrane forces and that Q_{yz} is perpendicular to the midplane of (e). Figure 17.5 provides a representation of these resultants.

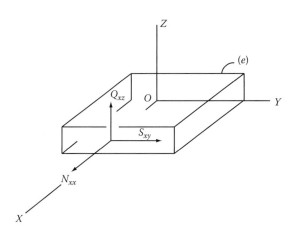

FIGURE 17.4 Stress resultants on the X-face of a plate element.

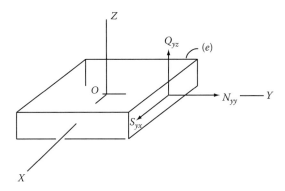

FIGURE 17.5 Stress resultants on the Y-face of a plate element.

Observe that since $\sigma_{xy} = \sigma_{yx}$ we have

$$S_{xy} = S_{yx} \tag{17.8}$$

Observe further that the stress resultants are not actual forces but instead they are entities of equivalent force systems (see Section 1.5.3). As such, the lines of action of the resultants can be placed through arbitrary points (in this case, the centers of the element faces) and then equivalency is ensured by calculating the moments about those points. The following section documents these moments.

17.4 BENDING AND TWISTING (WARPING) MOMENTS

Consider the modeling of the stress systems on the faces of a plate element by equivalent force systems consisting of stress resultants passing through the face centers together with stress couples. Consider now the moments of these stress couples: specifically, consider the moments of the stresses σ_{xx}, σ_{xy}, and σ_{xz}, acting on the X-face of a plate element, about the X- and Y-axes (see Figure 17.6). First, for σ_{xx}, from our experience with beam analysis (see Chapter 8), and with our assumptions of line elements normal to the undeformed midplate plane remaining straight and normal to the plane during bending, we expect σ_{xx} to vary linearly in the Z-direction, through the thickness of the plate. As such σ_{xx} will create a moment (a flexural moment) about the Y-axis. Following the notation of Ref. [1], we call this moment M_x and define it as

$$M_x = \int_{-h/2}^{h/2} z\sigma_{xx}dz \tag{17.9}$$

Observe that with the σ_{xx} stresses being directed along the X-axis, they will have no moment about the X-axis.

Next, for the shear stresses σ_{xy} we will have a "twisting" or "warping" moment about the X-axis which we call T_{xy}, defined as

$$T_{xy} = \int_{-h/2}^{h/2} z\sigma_{xy}dz \tag{17.10}$$

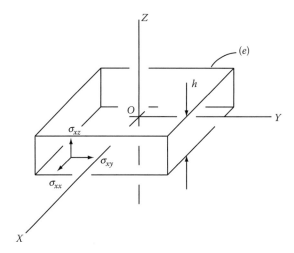

FIGURE 17.6 Plate element.

Observe that with the σ_{xy} being directed along the Y-axis, they will have no moment about the Y-axis.

Finally, for the vertical shear stresses σ_{xz}, we expect, from our experience in beam analysis, that the σ_{xz} will have a symmetric distribution across the face and then will not produce a moment about either the X- or Y-axes.

In a similar analysis for the stresses on the Y-face, we see that the normal stresses σ_{yy} produce a flexural moment M_y about the X-axis defined as

$$M_y = \int_{-h/2}^{h/2} z\sigma_{yy}dz \tag{17.11}$$

σ_{yy}, being parallel to the Y-axis, do not have any moment about the Y-axis.

The shear stresses σ_{yx} will produce a warping moment T_{yx} about the X-axis as

$$T_{yx} = \int_{-h/2}^{h/2} z\sigma_{yx}dz \tag{17.12}$$

σ_{yx}, being parallel to the X-axis, will have no moment about the X-axis.

Observe that with σ_{xy} being equal to σ_{yx} we have

$$T_{xy} = T_{yx} \tag{17.13}$$

Finally, the vertical shear stresses σ_{yz} being symmetrically distributed across the Y-face, will have no moments about either the X- or Y-axes.

17.5   EQUILIBRIUM FOR A PLATE ELEMENT

Recall from Equations 4.30 through 4.32 that for a body under loading, the equilibrium of a rectangular "brick" element within the body requires that the stresses satisfy the equilibrium equations

$$\frac{\partial \sigma_{xx}}{\partial x} + \frac{\partial \sigma_{xy}}{\partial y} + \frac{\partial \sigma_{xz}}{\partial z} = 0 \tag{17.14}$$

$$\frac{\partial \sigma_{yx}}{\partial x} + \frac{\partial \sigma_{yy}}{\partial y} + \frac{\partial \sigma_{yz}}{\partial z} = 0 \tag{17.15}$$

$$\frac{\partial \sigma_{zx}}{\partial x} + \frac{\partial \sigma_{zy}}{\partial y} + \frac{\partial \sigma_{zz}}{\partial z} = 0 \tag{17.16}$$

where the edges of the element are parallel to the coordinate axes.

We can use these equilibrium equations to obtain equilibrium equations for plate elements by integrating the equations through the plate thickness. To this end, consider again the rectangular plate element of Figure 17.2 and as shown in Figure 17.7. Recall that the simplifying assumptions of plate theory require the plate support loading to be perpendicular to its plane by flexure, that is, by forces and moments on the plate element faces normal to the *X*- and *Y*-axes. The first two equilibrium equations (Equations 17.14 and 17.15) involve stresses on these faces. Therefore, we will initially consider integration of these equations and reserve analysis of the third equation (Equation 17.16) until later.

First consider Equation 17.14: by integrating through the plate thickness, we have

$$\int_{-h/2}^{h/2} \frac{\partial \sigma_{xx}}{\partial x}\,dz + \int_{-h/2}^{h/2} \frac{\partial \sigma_{xy}}{\partial y}\,dz + \int_{-h/2}^{h/2} \frac{\partial \sigma_{xz}}{\partial z}\,dz = 0 \tag{17.17}$$

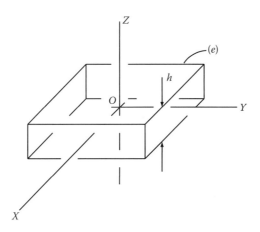

FIGURE 17.7   Plate element.

Since x, y, and z are independent variables, we have

$$\frac{\partial}{\partial x}\int_{-h/2}^{h/2}\sigma_{xx}dz+\frac{\partial}{\partial y}\int_{-h/2}^{h/2}\sigma_{xy}dz+\int_{-h/2}^{h/2}\frac{\partial\sigma_{xz}}{\partial z}dz=0 \tag{17.18}$$

In view of the definitions of Equations 17.4 and 17.5 we have

$$\frac{\partial N_{xx}}{\partial x}+\frac{\partial S_{xy}}{\partial y}+\sigma_{xz}\Big|_{-h/2}^{h/2}=0 \tag{17.19}$$

Regarding the last term, observe that with the elements of the stress matrix being symmetric, that is $\sigma_{ij}=\sigma_{ji}$, we have $\sigma_{xz}=\sigma_{zx}$. But σ_{zx} is zero on the plate surface since the loading is assumed to be directed perpendicular to the plane of the plate. Therefore, the third term of Equation 17.18 is zero. That is,

$$\sigma_{xz}\Big|_{-h/2}^{h/2}=\sigma_{zx}\Big|_{-h/2}^{h/2}=\sigma_{zx}(h/2)-\sigma_{zx}(-h/2)=0 \tag{17.20}$$

Thus Equation 17.19 takes the simple form

$$\frac{\partial N_{xx}}{\partial x}+\frac{\partial S_{xy}}{\partial y}=0 \tag{17.21}$$

Similarly, by integrating Equation 17.15 through the thickness we obtain

$$\frac{\partial S_{yx}}{\partial x}+\frac{\partial N_{yy}}{\partial y}=0 \tag{17.22}$$

Next, consider the moments of the stresses on the X- and Y-faces about the X- and Y-axes: If we multiply Equation 17.14 by z and integrate through the plate thickness we have

$$\int_{-h/2}^{h/2}z\frac{\partial\sigma_{xx}}{\partial x}dz+\int_{-h/2}^{h/2}z\frac{\partial\sigma_{xy}}{\partial y}dz+\int_{-h/2}^{h/2}z\frac{\partial\sigma_{xz}}{\partial z}=0 \tag{17.23}$$

In the first two terms with x, y, and z being independent, we can move the derivatives outside the integrals and in the third term we can integrate by parts, obtaining

$$\frac{\partial}{\partial x}\int_{-h/2}^{h/2}z\sigma_{xx}dz+\frac{\partial}{\partial y}\int_{-h/2}^{h/2}z\sigma_{xy}dz+z\sigma_{xz}\Big|_{-h/2}^{h/2}-\int_{-h/2}^{h/2}\sigma_{xz}dz=0 \tag{17.24}$$

In the third term (the integrated term) with the symmetry of the stress matrix and with the plate being loaded only in the direction normal to the plate, we see that the term is zero. Finally, by using the definitions of Equations 17.9, 17.10, and 17.6, Equations 17.23 and 17.24 become

$$\frac{\partial \mathbf{M}_{x}}{\partial x}+\frac{\partial \mathbf{T}_{xy}}{\partial y}-Q_{xz}=0 \tag{17.25}$$

By a similar analysis in integrating Equation 17.15 we obtain

$$\frac{\partial \mathbf{T}_{yx}}{\partial x} + \frac{\partial \mathbf{M}_y}{\partial y} - Q_{yz} = 0 \tag{17.26}$$

Then consider the third equilibrium equation (Equation 17.16): if we integrate through the plate thickness we have

$$\int_{-h/2}^{h/2} \frac{\partial \sigma_{zx}}{\partial x}\,\mathrm{d}z + \int_{-h/2}^{h/2} \frac{\partial \sigma_{zy}}{\partial y}\,\mathrm{d}z + \int_{-h/2}^{h/2} \frac{\partial \sigma_{zz}}{\partial z}\,\mathrm{d}z = 0 \tag{17.27}$$

Then by the independence of x, y, and z, the symmetry of the stress matrix, and in view of the definition of Equations 17.6 and 17.7, we have

$$\frac{\partial Q_{xz}}{\partial x} + \frac{\partial Q_{yz}}{\partial y} + \sigma_{zz} \Big|_{-h/2}^{h/2} = 0 \tag{17.28}$$

Recall that a simplifying assumption of plate theory is that the external loading is normal to the plane of the plate (assumption 4), and also in the interior of the plate, the stresses normal to the plane of the plate (σ_{xx}) are relatively small and can be neglected (assumption 9). Let the positive Z-axis designate the direction of positive loading. Then as a consequence of these assumptions we may regard the loading as being applied to either the upper or the lower plate surfaces, or equivalently to the mid plane. Let $p(x, y)$ be the loading. Then if we consider the loading in terms of the surface stresses, $p(x, y)$ may be expressed as

$$p(x, y) = \sigma_{zz} \Big|_{h/2} - \sigma_{zz} \Big|_{-h/2} \tag{17.29}$$

Therefore Equation 17.28 becomes

$$\frac{\partial Q_{xz}}{\partial x} + \frac{\partial Q_{yz}}{\partial y} + p(x,y) = 0 \tag{17.30}$$

Finally, consider integrating the moments of the stress derivatives in Equation 17.16

$$\int_{-h/2}^{h/2} z \frac{\partial \sigma_{zx}}{\partial x}\,\mathrm{d}z + \int_{-h/2}^{h/2} z \frac{\partial \sigma_{zy}}{\partial y}\,\mathrm{d}z + \int_{-h/2}^{h/2} z \frac{\partial \sigma_{zz}}{\partial z}\,\mathrm{d}z = 0 \tag{17.31}$$

It happens in view of the simplifying assumptions of plate theory, that each of these terms is either zero or negligible. Consider the first term: again due to the independence of x, y, and z, we have

$$\int_{-h/2}^{h/2} z \frac{\partial \sigma_{zx}}{\partial x}\,\mathrm{d}z = \frac{\partial}{\partial x} \int_{-h/2}^{h/2} z\sigma_{zx}\,\mathrm{d}z \tag{17.32}$$

Observe that, as in the theory of beam bending, the shear stress distribution across the plate thickness is expected to be parabolic or at least symmetric, that is, an even function symmetric about the midplane. Then with z being an odd function, the antiderivative will be even and with equal limits, the integral is zero.

A similar reasoning provides the same result for the second term.

In the third term, by integrating by parts, we have

$$\int_{-h/2}^{h/2} z \frac{\partial \sigma_{zz}}{\partial z} dz = z\sigma_{zz} \Big|_{-h/2}^{h/2} - \int_{-h/2}^{h/2} \sigma_{zz} dz$$

$$= (h/2)\sigma_{zz} \Big|_{h/2} -(-h/2) \Big|_{-h/2} \sigma_{zz} - \int_{-h/2}^{h/2} \sigma_{zz} dz \qquad (17.33)$$

Due to the assumption on the loading on the plate we have

$$\sigma_{zz} \Big|_{h/2} = -\sigma_{zz} \Big|_{-h/2} \qquad (17.34)$$

Thus the first two terms of Equation 17.33 cancel and the third term is insignificant in view of assumption 9 which states that the normal stresses σ_{zz} are small in the interior of the plate.

17.6 SUMMARY OF TERMS AND EQUATIONS

For reference purposes, it is helpful to summarize the foregoing results. The coordinate directions for a plate element are shown again in Figure 17.8. The plate thickness h is small compared with the in-plane dimensions of the plate but still sufficiently large that membrane forces, if they exist, do not affect the flexural (bending) forces. That is, loads applied normal to the plane of the plate are supported by flexural forces as opposed to membrane (or midplane) forces. Finally, in Figure 17.8 the Z-axis is normal to the plane of the plate and the X- and Y-axes are in the midplane of the plate.

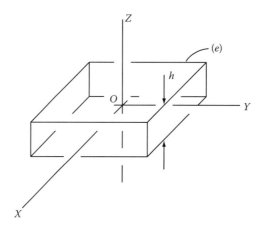

FIGURE 17.8 Plate element and coordinate directions.

17.6.1 IN-PLANE NORMAL (MEMBRANE) FORCES

$$N_{xx} \underset{(17.4)}{=} \int_{-h/2}^{h/2} \sigma_{xx}dz \quad N_{yy} \underset{(17.7)}{=} \int_{-h/2}^{h/2} \sigma_{yy}dz \qquad (17.35)$$

(Numbers under the equal sign refer to the original defining equation numbers.)

17.6.2 IN-PLANE SHEAR FORCES

$$S_{xy} = S_{yx} \underset{(17.5)}{=} \int_{-h/2}^{h/2} \sigma_{xy}dz \underset{(17.7)}{=} \int_{-h/2}^{h/2} \sigma_{yx}dz \qquad (17.36)$$

17.6.3 VERTICAL (Z-DIRECTION) SHEAR FORCES

$$Q_{xz} \underset{(17.6)}{=} \int_{-h/2}^{h/2} \sigma_{xz}dz \quad Q_{yz} \underset{(17.7)}{=} \int_{-h/2}^{h/2} \sigma_{yz}dz \qquad (17.37)$$

17.6.4 BENDING MOMENTS

$$M_x \underset{(17.9)}{=} \int_{-h/2}^{h/2} z\sigma_{xz}dz \quad M_y \underset{(17.11)}{=} \int_{-h/2}^{h/2} z\sigma_{yy}dz \qquad (17.38)$$

17.6.5 TWISTING MOMENTS

$$T_{xy} \underset{(17.10)}{=} \int_{-h/2}^{h/2} z\sigma_{xy}dz = T_{yx} \underset{(17.12)}{=} \int_{-h/2}^{h/2} z\sigma_{yx}dz \qquad (17.39)$$

17.6.6 LOADING CONDITIONS

Loads are applied in the direction normal to the plane of the plate. With the plate being thin, these loads may be regarded as equivalent to equally divided forces on the upper and lower surfaces of the plate as

$$p(x, y) \underset{(17.29)}{=} \sigma_{zz}\Big|_{h/2} - \sigma_{zz}\Big|_{-h/2} \qquad (17.40)$$

Alternatively, the loading may be regarded as being applied at the midplane.

17.6.7 EQUILIBRIUM EQUATIONS

(i) *In-plane (membrane) forces:*

$$\frac{\partial N_{xx}}{\partial x} + \frac{\partial S_{xy}}{\partial y} \underset{(17.21)}{=} 0, \quad \frac{\partial S_{yx}}{\partial x} + \frac{\partial S_{yy}}{\partial y} \underset{(17.22)}{=} 0 \tag{17.41}$$

(ii) *Vertical (Z-direction) forces:*

$$\frac{\partial Q_{xz}}{\partial x} + \frac{\partial Q_{yz}}{\partial y} + p(x, y) \underset{(17.20)}{=} 0 \tag{17.42}$$

(iii) *Moment equations:*

$$\frac{\partial M_x}{\partial x} + \frac{\partial T_{xy}}{\partial y} - Q_{xz} = 0, \quad \frac{\partial T_{yx}}{\partial x} + \frac{\partial M_y}{\partial y} - Q_{yz} = 0 \tag{17.43}$$

17.6.8 COMMENT

The normal stress σ_{zz} is assumed to be small in the interior of the plate, but on the surface the normal stress may be interpreted as the loading $p(x, y)$, as in Equation 17.40. Next, on the surface, the loading is assumed to be normal to the surface. Consequently the shear stresses on the surface are zero. This in turn means that the shear forces Q_{xz} and Q_{yz} are zero at the surface. Finally, as in beam theory, the shear forces are assumed to be quadratic across the plate thickness and the normal stresses σ_{zz} and σ_{yy} are assumed to be linear across the thickness.

17.7 STRESS–STRAIN–DISPLACEMENT RELATIONS

In Chapter 7, we developed stress–strain equations in Cartesian coordinates which, from Equations 7.49 through 7.54, can be expressed as

$$\varepsilon_{xx} = (1/E)\left[\sigma_{xx} - \nu(\sigma_{yy} + \sigma_{zz})\right] \tag{17.44}$$

$$\varepsilon_{yy} = (1/E)\left[\sigma_{yy} - \nu(\sigma_{zz} + \sigma_{xx})\right] \tag{17.45}$$

$$\varepsilon_{zz} = (1/E)\left[\sigma_{zz} - \nu(\sigma_{xx} + \sigma_{yy})\right] \tag{17.46}$$

$$\varepsilon_{xy} = (1/2G)\sigma_{xy} \tag{17.47}$$

$$\varepsilon_{yz} = (1/2G)\sigma_{yz} \tag{17.48}$$

$$\varepsilon_{zx} = (1/2G)\sigma_{zx} \tag{17.49}$$

where ε_{ij} $(i, j = x, y, z)$ are elements of the strain tensor, E is the elastic constant, G is the shear modulus and ν is Poisson's ratio. Earlier, in Chapter 5, we developed the strain–displacement relations:

$$\varepsilon_{xx} = \partial u/\partial x, \quad \varepsilon_{yy} = \partial v/\partial y, \quad \varepsilon_{zz} = \partial w/\partial z \tag{17.50}$$

$$\varepsilon_{xy} = (1/2)(\partial u/\partial y + \partial v/\partial x), \quad \varepsilon_{yz} = (1/2)(\partial v/\partial z + \partial w/\partial x), \quad \varepsilon_{zx} = 1/2(\partial w/\partial x + \partial u/\partial z) \tag{17.51}$$

(See Equations 5.10, 5.11, 5.12, 5.15, 5.16, 5.17, and 5.32.)

From our simplifying assumptions based upon the thinness of the plate, we have the displacements u, v, and w of the form

$$u = z\alpha(x, y), \quad v = z\beta(x, y), \quad w = w(x, y) \tag{17.52}$$

where α and β are constants (see Equations 17.1 through 17.3). Also, the normal stress σ_{zz} in the interior of the plate is small and negligible (assumption 9), so that for the purpose of analysis we have

$$\sigma_{zz} = 0 \tag{17.53}$$

By substituting Equations 17.52 and 17.53 into 17.50 and 17.51, we obtain:

$$\varepsilon_{xx} = \partial u/\partial x = z\partial\alpha/\partial x = (1/E)(\sigma_{xx} - \nu\sigma_{yy}) \tag{17.54}$$

$$\varepsilon_{yy} = \partial v/\partial y = z\partial\beta/\partial y = (1/E)(\sigma_{yy} - \nu\sigma_{xx}) \tag{17.55}$$

$$\varepsilon_{zz} = \partial w/\partial z = 0 = -(\nu/E)(\sigma_{xx} + \sigma_{yy}) \tag{17.56}$$

$$2\varepsilon_{xy} = \partial u/\partial y + \partial v/\partial x = z\partial\alpha/\partial y + z\partial\beta/\partial x = (1/G)\sigma_{xy} \tag{17.57}$$

$$2\varepsilon_{yz} = \partial v/\partial z + \partial w/\partial y = \beta + \partial w/\partial y = (1/G)\sigma_{yz} \tag{17.58}$$

$$2\varepsilon_{zx} = \partial w/\partial x + \partial u/\partial z = \partial w/\partial x + \alpha = (1/G)\sigma_{zx} \tag{17.59}$$

By integrating these equations through the thickness of the plate, we can obtain the constitutive equations (or reduced stress-displacement) equations for a plate.

17.8 INTEGRATION OF STRESS–STRAIN–DISPLACEMENT EQUATIONS THROUGH THE THICKNESS OF THE PLATE

Consider first Equation 17.54:

$$z\partial\alpha/\partial x = (1/E)(\sigma_{xx} - \nu\sigma_{yy}) \tag{17.60}$$

Multiplying by z and integrating we have

$$\int_{-h/2}^{h/2} z^2 \partial\alpha/\partial x\, dz = (1/E) \int_{-h/2}^{h/2} z\sigma_{xx}dz - (\nu/E) \int_{-h/2}^{h/2} z\sigma_{yy}dz \tag{17.61}$$

Then in view of Equation 17.38, we have

$$(h^3/12)\, \partial\alpha/\partial x = (1/E)\, (M_x - \nu\, M_y) \tag{17.62}$$

Next, recall the basic assumption of plate theory that line elements normal to the middle surface before loading remain straight and normal to the middle surface during and after loading (assumption 7). This means that locally the plate is not distorted during bending, which in turn means that the shear strains on surfaces normal to the Z-axis are zero. That is,

$$\varepsilon_{zx} = 0 \quad \text{and} \quad \varepsilon_{zy} = 0 \tag{17.63}$$

or

$$\partial u/\partial z + \partial w/\partial x = 0 \quad \text{and} \quad \partial v/\partial z + \partial w/\partial y = 0 \qquad (17.64)$$

or in view of Equations 17.1 and 17.2, we have

$$\partial w/\partial x = -\alpha \quad \text{and} \quad \partial w/\partial y = -\beta \qquad (17.65)$$

Then by substituting the first of these results into Equation 17.64 we obtain

$$(1/E)(M_x - \nu M_y) = -(\partial^2 w/\partial x^2)(h^3/12) \qquad (17.66)$$

Next consider Equation 17.55:

$$z\partial \beta/\partial y = (1/E)(\sigma_{yy} - \nu\sigma_{xx}) \qquad (17.67)$$

By an analysis similar to the foregoing we obtain

$$(1/E)(M_y - \nu M_x) = -(\partial^2 w/\partial y^2)(h^3/12) \qquad (17.68)$$

Thirdly, consider Equation 17.56:

$$\partial w/\partial z = 0 - (\nu/E)(\sigma_{xx} + \sigma_{yy}) \qquad (17.69)$$

Since this equation represents Z-displacement derivatives in the Z-direction, which are small, the terms do not contribute to flexural moments. Therefore, we can ignore the moment of this equation.
    Consider Equation 17.57:

$$z(\partial\alpha/\partial y) + z(\partial\beta/\partial x) = (1/G)\sigma_{xy} \qquad (17.70)$$

Multiplying by z and integrating, we have

$$\int_{-h/2}^{h/2} z^2(\partial\alpha/\partial y)\, dz + \int_{-h/2}^{h/2} z^2(\partial\beta/\partial x)\, dz = (1/G)\int_{-h/2}^{h/2} z\sigma_{xy}dz \qquad (17.71)$$

or in view of Equation 17.39 we have

$$(h^3/12)(\partial\alpha/\partial y + \partial\beta/\partial x) = (1/G)T_{xy} \qquad (17.72)$$

By using Equation 17.65 we can express $\partial\alpha/\partial y$ and $\partial\beta/\partial x$ in terms of second mixed derivatives of w so that Equation 17.72 takes the simplified form

$$\frac{(1+\nu)}{E}T_{xy} = -(h^3/12)\frac{\partial^2 w}{\partial x\,\partial y} \qquad (17.73)$$

where we have replaced G by $E/2(1+\nu)$ (see Equation 7.48).
    Finally, regarding Equations 17.58 and 17.59, we have already incorporated them into our analysis through the use of Equation 17.65.

Consider next the direct integration of the stress–displacement relations of Equations 17.54 through 17.59, which will involve the in-plane, or membrane force effects. For Equation 17.54, we have

$$\int_{-h/2}^{h/2} z\partial\alpha/\partial x\, dz = (1/E) \int_{-h/2}^{h/2} (\sigma_{xx} - \nu\sigma_{yy})\, dz \tag{17.74}$$

or in view of Equation 17.35 we have

$$(\partial\alpha/\partial x)(z^2/2) \Big|_{-h/2}^{h/2} = 0 = (1/E)N_{xx} - (\nu/E)N_{yy}$$

or

$$N_{xx} - \nu N_{yy} = 0 \tag{17.75}$$

Similarly, for Equation 17.55, we obtain

$$N_{yy} - \nu N_{xx} = 0 \tag{17.76}$$

Next, for Equation 17.56 we have

$$\int_{-h/2}^{h/2} \sigma_{xx} dz + \int_{-h/2}^{h/2} \sigma_{yy} dz = 0 \tag{17.77}$$

or

$$N_{xx} + N_{yy} = 0 \tag{17.78}$$

For Equation 17.57, we have

$$\int_{-h/2}^{h/2} z\frac{\partial\alpha}{\partial y} dz + \int_{-h/2}^{h/2} z\frac{\partial\beta}{\partial x} dz = (1/G) \int_{-h/2}^{h/2} \sigma_{xy} dz \tag{17.79}$$

or

$$(\partial\alpha/\partial y)z^2/2 \Big|_{-h/2}^{h/2} + (\partial\beta/\partial x)z^2/2 \Big|_{-h/2}^{h/2} = (1/G)S_{xy}$$

Since the integrated terms cancel to zero, we have

$$S_{xy} = 0 \tag{17.80}$$

Finally, for Equations 17.58 and 17.59, recall that the assumptions of plate theory require that there be no distortion within the plate during bending (assumption 7). This means that the shear strains ε_{xz} and ε_{yz} are zero (see Equations 17.63, 17.58, and 17.59) leading to

$$(1/G)\sigma_{yz} = 0 \quad \text{and} \quad (1/G)\sigma_{xz} = 0 \tag{17.81}$$

By integrating through the thickness we have

$$(1/G) \int_{-h/2}^{h/2} \sigma_{yz} dz = (1/G)Q_{yz} = 0 \quad \text{and} \quad (1/G) \int_{-h/2}^{h/2} \sigma_{xz} dz = (1/G)Q_{xz} = 0 \tag{17.82}$$

These expressions appear to present a contradiction in view of Equation 17.30, which states that the shear forces Q_{xz} and Q_{yz} must support the surface normal loading and thus cannot be zero. The explanation, or resolution, is that Equations 17.30 and 17.82 are both within the range of the approximations of plate theory. Specifically, no distortion implies an infinite shear modulus G, which satisfies Equation 17.82. An infinite value of G, however, implies an infinite elastic modulus E, which creates difficulties in other equations.

A better interpretation is that in the flexure of a plate the material near the surface provides the flexural strength. Also, since the external loading is normal to the plate surface, the shear stresses on the surface and consequently in the regions close to the surface are zero. In the midplane regions, however, the flexural support is minimal. But here the shear is not zero. Therefore Equation 17.82 may be viewed as approximately satisfying the flexural response for a plate, particularly in the surface regions of the plate. Alternatively, Equation 17.30 may be viewed as approximately satisfying the loading equilibrium of the plate, particularly in the interior, midplane region.

17.9   GOVERNING DIFFERENTIAL EQUATIONS

We can now obtain the governing differential equation for plate flexure by combining the equilibrium equations and the stress–strain (moment–slope) equations. To this end, it is helpful to list some principal relations from the foregoing sections:

17.9.1   EQUILIBRIUM EQUATIONS (SEE SECTION 17.5)

(1) *Moment–shear relations:*

$$\frac{\partial M_x}{\partial x} + \frac{\partial T_{xy}}{\partial y} - Q_{xz} \underset{(17.25)}{=} 0 \tag{17.83}$$

$$\frac{\partial T_{yx}}{\partial x} + \frac{\partial M_y}{\partial y} - Q_{yz} \underset{(17.26)}{=} 0 \tag{17.84}$$

(2) *Shear–loading relation:*

$$\frac{\partial Q_{xy}}{\partial x} + \frac{\partial Q_{yz}}{\partial y} + p(x, y) \underset{(17.30)}{=} 0 \tag{17.85}$$

(3) *In-plane (membrane) forces:*

$$\frac{\partial N_{xx}}{\partial x} + \frac{\partial S_{xy}}{\partial y} \underset{(17.21)}{=} 0 \tag{17.86}$$

$$\frac{\partial S_{yx}}{\partial x} + \frac{\partial N_{yy}}{\partial y} \underset{(17.22)}{=} 0 \qquad (17.87)$$

17.9.2 Displacement/Shear Assumptions

(1) *Displacements:*

$$\mathrm{u} \underset{(17.1)}{=} z\alpha(\alpha, y) \quad v \underset{(17.2)}{=} z\beta(x, y) \quad w \underset{(17.3)}{=} w(x, y) \qquad (17.88)$$

(2) *Shear strains (in surface regions):*

$$\varepsilon_{zx} \underset{(17.63)}{=} 0, \quad \varepsilon_{zy} \underset{(17.63)}{=} 0 \qquad (17.89)$$

(3) *Surface slopes:*

$$\frac{\partial w}{\partial x} \underset{(17.65)}{=} -\alpha, \quad \frac{\partial w}{\partial y} \underset{(17.65)}{=} -\beta \qquad (17.90)$$

17.9.3 Moment–Curvature and In-Plane Force Relations

(1) *Moment–curvature:*

$$(1/E)(M_x - \nu M_y) \underset{(17.66)}{=} -\left(\frac{h^3}{12}\right)\frac{\partial^2 w}{\partial x^2} \underset{(17.66)}{=} \left(\frac{h^3}{12}\right)\frac{\partial \alpha}{\partial x} \qquad (17.91)$$

$$(1/E)(M_y - \nu M_x) \underset{(17.68)}{=} -\left(\frac{h^3}{12}\right)\frac{\partial^2 w}{\partial y^2} \underset{(17.65)}{=} \left(\frac{h^3}{12}\right)\frac{\partial \beta}{\partial y} \qquad (17.92)$$

$$\left(\frac{1+\nu}{E}\right)\mathrm{T}_{xy} \underset{(17.73)}{=} -\left(\frac{h^3}{12}\right)\frac{\partial^2 w}{\partial x \partial y} \underset{(17.62)}{=} \left(\frac{h^3}{24}\right)\left(\frac{\partial \alpha}{\partial y} + \frac{\partial \beta}{\partial x}\right) \qquad (17.93)$$

(2) *In-plane force relations:*

$$N_{xx} - \nu N_{yy} \underset{(17.75)}{=} 0 \qquad (17.94)$$

$$N_{yy} - \nu N_{xx} \underset{(17.76)}{=} 0 \qquad (17.95)$$

$$N_{xx} + N_{yy} \underset{(17.78)}{=} 0 \qquad (17.96)$$

$$S_{xy} \underset{(17.80)}{=} 0 \qquad (17.97)$$

17.9.4 Governing Equation

We can solve Equations 17.91 through 17.93 for M_x, M_y, and T_{xy} as

$$M_x = -D\left(\frac{\partial^2 w}{\partial x^2} + \nu \frac{\partial^2 w}{\partial y^2}\right) \qquad (17.98)$$

$$M_y = -D\left(\frac{\partial^2 w}{\partial y^2} + \nu \frac{\partial^2 w}{\partial x^2}\right) \qquad (17.99)$$

$$T_{xy} = -(1 - \nu)D\frac{\partial^2 w}{\partial x \partial y} \qquad (17.100)$$

where D is defined as

$$D \stackrel{D}{=} Eh^3/12(1 - \nu^2) \qquad (17.101)$$

Then by substituting these results into the equilibrium equations (Equations 17.83 and 17.84) we obtain

$$-D\frac{\partial}{\partial x}\left(\frac{\partial^2 w}{\partial x^2} + \frac{\partial^2 w}{\partial y^2}\right) = Q_{xz} \qquad (17.102)$$

and

$$-D\frac{\partial}{\partial x}\left(\frac{\partial^2 w}{\partial x^2} + \frac{\partial^2 w}{\partial y^2}\right) = Q_{xz} \qquad (17.103)$$

Let the operator ∇^2 be defined as

$$\nabla^2() \stackrel{D}{=} \frac{\partial^2()}{\partial x^2} + \frac{\partial^2()}{\partial y^2} \qquad (17.104)$$

Then Equations 17.102 and 17.103 have the simplified forms:

$$-D\frac{\partial}{\partial x}\nabla^2 w = Q_{xz} \qquad (17.105)$$

and

$$-D\frac{\partial}{\partial x}\nabla^2 w = Q_{yz} \qquad (17.106)$$

Finally by substituting for Q_{xz} and Q_{yz} in Equation 17.85, we have

$$D\nabla^4 w = p(x, y) \qquad (17.107)$$

or more explicitly

$$\frac{\partial^4 w}{\partial x^4} + 2\frac{\partial^4 w}{\partial x^2 \partial y^2} + \frac{\partial^4 w}{\partial y^2} = p/D \qquad (17.108)$$

An advantage of the form of Equation 17.107, in addition to its simplicity is that we can readily express it in polar coordinates and then apply it with circular plates.

By solving Equations 17.94 through 17.97 for N_{xx}, N_{yy}, and S_{xy}, we immediately obtain

$$N_{xx} = 0, \quad N_{yy} = 0, \quad S_{xy} = 0 \qquad (17.109)$$

These results are consistent with the loading being directed normal to the plane of the plate, and with the small displacement so that in-plane (membrane) effects are independent of flexural effects. Vinson [1] shows that it is possible to have in-plane forces without violating the assumptions of plate theory. That is, with small displacements a plate can independently support loading normal to the plate surface and in-plane (membrane) forces. In other words, the flexural and membrane effects are decoupled (see Ref. [1] for additional details).

17.10 BOUNDARY CONDITIONS

Consider a rectangular plate and an edge perpendicular to the X-axis: the common support and end conditions are (1) simple support (zero displacement and zero moment, along the edge); (2) clamped (zero displacement and zero rotation); (3) free; and (4) elastic. The following paragraphs list the resulting conditions on the displacements for these conditions.

17.10.1 SIMPLE (HINGE) SUPPORT

In this case, the plate edge has restricted (zero) displacement, but it is free to rotate (about an axis parallel to the edge, the Y-axis). That is

$$w = 0 \quad \text{and} \quad M_x = 0 \tag{17.110}$$

From Equation 17.98, M_x is expressed in terms of the displacement ω as

$$M_x = -D\left(\frac{\partial^2 w}{\partial x^2} + v\frac{\partial^2 w}{\partial y^2}\right) = 0 \tag{17.111}$$

17.10.2 CLAMPED (FIXED OR BUILT-IN) SUPPORT

In this case, the edge displacement and rotation are zero. That is

$$w = 0 \quad \text{and} \quad \partial w/\partial y = 0 \tag{17.112}$$

17.10.3 FREE EDGE

In this case, there are no external restrictions on the movement of the edge. That is, there are no forces nor moments applied to the edge. Analytically, this means

$$Q_{xz} = 0, \quad M_x = 0, \quad T_{xy} = 0 \tag{17.113}$$

A difficulty with these equations, however, is that we now have three boundary conditions whereas the biharmonic operator ∇^4 of the governing equation (Equation 17.107) requires only two conditions per edge.* Therefore, to be consistent with the assumptions of plate theory, we need to combine the conditions of Equation 17.113, reducing the number from three to two. This can be accomplished using an ingenious analysis, attributed to Kirchoff: Recall that the stresses on the edge normal to the X-axis are σ_{xx}, σ_{xy}, and σ_{xz}. The shear stresses are the sources of the twisting moment T_{xy} and the shear force Q_{xz}. By examining the equilibrium of an element of the edge, we can approximately combine T_{xy} and Q_{xz} into an "effective" shear force V_{xz} defined as

$$V_{xz} = Q_{xz} + \partial T_{xy}/\partial y \tag{17.114}$$

---

* A fourth order equation in two dimensions requires eight auxiliary conditions, or two per edge for a rectangular plate.

To see this consider a representation of T_{xy} by a pair of equal magnitude but oppositely directed vertical forces as in Figure 17.9. As such T_{xy} is represented by a simple couple (see Section 1.5.2) and the directions of the two forces are irrelevant as long as they are parallel (that is, they may be vertical, as well as horizontal). Next, consider a representation of the twisting moment at a small distance Δy along the edge as in Figure 17.9. Using the first term of a Taylor series, the twisting moment at this location is approximately $T_{xy} + (\partial T_{xy}/\partial y)\Delta y$. Then by superposing adjoining forces we have an upward force of $(\partial T_{xy} + \partial y)\Delta y$ on an element of length Δy (see Figure 17.9). Hence there is a net vertical force V_{xz} on the element given by $Q_{xz} + \partial T_{xy}/\partial y$ as in Equation 17.114. By substituting from Equations 17.100 and 17.102 we see that V_{xz} may be expressed in terms of the displacement ω as

$$V_{xz} = Q_{xz} + \partial T_{xy}/\partial y = -D\left[\frac{\partial^3 w}{\partial x^3} + (2 - \nu)\frac{\partial^3 w}{\partial x \partial y^2}\right] \tag{17.115}$$

Then for a free edge the boundary conditions of Equation 17.113 are replaced by the conditions

$$M_x = 0 \quad \text{and} \quad V_{xz} = 0 \tag{17.116}$$

or in view of Equations 17.98 and 17.115

$$\frac{\partial^2 w}{\partial x^2} + \nu\frac{\partial^2 w}{\partial y^2} = 0 \quad \text{and} \quad \frac{\partial^3 w}{\partial x^3} + (2 - \nu)\frac{\partial^3 w}{\partial x \partial y^2} = 0 \tag{17.117}$$

17.10.4 ELASTIC EDGE SUPPORT

An "elastic edge" provides support proportional to the displacement and/or rotation. If, for example, an X-face is an elastic edge, the shear provided by the support is proportional to the Z-direction displacement w and/or the rotation, or slope, is proportional to the flexural moment. That is

$$V_{xz} = -k_d w \quad \text{and/or} \quad M_x = -k_r \partial w/\partial x \tag{17.118}$$

Then by substituting from Equations 17.98 and 17.115 we have

$$\frac{\partial^3 w}{\partial x^3} + (2 - \nu)\frac{\partial^3 w}{\partial x \partial y^2} = k_d w/D \quad \text{and} \quad \frac{\partial^2 w}{\partial x^2} + \frac{\partial^2 w}{\partial y^2} = k_r/D \tag{17.119}$$

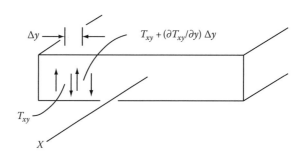

FIGURE 17.9 Representation of twisting moment along an edge normal to the X-axis.

17.11 INTERNAL STRESSES

Using Equations 17.98 through 17.103 we can immediately obtain expressions for the bending moments, the twist, and the shearing forces in terms of the displacement:

$$M_x \underset{(17.98)}{=} -D\left(\frac{\partial^2 w}{\partial x^2} + \nu\frac{\partial^2 w}{\partial y^2}\right) \tag{17.120}$$

$$M_y \underset{(17.99)}{=} -D\left(\frac{\partial^2 w}{\partial y^2} + \nu\frac{\partial^2 w}{\partial x^2}\right) \tag{17.121}$$

$$T_{xy} \underset{(17.102)}{=} -(1-\nu)D\frac{\partial^2 w}{\partial x \partial y} \tag{17.122}$$

$$Q_{xz} \underset{(17.102)}{=} -D\frac{\partial}{\partial x}\left(\frac{\partial^2 w}{\partial x^2} + \frac{\partial^2 w}{\partial y^2}\right) \tag{17.123}$$

$$Q_{yz} \underset{(17.103)}{=} -D\frac{\partial}{\partial y}\left(\frac{\partial^2 w}{\partial x^2} + \frac{\partial^2 w}{\partial y^2}\right) \tag{17.124}$$

where D is defined as

$$D \overset{D}{=} Eh^3/12(1-\nu^2) \tag{17.125}$$

Recall from beam theory that in the interior of the beam the axial stresses due to bending (the flexure) increase linearly across the cross section away from the neutral axis. That is, the stress is proportional to the distance, above or below, the neutral axis. Recall also in a beam that the shear stress has a quadratic (parabolic) distribution across the cross section. Since the assumptions of plate theory are analogous to those of beam theory, the stress distributions across the plate cross sections, about the midplane, are consequently analogous to those of beam theory. Specifically, for the stresses on the cross sections normal to the X- and Y-axes, we have

$$\sigma_{xx} = M_x z/(h^3/12), \quad \sigma_{xy} = T_{xy} z/(h^3/12), \quad \sigma_{yy} = M_y z/(h^3/12) \tag{17.126}$$

and

$$\sigma_{xz} = (3Q_{xz}/2h)\left[1 - (2z/h)^2\right], \quad \sigma_{yz} = (3Q_{yz}/2h)\left[1 - (2z/h)^2\right] \tag{17.127}$$

The procedure for determining these stresses is straight-forward: we solve the governing equation $\nabla^4 \omega = p/D$, Equation 17.107, for a given loading $p(x, y)$, subject to the boundary conditions (see Section 17.10) appropriate for a given plate support. Next, knowing the displacement, we can use Equations 17.120 through 17.124 to determine the bending moments, twist, and shear forces. Finally, Equations 17.126 and 17.127 provide the stresses.

17.12 COMMENTS

When compared with elementary beam theory, the assumptions of classical plate theory as in Sections 17.1 and 17.2 are considerably numerous and restrictive. The complexity of the geometry with bending in two directions, necessitates the simplifications provided by the assumptions. Even so, the resulting analysis is still not simple. Ultimately we need to solve a fourth-order partial differential equation (Equation 17.107) with varying degrees of boundary conditions. Closed form solutions are thus elusive or intractable except for the simplest of loading and boundary conditions.

In the following chapters, we will look at some of these elementary solutions. We will then consider problems of more practical importance in structural design and the ways of obtaining stress analyses for those cases.

SYMBOLS

D	$Eh^3/12(1 - \nu^2)$ (see Equation 17.101)
E	Elastic constant
(e)	Plate element
G	Shear modulus
h	Plate thickness
k_d, k_r	Shear and moment coefficients (see Equation 17.118)
M_x	Bending moment per unit edge length on the X-face (see Equation 17.9)
M_Y	Bending moment per unit edge length on the Y-face (see Equation 17.11)
N_{xx}	Membrane force per unit edge length in the X-direction (see Equation 17.4)
N_{yy}	Shear farce per unit edge length on the Y-face in the Y-direction (see Equation 17.7)
O	Origin of X, Y, Z coordinate axes
$p(x, y)$	Surface pressure; loading
Q_{xz}	Shear force per unit edge length on the X-face in the Z-direction (see Equation 17.6)
Q_{yz}	Shear force per unit edge length on the Y-face in the Z-direction (see Equation 17.7)
S_{xy}	Shear force per unit edge length, on the X-face in the Y-direction (see Equation 17.5)
S_{yx}	Shear force per unit edge length, on the Y-face in the X-direction (see Equation 17.7)
T_{xy}	Twisting moment per unit edge length on the X-face, about the X-face (see Equation 17.10)
T_{yx}	Twisting moment per unit edge length on the Y-face about the Y-face (see Equation 17.12)
u, v, w	Displacements in the X, Y, Z direction
V_{xz}	Effective shear (see Equation 17.114)
X, Y, Z	Rectangular (Cartesian axes)
x, y, z	Coordinates relative to X, Y, Z
α, β	Rotations of plate X-face, Y-face cross sections
ε_{ij} $(i, j = x, y, z)$	Strain matrix components
ν	Poisson's ratio
σ_{ij} $(i, j = x, y, z)$	Stress matrix components; stresses on the i-face in the j-direction

REFERENCES

1. J. R. Vinson, *The Behavior of Thin Walled Structures—Beams, Plates, and Shells*, Kluwer Academic, Dordrecht, 1989 (chaps. 1 and 2).
2. J. R. Vinson, *Structural Mechanics: The Behavior of Plates and Shells*, John Wiley & Sons, New York, 1974.
3. S. Timoshenko and S. Woinosky-Krieger, *Theory of Plates and Shells*, McGraw Hill, New York, 1959.
4. A. A. Armenakas, *Advanced Mechanics of Materials and Applied Elasticity*, CRC Taylor & Francis, Boca Raton, FL 2006 (chap. 17).

18 Plates: Fundamental Bending Configurations and Applications

18.1   REVIEW

In Chapter 17, we established the governing partial differential equation for plate deformation due to bending as a result of loading normal to the plate (Equation 17.107):

$$\nabla^4 w = p/D \tag{18.1}$$

where
  w is the displacement normal to the plate
  p is the loading function
  D is (Equation 17.91)

$$D = Eh^3/12(1 - \nu^2) \tag{18.2}$$

where
  h is the plate thickness
  E and ν are the elastic modulus and Poisson's ratio

In Cartesian coordinates the ∇^4 operator has the form

$$\nabla^4() = \frac{\partial^4()}{\partial x^4} + 2\frac{\partial^4()}{\partial x^2 \partial y^2} + \frac{\partial^4()}{\partial y^4} \tag{18.3}$$

In cylindrical coordinates the ∇^2 operator has the form [1]

$$\begin{aligned}
\nabla^2() &= \frac{\partial^2()}{\partial r^2} + \frac{1}{r}\frac{\partial()}{\partial r} + \frac{1}{r^2}\frac{\partial^2()}{\partial \theta^2} \\
&= \frac{1}{r}\frac{\partial}{\partial r}\left(r\frac{\partial()}{\partial r}\right) + \frac{1}{r^2}\frac{\partial^2()}{\partial \theta^2}
\end{aligned} \tag{18.4}$$

so that $\nabla^4()$ is then

$$\nabla^4() = \nabla^2\nabla^2() \tag{18.5}$$

In Cartesian coordinates, p is a function of x and y. In cylindrical coordinates, p is a function of r and θ, although for most circular plate problems of practical importance the loading is axisymmetric, that is $p = p(r)$.

For rectangular plates the boundary conditions are

1. *Simple (hinge) support* (parallel to Y-axis):

$$w = 0 \quad \text{and} \quad \frac{\partial^2 w}{\partial x^2} + \nu \frac{\partial^2 w}{\partial y^2} = 0 \quad (M_x = 0) \tag{18.6}$$

(See Equation 17.112.)

2. *Clamped (fixed) support* (parallel to Y-axis):

$$w = 0 \quad \text{and} \quad \partial w / \partial x = 0 \tag{18.7}$$

(See Equation 17.112.)

3. *Free edge* (parallel to Y-axis):

$$\frac{\partial^2 w}{\partial x^2} + \nu \frac{\partial^2 w}{\partial y^2} = 0 \quad \text{and} \quad \frac{\partial^3 w}{\partial x^3} + (2 - 0) \frac{\partial^3 w}{\partial x \partial y^2} = 0 \tag{18.8}$$

(See Equation 17.117.)

For circular plates the most common supports are simple support and clamped (built-in) support. For axisymmetric loading these may be expressed as [2]:

1. *Simple support:*

$$w = 0 \quad \text{and} \quad \frac{\partial^2 w}{\partial r^2} + \frac{\nu}{r} \frac{\partial w}{\partial r} = 0 \tag{18.9}$$

2. *Clamped support:*

$$w = 0 \quad \text{and} \quad \partial w / \partial r = 0 \tag{18.10}$$

The procedure for a given problem is to solve Equation 18.1 for w subject to the appropriate boundary conditions. Then knowing w, the bending moments and shears may be computed and from these the stresses may be evaluated. For rectangular plates, the moments, shears and stresses are given by Equations 17.120 through 17.127. For axisymmetrically loaded circular plates, the radial bending moment, shear, and stresses are [2]

$$M_r = -D \left[\frac{d^2 w}{dr^2} + \frac{\nu}{r} \frac{dw}{dr} \right] \tag{18.11}$$

$$Q_r = -D \frac{d}{dr} \left[\frac{1}{r} \frac{d}{dr} + \left(r \frac{dw}{dr} \right) \right] \tag{18.12}$$

$$\sigma_r = M_r z / (h^3 / 12) \tag{18.13}$$

$$\sigma_{rz} = \frac{3Q_r}{2h} \left[1 - \left(\frac{z}{h/2} \right)^2 \right] \tag{18.14}$$

In the following sections, we will review some elementary and fundamental plate loading problems and their solutions.

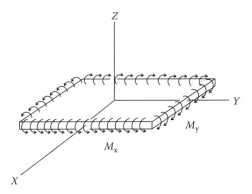

FIGURE 18.1 Pure bending of a rectangular plate.

18.2 SIMPLE BENDING OF RECTANGULAR PLATES [3]

Consider first a rectangular plate subjected to pure bending as represented in Figure 18.1.

Specifically let there be uniform moments applied along the edges as shown and let the twisting moment T_{xy} be zero. That is,

$$M_x = M_{x0}, \quad M_y = M_{y0}, \quad T_{xy} = 0 \tag{18.15}$$

Then from Equations 17.120, 17.121, and 17.122 the plate curvatures are

$$\frac{\partial^2 w}{\partial x^2} = -\frac{M_{x0} - \nu M_{y0}}{D(1 - \nu^2)} \tag{18.16}$$

$$\frac{\partial^2 w}{\partial y^2} = -\frac{M_{y0} - \nu M_{x0}}{D(1 - \nu^2)} \tag{18.17}$$

$$\frac{\partial^2 w}{\partial x \partial y} = 0 \tag{18.18}$$

From the third of these expressions, we immediately see that the displacement w has the form

$$w = f(x) + g(y) \tag{18.19}$$

Then

$$\frac{\partial^2 w}{\partial x^2} = \frac{d^2 f}{dx^2} = -\frac{M_{x0} - \nu M_{y0}}{D(1 - \nu^2)} \tag{18.20}$$

and

$$\frac{\partial^2 w}{\partial y^2} = \frac{d^2 g}{dy^2} = -\frac{M_{y0} - \nu M_{x0}}{D(1 - \nu^2)} \tag{18.21}$$

By integration we obtain

$$f(x) = -\frac{M_{x0} - \nu M_{y0}}{2d(1 - \nu^2)} x^2 + c_1 x + c_2 \tag{18.22}$$

and

$$g(y) = -\frac{M_{y0} - \nu M_{x0}}{2D(1 - \nu^2)}y^2 - c_3 y + c_4 \tag{18.23}$$

where $c_1, \ldots, c_4$ are constants. Therefore the displacement is

$$w = -\frac{M_{x0} - \nu M_{y0}}{2D(1 - \nu^2)}x^2 - \frac{M_{y0} - \nu M_{x0}}{2D(1 - \nu^2)}y^2 + c_1 x + c_3 y + c_2 + c_4 \tag{18.24}$$

To uniquely specify the displacement, we can eliminate rigid body movement by the conditions:

$$w(0,0) = 0, \quad \frac{\partial w}{\partial x}(0,0) = 0, \quad \frac{\partial w}{\partial y}(0,0) = 0 \tag{18.25}$$

The displacement then becomes

$$w(x,y) = -\frac{M_{x0} - \nu M_{y0}}{2D(1 - \nu^2)}x^2 - \frac{M_{y0} - \nu M_{x0}}{2D(1 - \nu^2)}y^2 \tag{18.26}$$

The plate surface then has the form of an elliptical paraboloid.

Finally, if the moments M_{x0} and M_{y0} are equal w has the simplified form:

$$w = -M_{x0}\frac{x^2 + y^2}{2D(1 + \nu)} \tag{18.27}$$

18.3   SIMPLY SUPPORTED RECTANGULAR PLATE

Consider a rectangular plate with dimensions a and b (along the X- and Y-axes) with hinged (pinned) edge supports. Let the origin of the axis system be placed at a corner, as in Figure 18.2. Let the loading on the plate be $p(x, y)$. The governing equation is then (see Equations 17.107 and 17.108)

$$\nabla^4 w = \frac{\partial^4 w}{\partial x^4} + 2\frac{\partial^4 w}{\partial x^2 \partial y^2} + \frac{\partial^4 w}{\partial y^4} = p(x, y)/D \tag{18.28}$$

where the boundary conditions are

$$w(0, y) = w(x, 0) = w(a, y) = w(x, b) = 0 \tag{18.29}$$

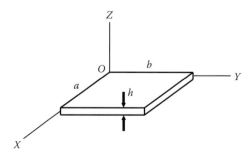

FIGURE 18.2   Rectangular plate and axis system.

$$\frac{\partial^2 w}{\partial x^2}(0, y) = \frac{\partial^2 w}{\partial y^2}(x, 0) = \frac{\partial^2 w}{\partial x^2}(a, y) = \frac{\partial^2 w}{\partial y^2}(x, b) = 0$$

The boundary conditions of Equations 18.29 will be satisfied if we can express the displacement w in terms of series of sine functions $m\pi x/a$ and $n\pi y/b$. This is feasible since these functions form a "complete" and "orthogonal" system, with their sum forming a Fourier series [4,5]. Therefore we seek a solution, $w(x, y)$ of Equation 18.28 in the form

$$w(x, y) = \sum_{m=1}^{\infty} \sum_{n=1}^{\infty} A_{mm} \sin\left(\frac{m\pi x}{a}\right) \sin\left(\frac{n\pi x}{b}\right) \tag{18.30}$$

By substituting into Equations 18.28 we obtain:

$$\sum_{m=1}^{\infty} \sum_{n=1}^{\infty} A_{mn} \left[\left(\frac{m\pi}{a}\right)^4 + 2\left(\frac{m\pi}{a}\right)^2 \left(\frac{n\pi}{b}\right) + \left(\frac{n\pi}{b}\right)^4\right] \sin\left(\frac{m\pi x}{a}\right) \sin\left(\frac{n\pi y}{b}\right) = p(x, y)/D \tag{18.31}$$

We can also express $p(x, y)$ in a double sine series as

$$p(x, y) = \sum_{m=1}^{\infty} \sum_{n=1}^{\infty} B_{mn} \sin\left(\frac{m\pi x}{a}\right) \sin\left(\frac{n\pi y}{b}\right) \tag{18.32}$$

where by Fourier expansion [4] the coefficients B_{mn} may be expressed as

$$B_{mn} = \left(\frac{4}{ab}\right) \int_0^a \int_0^b p(x, y) \sin\left(\frac{m\pi x}{a}\right) \sin\left(\frac{n\pi y}{b}\right) dxdy \tag{18.33}$$

By substituting from Equation 18.32 into Equation 18.31, we have

$$\sum_{m=1}^{\infty} \sum_{n=1}^{\infty} A_{mn} \left[\left(\frac{m\pi}{a}\right)^4 + 2\left(\frac{m\pi}{a}\right)\left(\frac{n\pi}{b}\right) + \left(\frac{n\pi}{b}\right)^4\right] \sin\left(\frac{m\pi x}{a}\right) \sin\left(\frac{n\pi y}{b}\right)$$

$$= (1/D) \sum_{m=1}^{\infty} \sum_{n=1}^{\infty} B_{mn} \sin\left(\frac{m\pi x}{a}\right) \sin\left(\frac{n\pi x}{b}\right)$$

or

$$\sum_{m=1}^{\infty} \sum_{n=1}^{\infty} \left\{A_{mn} \left[\left(\frac{m\pi}{a}\right)^2 + \left(\frac{n\pi}{b}\right)^2\right]^2 - (1/D) B_{mn}\right\} \sin\left(\frac{m\pi x}{a}\right) \sin\left(\frac{n\pi y}{b}\right) = 0 \tag{18.34}$$

This expression is identically satisfied by setting the coefficients of $\sin(m\pi x/a)\sin(n\pi y/b)$ equal to zero. Then we have

$$A_{mn} = (1/D) \frac{B_{mn}}{\left[\left(\frac{m\pi}{a}\right)^2 + \left(\frac{n\pi}{b}\right)^2\right]^2} \tag{18.35}$$

Finally, by substituting A_{mn} in Equation 18.30 with Equation 18.37, the displacement w is seen to be

$$w = (1/D) \sum_{m=1}^{\infty} \sum_{n=1}^{\infty} \frac{B_{mn}}{\left[\left(\frac{m\pi}{a} \right)^2 + \left(\frac{n\pi}{b} \right)^2 \right]^2} \sin\left(\frac{m\pi x}{a} \right) \sin\left(\frac{n\pi y}{b} \right) \tag{18.36}$$

where B_{mn} are given by Equation 18.33.

18.4 SIMPLY SUPPORTED RECTANGULAR PLATE WITH A UNIFORM LOAD

As an immediate application of the foregoing result, consider a simply supported rectangular plate with a uniform load p_0. Then from Equation 18.33, the coefficients B_{mn} are

$$B_{mn} = \frac{4}{ab} \int_0^a \int_0^b p_0 \sin\left(\frac{m\pi x}{a} \right) \sin\left(\frac{n\pi y}{b} \right) dx \, dy$$

or

$$B_{mn} = \frac{4p_0}{mn\pi^2} (\cos m\pi - 1)(\cos n\pi - 1) \tag{18.37}$$

Then from Equation 18.35, A_{mn} are seen to be

$$A_{mn} = \frac{4p_0}{\pi^6 D} \frac{(\cos m\pi - 1)(\cos n\pi - 1)}{mn \left[\left(\frac{m}{a} \right)^2 + \left(\frac{n}{b} \right)^2 \right]} \tag{18.38}$$

Consequently the displacement w may be written as

$$w = \frac{16p_0}{\pi^6 D} \sum_{m=1}^{\infty} \sum_{n=1}^{\infty} \frac{\sin\left(\frac{m\pi x}{a} \right) \sin\left(\frac{n\pi y}{b} \right)}{mn \left(\frac{m^2}{a^2} + \frac{n^2}{b^2} \right)^2} \tag{18.39}$$

where only the odd terms are used in the summations.

The presence of multiplied integers in the denominator of the expression for A_{mn} in Equation 18.38 provides for rapid convergence. To see this, consider a square plate ($a = b$): the first four A_{mn} are

$$A_{11} = \frac{4p_0 a^4}{\pi^6 D}, \quad A_{13} = A_{31} = \frac{4p_0 a^4}{75\pi^6 D}, \quad A_{33} = \frac{16p_0 a^4}{729\pi^6 D} \tag{18.40}$$

Observe that A_{33}/A_{11} is then only 5.48×10^{-3}.

18.5 SIMPLY SUPPORTED RECTANGULAR PLATE WITH A CONCENTRATED LOAD

Next consider a simply supported rectangular plate with a concentrated load, with magnitude p, at a point P, having coordinates (ξ, η) as represented in Figure 18.3. Timoshenko and Woinosky-Kreiger solve this problem in their treatise on plate theory [6]. Their procedure is to apply a uniform load

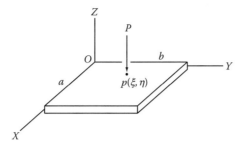

FIGURE 18.3   Concentrated point load on a simply supported rectangular plate.

over a rectangular region of the plate and then reduce the region to a point while simultaneously increasing the load. The resulting displacement w is

$$w = \frac{4P}{\pi^4 abD} \sum_{m=1}^{\infty} \sum_{n=1}^{\infty} \frac{\sin\left(\dfrac{m\pi\xi}{a}\right) \sin\left(\dfrac{n\pi\eta}{b}\right)}{\left(\dfrac{m^2}{a^2} + \dfrac{n^2}{b^2}\right)} \sin\left(\frac{m\pi x}{a}\right) \sin\left(\frac{n\pi y}{b}\right) \tag{18.41}$$

Observe that this result could have been obtained by an analysis of Equations 18.33 and 18.36 by using a two-dimensional singularity (or impulse) function (see Chapter 10).

18.6   COMMENTS

The solutions presented in Equations 18.36, 18.39, and 18.41 are the most elementary of the many possible solutions of rectangular plate problems. References [6–8] provide many other solutions and the listings in Roark and Young [9] and Pilkey [10] provide additional solutions.

Although the solutions of Equations 18.36, 18.39, and 18.41 are relatively simple in their forms and formulation, they nevertheless have double infinite series. Even though convergence is relatively rapid, as seen in Section 18.4, for computational purposes it is sometimes helpful to look for simpler forms of solutions. By insightful analysis [2,6], it is seen that these solutions may be expressed in a single series. This in turn has produced a number of results of practical importance [6,9,10].

Finally, a feature of the solution of simply supported rectangular plates is that the surface may become anticlastic. This may be simulated by forces concentrated at the corners. At one time, this feature was used in experiments to verify the basic theory of plate bending. For example, the corners of a uniformly downward loaded, simply supported square plate have a tendency to rise.

18.7   CIRCULAR PLATES

Circular plates are used in virtually all kinds of structural applications. For the most part, the loading and support are axisymmetric.

We can obtain the governing equations for circular plates by following the same procedures as in Chapter 17. Alternatively, we can simply make a coordinate transformation from rectangular coordinates (x, y, z) to cylindrical coordinates (r, θ, z). Recall from Equations 17.107 and 18.1 that the governing equation for the plate deformation w is

$$\nabla^4 w = p/D \qquad (18.42)$$

The operator $\nabla^4( )$ may be expressed as $\nabla^2 \nabla^2( )$. In cylindrical coordinates $\nabla^2( )$ is [1]

$$\nabla^2( ) = \frac{\partial^2( )}{\partial r^2} + \frac{1}{r}\frac{\partial( )}{\partial r} + \frac{\partial^2( )}{\partial \theta^2} \qquad (18.43)$$

Correspondingly, the equilibrium equations are [2,6]

$$\frac{\partial Q_r}{\partial r} + \frac{1}{r}\frac{\partial Q_\theta}{\partial \theta} + \frac{1}{r}Q_r + p(r, \theta) = 0 \qquad (18.44)$$

$$\frac{\partial M_r}{\partial r} + \frac{1}{r}\frac{T_{r\theta}}{\partial \theta} + \frac{M_r - M_\theta}{r} - Q_r = 0 \qquad (18.45)$$

$$\frac{\partial T_{r\theta}}{\partial r} + \frac{1}{r}\frac{\partial M_\theta}{\partial \theta} + \frac{2}{r}M_{r\theta} - Q_\theta = 0 \qquad (18.46)$$

where

    Q_r and Q_θ are the shear forces per unit length on the radial and circumferential faces of an interior element

    M_r and M_θ are the bending moments per unit length on the radial and circumferential faces

    $T_{r\theta}$ is the twisting moment

In terms of the displacement w, the moments and twist are [2]

$$M_r = -D\left[\frac{\partial^2 w}{\partial r^2} + \frac{\nu}{r}\frac{\partial w}{\partial r} + \frac{\nu}{r^2}\frac{\partial^2 w}{\partial \theta^2}\right] \qquad (18.47)$$

$$M_\theta = -D\left[\frac{1}{r^2}\frac{\partial^2 w}{\partial \theta^2} + \frac{1}{r}\frac{\partial w}{\partial r} + \nu\frac{\partial^2 w}{\partial \theta^2}\right] \qquad (18.48)$$

$$T_{r\theta} = -D(1 - \nu)\left[\frac{1}{r}\frac{\partial^2 w}{\partial r \partial \theta} - \frac{1}{r^2}\frac{\partial w}{\partial \theta}\right] \qquad (18.49)$$

As with rectangular plates we assume that the in-plane (membrane) forces are either zero or sufficiently small that they do not affect the shears, moments, or displacements due to bending.

When the loading and support are axisymmetric, the foregoing equations simplify considerably: $\nabla^2( )$ and $\nabla^4( )$ are

$$\nabla^2( ) = \frac{d^2( )}{dr^2} + \frac{1}{r}\frac{d( )}{dr} = \frac{1}{r}\frac{d}{dr}\left[r\frac{d( )}{dr}\right] \qquad (18.50)$$

and

$$\nabla^4( ) = \nabla^2\nabla^2( ) = \frac{1}{r}\frac{d}{dr}\left\{r\frac{d}{dr}\left[\frac{1}{r}\frac{d}{dr}\left(r\frac{d( )}{dr}\right)\right]\right\} \qquad (18.51)$$

The equilibrium equations (Equations 18.44, 18.45, and 18.46) then become

$$\frac{dQ_r}{dr} + \frac{1}{r}Q_r + p(r) = 0 \tag{18.52}$$

$$\frac{dM_r}{dr} + \frac{M_r - M_\theta}{r} - Q_r = 0 \tag{18.53}$$

where Equation 18.46 is identically satisfied.

Similarly, the moment–displacement equations (Equations 18.47, 18.48, and 18.49) become

$$M_r = -D\left[\frac{d^2 w}{dr^2} + \frac{\nu}{r}\frac{dw}{dr}\right] \tag{18.54}$$

$$M_\theta = -D\left[\frac{1}{r}\frac{dw}{dr} + \nu\frac{d^2 w}{dr^2}\right] \tag{18.55}$$

$$T_{r\theta} = 0 \tag{18.56}$$

Finally, by solving Equations 18.45 and 18.46 for Q_r and Q_θ and using Equations 18.54, 18.55, and 18.56 we have

$$Q_r = -D\left[\frac{d^3 w}{dr^3} + \frac{1}{r}\frac{d^2 w}{dr^2} - \frac{1}{r^2}\frac{dw}{dr}\right] \tag{18.57}$$

and

$$Q_\theta = 0 \tag{18.58}$$

18.8 SOLUTION OF THE GOVERNING EQUATION FOR CIRCULAR PLATES

From Equations 18.42 and 18.51, the governing equation for an axisymmetrically loaded and axisymmetrically supported circular plate is

$$\frac{1}{r}\frac{d}{dr}\left\{ r\frac{d}{dr}\left[\frac{1}{r}\frac{d}{dr}\left(r\frac{dw}{dr}\right)\right]\right\} = p(r)/D \tag{18.59}$$

Thus if we know $p(r)$, we can integrate four times to obtain $w(r)$ and then we can compute the bending moments and shear forces using Equations 18.54 through 18.58. The radial and circumferential flexural stresses are then simply [2]

$$\sigma_{rr} = \frac{M_r z}{h^3/12} \quad \text{and} \quad \sigma_{\theta\theta} = \frac{M_\theta z}{h^3/12} \tag{18.60}$$

Similarly the shear stresses are [2]

$$\sigma_{r\theta} = 0, \quad \sigma_{\theta z} = 0, \quad \sigma_{rz} = \frac{3Q_r}{2h}\left[1 - \left(\frac{z}{h/2}\right)^2\right] \tag{18.61}$$

Upon integrating Equation 18.59 four times, we obtain four constants of integration, which may be evaluated from the support and symmetry conditions. As an illustration, suppose $p(r)$ is a uniform load P_0: then the four integrations of Equation 18.59 leads to

$$w = \frac{P_0 r^4}{64D} + c_1 r^2 \ell nr + c_2 r^2 + c_3 \ell nr + c_4 \tag{18.62}$$

where $c_1, \ldots, c_4$ are the integration constants. From Equations 18.54 through 18.58, the bending moment and shear forces are then seen to be

$$M_r = -D[(3 + \nu)\frac{P_0 r^2}{16D} + 2c_1(1 + \nu)\ell nr + (3 + \nu)c_1$$
$$+ 2c_2(1 + \nu) + c_3(\nu - 1)/r^2] \tag{18.63}$$

$$M_\theta = -D[(1 + 3\nu)\frac{P_0 r^2}{16D} + (1 + \nu)2c_1 \ell nr + (1 + 3\nu)c_1$$
$$+ 2(1 + \nu)c_2 + (1 - \nu)c_3/r^2] \tag{18.64}$$

$$T_{r\theta} = 0 \tag{18.65}$$

$$Q_r = -D\left(\frac{P_0 r}{2D} + 4\frac{c_1}{r}\right) \tag{18.66}$$

$$Q_\theta = 0 \tag{18.67}$$

For finite displacement, finite shear, and finite bending moment at the origin (plate center), we must have

$$c_1 = 0 \quad \text{and} \quad c_3 = 0 \tag{18.68}$$

We can use the support conditions to evaluate c_2 and c_4. Consider the two common support cases: (1) simple support and (2) clamped (or fixed) support.

18.8.1 SIMPLY SUPPORTED, UNIFORMLY LOADED, CIRCULAR PLATE

In this case the support conditions are
When $r = a$

$$w = 0 \quad \text{and} \quad M_r = 0 \tag{18.69}$$

where a is the plate radius. From Equations 18.62 and 18.68, the second boundary conditions lead to

$$M_r(a) = 0 = (3 + \nu)\frac{P_0 a^2}{16D} + 2c_2(1 + \nu)$$

or

$$c_2 = -\frac{(3 + \nu)}{2(1 + \nu)}\frac{P_0 a^2}{16D} \tag{18.70}$$

From Equations 18.62 and 18.68 the first condition of Equation 18.69 then becomes

$$w(a) = 0 = \frac{P_0 a^4}{64D} - \frac{3+\nu}{2(1+\nu)} \frac{P_0 a^4}{16D} + c_4$$

or

$$c_4 = \frac{P_0 a^4}{64D} \left(\frac{5+0}{1+\nu} \right) \tag{18.71}$$

Therefore the displacement w of Equation 18.62 becomes

$$w = \frac{P_0}{64D} \left[r^4 - \frac{2(3+\nu)}{1+\nu} a^2 r^2 + \frac{5+\nu}{1+D} a^4 \right] \tag{18.72}$$

18.8.2 Clamped Uniformly Loaded Circular Plate

In this case the support conditions are
    when $r = 1$

$$w = 0 \quad \text{and} \quad \frac{dw}{dr} = 0 \tag{18.73}$$

From Equations 18.62 and 18.68 dw/dr is seen to be

$$\frac{dw}{dr} = \frac{P_0 r^3}{16D} + 2c_2 r \tag{18.74}$$

Then the second boundary condition becomes

$$\frac{dw}{dr}(a) = 0 = \frac{P_0 a^3}{16D} + 2c_2 a$$

or

$$c_2 = -\frac{P_0 a^2}{32D} \tag{18.75}$$

From Equations 18.62 and 18.68, the first condition of Equation 18.73 then becomes

$$w(a) = 0 = \frac{P_0 a^4}{64D} - \frac{P_0 a^4}{32D} + c_4$$

or

$$c_4 = \frac{P_0 a^4}{64D} \tag{18.76}$$

Therefore the displacement w of Equation 18.62 becomes

$$w = \frac{P_0}{64D}(r^4 - 2a^2 r^2 + a^4) \tag{18.77}$$

18.9 CIRCULAR PLATE WITH CONCENTRATED CENTER LOAD

Centrally loaded circular plates are common structural components. We can study them in the same way as we did for rectangular plates with concentrated loads. We can apply a uniform load over a central circular region of the plate with radius b ($b < a$), with a being the plate radius. Then as b is reduced to zero with the overall load remaining the same, we have the concentrated load configuration. Timoshenko and Woinowsky-Krieger [6] present the details of this analysis. The results for a simple supported and clamped plate are summarized in the following sections.

18.9.1 SIMPLY SUPPORTED CIRCULAR PLATE WITH A CONCENTRATED CENTER LOAD

The deflection w at any point of a distance r from the plate center is [6]

$$w = \frac{P}{16\pi D}\left[\left(\frac{3+\nu}{1+\nu}\right)(a^2 - r^2) + 2r^2\ell n(r/a)\right] \tag{18.78}$$

where P is the magnitude of the concentrated center load. The maximum deflection, occurring at $r = 0$, is then

$$w_{\max} = \frac{Pa^2}{16\pi D}\left(\frac{3+\nu}{1+\nu}\right) \tag{18.79}$$

18.9.2 CLAMPED CIRCULAR PLATE WITH A CONCENTRATED CENTER LOAD

The deflection w at any point of a distance r from the plate center is [6]

$$w = \frac{P}{16\pi D}[(a^2 - r^2) + 2r^2\ell n(r/a)] \tag{18.80}$$

where again P is the magnitude of the concentrated center load. The maximum deflection, occurring at $r = 0$, is then

$$w_{\max} = \frac{Pa^2}{16\pi D} \tag{18.81}$$

18.10 EXAMPLE DESIGN PROBLEM

Consider a clamped circular plate with radius a with a uniform load as in Figure 18.4. Suppose the center deflection δ is equal to the thickness h of the plate. Determine the flexural stresses on the surface at the center and at the rim support.

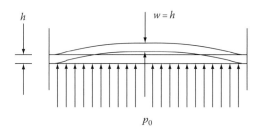

FIGURE 18.4   Uniformly loaded clamped circular plate.

SOLUTION

From Equation 18.77, the displacement is

$$w = \frac{P_0}{64D}(r^4 - 2a^2r^2 + a^4) \tag{18.82}$$

The deflection δ at the center ($r = 0$) is

$$\delta = \frac{P_0 a^4}{64D} \tag{18.83}$$

From Equation 18.60 the upper surface flexural stresses are

$$\sigma_{rr} = \frac{M_r(h/2)}{(h^3/12)} = 6M_r/h^2 \quad \text{and} \quad \sigma_{\theta\theta} = \frac{M_\theta(h/2)}{(h^3/12)} = 6M_\theta/h^2 \tag{18.84}$$

From Equations 18.63, 18.64, 18.68, 18.75, and 18.76 M_r and M_θ are seen to be

$$M_r = -\frac{(3 + \nu)p_0 r^2}{16} + \frac{(1 + \nu)p_0 a^2}{16} \tag{18.85}$$

and

$$M_\theta = -\frac{(1 + 3\nu)p_0 r^2}{16} + \frac{(1 + \nu)p_0 a^2}{16} \tag{18.86}$$

Then at the center, with $r = 0$, the stresses are

$$\sigma_{rr} = \frac{3(1 + \nu)}{8}p_0(a^2/h^2) \quad \text{and} \quad \sigma_{\theta\theta} = \frac{3(1 + \nu)}{8}p_0(a^2/h^2) \tag{18.87}$$

Suppose now, that according to the example statement δ, as given by Equation 18.83, is equal to the thickness, h of plate, then the corresponding loading p_0 is

$$p_0 = \frac{64Dh}{a^4} = \frac{16Eh^4}{3(1 - \nu^2)a^4} \tag{18.88}$$

The stresses at the center are then

$$\sigma_{rr} = \sigma_{\theta\theta} = \frac{2Eh^2}{a^2(1 - \nu)} \tag{18.89}$$

Similarly, from Equations 18.63 at the plate rim where $r = a$, the bending moments are

$$M_r = -p_0 a^2/8 \quad \text{and} \quad M_\theta = -p_0 \nu a^2/8 \tag{18.90}$$

From Equation 18.60 the upper surface flexural stresses are

$$\sigma_{rr} = -3p_0 a^2/4h^2 \quad \text{and} \quad \sigma_{\theta\theta} = -3p_0 \nu a^2/4h^2 \tag{18.91}$$

Then with p_0 given by Equation 18.88, σ_{rr} and $\sigma_{\theta\theta}$ become

$$\sigma_{rr} = -\frac{4Eh^2}{(1-\nu^2)a^2} \quad \text{and} \quad \sigma_{\theta\theta} = -\frac{4\nu Eh^2}{(1-\nu^2)a^2} \tag{18.92}$$

Observe that the stresses on the upper plate surface are positive in the center of the plate (tension) and negative at the rim (compression).

18.11   A FEW USEFUL RESULTS FOR AXISYMMETRICALLY LOADED CIRCULAR PLATES

By similar analyses we can obtain results for other problems of practical importance. Table 18.1 provides a listing of a few of these for the case where Poisson's ratio is 0.3. Specifically, the maximum displacement w_{max} and the maximum stress σ_{max} and their locations are given.

TABLE 18.1

A Few Useful Formulas for Axisymmetrically Loaded Circular Plates

1. Simple rim support, central uniform load

$$\sigma_{max} = \frac{px^2}{h^2}\left[1.5 + 1.95\ell n(r_0/x) - 0.263(x/r_0)^2\right]\text{(at the center)}$$

$$w_{max} = \frac{px^2}{Eh^2}\left[1.733r_0^2 - 0.683x^2\ell n(r_0/x) - 1.037x^2\right]$$

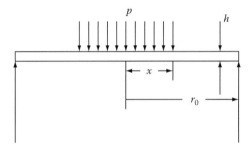

2. Simple rim support, central ring load

$$\sigma_{max} = \frac{p}{h^2}\left[0.167 + 0.621\ell n(r_0/x) - 0.167(x/r_0)^2\right]\text{(at the center)}$$

$$w_{max} = \frac{p}{Eh^3}\left[0.551(r_0^2 - x^2) - 0.434x^2\ell n(r_0/x)\right]$$

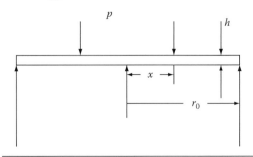

TABLE 18.1 (continued)
A Few Useful Formulas for Axisymmetrically
Loaded Circular Plates

3. Clamped rim support, central uniform load

$$\sigma_{max} = \frac{px^2}{h^2}\left[1.5 - 0.75(x/x_0)^2\right] \quad \text{at the rim} \quad \text{for } x > 0.58r_0$$

$$\sigma_{max} = \frac{px^2}{h^2}\left[1.95\ell n(r_0/x) + 0.488(x/r_0)^2\right] \quad \text{at the rim} \quad \text{for } x < 0.58r_0$$

$$w_{max} = \frac{px^2}{Eh^3}\left[0.683r_0^2 - 0.683x^2\ell n\right](r_0/x) - 0.512x^2$$

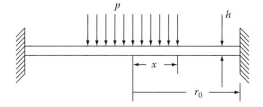

4. Clamped rim support, central ring load

$$\sigma_{max} = \frac{0.477p}{h^2}\left[1 - (x/r_0)^2\right] \quad \text{at the rim} \quad \text{for } x > 0.31r_0$$

$$\sigma_{max} = \frac{0.31p}{h^2}\left[2\ell n(r_0/x) + (x/r_0)^2 - 1\right] \quad \text{at the center} \quad \text{for } x < 0.31r_0$$

$$w_{max} = \frac{p}{Eh^3}\left[0.217(r_0^2 - x^2) - 0.434x^2\ell n(r_0/x)\right]$$

18.12 COMMENTS

All the foregoing analyses and examples have simple loadings (uniform load or concentrated load) and simple (pinned) or clamped (fixed or built-in) supports. Nevertheless, they represent many structural applications, particularly when plates are used as closures or covers.

If the loadings or support are more complex we have several options:

1. We can attempt to approximately solve the governing partial differential equation. After all, the governing equation is itself an approximation based upon numerous simplifying assumptions (see Chapter 17).
2. We can consult the several fine handbooks of solutions and approximate solutions to various plate loading and support configurations [9,10].
3. We can seek a finite element solution. This is a useful approach if software and computer hardware are available.
4. We can approximate a given structure or loading with simpler models.

In Chapter 19, we examine modeling and approximations for flanges, brackets, and panels.

SYMBOLS

A	Circular plate radius
a, b	Plate edge dimensions
A_{mn}, B_{mn}	Fourier coefficients (see Equations 18.30, 18.32, 18.33, and 18.35)
D	$Eh^3/12(1 - \nu^2)$ (see Equation 18.2)
E	Elastic modulus
h	Plate thickness
M_r	Radial bending moment
M_x, M_y	Edge bending moments
M_{xo}, M_{yo}	Uniform values of M_x, M_y
M_θ	Tangential bending moment
p	Loading normal to the plate surface
p_O	Uniform loading
Q_r	Radial shear force
Q_θ	Tangential shear force
r	Radial (polar) coordinate
$T_{xy}, T_{r\theta}$	Twisting moments
w	Plate displacement, normal to the plane
X, Y, Z	Rectangular (Cartesian) coordinate axes
x, y, z	Coordinates relative to X, Y, Z
θ	Angular (polar) coordinate
δ	Center displacement of a circular plate
ν	Poisson's ratio
σ_r, σ_{rr}	Radial stress
σ_{rz}	Shear stress
$\sigma_{r\theta}$	Shear stress
$\sigma_{\theta z}$	Shear stress
$\sigma_{\theta\theta}$	Tangential stress

REFERENCES

1. *CRC Standard Mathematical Tables*, 20th ed., The Chemical Rubber Co., Cleveland, OH, 1972, p. 551.
2. J. R. Vinson, *The Behavior of Thin Walled Structures, Beams, Plates, and Shells*, Kluwer Academic, Dordrecht, the Netherlands, 1989.
3. M. A. Brull, *Lecture Notes on Plate Theory*, University of Pennsylvania, Philadelphia, PA, 1961.
4. C. R. Wiley, *Advanced Engineering Mathematics*, 4th ed., McGraw Hill, New York, 1975, pp. 353–367.
5. F. B. Hildebrand, *Advanced Calculus for Application*, 2nd ed., Prentice Hall, Englewood Cliffs, NJ, 1976, p. 208.
6. S. Tinoshenko and S. Woinowsky-Krieger, *Theory of Plates and Shells*, McGraw Hill, New York, 1959.
7. V. Panc, *Theories of Elastic Plates*, Noordhoff International Publishing, Leyden, the Netherlands, 1975.
8. E. Ventsel and T. Krauthammer, *Thin Plates and Shells*, Marcel Dekker, New York, 2001.
9. W. C. Young and R. G. Budynas, *Roark's Formulas for Stress and Strain*, 7th ed., McGraw Hill, New York, 2002 (chap. 11).
10. W. D. Pilkey, *Formulas for Stress, Strain, and Structural Matrices*, Wiley-Interscience, New York, 1994 (chap. 18).

19 Panels and Annular Plate Closures

19.1 PROBLEM DEFINITION

Some of the fundamentals outlined in the previous chapters point to the degree of complexity of the various plate solutions. When plate applications arise, a good deal of specialization is required, backed up by experimental work. This type of information is not easy to obtain and the designer has to fall back on the classical solutions and the conservative assumptions of elasticity. In this chapter on panels, we will attempt to summarize some of the more basic practical data related to those plate configurations that occur most frequently and which can be used as approximate models for more complex solutions.

A typical structural panel may be defined as a flat material, usually rectangular, elliptical, or similar in shape, which forms a part of the surface of a wall, door, cabinet, duct, machine component, fuselage window, floor, or similar component. The panel boundaries illustrated in Figure 19.1 may involve some degree of fixity or freedom when a given panel is subjected to uniform loading. A difficult consideration in estimating the panel strength and rigidity is the choice of the correct boundary condition. This process depends entirely on a knowledge of loading and support, which varies from problem to problem. The boundary conditions can vary from being completely built-in to having a simple roller-type support, allowing full freedom of rotation. In the majority of practical configurations, some intermediate conditions exist, requiring engineering judgment in selecting the most realistic model for panel support. The design criteria for uniform transverse loading can be governed by either the maximum bending strength or the allowable maximum deflection. Our purpose is to provide a set of working equations and charts suitable for design.

19.2 DESIGN CHARTS FOR PANELS

Simple rectangular panels are often supported by structural shapes whose bending stiffness is relatively high compared with that of the panels themselves. Under these conditions, fixed edges can be assumed in the calculations. However, when the supporting shapes are such that a finite slope can develop in the plane perpendicular to the panel, the design should be based on a simple support criterion. Table 19.1 provides a summary of some of the more commonly used design equations for the rectangular and the elliptical panels.

In Figures 19.2 and 19.3, we graph the design factors A_1 through A_8 against the panel length ratio a/b. For a rectangular panel, a and b denote the smaller and larger sides, respectively. For an elliptical geometry, a and b are the minor and major axes, respectively. While the maximum bending stress is found at the center for the simply supported rectangular and elliptical panels, built-in panels are stressed more at the supports. For a rectangular built-in panel, this point is at the midpoint of the longer edge. In the built-in elliptical panel, the maximum bending stress is at the ends of the minor axis a. The example problem of Section 19.4 shows that the maximum deflection is a function of the a/b ratio. By taking $b = 2a$, we find A_8 to be: $3(1 - \nu^2)/118$. This compares well with the value obtained from the graph of Figure 19.3.

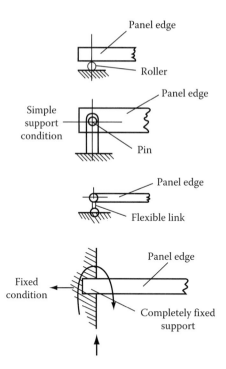

FIGURE 19.1 Examples of edge conditions for panel design.

19.3 SIMILARITIES OF RECTANGULAR AND ELLIPTICAL PANELS

The charts given in Figures 19.2 and 19.3 indicate a definite correlation between rectangular and elliptical panels in their structural behavior. For this reason, a great number of panel shapes that fall between the rectangular and elliptical boundaries can be designed with the help of the charts given in Figures 19.2 and 19.3. For example, the arbitrary profile shown in Figure 19.4 should display strength and rigidity characteristics, which might be termed as intermediate between those of the elliptical and rectangular configurations, provided that the overall a and b dimensions remain the same.

The design engineer concerned with such a problem can develop an individual method of interpolation between the relevant results. For instance, the ratio of the unused corner area F to the total area difference between the rectangular and elliptical geometries can be used as a parameter. In terms of the dimensions indicated in Figure 19.4, this parameter may be defined as $16F/ab(4 - \pi)$.

TABLE 19.1

Design Equations for Simple Panels under Uniform Loading

Type of Panel	Maximum Stress	Maximum Deflection
Rectangular simply supported	$S = \dfrac{qa^2 A_1}{t^2}$	$\delta = \dfrac{qa^4 A_2}{Et^3}$
Rectangular built-in	$S = \dfrac{qa^2 A_3}{t^2}$	$\delta = \dfrac{qa^4 A_4}{Et^3}$
Elliptical simply supported	$S = \dfrac{qa^2 A_5}{t^2}$	$\delta = \dfrac{qa^4 A_6}{Et^3}$
Elliptical built-in	$S = \dfrac{qa^2 A_7}{t^2}$	$\delta = \dfrac{qa^4 A_8}{Et^3}$

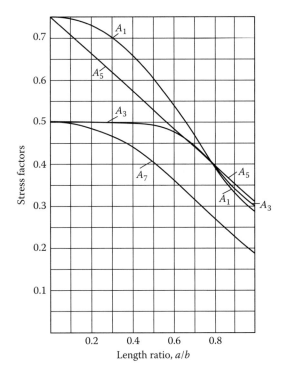

FIGURE 19.2 Stress chart for simple panels.

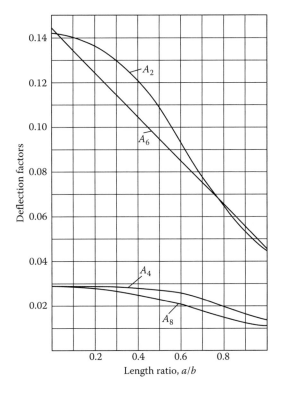

FIGURE 19.3 Deflection chart for simple panels.

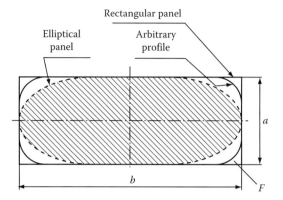

FIGURE 19.4 Comparable plate configurations.

It should be emphasized, however, that such a linear interpolation can be justified only because of the inherent similarities in the structural behavior of the rectangular and elliptical configurations. The error introduced by this procedure is expected to be relatively small and certainly acceptable within the scope of the preliminary design, which under normal conditions, involves ample margins of safety.

19.4   EXAMPLE DESIGN PROBLEM

Figure 19.5 depicts a pressure plate of rectangular geometry with rounded-off corners. It is simply supported and carries a uniform transverse loading of 200 psi. Assuming the dimensions shown in the figure and steel as the material, calculate the maximum stresses and deflections using the interpolation method described in Section 19.3.

SOLUTION

From Figure 19.5 the unused corner area A is

$$A = 1 - \frac{\pi}{4} = 0.215 \text{ in.}^2 \tag{19.1}$$

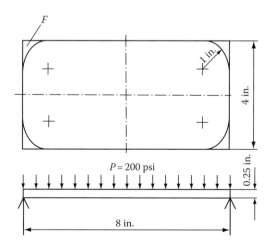

FIGURE 19.5 Panel of arbitrary profile.

The total unused corner area between the rectangular and elliptical boundaries is

$$ab\frac{(4-\pi)}{16} = 8 \times \frac{4(4-\pi)}{16} = 1.717\,\text{in.}^2 \tag{19.2}$$

The dimensionless ratio is then

$$\frac{0.215}{1.717} = 0.125 \tag{19.3}$$

For $a/b = 4/8 = 0.5$, Figure 19.2 gives approximately

$$A_1 = 0.61, \quad A_5 = 0.53 \tag{19.4}$$

The equation for interpolating the required stress factor can now be set up as follows:

$$A_1 - \frac{(A_1 - A_5)16A}{ab(4-\pi)} \tag{19.5}$$

Hence,

$$0.61 - (0.61 - 0.53) \times 0.125 = 0.600$$

and using the formula for a rectangular plate from Table 19.1 gives

$$S = \frac{200 \times 16 \times 0.600}{09.25 \times 0.25} = 30{,}720\,\text{psi}\,(212\,\text{N/mm}^2) \tag{19.6}$$

From Figure 19.3

$$A_2 = 0.11, \quad A_6 = 0.096 \tag{19.7}$$

Again, the interpolation formula for this case is

$$A_2 - \frac{(A_2 - A_6)16A}{ab(4-\pi)} \tag{19.8}$$

and since the parameter $16A/ab(4-\pi) = 0.125$, as before, we get

$$0.110 - (0.110 - 0.096)0.125 = 0.1083$$

Hence, using the plate deflection formula from Table 19.1 yields

$$\delta = \frac{200 \times 256 \times 0.1083}{30 \times 10^6 \times 0.25^3} = 0.012\,\text{in.}\,(0.30\,\text{mm}) \tag{19.9}$$

19.5 ANNULAR MEMBERS

Circular plates with centered round holes form a large class of problems related to flanges, rings, and circular closures, with numerous structural applications. For axisymmetric loading and support, the governing equations for the displacement, bending moments, twisting moments, and shear forces are given by Equations 18.62 through 18.67 as

$$w = \frac{p_o r^4}{64D} + c_1 r^2 \ell n r + c_2 r^2 + c_3 \ell n r + c_4 \tag{19.10}$$

$$M_r = -D\left[(3+v)\frac{p_0 r^2}{16D} + 2c_1(1+v)\ell nr + (3+v)c_1\right.$$

$$\left. + 2c_2(1+v) + c_3(v-1)/r^2\right] \tag{19.11}$$

$$M_\theta = -D\left[(1+3v)\frac{p_0 r^2}{16D} + (1+v)2c_1\ell nr + (1+3v)c_1\right.$$

$$\left. + 2(1+v)c_2 + (1-v)c_3/r^2\right] \tag{19.12}$$

$$T_{r\theta} = 0 \tag{19.13}$$

$$Q_r = -D\left(\frac{p_0 r}{2D} + 4\frac{c_1}{r}\right) \tag{19.14}$$

$$Q_\theta = 0 \tag{19.15}$$

where, $c_1, \ldots, c_4$ are integration constants arising in the integration of Equation 18.59, and where the notation is the same as that in Chapter 18. As before, the constants $c_1, \ldots, c_4$ are to be evaluated using the support conditions.

To illustrate the procedure, consider a plate with a central opening and uniformly distributed edge moments as in Figure 19.6. If there is no transverse loading, the pressure p_0 and the shear Q_r are zero. That is

$$p_0 = 0 \quad \text{and} \quad Q_r = 0 \tag{19.16}$$

Then from Equation 19.14 we see that c_1 is zero.

$$c_1 = 0 \tag{19.17}$$

Then from Equation 19.12 M_r is

$$M_r = -D\left[2c_2(1+v) + (c_3/r^2)(v-1)\right] \tag{19.18}$$

From the loading depicted by Figure 19.6 the edge (rim) conditions are

$$\text{At } r = r_i: \ M_r = M_i \quad \text{and at} \quad r = r_0: \ M_r = M_0 \tag{19.19}$$

or

$$M_i = -D\left[2c_2(1+v) + (c_3/r_i^2)(v-1)\right] \tag{19.20}$$

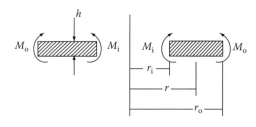

FIGURE 19.6   Annular plate with uniform edge moments.

and

$$M_o = -D\left[2c_2(1 + v) + (c_3/r_o^2)(v - 1)\right] \tag{19.21}$$

By solving c_2 and c_3, we obtain

$$c_2 = \frac{M_i r_i^2 - M_o r_o^2}{2D(1 + v)(r_o^2 - r_i^2)} \quad \text{and} \quad c_3 = \frac{M_o - M_i}{D(v - 1)(r_o^2 - r_i^2)} \tag{19.22}$$

If the plate is supported at its outer rim such that

$$w = 0 \quad \text{when} \quad r = r_o \tag{19.23}$$

Then from Equations 19.10, 19.16, and 19.17 we have

$$0 = c_2 r_o^2 + c_3 \ell n r_o + c_4 \tag{19.24}$$

From Equation 19.22 it is obvious that, c_4 is

$$c_4 = \frac{(M_o r_o^2 - M_i r_i^2) r_o^2}{2D(1 + v)(r_o^2 - r_i^2)} + \frac{(M_i - M_o) r_i^2 r_o^2 \ell n r_o}{D(v - 1)(r_o^2 - r_i^2)} \tag{19.25}$$

Finally, by substituting for $c_1, \ldots, c_4$ in Equations 19.10 and 19.11, the displacement and radial bending moment are

$$w = \frac{1}{D(r_o^2 - r_i^2)}\left[\frac{(M_o r_o^2 - M_i r_i^2)(r_o^2 - r^2)}{2(1 + v)} + \frac{(M_o - M_i) r_o^2 r_i^2 (\ell n r_o - \ell n r)}{1 - v}\right] \tag{19.26}$$

and

$$M_r = \frac{1}{(r_o^2 - r_i^2)}\left[r_o M_o^2 - r_i^2 M_i^2 - (M_o - M_i) r_i^2 r_o^2 / r^2\right] \tag{19.27}$$

As a second illustration, suppose the inner rim of the plate is restricted from displacement and rotation as represented in Figure 19.7. Then with a radial moment M_0 at the outer rim and an absence of loading on the surface, the edge (rim) conditions are

$$\text{At } r = r_i: \ w = 0 \quad \text{and} \quad dw/dr = 0 \tag{19.28}$$

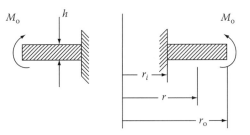

FIGURE 19.7 Annular plate with fixed inner rim and movement at the outer rim.

and

$$\text{At } r = r_0: \ M_r = M_t \tag{19.29}$$

Using a similar analysis, the displacement and radial bending moment become

$$w = \frac{M_o r_o^2 \left[r_i^2 - r^2 + 2r_i^2 \ell n(r/r_i) \right]}{2D \left[r_o^2 (1 + \nu) + r_i^2 (1 - \nu) \right]} \tag{19.30}$$

and

$$M_r = \frac{M_o r_o^2 \left[1 + \nu + (1 - \nu)(r_i^2/r^2) \right]}{r_o^2 (1 + \nu) + r_i^2 (1 - \nu)} \tag{19.31}$$

19.6   SELECTED FORMULAS FOR ANNULAR PLATES

Table 19.2 provides a listing for the maximum stress σ_{max} and the maximum displacement w_{max} for several support and loading conditions of axisymmetrically loaded annular plates. Figures 19.8 through 19.11 provide values of the parameters $F_1, \ldots, F_8$ and $B_1, \ldots, B_8$ for Poisson ratio ν: 0.3. As before, h is the plate thickness.

TABLE 19.2

Maximum Stress and Displacement Values for Axisymmetrically Loaded Circular Plates for Various Support Conditions

Simple support at the outer rim and ring loaded at the inner rim

$$\sigma_{max} = \frac{PF_1}{h^2} \qquad w_{max} = \frac{Pr_o^2 B^1}{Eh^3}$$

Simple support at the inner rim and uniform load on the plate

$$\sigma_{max} = \frac{pr_o F_2}{h^2} \qquad w_{max} = \frac{pr_o^4 B_2}{Eh^3}$$

TABLE 19.2 (continued)
Maximum Stress and Displacement Values for Axisymmetrically Loaded Circular Plates for Various Support Conditions

Clamped inner rim and uniform load on the plate

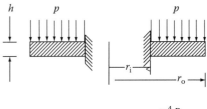

$$\sigma_{max} = \frac{pr_o^2 F_3}{h^2} \qquad w_{max} = \frac{pr_o^4 B_3}{Eh^3}$$

Simple support at the outer rim, horizontal slope at the inner rim, and uniform load on the plate

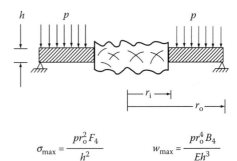

$$\sigma_{max} = \frac{pr_o^2 F_4}{h^2} \qquad w_{max} = \frac{pr_o^4 B_4}{Eh^3}$$

Clamped outer rim, horizontal slope at the inner rim, ring load at the inner rim

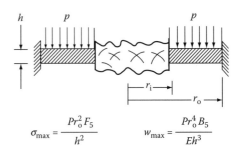

$$\sigma_{max} = \frac{Pr_o^2 F_5}{h^2} \qquad w_{max} = \frac{Pr_o^4 B_5}{Eh^3}$$

Clamped outer rim, horizontal slope at the inner rim, ring load at the inner rim

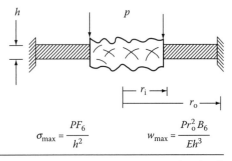

$$\sigma_{max} = \frac{PF_6}{h^2} \qquad w_{max} = \frac{Pr_o^2 B_6}{Eh^3}$$

(continued)

TABLE 19.2 (continued)

Maximum Stress and Displacement Values for Axisymmetrically Loaded Circular Plates for Various Support Conditions

Simple support at the outer rim and uniform load
 on the plate

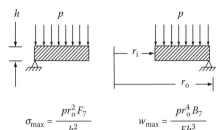

$$\sigma_{max} = \frac{pr_o^2 F_7}{h^2} \qquad w_{max} = \frac{pr_o^4 B_7}{Eh^3}$$

Simple support at the outer rim, horizontal slope at
 the inner rim, and ring load at the inner rim

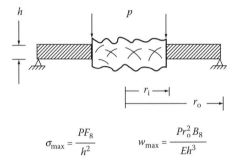

$$\sigma_{max} = \frac{PF_8}{h^2} \qquad w_{max} = \frac{Pr_o^2 B_8}{Eh^3}$$

Note: Figures 19.8 through 19.11 provide values of F_i and B_i $(i = 1, \ldots, 8)$ for Poisson ratio $\nu = 0.3$.

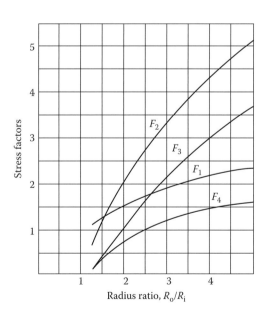

FIGURE 19.8 Stress factors F_1 through F_4 for the annular plates of Table 19.2.

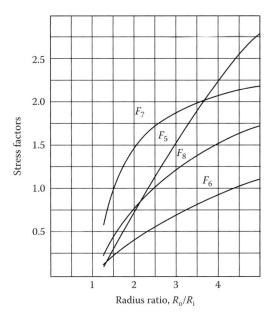

FIGURE 19.9 Stress factors F_5 through F_8 for the annular plates of Table 19.2.

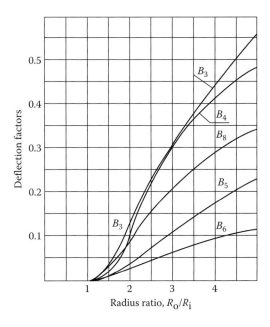

FIGURE 19.10 Plate displacement factors B_3, B_4, B_5, B_6, and B_8 for the annular plates of Table 19.2.

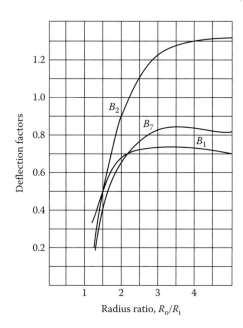

FIGURE 19.11 Plate displacement factors B_1, B_2, and B_7 for the annular plates of Table 19.2.

SYMBOLS

A	Corner area
A_1 through A_8	Factors for panels
a	Smaller side or minor axis
B_1 through B_8	Plate deflection factors
b	Larger side or major axis
c_1, c_2, c_3	Integration constants
D	Plate flexural rigidity; $Eh^3/12(1 - \nu^2)$ (see Equation 18.2)
E	Modulus of elasticity
F	Corner area
F_1 through F_8	Plate stress factors
h	Plate thickness
M_1, M_i, M_o	Bending moments
M_i, M_o	Inner and outer bending moments
M_r, M_θ	Radial and transverse bending moment
p_O	Uniform load
q	Uniform load
Q_r, Q_θ	Radial and transverse shear
R_i	Inner radius of plate
R_o	Outer radius of plate
r	Radial coordinate
r_i, r_o	Inner and outer radii
S	Stress
S_y	Yield strength
t	Thickness of plate
$T_{r\theta}$	Twisting moment

w	Transverse displacement
W	Total load on plate
x	Arbitrary distance
δ	Maximum deflection
ν	Poisson's ratio
ϕ	Slope, rad

20 Flanges

20.1 INTRODUCTORY REMARKS

Flanges and support rings have various configurations. They present an extensive variety of analysis and design problems. Forces acting on these members can arise from any direction. Those acting out of plane are the most difficult to describe analytically.

As a result of these difficulties and complexities, problems involving circular flanges with reinforced gussets are seldom found in the open literature. The configurations and loadings on these structural components necessitate a detailed three-dimensional analysis. Even with the help of finite element methods (FEM), the design of these components represents a tedious and costly procedure. Thus a design engineer may have to make difficult decisions in practical cases because of the lack of a well-established design methodology and a lack of readily available published results.

These issues can lead to gross overdesign and excessive cost, especially where large-diameter pipes with great pipe lengths and large flanges are involved. In conventional conservative design applications, the trend appears to be toward greater depth and larger overall sizes of the components. A review of flange design theory and practice is therefore useful. In this chapter, we present such a review.

As evidenced by recent theoretical and experimental work, flange analysis can be very time-consuming even in the case of simpler flange configurations and simple loadings. It is helpful, therefore, not only to suggest a simplified approach to the problems at hand but also to review some of the more commonly accepted theoretical concepts and formulas.

The objective of our review is to develop usable procedures for practical applications. In this regard, the theoretical information concerned with generic flange design is useful for developing a simplified approach and philosophy for designing rib-stiffened configurations. We base our review upon selected references and flange design standards of the United States, the United Kingdom, and Germany [1–18].

20.2 STRESS CRITERIA

One of the key messages included in this section concerns the idea of elastic versus plastic stresses. Since the great majority of practising engineers have been taught in the tradition of the theory of elasticity and the concept of the elastic strength of materials, it is relatively easy to misinterpret the true meaning of computed stresses. In fact, it is often presumed that the calculated values exceeding the elastic limit must necessarily be dangerous. This seems to be particularly misleading where the design formulas give the sum of the bending and membrane stresses without due allowance for material ductility and stress redistribution.

In a typical integral-type flange, that is, where the flange is butt-welded to the wall of the pipe, the adjacent portion of the wall is considered to act as a hub. The accepted design practice calls for calculation of the three major stresses: maximum axial stress in the hub, radial stress in the flange ring at its inside diameter, and the corresponding tangential stress at the same point. The theoretical and experimental evidence indicates that the axial stress in the hub is frequently by far the highest and it is often used as the basic design criterion for sizing the wall thickness. Some applications of this general rule are considered here in evaluating the maximum theoretical stress in the hub of a rib-stiffened flange.

The selection of a suitable design criterion and the corresponding calculation procedure depends in general upon the flange geometry and the materials involved. Various theories and design methods in the past utilized straight beam, cantilever, circular ring, and plate model approaches for the purpose of checking the flange stresses. The method of rib sizing, proposed in this section, is based on the theory of beams on elastic foundation.

20.3   EARLY DESIGN METHODS

The development of pressure vessels having increasingly higher pressures and temperatures has been a stimulus of increased interest in flange stress formulas in the West [2–4]. Early flange design involved hubs of approximately uniform thickness and the designs were checked by calculating the tangential stress at the inner diameter of the flange, ignoring entirely the possibility of the hub stresses. Further limitations of the early methods involved their narrow range of applicability, as they were developed for specific types and proportions of the flanges. This predicament persisted until publication of the Waters-Taylor formulas [9], which were based on theoretical and experimental results. This classical paper marked the start of extensive deliberations of various approaches to flange design.

20.4   THIN HUB THEORY

When the hub is relatively thin and a critical section is assumed to exist along one of the flange diameters, the maximum stress can be calculated from a simple beam formula. This approach, which is probably one of the earliest and best known, is illustrated in Figure 20.1, where we assume that the flange is clamped along this radial cross section. The design is based on bending due to the external moment obtained by lumping together all bolt loads and utilizing the concept of a moment

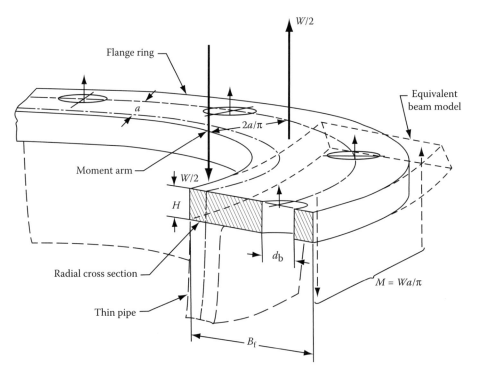

FIGURE 20.1   Flange fixed along radial cross section.

arm. In effect, a simple beam model is postulated where the net cross section is found by subtracting the projected areas of bolt holes. Using the notation indicated in Figure 20.1, the available section modulus for the flange ring becomes

$$Z = \frac{(B_f - d_b)H^2}{3} \tag{20.1}$$

Utilizing the moment arm shown in Figure 20.1, the available section modulus for the Figure 20.1 we obtain the maximum bending stress σ_b from the elementary beam formula. Note that the term $2a/\pi$ follows from a consideration of the centers of gravity for the two concentric, semicircular arcs. Hence

$$\sigma_b = 0.95 \frac{Wa}{(B_f - d_b)H^2} \tag{20.2}$$

Obviously, Equation 20.2 is only approximate since the curvature of the flange ring and the effect of the pipe wall have been ignored. Nevertheless, the method is a rather ingenious use of the theory of straight beams and it gives surprisingly good results when applied to loose flanges or flanges welded to thin pipes. The effect of radial stresses in such flanges can, of course, be neglected.

20.5 FLANGES WITH THICK HUBS

When a pipe is relatively thick and the circumferential stresses are ignored by assuming a number of radial slots, a cantilever beam method is sometimes employed. The corresponding notation and configurational details for this analytical model are given in Figure 20.2. This method of calculation

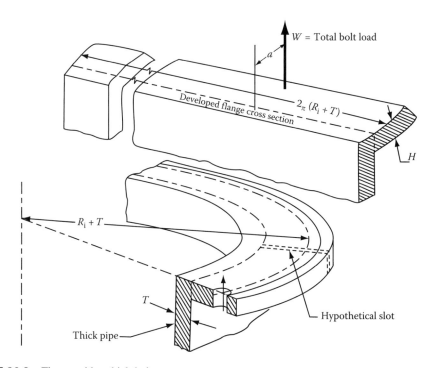

FIGURE 20.2 Flange with a thick hub.

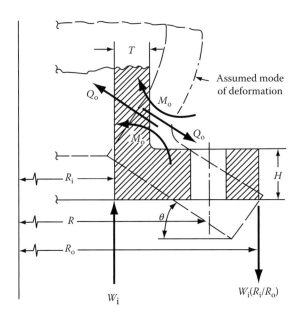

FIGURE 20.3   Flange treated as a circular ring.

yields only radial stresses and it is expected to have a rather limited field of application. In reality, it can be used only in the case of a relatively thin flange made integral with the thick pipe of a large diameter. Under these conditions the maximum radial stress σ_r becomes

$$\sigma_r = 0.95 \frac{Wa}{(R_i + T)H^2} \tag{20.3}$$

When radial stresses are expected to be relatively small, a significant refinement is achieved by utilizing the theory of rings [6]. The corresponding mode of deformation and the basic notation are given in Figure 20.3, where the cross section of the flange ring is assumed to rotate through angle θ, shown in an exaggerated manner. The cross-sectional dimensions of the flange ring are relatively small compared to the ring diameter and it is assumed the rectangular shape of the cross section does not change under stress. The latter assumption is consistent with the idea of neglecting radial stresses, which suggests that this theory applies to flanges attached to relatively thin pipes.

20.6   CRITERION OF FLANGE ROTATION

In establishing the equations for calculating the bending moment and the shearing force per unit length of the inner circumference of the pipe, where the flange ring and the pipe are joined, radial deflection is assumed to be zero and the angle of rotation of the edge of the pipe is made equal to the angle of rotation of the flange cross section. In Figure 20.3, this angle is denoted by θ and has been the theory of local bending and discontinuity stresses in thin shells [11] used, together with the theory of a circular ring subjected to toroidal deformation.

Following the formulation and analysis of Timoshenko [6], the maximum bending stress σ_b in the pipe using this theory and the notation of Figure 20.3 is

$$\sigma_b = 6M_0/T^2 \tag{20.4}$$

where M_0 is the bending moment per unit length of the inner circumference of the flange with radius R_i, and is given by

$$M_0 = \frac{W_i(R_o - R_i)}{(1 + (\beta_s H/2)) + (1 - v^2 2\beta_s R_i)(H/T)^3 \log_e (R_0/R_i)} \qquad (20.5)$$

and the corresponding shear force is

$$Q_0 = \beta_s M_0 \qquad (20.6)$$

where β_s is

$$\beta_s = \frac{1.285}{(R_i T)^{1/2}} \qquad (20.7)$$

The parameter β_s is useful in the analysis of beams on an elastic foundation indicating the extent of stress-affected zones in the vicinity of edge or local loading.

In Equation 20.5, W_i denotes the force per unit length of the inner circumference of the flange corresponding to radius R_i. The external bending moment applied to the flange involves the moment arm, which in Timoshenko's case is defined as $R_o - R_i$. A brief comparison with other methods indicates that the assumption of different moment arms is bound to significantly affect the calculated results. It is quite likely that Equation 20.5 will always overestimate the bending moment M_0 because of the maximum moment arm used. Under the actual conditions, the loading may be found to be significantly removed from the inner and outer edges of the flange. Nevertheless, integral flanges with relatively thick pipes have been used successfully with and form the basis of some of the existing design standards in industry.

20.7 USE OF PLATE THEORY WITH FLANGES

Further refinement of and insight into flange analysis may be obtained by applying plate theory, where radial and circumferential stresses are taken into account. Radial stresses may be of importance in flanges integral with thick pipes, which can resist the angle of tilt much better. This angle is shown in Figure 20.4.

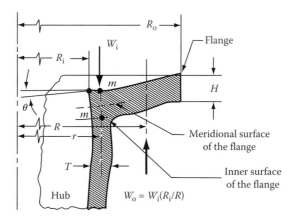

FIGURE 20.4 Flange treated as a circular plate.

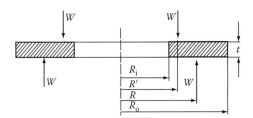

FIGURE 20.5 Simply supported plate under concentric loading.

In applying plate theory to the solution of a flange problem or to the stresses in a cylinder with rigidly attached flat heads [8–10], a strip may be cut out of the cylinder and treated as a beam on an elastic foundation while the flange is regarded as a flat plate with a central hole. The slopes and deflections at the end of the cylinder can be expressed in terms of the unknown moments and shear forces. These displacements are then equated to the slopes and deflections similarly determined for the plate. In this manner, we establish a sufficient number of equations, which are solved simultaneously to yield the unknown reactions and displacements.

The classical approach to the flange problem can be based on the premise that the bending moment existing at the root of the hub acts at the meridian plane of the flange instead of at its inner plane. When the flange portion is deformed, these planes become curved, as shown in Figure 20.4. Furthermore, it can be assumed that the cylindrical surface containing points $m - m$ does not alter its original curvature and that the expansion of the hub due to the internal pressure can be ignored.

The solutions for flange problems using plate theory assume the flange to be a circular plate with a central hole either loaded at the edge or loaded uniformly over the whole surface [9,10]. When a circular plate is loaded and supported in the manner shown in Figure 20.5, the maximum hoop stress is developed at the inside corners of the plate. According to the flat ring theory of Waters [9], the corner stress σ_h becomes

$$\sigma_h = \frac{W(R - R')}{(R_o - R_i)t^2}\left[\frac{1.242R_o^2 \log_e (R_o/R_i)}{(R_o^2 - R_i^2)} + 0.335\right] \tag{20.8}$$

The corresponding Holmberg-Axelson formula for this stress is

$$\sigma_h = \frac{W}{(R_o^2 - R_i^2)}\left[0.35(R^2 - R'^2) + 1.195R_o^2 \log_e \left(\frac{R}{R'}\right)\right] \tag{20.9}$$

Examination of Equation 20.8 shows certain natural limitations when the R_o/R_i ratio is large, since such a case would correspond to the theory of plates rather than that for flat rings on which Equation 20.8 is based. It appears that Equation 20.9 is correct for all values of R_o/R_i. The results of the study also indicate that a plate pierced by a small hole in the center has the maximum circumferential stress twice as large as if the plate had been solid. This finding may be of special importance in those cases where the plate is not made of fracture-tough material.

It appears that in treating certain plate configurations, which resemble machine and pipe flanges rather than circular closures, the analyst has the choice of following either the treatment of flat rings or flat plates. Probably the best method demonstrated in industry so far involves breaking down the flange or plate structures into a series of concentric rings, each of which has a simple loading [10]. The boundary conditions are solved by making the slope and radial moments continuous. Since, in this type of treatment, the number of constants is always equal to twice the number of rings, a large number of simultaneous equations may be involved. This, however, will not pose special problems where large electronic computers and finite element methods are available.

20.8 FORMULA FOR HUB STRESS

Hub stress arising from bolt loading is a major concern in flange analyses. In this section, we present a simplified formula for the hub stress. The formula is based upon analyses of compressor casing flanges in jet engines.

Figure 20.6 defines the notation. The maximum bending stress σ_{max} is then approximately

$$\sigma_{b\,max} = \frac{0.48 W \beta_s T (R_o + R_i)}{\phi_0 B_s R_i T^3 (2 + \beta_s H) + H^3} \tag{20.10}$$

where as before β_s is (see Equation 20.7)

$$\beta_s = \frac{1.285}{(R_i T)^{1/2}} \tag{20.11}$$

and where for a typical Poisson ratio ν of 0.3, Figure 20.7 provides values of the flange ratio ϕ_o as a function of the ring ratio R_o/R_i. As noted in the figure, ϕ_o may also be calculated using the expression

$$\phi_o = \frac{0.77 + 1.43 k^2}{k^2 - 1} \tag{20.12}$$

where k denotes the ring ratio R_o/R_i.

By substituting from Equations 20.11 and 20.12 into Equation 20.10, we obtain the approximation

$$\frac{\sigma_{bmax} T^2}{W} = \frac{0.614(k+1)(k^2-1)(m)^{1/2}}{(0.77 + 1.43 k^2)[2.57 m^{1/2} + 1.65 n] + n^3 (k^2 - 1)} \tag{20.13}$$

where m and n are the dimensionless parameters:

$$m = R_i/T \quad \text{and} \quad n = H/T \tag{20.14}$$

Equation 20.13 provides an "apparent" instead of an "actual" stress, but it is conservative particularly for ductile materials. That is, the calculated stresses will be higher than those actually occurring.

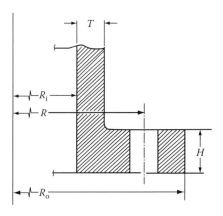

FIGURE 20.6 Simplified notation for the analysis of straight flanges.

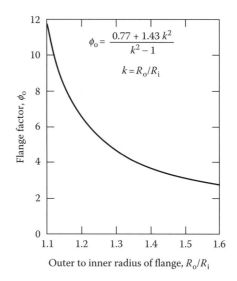

FIGURE 20.7 Flange factor ϕ_0.

20.9   GERMAN AND AMERICAN FLANGE DESIGN PRACTICE

According to machine design practice in Germany [4], the calculation of maximum bending stresses in the pipe wall adjacent to the flange ring can be accomplished in a straightforward manner. This procedure recognizes the two basic modes of failure depending on the relative thickness of the flange ring and the pipe, and it can be applied to tapered as well as to straight hubs. When the flange ring is thicker than the pipe, failure is expected in the hub. The reverse is true for heavy pipe walls. In both instances, the flange ring surface under strain is assumed to conform to a spherical shape.

Figure 20.8 shows the notation and overall proportions of a flanged section used in German and American flange standards. The pipe section typically fails (fractures) at the inclination α as shown where α ranges between 20° and 30°. The depth s_1 of the fractured surface can be related to the nominal thickness s_0 of the pipe as

$$s_0 = s_1 \cos \alpha \qquad (20.15)$$

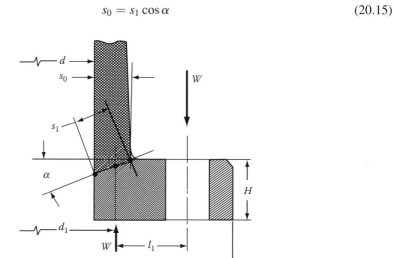

FIGURE 20.8 Flange section notation.

In terms of the notation of Figure 20.8 we have the following approximate relations:

$$d_1 = 2R_i + T \tag{20.16}$$

$$s_1 = T/\cos\alpha \tag{20.17}$$

$$\ell_1 = (R_o - R_i)/2 \tag{20.18}$$

where the radii R_i and R_o are the half-diameters $d_i/2$, and T is the nominal hub or pipe thickness (see Figure 20.4). Using a mean value of α as 25°, the maximum bending stress σ_{bmax} in the pipe, adjacent to the flange, is (German standard)

$$\sigma_{bmax} = \frac{W(k-1)[0.228 + 0.035n\sin^2(\pi/n)]}{nT^2} \tag{20.19}$$

where, as before, we employ the dimensionless parameters k, m, and n as

$$k = R_o/R_i, \quad m = R_i/T, \quad n = H/T \tag{20.20}$$

The derivation of this equation involves a number of simplified steps and symbols consistent with other formulas given in this chapter [16].

If the flange design is made according to American practice, the maximum bending stress σ_{bmax} is (Waters-Taylor formula):

$$\sigma_{bmax} = \frac{W}{T^2}\left\{\frac{0.25(m)^{1/2}[k^2(1+8.55\log k)-1]}{(1.05+1.94k^2)[(m)^{1/2}+0.64n]+0.53n^3(k^2-1)}\right\} \tag{20.21}$$

where the logarithmic term is calculated to base 10. The formula consistent with the German code should be applicable to all cases for which $n \neq 0$ and $k < 1.8$. Parametric studies also indicate that Equations 20.13 and 20.19 yield lower numerical values of hub stresses than those which can be predicted on the basis of Equation 20.21.

20.10 CIRCUMFERENTIAL STRESS

It may be of interest to note a simplified method of predicting the circumferential stresses in a standard flange ring in terms of the maximum bending hub stress [3]. In this type of analysis, we assume that the radial expansion of the flange ring due to the discontinuity shear force and the bulging of the pipe due to any internal pressure can be ignored. Because of the flange rotation, one can expect to find the maximum tensile stress on the outside of the pipe wall, and pipe bulging due to internal pressure. Hence, the estimate should be conservative. Also, we assume that the pipe in the vicinity of the flange ring can be stressed to the yield point of the material. Under these conditions, the circumferential flange stress σ_f may be expressed in terms of the maximum hub bending stress σ_b as

$$\sigma_f = \frac{n(2m+1)}{3.64mn + 4m(2m+1)^{1/2}}\sigma_b \tag{20.22}$$

where as before m is the radius/thickness ratio R_i/T and n is the flange/hub thickness ratio H/T (see, for example, Figure 20.6).

In Equation 20.22, σ_f denotes the flange-ring circumferential stress produced by the deformation due to the bolt loading and σ_b is set equal to the yield stress. For large values of the radius/thickness ratio m, Equation 20.22 has the graphical form shown in Figure 20.9.

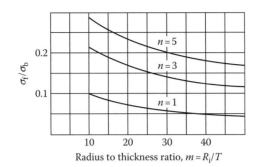

FIGURE 20.9   Ratio of flange to hub stress.

The theoretical limits for the stress ratio σ_f/σ_b are 0.55 and 0.0 for zero and infinite values of the radius/thickness ratio m. The intermediate range of values show that the maximum circumferential stresses in the flange ring are always considerably smaller than the hub stresses. Conversely, the theory indicates that if yielding of the pipe in the vicinity of the flange is to be avoided, the flange ring would have to be extremely thick and therefore unacceptable for all practical purposes. The major conclusion drawn from this finding is that, for truly economic design in ductile materials, plastic deformation of the pipe in the vicinity of the flange ring can be permitted. The reserve of strength beyond the onset of yield can be quite significant. For example, the theoretical collapse load of a beam of rectangular cross section is 1.5 times the load causing yield in the outer fibers, assuming a rigid-plastic, stress–strain characteristic for the material.

20.11   APPARENT STRESS CRITERIA—A DISCUSSION

A principal issue in mechanical design is establishing the maximum allowable stress level. Section 20.10 shows that if we simply restrict the maximum stress to be less than the yield stress, we will not obtain the optimal design. For ductile materials, the yield stress is considerably less than the rupture (or fracture) stress.

Generally speaking, the hoop stress in a flange ring, or the radial stress across the junction of a hub and ring, may have moderate values. However, the longitudinal bending stress in the hub is usually much larger than either of these.

In a standard flange application, the distributed loading consists of bolt load, gasket pressure, hydrostatic pressure of the flange leakage area, and the hydrostatic end force. All the loading is represented by an equivalent bending couple consisting of two equal and opposite loads. When the internal pressure is absent and a flange gasket is relatively close to the bolt circle, these loads can be appreciably removed from the inner and outer edges of the flange. Equation 20.13 corresponds to the loads, which are placed sufficiently far from the flange edges.

The theoretical hub stresses calculated from Equation 20.13, 20.19, or 20.21 are based on the elastic behavior and should be considered as "apparent" rather than actual stresses. Generally, this situation persists in many areas of engineering analysis and should be reviewed continually with reference to practical design requirements. The numerical values of apparent stresses, interpreted as strains multiplied by the relevant moduli of elasticity, often bear little relation to the actual material stresses and can be evaluated only with special regard to the stress–strain curve. The concept of apparent stresses is a very real one and is fully supported by practical experience. It is generally recognized that in many actual flange designs, the calculated hub stresses are extremely high and yet the flanges are satisfactory. In such circumstances, of course, the classical elastic formulas show limited validity and flange design by test is recommended [12]. Such formulas are basically correct, but the relevant interpretation of the numerical results is misleading if the yield strength and the reserve of plastic strength in a flange, or other machine part, are ignored.

The fact that calculated apparent stress exceeds the proportional limit of a material does not necessarily mean that failure is imminent. Engineers and designers have often been inclined to think in terms of the vertical axis of a stress–strain diagram and have paid less attention to the problem of strain. It may be fitting in many areas of design analysis to think in terms of the maximum allowable strain rather than in terms of maximum allowable stress.

Therefore, in flange design, regarding the level of hub stresses, it is advisable to recognize the existence of local yielding as long as there is sufficient reserve of strength in the adjacent section of the flange, to take care of the increased loading placed upon these sections by the region of local yielding.

Finally, in the vast majority of flange applications significant dynamic loading is usually absent. that is, the static loading due to bolt loading and pipe pressure has little, if any, fluctuation. Thus fatigue from multiple stress cycling does not occur. This also allows for an optimal design where the maximum stress may exceed the yield stress.

20.12 PLASTIC CORRECTION

Consider the stress–strain graphic of Figure 20.10. This figure is useful for obtaining insight into the problem of apparent stress and to suggest an approximate plastic correction for interpreting calculated stresses higher than the yield strength of the material. The stress–strain curve shown in Figure 20.10 is typical of a low-strength ductile material, although the strain at yield ε_y, has been exaggerated for the purpose of clarity. Hence, the slope of the curve below the proportional limit should not be used for direct numerical evaluation of the modulus of elasticity. The values of stresses and strains are merely to indicate that the material characteristics discussed are close to those found in low-carbon steel, aluminum alloy, or other materials that may behave similarly.

Let us first assume that the stress–strain diagram has been established experimentally and that the curve obtained can be approximated by the two straight lines from zero to the yield point and from the yield region to the highest point on the curve, as shown by the line $x–x$. Let the corresponding elastic moduli be E and E_0, respectively. In the study of plasticity of materials, a

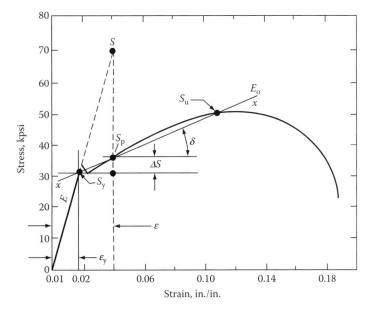

FIGURE 20.10 Typical stress–strain curve for a low-strength ductile material.

bilinear approximation of this kind is well known and indicates that the material in question conforms to the law of linear strain hardening [13].

In terms of the calculated stress S, shown as 70,000 psi in Figure 20.10, Hooke's law gives

$$\varepsilon = \frac{S}{E} \tag{20.23}$$

The corresponding strain at yield stress S_y is

$$\varepsilon_y = \frac{S_y}{E} \tag{20.24}$$

By the definition of the elastic modulus

$$\tan \delta = E_0 = \frac{\Delta S}{\varepsilon - \varepsilon_y} \tag{20.25}$$

Denoting the actual stress by S_p, we have

$$\Delta S = S_p - S_y \tag{20.26}$$

Hence, substituting Equation 20.26 into Equation 20.25, the elastic modulus for the $x-x$ portion of the curve becomes

$$E_0 = \frac{S_p - S_y}{\varepsilon - \varepsilon_y} \tag{20.27}$$

Solving for S_p and eliminating strains with the aid of Equations 20.23 and 20.24 gives

$$S_p = S\frac{E_0}{E} + S_y\left(1 - \frac{E_0}{E}\right) \tag{20.28}$$

It is evident that when E_0 tends to zero, a stress–strain curve representing ideal plastic material is obtained, and $S_p = S_y$ for all values of strain. Alternatively, when $E_0 = E$, there is only elastic behavior, and the stress–strain curve represents purely elastic action up to the point of failure. The concept of ideal plastic material is often employed in the theory of plasticity because a typical mild steel stress–strain diagram is close to that of a perfect plastic material. With various alloying elements, the material still exhibits desirable ductility, but, in addition, there may be some strain hardening.

Considering the strain-hardening characteristic to be the one illustrated in Figure 20.10, the following appraisal of the apparent versus actual stress can be made. Assume that the calculated hub stress in a flange is 70,000 psi, as shown in Figure 20.10. If material knowledge gives the yield strength $S_y = 31,000$ psi, and the ultimate strength $S_u = 50,000$ psi, but the shape of the stress–strain curve is unknown, then on the basis of our calculation we may conclude that the pipe section will fail. Specifically, the safety factors 0.44 and 0.71 are based on the given values of yield and ultimate strength, respectively. To increase these factors, it would appear that either a better material is needed or that the nominal wall thickness of the pipe should be increased. Either approach would not be in the interest of economy and could lead to gross overdesign.

Assume now some knowledge of the actual stress–strain curve, and take $E_0/E = 0.25$, not an unusual ratio for a material with linear strain hardening, so that using Equation 20.27 yields

$$S_p = 0.25 \times 70,000 + 31,000(1 - 0.25) = 40,700 \, \text{psi} \tag{20.29}$$

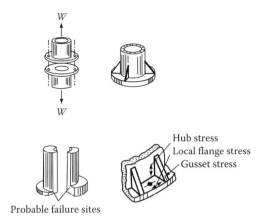

Hub stress
Local flange stress
Gusset stress

Probable failure sites

FIGURE 20.11   Typical flanges and reinforcements.

The corresponding safety factors are then changed to 0.76 and 1.23, respectively. If the flange in question is not subject to fatigue, as is normally the case with many flanged configurations, the pipe failure becomes less likely.

If the material assumed in the above numerical illustration can be considered perfectly plastic, then $E_0/E = 0$ and the maximum factor of safety becomes unity, because the actual hub stress cannot exceed the value of yield. The results of this discussion point clearly to the need for a realistic approach to the interpretation of calculated elastic stresses in conjunction with the stress–strain characteristics of the materials involved. Hence if the maximum bending stress in the hub is restricted to the pipe surface any higher stress than the yield strength of the material at that point will cause stress redistribution.

Figure 20.11 shows several examples of flanges used to connect piping and tubular structural members [14,15] with the features of progressive complexity and the regions of more significant stress components. The correction for plastic action described by Equation 20.28 can be applied to any of the stress components, provided their elastic values can be estimated. The most difficult problem, however, is the first estimate of the elastic response. This situation has not changed markedly despite the significant progress in numerical techniques.

20.13 HEAVY-DUTY FLANGES

In the development of heavy-duty pipe flanges and similar hardware components, it has been customary to utilize the concept of a compound flanging where the two concentric flanged rings on a pipe are joined by external ribs parallel to the pipe axis. Although this is not an unusual concept where the overall flange rigidity against the toroidal deformation is required, a rigorous analytical approach to the design of such a three-dimensional structure must be highly involved. A typical approach to the theoretical problem would be through finite-element techniques and rather lengthy experimental verification of the stress picture using strain gages, photostress, or three-dimensional photoelasticity. If, for economic or scheduling reasons, a more fundamental approach to this problem is not feasible, a relatively simple approximate solution can be developed utilizing the existing knowledge of structural mechanics. Such solutions appear to be on the conservative side and are not far removed from reality as shown by theoretical and experimental investigations conducted by Werne [17].

To reach a rational compromise as to the selection of flange thickness in the case of double-ring ribbed design, consider Figure 20.12. The action of a double-flanged ring may be simulated by a twist of a circular ring loaded by couples M_0, which are uniformly distributed along the center line

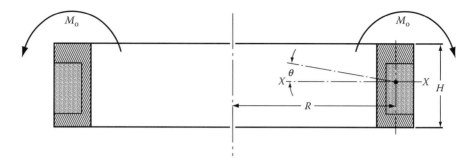

FIGURE 20.12   Model of a twisted flange ring.

of the ring. To simplify the calculation, it is assumed that the cross-sectional dimensions of the ring are relatively small when compared with the mean radius R. Timoshenko [6] shows that the angle of rotation of flange cross section θ and the normal stress are inversely proportional to the moment of inertia about the ring axis $x-x$.

20.14   EQUIVALENT DEPTH FORMULA

The assumption of the cumulative value H, sometimes used in the calculations, cannot be accurate because the material, indicated as the dotted areas of Figure 20.12 is really nonexistent except in the regions of the ribs. However, to deal with this characteristic dimension, the equivalent depth H_e may have to be calculated with reference to Figure 20.13 to provide a solid rectangular cross section that would yield the same angle of rotation as the original channel section of depth H. The possibility of some local distortion of the channel section in between the ribs exists; however, in the simplified approach, this feature will be excluded.

Although the composite channel section indicated in Figure 20.13 is not symmetrical about the axis $x-x$, the moment of area can be found directly without calculating the position of the center of gravity of the section [5]. For this purpose, a convenient baseline, $m-m$, is selected as shown in Figure 20.13, leading to the equation

$$I_x = I_b + I_g - \frac{J^2}{A_t} \tag{20.30}$$

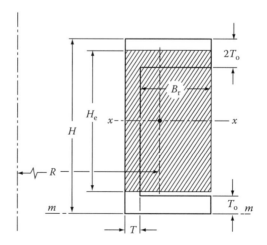

FIGURE 20.13   Section details for equivalent depth calculation.

where

> J and I_b denote the first and second moments of area with respect to the baseline $m-m$
>
> I_g stands for the sum of the moments of all component areas about their own centroids
>
> A_t defines the total cross-sectional area

The problem is handled in the usual way by breaking the whole area into geometrically convenient component areas and applying Equation 20.30. Invoking this rule and adhering to the notation shown in Figure 20.13 gives

$$I_x = \frac{H^4 T^2 + 36 B_r H T T_0^3 - 30 B_r H^2 T T_0^2 + 12 B_r H^3 T T_0 + 81 B_r^2 T_0^4}{12(HT + 3 B_r T_0)}$$
$$+ \frac{2 B_r^2 H^2 T_0^2 - 6 B_r^2 H T_0^3}{HT + 3 B_r T_0} \tag{20.31}$$

Hence, the equivalent depth of the ribbed flange consisting of the flange and backup rings is seen to be

$$H_e = 2.29 \left(\frac{I_x}{B_r + T} \right)^{1/3} \tag{20.32}$$

Figure 20.14 provides a view of the compound flange with the main flange ring, backup ring, rectangular ribs, and the triangular gussets. Appropriate size and location of the flange reinforcement should be determined on the basis of hub, gusset, and flange stresses. The equivalent depth of the ribbed flange H_e can be used in conjunction with such formulas as those given by Equations 20.13, 20.19, or 20.21. When the backup ring is not used and $2T_0 = h$, the formula for calculating the equivalent depth of the ribbed flange becomes

$$H_e = \left[\frac{HT^3(HT + 4hB_r) - 3HTh^2 B_r(2H - h)}{(B_r + T)(HT + hB_r)} \right]^{1/3} \tag{20.33}$$

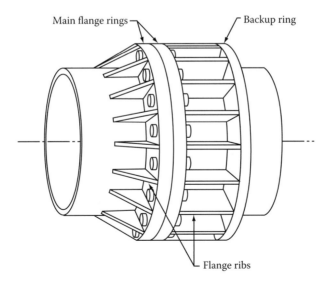

Main flange rings ⎯ Backup ring

Flange ribs

FIGURE 20.14 Pipe flange geometry with rectangular and triangular ribs.

20.15 LOAD SHARING IN RIBBED FLANGES

Figure 20.15 illustrates a partial view of a rib-stiffened main flange ring. In evaluating the moment carrying capacity of a stiffening rib in relation to that of the main flange ring, we can use the theory of beams on elastic foundations and toroidal deformation of a circular ring. If the moment arm is taken as $R - r$, the total external moment due to the bolt load, referring to unit length along the bolt circle, may be defined as

$$M = \frac{W(R - r)}{2\pi R} \tag{20.34}$$

If N denotes the total number of ribs supporting the flange ring, the external bending moment per length of the circumference corresponding to one rib spacing is

$$M_c = \frac{2\pi R M}{N} \tag{20.35}$$

Substituting from Equation 20.34 into Equation 20.35 gives

$$M_c = \frac{W(R - r)}{N} \tag{20.36}$$

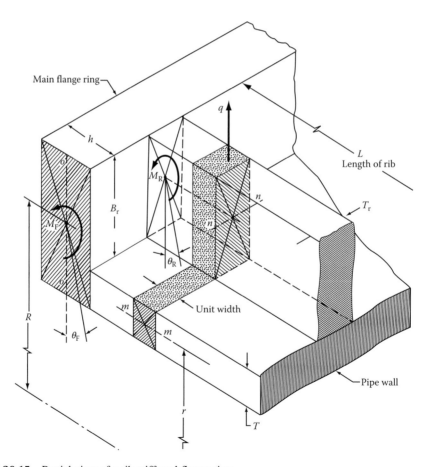

FIGURE 20.15 Partial view of a rib-stiffened flange ring.

Denoting the bending moments M_F and M_R carried by the main flange ring and rib, respectively, gives

$$M_c = \frac{2\pi R M_F}{N} + M_R \tag{20.37}$$

Hence Equations 20.36 and 20.37 yield

$$W(R - r) = 2\pi R M_F + N M_R \tag{20.38}$$

Although the foregoing algebraic operations are rather elementary, it is advisable to follow the basic derivations to assure correct dimensional identities before developing subsequent working formulas. We note, therefore, that M_F is expressed in lb-in./in. and M_R is given in lb-in., in the English system.

The angle of twist of an elastic ring undergoing a toroidal deformation under the action of a twisting moment M_F may be expressed by the classical formula as

$$\theta_F = \frac{M_F R^2}{E I_0} \tag{20.39}$$

The bending slope of the stiffening rib at the rib–flange junction, treated as a beam of finite length resting on the elastic foundation and acted upon by a concentrated end moment M_R, can be calculated from the following relations: let Y denote the deflection of a beam on an elastic foundation with hinged ends that is bent by a couple M_R, applied at the end as shown in Figure 20.16. Then the general expression for the deflection line is

$$Y = \frac{2 M_R \beta^2 [\cosh \beta L \sin \beta x \sinh \beta(L - x) - \cos \beta L \sinh \beta x \sin \beta(L - x)]}{K(\cos^2 \beta L - \cos^2 \beta L)} \tag{20.40}$$

The required slope is found by calculating dy/dx from Equation 20.40 and making $x = 0$. This gives

$$\theta_R = \frac{2 M_R \beta^3 (\cos \beta L \sin \beta L - \cos \beta L \sin \beta L)}{K(\cosh^2 \beta L - \cos^2 \beta L)} \tag{20.41}$$

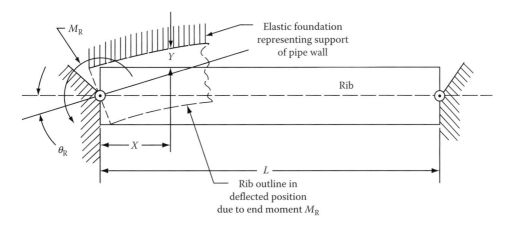

FIGURE 20.16 Mode of rib deformation as a beam on an elastic foundation.

where

$$\beta = \left(\frac{K}{4EI_n}\right)^{1/4}$$

(20.42)

Eliminating K with the aid of Equation 20.42 and introducing an auxiliary dimensional function characterizing the above beam on elastic foundation yields

$$\theta_R = \frac{M_R \phi(\beta, L)}{EI_n}$$

(20.43)

where $\phi(\beta, L)$ is

$$\phi(\beta, L) = \frac{\cosh \beta L \sinh \beta L - \cos \beta L \sin \beta L}{2\beta(\cosh^2 \beta L - \cos^2 \beta L)}$$

(20.44)

To compute this value, it is first necessary to calculate the parameter β from Equation 20.42, which contains term K, defined as the modulus of the elastic foundation. As the ribs are supported by the pipe wall, we can consider a slice of the pipe wall together with the rib as shown in Figure 20.15. A partial view of the slice is shown in the figure as a dotted area for the sake of clarity. The complete circumferential slice can then be represented by a mathematical model of a radially loaded circular ring as illustrated in Figure 20.17. Radial load intensity q is shown in Figures 20.15 and 20.17. For a slice of unit width load intensity, q is numerically equal to the load per linear inch of the rib acting upon the ring at N equidistant points. The modulus of the foundation then can be interpreted directly as a spring constant K, since the load–deflection relation for a particular ring loading [7] may be given as

$$y = \frac{qr^3(\theta_p^2 + \theta_p \sin \theta_p \cos \theta_p - 2\sin^2 \theta_p)}{4EI_m \theta_p \sin^2 \theta_p}$$

(20.45)

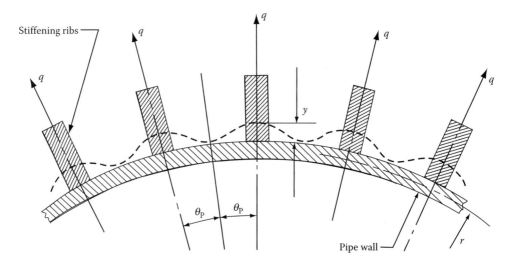

FIGURE 20.17   Circular ring model for modulus of foundation.

Since by definition $K = q/y$ and $\theta_p = \pi/N$, substituting for the modulus of the foundation in Equation 20.45 yields

$$\beta = \left\{ \frac{2\pi N I_m \sin^2 (\pi/N)}{r^3 I_n [2\pi^2 + \pi N \sin (2\pi/N) - 4N^2 \sin^2 (\pi/N)]} \right\}^{1/4} \tag{20.46}$$

A parametric investigation of Equation 20.46 within the range $8 < N < 100$ leads to the simplification

$$\beta = 0.69 \left(\frac{N}{r} \right) \left(r \frac{I_m}{I_n} \right)^{1/4} \tag{20.47}$$

Assuming next that the angle of twist of the main flange ring is equal to the bending slope of the rib at $x = 0$, Equations 20.39 and 20.43 give

$$\omega = \frac{M_R}{M_F} = \frac{I_n R^2}{I_0 \phi (\beta, L)} \tag{20.48}$$

Hence, solving Equations 20.38 and 20.48 simultaneously yields

$$M_R = \frac{W(R - r)\omega}{2\pi R + \omega N} \tag{20.49}$$

and

$$M_F = \frac{W(R - r)}{2\pi R + \omega N} \tag{20.50}$$

The main flange ring can now be viewed as consisting of N equal sectors in which the load distribution between the ribs and the corresponding sectors of the flange can be calculated from a load ratio that follows directly from Equation 20.38 and Figure 20.15

$$f = \frac{0.16N \, RT_r B_r^3}{(B_r + T)h^3 \phi(\beta, L)} \tag{20.51}$$

Since the auxiliary function given by Equation 20.44 is a dimensional quantity expressed in inches, the load-sharing equation above is nondimensional, in contrast with the ratio ω defined by Equation 20.48.

20.16 STRENGTH OF FLANGE RIBS

The analysis of load sharing between the stiffening ribs and the main flange ring must necessarily be considered as conservative because of the assumed pin-jointed supports (Figure 20.16) and because the shearing stresses between the pipe wall and the rib have been neglected. Indeed, the analysis appears to indicate that the weld between the rib and the wall might possibly be omitted in the double flange ring design for several manufacturing reasons. Some of these reasons include cost reduction, metallurgical control of welding procedure, reliability of weld inspection, and residual stress effects due to welding.

While sizing a new compound flange, Equation 20.51 may be useful in establishing the first criterion for sharing of the bending moment between the rib and the corresponding sector of the main flange. Combining Equations 20.51 and 20.38 yields

$$M_R = \frac{fW(R-r)}{N(1+f)} \tag{20.52}$$

Hence, the corresponding bending stress in the rib becomes

$$\sigma_{bR} = \frac{6fW(R-r)}{NT_r(1+f)(B_r+T)^2} \tag{20.53}$$

Equation 20.53 is applicable to the stiffening ribs sharing the external load with the main flange ring on the premise that the weld between the rib and the pipe wall is sufficiently strong. Here again, our criterion should be conservative because the additional cross-sectional area $T_r \times T$ included in Equation 20.53 does not truly represent the effect of the pipe wall at the welded junction. However, for the purpose of the preliminary design, Equation 20.35 is satisfactory, provided the maximum computed stress is elastic. Further interpretation of the maximum stress value may be made utilizing the concept of plastic correction.

Although the load-sharing capacity for a typical stiffening rib has been established only on the basis of flange ring rotation and beam bending due to a couple applied at its end, the effect of direct tension on the maximum stress can be included. The maximum bending stress and the tensile stress in the rib may be added directly. Hence, the total stress is as follows: in direct tension, load sharing between the rib and the pipe wall will be established in direct proportion to the working areas. The portion of the tensile load carried by the rib, denoted by W_r, as illustrated in Figure 20.18 gives

$$W_R = \frac{WB_rT_r}{NB_rT_r + 2\pi rT} \tag{20.54}$$

Since the simple tensile stress is W_R/B_rT_r, combining Equations 20.53 and 20.54 yields

$$\sigma_{TR} = \frac{6fW(R-r)}{NT_r(1+f)(B_r+T)^2} + \frac{W}{NB_rT_r + 2\pi rT} \tag{20.55}$$

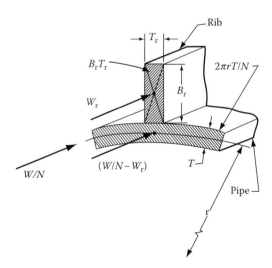

FIGURE 20.18   Tensile load-sharing diagram for a rib and pipe sector.

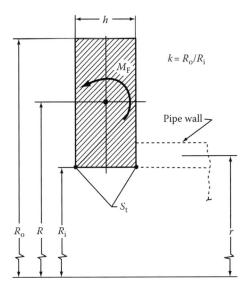

FIGURE 20.19 Flange ring notation for a stress check under toroidal moment.

The estimate of load-sharing capability based on Equation 20.41 indicates that in the majority of design situations involving rib-stiffened flanges of the usual proportions, only a limited toroidal moment will be expected to be carried by the main flange ring, and in many such cases the corresponding stresses can be ignored. However, in those instances where the calculated number f from Equation 20.51 is found to be rather small (of the order of 5 or less), the maximum flange stress due to the twist may be calculated from the simplified expression

$$\sigma_t = \frac{0.96W(R-r)R}{h^2 r^2 (1+f) \log_e k} \tag{20.56}$$

where the flange ring notation is shown in Figure 20.19.

20.17 LOCAL BENDING OF FLANGE RINGS

When the rib system is relatively rigid and the resulting toroidal deformation of the main flange ring is limited, it is recommended for design purposes that the order of the local stresses likely to exist in the flange ring under the individual bolt loads be evaluated. Due to the symmetry, only one portion of the flange, held by the two consecutive ribs and the corresponding portion of the pipe wall, needs to be examined. Basically, this is a three-dimensional problem where an experimental analysis can provide a reliable answer. As such, test data appear to be very scarce, and since a formal three-dimensional solution to this problem would be very lengthy, a compromise is suggested here, based on the elastic theory of plates. This method may be used in preliminary design and data reduction in support of experimental work.

Utilizing as far as practicable the notation already employed, Figure 20.20 illustrates the proposed mathematical model. Some theoretical solutions are available [8], which involve a rectangular plate having one long edge free, and both short edges simply supported. The boundary conditions along the shorter edges are likely to be closer to those of a built-in character, and therefore the stresses calculated according to the model illustrated in Figure 20.20 should be considered as approximate at best.

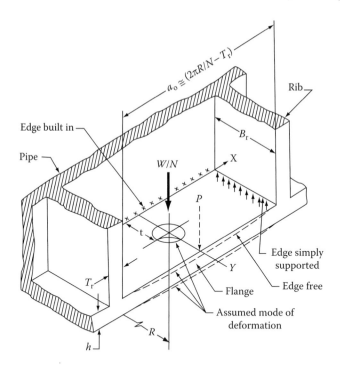

FIGURE 20.20   Approximate plate model for main flange ring analysis.

This analysis is based on the deflection of the free edge of a relatively long plate under the action of a concentrated force P. From Figure 20.20, the edge force may be taken as proportional to Wt/NB_r. Hence, the edge deflection Y_p becomes

$$Y_p = 1.83 \frac{WtB_r}{ENh^3} \tag{20.57}$$

The edge deflection Y_q for a uniformly loaded plate for the same boundary conditions is

$$Y_q = 1.37 \frac{WB_r^3}{a_0 ENh^3} \tag{20.58}$$

The maximum bending stress in the flange ring (Figure 20.21) should be at the middle of the built-in edge (i.e., where $x = y = 0$), as shown in the sketch. If the bolt load W/N is first assumed to be uniformly distributed, the maximum bending moment at the midpoint of the built-in edge in a classical solution varies as a function of B_r/a_0, as shown in Figure 20.21. Here M_y denotes the moment acting about the built-in edge and it is expressed in lb-in. per inch of circumference. Hence, the parameter $V = a_0 N M_y / WB_r$ must be nondimensional. The dimensional quantity a_0 is defined in Figure 20.21.

The bending stress due to the equivalent uniform load W/Na_0B_r may be given as

$$\sigma_b = \frac{6VWB_r}{Na_0h^2} \tag{20.59}$$

To make a conservative correction for the effect of a concentrated loading W/N, the bending stress can be assumed to be roughly proportional to Y_p/Y_q as determined by Equations 20.52 and 20.53. The criterion for the flange stress, located at the midpoint between the two adjoining ribs and very close to the outer surface of the pipe, can now be defined as

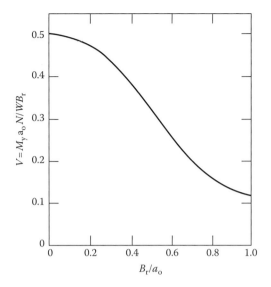

FIGURE 20.21 Maximum moment for a rectangular plate simulating ribbed flange.

$$\sigma_b = \frac{8VWt}{NB_r h^2} \qquad (20.60)$$

When the outer radius R_o of the flange is relatively large compared with the mean radius of the pipe, the approximate stress in the flange ring becomes

$$\sigma_b = \frac{0.95Wt}{NB_r h^2} \qquad (20.61)$$

The above stresses are shown to decrease with an increase in the number of gussets.

20.18   CORRECTION FOR TAPERED GUSSETS

When gussets have a linear change in depth from B_r to A_r as shown in Figure 20.22, the equations derived so far for the rectangular gussets can also be applied to the tapered geometry by determining

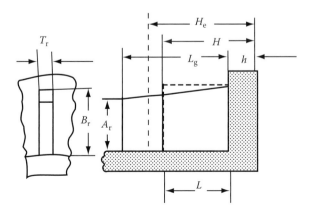

FIGURE 20.22 Tapered gusset.

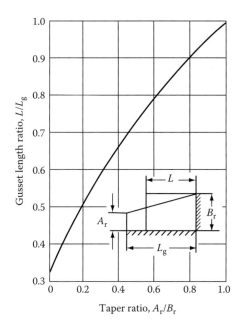

FIGURE 20.23 Design chart for length of tapered gussets.

an equivalent length L for a gusset of total length L_g. This equivalent length L can be found from Figure 20.23 for a given value of the ratio A_r/B_r. Once L is known, the overall depth H can be taken as $L + h$. Then H_e and the stresses can be estimated in the same way in which they were calculated for rectangular gussets. The calculation can now proceed with Equations 20.32 and 20.33 using the parameter $n = H_e/T$.

The development of the equivalent gusset length, Figure 20.23, is based on the resolution of the external forces acting on the gusset. The assumption is also made that the average shearing stresses, applied along the gusset and perpendicular to it, were numerically equal. The derivation involved the summation of the external moments for the tapered and "equivalent" rectangular ribs.

The foregoing analysis of a compound circular flange suggests bending rather than shear as the basic mechanism of potential structural failure. The approach then seems to be particularly sensitive with regard to the application of the principles of fracture-safe design.

SYMBOLS

A	Area of cross section
A_r	Depth of tapered rib
A_t	Total cross-sectional area
a	Moment arm; depth of tapered plate
a_0	Mean length of flange sector
B	Width of bracket
B_r	Depth of rib
c	Tapered plate parameter
d	Distance to loaded point
d_1	Mean pipe diameter
d_b	Bolt hole diameter
E	Elastic modulus
E_0	Reduced modulus of elasticity
E_t	Tangent modulus

e	Eccentricity of load application
F	Load on weld seam
F_f	Width of flange cross-section
$\underline{F}_n$	Normal force component
F_s	Shear force component
f	Load-sharing ratio
FEM	Finite element method
H	Depth of standard flange; maximum depth of bracket
H_e	Equivalent depth of flange
h	Thickness of flange ring; size of weld leg
I_b	Second moment of area
I_g	Moment of inertia of a component section
I_m	Moment of inertia of wall element of unit width
I_n	Moment of inertia of a rib cross section
I_0	Moment of inertia of main flange section
I_x	Moment of inertia about central axis
J	First moment of area
K	Modulus of elastic foundation
K_b	Buckling coefficient
$k = R_o/R_i$	Flange ring ratio
L	Length of rib of constant depth; length of bracket
L_g	Length of tapered rib
ℓ_1	Moment arm
M	General symbol for bending moment
M_1, M_2	Bending moment components
M_c	Bending moment per one rib spacing
M_f	Toroidal moment on flange ring
M_0	Discontinuity bending moment
M_R	Bending moment on rib
M_y	Bending moment about longer edge of plate
$m = R_i/T$	Ratio of inner radius to wall thickness
N	Number of ribs
$n = H/T$	Ratio of flange to pipe thickness
P	Tensile load on bracket; edge force on plate
Q_0	Discontinuity shearing force
q	Radial load intensity
R	Radius to bolt circle; mean flange radius
R_i	Inner radius of pipe
R_o	Outer radius of flange
r	Mean radius of pipe
S	General symbol for stress
S_1, S_2	Weld stress components
S_{bR}	Rib bending stress
S_c	Compressive stress
S_{Cr}	Critical compressive stress
S_F	Flange dishing stress
S_N	Normal stress
S_p	Plastic stress
S_R	Radial stress in flange ring
S_s	Shear stress
S_t	Tensile stress; toroidal stress in flange

S_{TR}	Total stress in rib
S_u	Ultimate strength
S_y	Yield strength
S_{max}	Maximum principal stress
s_0	Wall thickness
s_1	Depth of section at failure
T	Thickness of pipe; thickness of plate
t	Distance from bold circle to outer pipe surface; thickness of fillet weld
T_r	Thickness of rib
T_0	Thickness of backup ring
V	Moment factor in plate analysis
W	Total bolt load; external load on bracket
W_i	Load per inch of pipe circumference
W_0	Load per inch of bolt circle
W_R	Tensile load on rib
W_{p1}	Plastic load on bracket
x	Arbitrary distance
Y	Deflection of beam on elastic foundation
Y_p	Plate edge deflection under concentrated load
Y_q	Plate edge deflection under uniform load
y	Coordinate; ring deflection
Z	Section modulus
α	Angle of fractured part; bracket angle, rad
β	Elastic foundation parameter
β_s	Shell parameter
δ	Slope of stress–strain curve, rad
ε	Strain
ε_y	Uniaxial strain at yield
η	Modulus ratio
θ	Angle of twist; angle in weld analysis, rad
θ_F	Angle of twist of main flange ring, rad
θ_P	Rib half-angle, rad
θ_R	Bending slope at end of rib, rad
μ	Poisson's ratio
σ, S	Stress
σ_b, S_b	Bending stress
τ	Shear stress component
τ_{max}	Principal shear stress
ϕ	Plate angle (rad)
ϕ_0	Flange factor
$\phi(\beta, L)$	Auxiliary function for a beam on elastic foundation
ω	Ratio of rib to flange moment

REFERENCES

1. E. O. Holmberg and K. Axelson, Analysis of stresses in circular plates and rings, Paper APM-54-2, Transactions, American Society of Mechanical Engineers, New York, Vol. 43, 1931.
2. E. O. Waters, D. B. Wesstrom, D. B. Rossheim, and F. S. G. Williams, Formulas for stresses in bolted flanged connections, *Transactions*, American Society of Mechanical Engineers, New York, Vol. 59, 1937, pp. 161–169.
3. H. J. Bernhard, Flange theory and the revised standard: B. S. 10: 1962—Flanges and bolting for pipes, valves and fittings, Proceedings, Institute of Mechanical Engineers, London, Vol. 175, No. 5, 1963–1964.

4. F. Sass, Ch. Bounche, and A. Leitner, *Dubbels Taschenbuch für den Maschinenbau*, Springer-Verlag, Berlin, 1966.
5. A. Blake, *Design of Curved Members for Machines*, Robert E. Krieger, Huntington, NY, 1979.
6. S. Timoshenko, *Strength of Materials, Part II Advanced Theory and Problems*, D. van Nostrand, New York, 1956.
7. W. C. Young and R. G. Budynas, *Roark's Formulas for Stress and Strain*, 7th ed., McGraw Hill, New York, 2002.
8. S. Timoshenko and S. Woinowsky-Krieger, *Theory of Plates and Shells*, 2nd ed., McGraw Hill, New York, 1959.
9. E. O. Waters and J. H. Taylor, The strength of pipe flanges, *Mechanical Engineering*, Vol. 49, American Society of Mechanical Engineers, New York, 1927, pp. 531–542.
10. A. M. Wahl and G. Lobo, Stresses and deflections in flat circular plates with central holes, Transactions, American Society of Mechanical Engineers, New York, Vol. 51, 1929, pp. 1–13.
11. R. H. Johns and T. W. Orange, Theoretical elastic stress distributions arising from discontinuities and edge loads in several shell type structures, NASA, Report TRR-103, Lewis Research Center, Cleveland, OH, 1961.
12. E. O. Waters and F. S. G. Williams, Stress conditions for flanged joints for low-pressure service, Transactions, American Society of Mechanical Engineers, New York, Vol. 64, 1952, pp. 135–148.
13. A. Phillips, *Introduction to Plasticity*, Ronald Press, New York, 1956.
14. A. Blake, Flanges that won't fail, *Machine Design*, 46, 1974, pp. 45–48.
15. A. Blake, Stress in flanges and support rings, *Machine Design*, 1975.
16. A. Blake, Design considerations for Rib-stiffened flanges, Report UCRL-50756, Lawrence Livermore Laboratory, Livermore, CA, 1969.
17. R. W. Werne, Theoretical and experimental stress analysis of rib-stiffened flanges, EG&G Technical Report No. EGG-1183-4054, San Ramon, CA, 1972.
18. A. Blake, *Design of Mechanical Joints*, Marcel Dekker, New York, 1985.

21 Brackets

21.1 INTRODUCTION

Brackets have manifold applications in structures and machinery. They serve as pipe supports, motor mounts, connecting joints, fasteners, and seats of various types. [They involve rolled shapes, plate components, and prefabricated structural elements to meet requirements like strength, rigidity, appearance, and low manufacturing cost.]

Due to their many applications, brackets have a wide variety of geometry and loading configurations. The loads may be dynamic with changing directions. Optimal design is thus elusive. Nevertheless, in this chapter, we offer a methodology and design philosophy for safe, reliable, and economical designs.

As with the design of all structural components, bracket design criteria must also be based upon the fundamentals of strength of materials, elasticity theory, and elastic stability. For brackets, the designation of critical dimensions may also be governed by the elastoplastic response and the local buckling resistance. Thus, experimental stress analysis results may be useful. All these considerations may be important in evaluating brackets and in predicting their structural safety and performance. There is, however, relatively little information available on bracket design in the open literature. One of the reasons for this, as noted earlier, is the inherent diversity of configurations, loading conditions, and safety of individual bracket applications.

21.2 TYPES OF COMMON BRACKET DESIGN

We focus upon generic designs that are commonly employed in structural systems. Specialized brackets for a given application are expected to be similar to those described here.

Brackets in general have a bearing plate to distribute a load together with an edge-loaded plate, or plates, to act as a stiffener or gusset. The bearing and stiffening plates are usually attached or bonded by welding.

Figures 21.1 through 21.7 show a number of typical bracket designs, as previously documented by Blake [1]. These designs do not exhaust all possibilities but they illustrate some of the more important structural features that affect the design choice and methods of stress analysis. The examples selected indicate welded configurations, which, with modern fabricating techniques are likely to be reliable and economic. However, this statement is not intended to imply that welding processes never cause problems. Despite significant progress during the past years, strict quality control of welding should be maintained at all times. Fracture-safe design, for instance, can easily be compromised by a change in material properties in the head-affected zone due to welding, flame cutting, or other operation.

The mechanical characteristics of the various support brackets can be summarized as follows. The short bracket shown in Figure 21.1 is made of a standard angle with equal legs. This component can be designed on the basis of bending and transverse shear. When loading arm d is relatively short, the structural element is rigid and the effect of bending may be neglected.

A box-type support bracket (Figure 21.2) can be made out of two channels using butt-welding techniques. The strength check here is performed using a simple beam model under bending and shear.

Rugged bracket construction is illustrated in Figure 21.3, where heavy loads have to be supported. Because of the frame-type appearance and mechanics of this type of a support, external loading can be

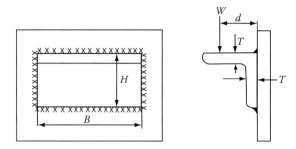

FIGURE 21.1 Shear-type bracket.

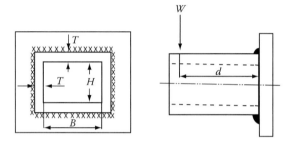

FIGURE 21.2 Box-type support bracket.

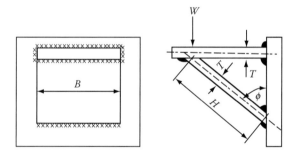

FIGURE 21.3 Heavy-duty plate bracket.

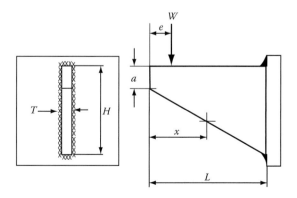

FIGURE 21.4 Tapered-plate bracket.

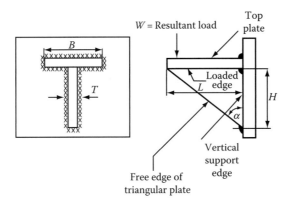

FIGURE 21.5   T-section bracket.

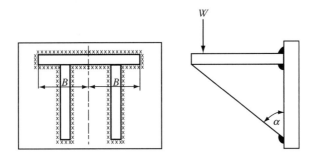

FIGURE 21.6   Double-T section bracket.

resolved into tensile and compressive forces for design purposes. In this design, the cross sections of the tensile and compression members are large enough to carry substantial loads.

A simple and light construction is illustrated in Figure 21.4. When the plate is relatively long, the bracket must be designed to resist bending, shear, and local buckling loads.

A more conventional type of bracket design is shown in Figure 21.5. This bracket can be made either by flame cutting and welding separate plate members or by cutting standard rolled shapes such as I or T beams.

For larger loads, a double-T configuration bracket design, shown in Figure 21.6, may be recommended. The design should be checked, however, for bending effects, shear strength, and stability of the free edges due to the compressive stresses.

Yet another version, shown in Figure 21.7, can be flame-cut from a standard channel and welded to the base plate to form a solid unit. The design analysis in this case is similar to that employed for the configuration given in Figure 21.6.

21.3   WELD STRESSES

With welding being a bonding agent between the plates forming brackets, it is essential that stresses in the welds be considered in overall stress analyses of brackets. In this section, we present a review of formulas for calculating welding stresses. The major findings are documented by the American Welding Society. For additional details, refer to welding handbooks, publications of the Welding Research Council, and of texts on materials science (see, for example, Ref. [2]).

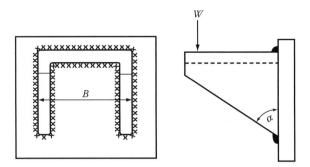

FIGURE 21.7　Channel-type heavy-duty bracket.

In reviewing the designs illustrated by Figures 21.1 through 21.7, we see that we have to consider both transverse and parallel welds subjected to bending moments. To examine the principles involved, consider the case shown in Figure 21.8: for the fillet weld shown, the size of the weld leg is h. The overall linear dimensions of the weld are B and H for the transverse and longitudinal welds respectively. The bending moment M_1 on the transverse welds can be imagined to be a couple consisting of two equal forces F acting at the center of the weld legs, as shown. Since it is standard practice to calculate the stresses on the basis of a weld-throat section, the area on which the component force F is acting must be approximately equal to $Bh/\sqrt{2}$. This is somewhat conservative because of the additional weld material found at the corner, which is not accounted for in calculating the weld area. Thus, we have

$$M_1 = F(H + h) \tag{21.1}$$

and the tensile stress across the throat section is

$$\sigma_1 = \frac{F\sqrt{2}}{Bh} \tag{21.2}$$

Combining these expressions gives

$$\sigma_1 = \frac{\sqrt{2}M_1}{Bh(H + h)} \tag{21.3}$$

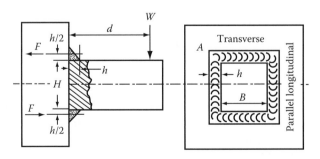

FIGURE 21.8　Example of fillet weld in bending.

The effect of the external load W on the parallel welds can be treated with the help of simple beam theory. The section modulus z of the parallel weld throat is approximately equal to

$$z = \frac{bH^2}{6\sqrt{2}} \tag{21.4}$$

Since both longitudinal sections are involved in resisting M_2, we have

$$\sigma_2 = \frac{3\sqrt{2}M_2}{hH^2} \tag{21.5}$$

As noted earlier, the stress at a common point must be the same for both the transverse and longitudinal welds. That is,

$$\sigma_1 = \sigma_2 \tag{21.6}$$

Then, from Equations 21.1 and 21.5, we obtain a relation between the bending moments as

$$M_1 = \frac{3B(H + h)M_2}{H^2} \tag{21.7}$$

Since M is $M_1 + M_2$, this equation provides an expression for M as

$$M = M_2 \left[1 + \frac{3B(H + h)}{H^2}\right] \tag{21.8}$$

Finally, from Equations 21.5 and 21.7 and by observing further in Figure 21.8 that M is Wd, we obtain the bending stress σ in terms of the load W as

$$\sigma = \frac{3\sqrt{2}Wd}{h[H^2 + 3B(H + h)]} \tag{21.9}$$

Correspondingly, the average shear stress τ due to the load W is

$$\tau = \frac{\sqrt{2}W}{2h(H + h)} \tag{21.10}$$

Consider the bracket loaded in tension as in Figure 21.9. At the plate junction (or "throat"), the nominal stress S on the fillet weld is simply

$$S = P/2Bh \tag{21.11}$$

where B is the weld length. For design purposes, it is customary to use the more conservative stress estimate:

$$S = \sqrt{2}P/2Bh \tag{21.12}$$

where the $\sqrt{2}$ factor is included since in actual welding practice, the effective weld area is generally smaller than $2Bh$.

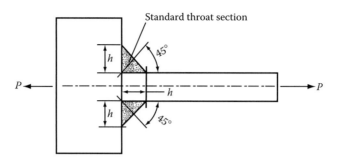

FIGURE 21.9 Symmetrical fillet weld in tension.

To look into this further, consider the double fillet weld represented in Figure 21.10 where a sketch of the probable outline of an actual weld is given. Thus, if the effective area is reduced due to the welding, the thickness of the weld is probably greater than that used in the mathematical model. To explore the theoretical model in more detail, consider an arbitrary section of the weld designated by the angle θ as in Figure 21.10. Let F_n and F_s be the normal and shear forces on the section respectively and let t be the thickness of the section as shown. From Figure 21.10, we see that t may be expressed in terms of the weld height h and angle θ as

$$t = h/(\sin\theta + \cos\theta) \tag{21.13}$$

To see this, consider an enlarged view of, say, the upper weld profile of Figure 21.10 as shown in Figure 21.11. By focusing upon triangle ABC and by using the law of sines we have

$$\frac{t}{\sin\pi/4} = \frac{h}{\sin\phi} \tag{21.14}$$

By noting that angle ϕ is $(\pi/4 + \theta)$ we see that $\sin\phi$ is

$$\sin\phi = \sin[\pi - (\pi/4 + \theta)] = \sin(\pi/4 + \theta)$$

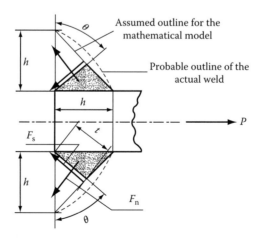

FIGURE 21.10 Free-body diagram of fillet weld.

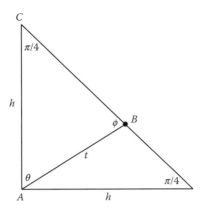

FIGURE 21.11 Bracket weld profile.

or

$$\sin \phi = \sin \pi/4 \cos \theta + \cos \pi/4 \sin \theta = \left(\sqrt{2}/2\right)(\cos \theta + \sin \theta) \qquad (21.15)$$

Then, by substituting into Equation 21.14 we have the result of Equation 21.13. That is,

$$t = h \sin (\pi/4)/ \sin \phi = h/(\sin \theta + \cos \theta) \qquad (21.16)$$

Next, referring again to Figure 21.10, consider a free-body diagram of the shaded portion of the lower weld of the bracket as in Figure 21.12.

By adding forces horizontally and vertically we have

$$P/2 - F_s \sin \theta - F_n \cos \theta = 0 \qquad (21.17)$$

and

$$F_s \cos \theta - F_n \sin \theta = 0 \qquad (21.18)$$

Then, by solving these expressions for F_s and F_n, we have

$$F_s = (P/2) \sin \theta \quad \text{and} \quad F_n = (P/2) \cos \theta \qquad (21.19)$$

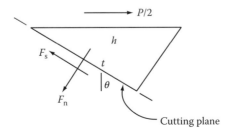

FIGURE 21.12 Free-body diagram of a weld segment.

Finally, for a weld length B, the shear and normal stresses on the cutting plane surface of Figure 21.12 are

$$\sigma_s = F_s/2Bt \quad \text{and} \quad \sigma_n = F_n/2Bt \tag{21.20}$$

By substituting for t, F_s, and F_n from Equations 21.16 and 21.19, the stresses become

$$\sigma_s = \frac{(P\sin\theta)(\sin\theta + \cos\theta)}{2Bh} \tag{21.21}$$

and

$$\sigma_n = \frac{(P\cos\theta)(\sin\theta + \cos\theta)}{2Bh} \tag{21.22}$$

An examination of Equations 21.21 and 21.22 shows that at no point of the weld do these theoretical stress values exceed the value estimated by Equation 21.12. Therefore, Equation 21.12 may be viewed as a safe upper bound on the stresses, and that the actual weld stresses are likely to be considerably smaller.

21.4 STRESS FORMULAS FOR VARIOUS SIMPLE BRACKET DESIGNS

Consider the simple shear bracket of Figure 21.1 and shown again in Figure 21.13. This bracket is simple both in design and manufacture. If the line of action of the load is a distance d from the main plate, the bending (σ_b) and shear (σ_s) stresses on the bracket plate may be computed as

$$\sigma_b = 6W(d - T)/BT^2 \tag{21.23}$$

and

$$\sigma_s = W/BT \tag{21.24}$$

The corresponding weld stresses are

$$\sigma_b = \frac{4.24Wd}{h(H^2 + 3BH + 3Bh)} \tag{21.25}$$

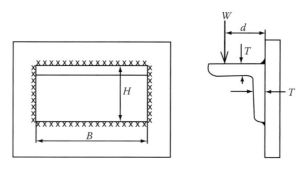

FIGURE 21.13 Shear-type bracket.

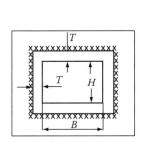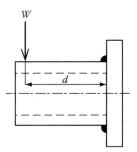

FIGURE 21.14   Box-type support bracket.

and

$$\sigma_s = \frac{0.7071W}{h(H+h)} \tag{21.26}$$

Next, consider the box-type bracket of Figure 21.2 and shown again in Figure 21.14. Here, the bending shear stresses are

$$\sigma_b = \frac{3Wd}{HT(H+2B+4T)} \tag{21.27}$$

and

$$\sigma_s = \frac{W}{2(H+2T)} \tag{21.28}$$

The corresponding weld stresses are

$$\sigma_b = \frac{4.24Wd}{h[H(H+4T)+3(B+2T)(H+h)]} \tag{21.29}$$

and

$$\sigma_s = \frac{0.7071W}{h(H+2T+h)} \tag{21.30}$$

For the heavy-duty plate bracket of Figure 21.3 and shown again in Figure 21.15, we can perform a stress analysis by assuming that the load W may be resolved into two components acting along the central planes of the plates. The stress is then tensile in the horizontal member and compressive in the inclined member. These stresses are

$$\sigma_t = \frac{W\sin\phi}{BT\cos\phi} \tag{21.31}$$

and

$$\sigma_c = \frac{W}{BT\cos\phi} \tag{21.32}$$

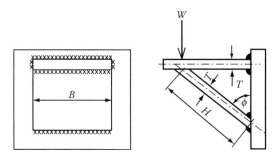

FIGURE 21.15 Heavy-duty plate bracket.

The corresponding weld stresses are approximately

$$\sigma_b = \frac{0.7071W\tan\phi}{Bh} \tag{21.33}$$

and

$$\sigma_s = \frac{0.5W\tan\phi}{Bh} \tag{21.34}$$

It is not practical to use large angles ϕ because the corresponding plate forces become relatively high, as can be seen from the foregoing expressions. In addition, the bracket having high ϕ loses its frame character of structural behavior and tends to become a cantilevered member for which even small transverse loads can cause substantial bending stresses.

A bracket angle ϕ of 45° is often selected in practical design. With the typical proportions of plate members in use, Equations 21.31 and 21.32 suffice for sizing calculations. However, it should be appreciated that the compressive member of the bracket can become elastically unstable if its thickness is drastically reduced. Since in the angle brace of Figure 21.4, the two edges of the plate are free to deform, we have the case of buckling of a relatively wide beam subjected to axial compression. Denoting the width and length of this beam by B and H, respectively, and assuming end fixity due to welding, the following expression for the critical buckling stress can be used

$$S_{c_r} = \frac{3.62ET^2}{H^2} \tag{21.35}$$

This formula is limited to elastic behavior, and therefore the yield strength of the material S_y can be used to determine the maximum allowable value of H/T to avoid failure by buckling. The corresponding critical ratio is

$$\frac{H}{T} = 1.9\left(\frac{E}{S_y}\right)^{1/2} \tag{21.36}$$

The term E/S_y may be called the inverse strain parameter because it follows directly from Hooke's law. For the conventional metallic materials, the ratio E/S_y varies between 100 and 1000 for high-strength and low-strength materials, respectively.

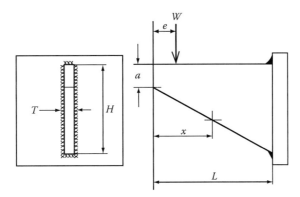

FIGURE 21.16 Tapered-plate bracket.

21.5 STRESS AND STABILITY ANALYSES FOR WEB-BRACKET DESIGNS

Consider the tapered plate bracket of Figure 21.4 shown again in Figure 21.16. This design, in effect, is a cantilever plate loaded on edge. The normal stresses on a section, say at x, must vary from tension to compression. The maximum bending stress σ_b depends upon the taper. It can be calculated using the expression

$$\sigma_b = \frac{6WL^2(x - e)}{T[aL + x(H - a)]^2} \tag{21.37}$$

The distance x at which the highest bending stresses develop can be found form the condition $d\sigma_b/dx = 0$, calculated from Equation 21.37. This yields

$$x = e + (e^2 + c)^{1/2} \tag{21.38}$$

where c is

$$c = \frac{aL[23(H - a) + aL]}{(H - a)^2} \tag{21.39}$$

The procedure is to compute x from Equations 21.38 and 21.39, and to substitute this value into Equation 21.37 to obtain the maximum stress value. With the usual proportion of brackets found in practice, the aspect ratio H/L can be used to make a rough estimate of the relevant buckling coefficient K_b from Figure 21.17. This coefficient is then used in calculating the critical elastic stress of the free edge of the bracket using the expression

$$\sigma_{Cr} = K_b E \left(\frac{T}{H}\right)^2 \tag{21.40}$$

The plate buckling coefficient K_b given in Figure 21.17 can be determined experimentally for each case of plate proportions, boundary conditions, and type of stress distribution. It represents the tendency of a free edge of the plate element to move toward local instability when the

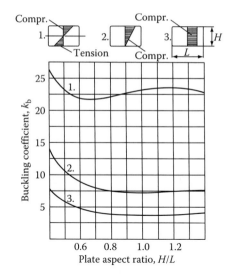

FIGURE 21.17 Buckling coefficients for simply supported plates under nonuniform longitudinal stresses.

compressive stresses reach a certain critical value. The consequence of local buckling may then be interpreted in two ways:

1. Overall collapse by rendering the plate element less effective in the postbuckling region of structural response
2. Detrimental stress redistribution influencing the load-carrying capacity of the system

The design values given in Figure 21.17 depend largely on the type of stress distribution in compression. Although K_b values are sensitive to the type of stress distribution and vary in a nonlinear fashion, their dependence on the aspect ratio H/L is only moderate.

When the actual compressive stress given by Equation 21.37 exceeds that given by Equation 21.40, it is customary to assume that the free edge of the bracket is susceptible to local elastic buckling. To make a conservative allowance for the critical buckling stress in the plastic range, the following set of design formulas may be used

$$S_{Cr} = K_b E \eta \left(\frac{T}{H}\right)^2 \tag{21.41}$$

where η is

$$\eta = \left(\frac{E_t}{E}\right)^{1/2} \tag{21.42}$$

and E_t is

$$E_t = \frac{dS}{d\varepsilon} \tag{21.43}$$

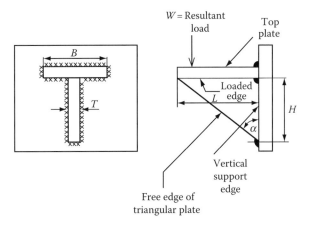

FIGURE 21.18 T-section bracket.

In Equation 21.43, the terms S and ε denote the normal stress and uniaxial strain, respectively. Therefore, Equation 21.43 defines the tangent modulus of the stress–strain characteristics of the material at a specified level of stress.

The strength of the weld in bending is estimated as follows:

$$\sigma_b = \frac{4.24W(L - e)}{h(H^2 + 3HT + 3hT)} \tag{21.44}$$

The numerical value of shear stress for this case can be obtained from Equation 21.26.

Figures 21.5 through 21.7 reproduced here as Figures 21.18 through 21.20 show various common designs of tapered plate brackets. In spite of their common use, comparatively few stress formulas for these brackets are available.

The design shown in Figure 21.18 contains two basic elements of structural support: the top support plate which helps to distribute the load; and the triangular plate loaded on edge and designed to carry the major portion of the load. The two plates acting together form a relatively rigid "tee" configuration.

Experience indicates that the free edge carries the maximum compressive stress X_{max}, which depends on the aspect ratio L/H. Practical design situations give aspect ratios somewhere between

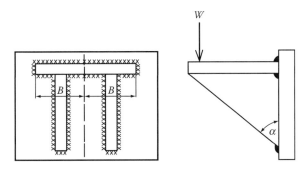

FIGURE 21.19 Double-T section bracket.

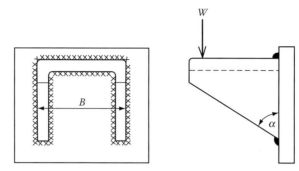

FIGURE 21.20   Channel-type heavy-duty bracket.

0.5 and 2.0 [3]. For this particular case, the maximum allowable total load W near the center of the upper plate can be estimated as

$$W = S_{max}(0.60H - 0.21L)\frac{TL}{H} \qquad (21.45)$$

When the working load W is specified, the maximum corresponding stress S_{max} can be calculated from Equation 21.45. It is then customary to make $S_{max} < S_y$, where S_y denotes the yield compressive strength of the material. The design condition for the critical values of L/T can be represented by the following criteria.
   For $0.5 \leq L/H \leq 1.0$,

$$\frac{L}{T} \leq \frac{180}{S_y^{1/2}} \qquad (21.46)$$

For $1.0 \leq L/H \leq 2.0$,

$$\frac{L}{T} \leq \frac{60 + 120\left(\frac{L}{H}\right)}{S_y^{1/2}} \qquad (21.47)$$

Equations 21.45 through 21.47 are valid when the resultant load W is located reasonably close to the center of the top plate and when the yield strength S_y is expressed in ksi. However, when this load moves out toward the edge of the plate, the analytical method described above loses its degree of conservatism and an alternative approach based on the concept of increased eccentricity should be designed. The strength of a welded connection in this design may be checked from Equations 21.33 and 21.34.

   Some of the specific features of the triangular-plate bracket can be analyzed with reference to Figure 21.21. The maximum stress at the free edge of the triangular part may be calculated on the basis of elementary beam theory, by combining the stresses due to the bending moment $W \times e$ and the compressive load equal to $W/\cos \alpha$. This gives

$$S_{max} = \frac{W(L + 6e)}{TL^2 \cos^2 \alpha} \qquad (21.48)$$

A conservative check on free-edge stability can be made by assuming that the shaded portion of the plate acts as a column with a cross section equal to $(TL \cos\alpha)/4$ and length equal to $H/\cos \alpha$.

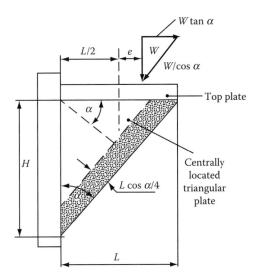

FIGURE 21.21 Approximate model for a triangular plate bracket.

When a relatively small value of L/T must be used, there is little danger of elastic instability and the bracket can be designed to undergo a certain amount of local yielding. For $0.5 \leq L/H \leq 2.0$, the recommended L/T ratio is

$$\frac{L}{T} \leq \frac{48 + 24\left(\frac{L}{H}\right)}{S_y^{1/2}} \tag{21.49}$$

Here, the yield strength S_y is expressed in ksi [3].

The maximum permissible load on the bracket under fully plastic conditions can be calculated from the expression:

$$W_{pl} = TS_y \cos^2 \alpha \left[(L^2 + 4e^2)^{1/2} - 2e\right] \tag{21.50}$$

The cross-sectional area of the top plate should be designed for the horizontal component of the external load

$$A = W_{pl} \frac{\tan \alpha}{S_y} \tag{21.51}$$

The design formulas given by Equations 21.23 through 21.51 are applicable to various practical situations wherever a particular structure can be modeled as a support bracket similar to one of the configurations illustrated in Figure 21.13 through 21.20. By checking the weld strength, beam strength, and stability the structural integrity of a bracket can be assured, provided that material and fabrication controls are not compromised.

SYMBOLS

A	Area of cross section
A_r	Depth of tapered rib
A_t	Total cross-sectional area
a	Moment arm; depth of tapered plate

a_0	Mean length of flange section
B	Width of bracket
B_r	Depth of rib
c	Tapered plate parameter
d	Distance to loaded point
d_1	Mean pipe diameter
d_b	Bolt hole diameter
E	Elastic modulus
E_0	Reduced modulus of elasticity
E_t	Tangent modulus
e	Eccentricity of load application
F	Load on weld seam
F_n	Normal force component
F_s	Shear force component
F_t	Width of flange cross section
f	Load-sharing ratio
H	Depth of standard flange; maximum depth of bracket
H_e	Equivalent depth of flange
h	Thickness of flange ring; size of weld leg
I_b	Second moment of area
I_g	Moment of area of a component section
I_m	Moment of area of wall element of unit width
I_n	Moment of area of a rib cross section
I_0	Moment of area of main flange section
I_x	Moment of area about central axis
J	First moment of area
K	Modulus of elastic foundation
K_b	Buckling coefficient
$k = R_o/R_i$	Flange ring ratio
L	Length of rib of constant depth; length of bracket
L_g	Length of tapered rib
ℓ_1	Moment arm
M	General symbol for bending moment
M_1, M_2	Bending moment components
M_c	Bending moment per one rib spacing
M_F	Toroidal moment on flange ring
M_0	Discontinuity bending moment
M_R	Bending moment on rib
M_y	Bending moment about longer edge of plate
$m = R_i/T$	Ratio of inner radius to wall thickness
N	Number of ribs
$n = H/T$	Ratio of flange to pipe thickness
P	Tensile load on bracket; edge force on plate
Q_0	Discontinuity shearing force
q	Radial load intensity
R	Radius to bolt circle; mean flange radius
R_i	Inner radius of pipe
R_o	Outer radius of flange
r	Mean radius of pipe
S	General symbol for stress
S_1, S_2	Weld stress components

S_{bR}	Rib bending stress
S_c	Compressive stress
S_{Cr}	Critical compressive stress
S_F	Flange dishing stress
S_n	Normal stress
S_p	Plastic stress
S_R	Radial stress in flange ring
S_s	Shear stress
S_t	Tensile stress; toroidal stress in flange
S_{TR}	Total stress in rib
S_u	Ultimate strength
S_y	Yield strength
S_{max}	Maximum principal stress
s_0	Wall thickness
s_1	Depth of section at failure
T	Thickness of pipe; thickness of plate
t	Distance from bolt circle to outer pipe surface; thickness of fillet weld
T_r	Thickness of rib
T_0	Thickness of backup ring
V	Moment factor in plate analysis
W	Total bolt load; external load on bracket
W_i	Load per inch of pipe circumference
W_0	Load per inch of bolt circle
W_R	Tensile load on rib
W_{pl}	Plastic load on bracket
x	Arbitrary distance
Y	Deflection of beam on elastic foundation
Y_p	Plate edge deflection under concentrated load
Y_q	Plate edge deflection under uniform load
y	Coordinate; ring deflection
Z	Section modulus
α	Angle of fractured part; bracket angle, rad
β	Elastic foundation parameter
β_s	Shell parameter
δ	Slope of stress–strain curve, rad
ε	Strain
ε_y	Uniaxial strain at yield
η	Modulus ratio
θ	Angle of twist; angle in weld analysis, rad
θ_F	Angle of twist of main flange ring, rad
θ_p	Rib half-angle, rad
θ_R	Bending slope at end of rib, rad
μ	Poisson's ratio
σ, S	Stress
σ_b, S_b	Bending stress
τ	Shear stress component
τ_{max}	Principal shear stress
ϕ	Plate angle, rad
ϕ_0	Flange factor
$\phi(\beta, L)$	Auxiliary function for a beam on elastic foundation
ω	Ratio of rib to flange moment

REFERENCES

1. A. Blake, Design of welded brackets, *Machine Design*, 1975.
2. D. R. Askeland, *The Science and Engineering of Materials*, 3rd ed., PWS Publishing Co., Boston, MA, 1989.
3. L. Tall, L. S. Beedle, and T. V. Galambos, *Structural Steel Design*, Ronald Press, New York, 1964.

22 Special Plate Problems and Applications

22.1 INTRODUCTION

The many applications of plates in structures and machines, and the consequent varied geometries and loadings lead to diverse and special problems which are seldom listed or discussed in textbooks and handbooks. In this chapter, we look at some of these special configurations which may be of interest and use to analysts and designers. Specifically, we consider plates with large deflections, perforated plates, reinforced plates, pin loaded plates, and washers.

22.2 LARGE DISPLACEMENT OF AXISYMMETRICALLY LOADED AND SUPPORTED CIRCULAR PLATES

The fundamentals of plate theory of Chapter 18, and the application with panels, flanges, and brackets of Chapters 19 through 21 assume that the plate components have relatively small thicknesses whose deflections do not exceed the magnitude of the thicknesses. If, however, the deflections are large exceeding the plate thickness, the analysis should include the effect of the strain in the middle plane of the plate [1,2]. But this leads to a set of nonlinear differential equations that are difficult to solve even numerically. Nevertheless an approximate solution for circular plates with axisymmetric loading and support may be obtained using Nadai's equations as recorded by Timoshenko and Woinowsky-Kreiger [1]:

$$\frac{d^2u}{dr^2} + \frac{du}{dr}\bigg/4 - u/r^2 = -(1-\nu)\left(\frac{dw}{dr}\right)^2\bigg/2r - \left(\frac{dw}{dr}\right)\left(\frac{d^2w}{dr^2}\right) \tag{22.1}$$

and

$$\frac{d^3w}{dr^3} + \frac{d^2w}{dr^2}\bigg/r - \frac{dw}{dr}\bigg/r^2 = (12/h^2)\left(\frac{dw}{dr}\right)\left[\frac{du}{dr} + \nu u/r + \left(\frac{dw}{dr}\right)^2\bigg/2\right]$$
$$+ (1/Dr)\int_0^r q(r)rdr \tag{22.2}$$

where
    r is the radial coordinate
    u is the radial displacement
    w is the midplane displacement normal to the plate
    D is $Eh^3/12(1-\nu^2)$
    h is the plate thickness
    E is Young's modulus of elasticity
    ν is Poisson's ratio
    $q(r)$ is the axisymmetric loading.

An approximate iterative solution to Equations 22.1 and 22.2 may be obtained by assuming a reasonable first solution $w(r)$. Then by substituting into the right-hand side of Equation 22.1 we obtain a linear equation for u, which can be integrated to obtain a first solution for u.

Next, by substituting for u and w in the right-hand side of Equation 22.2 we obtain a linear equation for w, which after integration gives a second approximation for $w(r)$. This in turn, may be substituted into Equation 22.1 to obtain further refined values for u and w.

This procedure may be used to obtain the maximum (center) displacement $w_{\max}$ of a circular membrane with fixed edges and Poisson's ratio $\nu = 0.25$, as

$$(w_{\max}/h) + 0.583(w_{\max}/h)^3 = 0.176(q/E)(r_0/h)^4 \tag{22.3}$$

For large displacement, the first term is relatively small compared with the second, and thus $w_{\max}$ is approximately

$$w_{\max} = 0.671 r_0 (q r_0 / Eh)^{1/3} \tag{22.4}$$

The corresponding tensile membrane stresses at the center and boundary of the membrane are then approximately

$$S_{\text{center}} = 0.423 (Eq^2 r_0^2 / h^2)^{1/3} \tag{22.5}$$

and

$$S_{\text{boundary}} = 0.328 (Eq^2 r_0^2 / h^2)^{1/3} \tag{22.6}$$

It may be of interest to note that the deflection of Equation 22.4 is not directly proportional to the load intensity q but instead varies with the cube root of q. To make the deflection proportional to q, it would be necessary to have a corrugated membrane as is frequently done in the field of instrumentation.

22.3 DESIGN CHARTS FOR LARGE DEFLECTION OF CIRCULAR PLATES

The approach of the foregoing section where an iterative procedure is used to predict a large plate displacement, has led to a number of practical formulas and rules for design. Many of these involve procedures where the beginning value of normal displacement w is assumed using the small-deflection theory. Since many large-deflection formulas [3] are expressed in terms of w/h ratios of values somewhat smaller than that which is compatible with the elastic deformation, a more accurate value of w/h can be obtained using relatively few successive approximations.

It should be pointed out that when the deflection of a plate becomes comparatively large, the middle surface is additionally strained due to the membrane plate support. The total stress will be equal to the sum of the maximum stress due to flexure and membrane tension. The load–deflection and load–stress relations, in this particular case, become nonlinear.

The majority of practical design applications involve circular plates and diaphragms under uniform pressure. The relevant deflection criterion can be expressed in dimensionless form as

$$\frac{q r_0^4}{E h^4} = H(w, t) \tag{22.7}$$

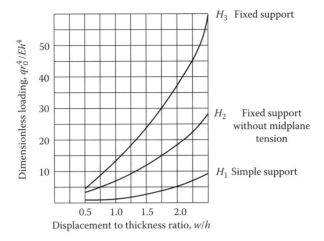

FIGURE 22.1   Design chart for large deflections of circular plates.

Figure 22.1 provides a graphical representation of H for three usual support conditions:

1. H_1: Simple support
2. H_2: Edge restrained but no membrane (midplane) tension
3. H_3: Fixed support with full membrane tension

Figure 22.2 provides a graphical representation of a dimensionless stress parameter F as a function of the deflection to thickness ratio w/h, where F is defined as

$$F = Sr_0^4/Eh^2 \tag{22.8}$$

where S is the estimated actual stress.

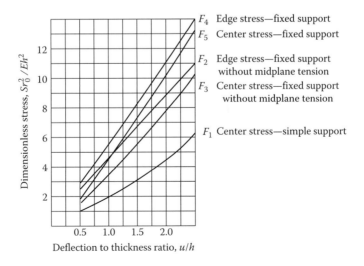

FIGURE 22.2   Design chart for large deflections of circular plates.

Figure 22.2 represents five curves depending upon the stress location (center or edge) and the edge support. These are

1. F_1: Center stress—simple support
2. F_2: Edge stress—edge support but no membrane (midplane) tension
3. F_3: Center stress—edge support but no membrane (midplane) tension
4. F_4: Edge stress—fixed support
5. F_5: Center stress—fixed support

The design problem is thus reduced to a simple procedure: given the loading, support conditions, and the geometrical and physical parameters, Figure 22.1 provides the displacement. Then knowing the displacement, Figure 22.2 provides the stress. Alternatively, given a stress limit, Figure 22.2 provides the displacement and then Figure 22.1 provides the loading limit. Still more, for given stress or displacement limits, the figures provide design criteria for the geometric and physical parameters.

The following example illustrates the procedure.

22.4   DESIGN EXAMPLE FOR A LARGE DISPLACEMENT OF CIRCULAR PLATE

Consider an aluminum plate with thickness h of 0.375 in. and radius r_0 of 24 in. Let the plate have a fixed (built-in) rim. Suppose the plate is loaded by a pressure q of 6 psi. Calculate the maximum displacement and stress.

Solution

The elastic modulus E for aluminum is approximately 10×10^6 psi. Then, from the given data the dimensionless loading H is

$$H = H_3 = \frac{qr_0^4}{Eh^4} = \frac{(6)(24)^4}{(10)^7(0.375)^4} = 10.066 \tag{22.9}$$

From Figure 22.1, the displacement/thickness ratio w/h is then approximately

$$w/h = 1.075 \tag{22.10}$$

Hence, the center displacement w_{max} is

$$w_{max} = 1.075h = 0.403 \text{ in.} = 0.0102 \text{ m} \tag{22.11}$$

From Figure 22.2, the dimensionless stress F for w/h of 1.075 is approximately

$$F = F_4 = \frac{Sr_0^4}{Eh^4} = 5.3 \tag{22.12}$$

Hence, the maximum stress S_{max} (occurring at the plate edge) is

$$S_{max} = 5.3 \frac{Eh^2}{r_0^2} = 5.3 \frac{(10)^7(0.375)^2}{(24)^2} = 12.939 \text{ psi} = 89.2 \text{ Pa} \tag{22.13}$$

22.5 LARGE DISPLACEMENT OF RECTANGULAR PLATES

The theoretical problem of large deflection of a rectangular plate has been the subject of numerous investigations. Various solutions obtained have been expressed in terms of dimensionless parameters similar to those used in conjunction with circular plates. Many solutions have also been compared with experimental data during the various phases of research sponsored by the National Advisory Committee for Aeronautics. A convenient summary of deflections and stresses for rectangular plates under uniform load is given by Roark [4]. The boundaries for these plates include simple supports, riveting constraints, and completely fixed conditions. The plate-loading parameter qa^4/Eh^4 covers the range 0–250, where a is the long dimension. The maximum ratio of displacement to thickness in Roark's summary is 2.2.

22.6 PERFORATED PLATES

A perforated plate is a plate with numerous small holes. We can easily extend our plate analysis to perforated plates provided the perforated plates satisfy certain conditions. These are as follows:

1. The holes are numerous (20 or more) and circular
2. The holes are regularly positioned into either square or equilateral-triangular arrays (see Figure 22.3)
3. The plate thickness is more than twice the hold pitch (center to center distance, p_0)
4. The "ligament efficiency," h_0/p_0 is greater than 5%
5. Local reinforcement effects are included in the ligament efficiency

The "ligament efficiency" is a principal parameter in the analysis of perforated plates. It is a dimensionless parameter defining the geometric spacing of the holes as (see Figure 22.3):

$$\text{Ligament efficiency} = h_0/p_0 \qquad (22.14)$$

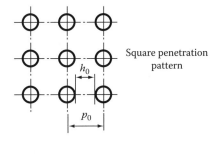

Square penetration pattern

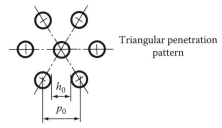

Triangular penetration pattern

FIGURE 22.3 Typical hole patterns for perforated plates.

The concept of ligament efficiency and its importance have been the subjects of various theoretical and experimental investigations [5]. In principle, we can design and analyze perforated plates by simply replacing the elastic constants E and v by modified constants E^* and v^* which are dependent upon the ligament efficiency.

The modified constants are considered to be functions of the ligament efficiency within the range 0.05–1.0, and specific detailed design charts that feature stress multipliers as functions of hole orientation angle in relation to the direction of loading are available. The actual plate stresses are obtained by multiplying the design factors by the nominal stresses calculated for the equivalent solid plate. All conditions treated by the codes are axisymmetric. The effects of temperature are included in consideration of structural interaction with the adjacent members. Where thin or irregular ligament patterns are involved, the code recommends using the average stress intensities.

In practice, triangular patterns appear to be more widely used, particularly in the construction of boiler feedwater heaters, steam generators, heat exchangers, and similar systems. In some of these systems, a perforated plate may house tubes which in turn could increase the plate rigidity. In such cases, the concept of virtual ligament efficiency can be defined as the actual ligament plus an effective portion of the tube wall divided by the tube hold pitch [7].

Figure 22.4 provides an example of how elastic constants should be modified for plates with triangular pattern holes, according to the American Society of Mechanical Engineers (ASME) code [6], where t is the plate thickness. The graph of Figure 22.4 has been developed for a Poisson's ratio of 0.3 and plate-thickness/hole-pitch ratios (t/p_0) greater than 2. Design curves for other values of Poisson's ratio are also available [5]. The ASME code provides various stress formulas for use with the modified elastic constants E^* and v^*.

Some analysts suggest that ligament efficiency might better be estimated using the hole area A as

$$\text{Ligament efficiency} = (p_0^2 - A)/p_0^2 \tag{22.15}$$

Furthermore, this and other formulas for ligament efficiency appear to be equally applicable to square and triangular hole patterns. The advantage of Equation 22.15 is that the relevant magnitudes

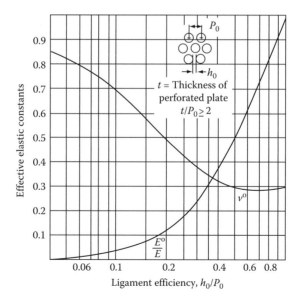

FIGURE 22.4 Modified elastic constants for triangular patterns ($v = 0.3$). (From *ASME Boiler and Pressure Vessel Code*, Section VIII, American Society of Mechanical Engineers, New York, 1971.)

of efficiency fall closer between 0.5 and 0.8, which seems to be more realistic than the smaller values resulting from h_0/p_0. A preliminary estimate of the maximum stress in the perforated plate can be obtained by dividing the maximum stress in the homogeneous plate by the ligament efficiency. Similarly, the deflections of the perforated plate can be calculated if the flexural rigidity of the homogeneous plate D is multiplied by the ligament efficiency. Since the deflection is inversely proportional to D, ligament correction tends to increase the magnitude of the deflection for a perforated plate compared to that of a solid plate.

22.7 REINFORCED CIRCULAR PLATES

We can reinforce circular plates by attaching ribs to the plate. In practice, these ribs being radial and/or concentric, are usually attached axisymmetrically. They may or may not be symmetric relative to the middle plane of the plate.

We may approximately model a stiffened plate by increasing the value of the flexural rigidity D. For a symmetric rib system, we may express the rigidity as

$$D = \frac{Eh^3}{12(1 - \nu^2)} + \frac{EI}{d} \qquad (22.16)$$

where
   I is the second moment of area of the rib with respect to the middle axis of the plate
   d is the average rib spacing

For a nonsymmetric rib system such as a T-shaped rib and plate combination (rib on one side of the plate only), the plate rigidity D may also be expressed as in Equation 22.16, but here I is the average second moment of area of the T-section about its centroid.

Grillage-type plates are used in the nuclear core reactor vessels and other applications where support and cover plate size requirements are such that these members cannot be procured as solid plates. The fabrication is accomplished by welding together a complex web system.

In some cases, the plates can be reinforced by the use of a concentric stiffening ring which reduced the stresses and deflection due to its toroidal stiffness. Such a concentric ring has the tendency to turn inside out as the plate deflects under the load. Working design charts have been developed for the effect of such a reinforcing ring on the maximum deflection and radial stress for a circular plate with a built-in edge and the transverse uniform load [8]. The most effective location of a ring of a specified size, is not the same for the stress and deflection criteria, and a design compromise may be needed. Generally, placing the ring at about 0.6 value of the radius measured from the plate center, represents a satisfactory compromise.

Where the major design criterion is deflection rather than stress, an alternative to a thick cover plate would be a relatively thin plate heavily reinforced. One such design, for instance, involves a system of straight radial ribs radiating to the outer edge of the plate as shown in Figure 22.5. While a relatively exact mathematical model can be applied to the bending of plate with orthogonal ribs, no flexural theory has yet been established for accurate calculation of the deflection of a plate with radial reinforcement [9]. Radial ribs may be of constant depth or tapered geometry with gradually diminishing depth toward the outer edge of the plate. Because of this, such a reinforcement can only be analyzed with the aid of the three-dimensional theory of elasticity, presenting almost unmanageable boundary conditions. Experimental evaluation of the stresses and deflections can be made but many models have to be tested prior to determining the optimum criteria of strength and rigidity. In such cases, however, where the rib system is relatively stiff compared with the plate to which these ribs are attached, the approximate solutions may be possible on the basis of flexible sectorial plates.

In the case of a cover plate shown in Figure 22.5, individual radial rib can be regarded as a simple beam subjected to a bending moment at the junction with the circular stiffener, a supporting reaction at the other end, and a distributed load along the rib length according to a linear function.

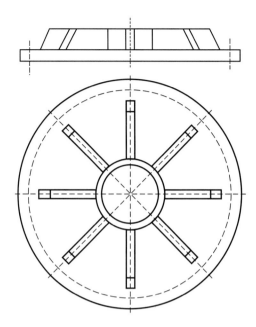

FIGURE 22.5   Cover plate with radial stiffeners.

The basic difficulty in the analysis, however, is the fact that ribs behave as T-beams of variable stiffness and that the portions of the adjacent plate act as flanges. Because of these constraints the only tractable approach has been so far, through experimental techniques [9]. Ribbed configurations in this type of an experiment are machined from a solid flat plate, and in the sequence of tests, the depth of the webs is progressively reduced providing the test samples with different combinations of web shape and size. All plates in a quoted experiment had eight ribs, and the deflections under uniform load were measured by optical means involving an interferometer. The plate models were about 8 in. in diameter, and the measured deflections were assessed in relation to the theoretical deflections of solid plates of equal weight calculated for a simply supported boundary. The results of this research indicated that the most effective use of material could be achieved with deep and slender ribs in such a way as to make the cumulative mass of the rib system equal to about 40% of the total mass of the plate. An empirical formula for the central deflection of the eight-ribbed design shown in Figure 22.5 was established as follows:

$$Y = \frac{21.6qa^{10}\lambda^3}{EW^3} \tag{22.17}$$

where the Poisson's ratio was assumed to be equal to 0.3, which is good for a majority of metallic materials. For the more heavily reinforced ribs, 40% may not be possible to achieve. This type of reinforcement, as well as the elimination of any potential waviness of the plate boundary could be achieved by increasing the number of ribs. Although further experimental analysis is required for finding the precise effect of the number of ribs on the deflection and local stresses, the eight-rib system analyzed so far can be used as a rough guide for establishing the deflection criteria for other designs. For instance, a plate of 12 ribs of the same size relative to the basic plate and one of the eight-rib type are expected to have similar deflection ratios.

22.8   PIN-LOADED PLATES

In a riveted and bolted connection, a plate can be loaded through a pin in the hole which causes a complex stress distribution [10]. Such a stress pattern may be of special interest in the determination

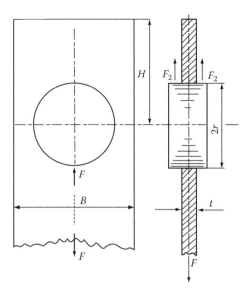

FIGURE 22.6 Notation for a pin-loaded plate.

of the fatigue strength of a joint made of high-strength alloy steels or lightweight alloys. For these materials, the endurance limits are well within the elastic range, in contrast to mild steel which has the endurance limit close to the yield strength of the material.

In the study quoted here [10], strain gage and photoelastic measurements were made to determine the stress concentration factor for various ratios of hole diameter to the width of plate, ranging from 0.086 to 0.76.

Figure 22.6 shows the configuration and notation for the problem. Let S designate the maximum tensile stress at the edge of the plate hole and let σ designate the average bearing stress on the pin. Let K_b be the ratio of these stresses defined as

$$K_b = S/\sigma \tag{22.18}$$

Then K_b represents a stress concentration factor.

From Figure 22.6 we see that K_b is

$$K_b = 2rtS/F \tag{22.19}$$

Figure 22.7 provides an experimentally determined graphical representation of K_b in terms of the ratio of hole diameter to plate width. A problem with the empirical data, however, is that an extrapolation of the data may be in error. For example, we should not expect the representation for K_b in Figure 22.7, to be accurate beyond $2r/B = 0.8$. Nevertheless, the greater the ratio of the hole diameter to the plate width, the more the stress concentration.

Clearance between the pin and the plate also increases stress concentration. Moreover, clearance effects increase with larger pin size. For example, smaller the ratio H/B of Figure 22.6, greater the effect of clearance. Clearance increases the stress by promoting changes in curvature. Experiments show that for neat-fitting pins, the maximum stresses occur at the ends of the horizontal diameter. These stresses move away from there as the clearance increases.

Finally, it should be noted that stress concentrations are highest when there is only a single pin in the plate. That is, if the load is divided between, say, two pins having individual diameters equal to one-half of the single pin diameter, the stress concentration is reduced by about 20%. Pin lubrication also produces a small decrease in K_b values.

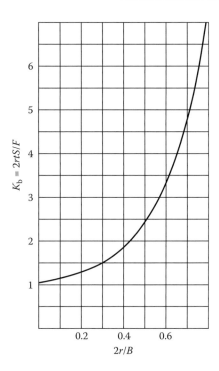

FIGURE 22.7   Stress concentration factor for a pin-loaded plate. (From Frocht, M. M. and Hill, H. N., *J. Appl. Mech.*, 7, 1940.)

22.9 BELLEVILLE WASHERS

A Belleville washer is an annular conical disk which is commonly used as a spring. Figure 22.8 provides a sketch of the device together with notation, which we can use in our stress and displacement analyses. A Belleville washer is not a flat plate. Instead it is more like a plate ring or pipe flange in its response to loadings. As such it is often called a "Belleville spring."

The Belleville spring is an important machine element where, among other features, space limitation, high load, and relatively small deflections are required. Although in practice, calculations of load–deflection characteristics by a majority of available methods generally show satisfactory correlation, the problems of stress still remain speculative in various design applications.

A rigorous analytical or mathematical analysis of Belleville springs is elusive as in practice, the loading and consequent displacement create in effect a large, elastic–plastic displacement of a

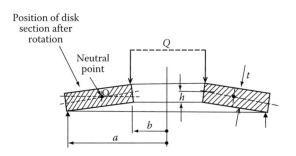

FIGURE 22.8   A Belleville washer.

conical shell. Nevertheless, in 1936, Almen and Laszlo [11] developed useful design formulas for maximum stresses at the inner edges of the washer, and their method has been widely accepted in industrial applications. Their derivation is based upon an elastic analysis assuming that radial stresses are negligible. The disk cross section is assumed to rotate about the neutral point O shown in Figure 22.8.

In 1946, Ashworth [12] conducted further analysis of the springs, resulting in expressions for tensile and compressive stresses which agreed well with the results of Almen and Laszlo. Ashworth also reported on tests conducted in Germany, according to which there was a good correlation between the theory and the experiment with regard to the maximum compressive stresses. The same experiments confirmed that the radial strains measured at various points of the disk spring, were rather small in comparison with the circumferential strains. A more recent investigation by Wempner [13] resulted in a proposed refinement of the Almen and Laszlo solution. In Wempner's work, the effect of radial strains was included. The relevant stresses, however, appeared to be only some 10% higher. Hence, for all practical purposes and despite some of the limitations, the Almen and Laszlo theory remained essentially unchanged. Their key representation for stress S and the rim displacement Y is

$$S = \frac{YE}{2(1 - \nu^2)a^2 C_1}[C_2(2h - Y) + 2tC_3] \tag{22.20}$$

where, as before, E and ν are the elastic constants and Poisson ratio and where Y may be determined from the load formula:

$$Q = \frac{YE}{(1 - \nu^2)a^2 C_1}\left[(h - Y)\left(h - \frac{Y}{2}\right)t + t^3\right] \tag{22.21}$$

where Q is the load as illustrated in Figure 22.8 and where C_1, C_2, and C_3 are dimensionless geometric parameters given by

$$C_1 = \frac{6\left(\frac{a}{b} - 1\right)^2}{\pi \log_e\left(\frac{a}{b}\right)} \tag{22.22}$$

$$C_2 = \frac{6}{\pi \log_e\left(\frac{a}{b}\right)}\left[\frac{\frac{a}{b} - 1}{\log_e\left(\frac{a}{b}\right)} - 1\right] \tag{22.23}$$

and

$$C_3 = \frac{3\left[\left(\frac{a}{b}\right) - 1\right]}{\pi \log_e\left(\frac{a}{b}\right)} \tag{22.24}$$

Figure 22.8 also illustrates the values of a, b, h, and t, where t, is the washer thickness.

It happens that for metals with Poisson ratio ν being approximately 0.3, there is relatively little error by replacing the stress/displacement and load/displacement equations (Equations 22.20 and 22.21) by the simpler expressions:

$$S = \frac{CYE(2h + 2t - Y)}{a^2} \tag{22.25}$$

and

$$Q = \frac{GYE\left[(h - Y)\left(h - \frac{Y}{2}\right)t + t^3\right]}{a^2} \tag{22.26}$$

where C and G are dimensionless parameters called the "stress factor" and the "displacement factor," respectively. Figures 22.9 and 22.10 provide graphical representations of C and G in terms of the outer/inner rim radius ratio a/b.

Observe the nonlinear relation between the stress S and displacement Y in Equations 22.20 and 22.21 and even also in the simplified expressions of Equations 22.25 and 22.26. For example, if we know the physical and geometrical parameters and the load Q, we need to solve either Equation 22.27 or Equation 22.26 for the displacement Y, and then by substituting into either Equation 22.20 or Equation 22.25 to obtain the stress S.

In many applications where a metal (typically steel) spring is used, the spring is completely flattened out so that $Y = h$. On these occasions, the stress and load equations (Equations 22.25 and 22.26) simply become

$$S + S_{\text{flat}} = \frac{CEh(h + 2t)}{a^2} \tag{22.27}$$

and

$$Q = Q_{\text{flat}} = \frac{GEht^2}{a^2} \tag{22.28}$$

We can develop a useful design equation by dividing the load Q of Equation 22.26 by the flattening load Q_{flat} of Equation 22.28, yielding

$$Q/Q_{\text{flat}} = (\eta m^2/2)(1 - \eta)(2 - \eta) + \eta \tag{22.29}$$

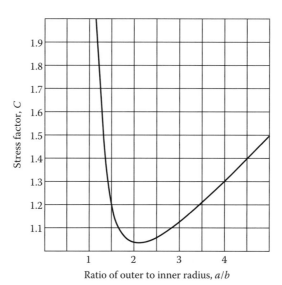

FIGURE 22.9 Stress factor/radius ratio chart.

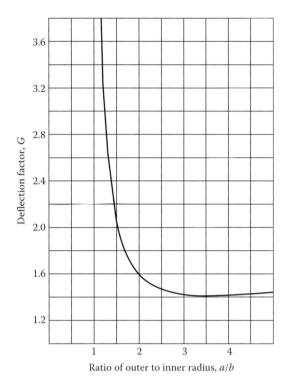

FIGURE 22.10 Deflection factor/radius ratio chart.

where η and m are the dimensionless displacement and free height/thickness ratio given by

$$\eta = Y/h \text{ and } m = h/t \qquad (22.30)$$

Equation 22.29 represents a family of curves known as the load–deflection characteristics of conical-disk springs of various proportions. Such curves are given in engineering handbooks and spring design manuals for direct design applications. When $0 < m < 1.4$, the spring rate is always positive. For the interval $0.6 < \eta < 1.2$ with $1.4 < m < 2.8$, the spring rate is negative and the actual load decreases as the deflection increases. The Belleville washer becomes unstable for $m > 2.8$; that is, at a particular compressive load Q the conical disk shape snaps into a new position. Also for $m \cong 1.6$, a zero spring rate can be achieved for an appreciable range of η.

Similarly, we can obtain a dimensionless stress ratio by dividing stress at flattening S_{flat} of Equation 22.27 by the stress S at a less-than-flattening load, of Equation 22.25. That is,

$$S_{\text{flat}}/S = \frac{m+2}{\eta(2 + 2m - \eta m)} \qquad (22.31)$$

Figure 22.11 displays this stress ratio for a few typical values of the free height/thickness ratio m. The curves show that the stress ratio is sensitive to changes in the displacement as we would expect.

In practice, many Belleville springs operate at deflections smaller than those corresponding to the load at solid. The amount of permanent set that can be tolerated is about 2% of the maximum working deflection. If it is expected that permanent set can exceed 2%, manufacturers specify a setting-out operation. This consists of loading the washer to solid and noting the amount of decrease in the free

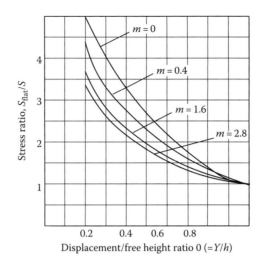

FIGURE 22.11   Stress/displacement relations.

height after unloading. The experiment shows that the first loading cycle causes the maximum amount of permanent deformation and it considerably stabilizes the washer geometry (see Ref. [13]).

Finally we cite Fortini [14] who has developed simplified tables for single washer and nests of washer designs. These tables may help reduce the trial-and-error procedures in using Equations 22.20 and 22.21.

SYMBOLS

A	Area
a	Plate radius
B	Width of rectangular plate
b	Inner radius
C	Approximate stress factor
C_1, C_2, C_3	Almen and Laszlo factors (see Equation 22.20)
D	$Eh^3/12(1 - \nu^2)$
d	Rib spacing dimension
E	Modulus of elasticity
F	Force on pin
$F_1, \ldots, F_5$	Stress function for design (see Figure 22.2)
G	Deflection factor
h	Plate thickness; free height of Belleville spring (see Figure 22.8)
h_0	Hole separation dimension (see Figure 22.3)
H	Load/displacement function for large deflection of circular plates (see Figure 22.1); head distance from edge of hole
I	Second moment of area
K_b	Stress concentration factor (see Equation 22.18)
$m = h/t$	Free height to thickness ratio (see Equation 22.30)
p_0	Hole separation dimension (see Figure 22.3)
Q	Load on Belleville spring (see Figure 22.7)
$q(r)$	Loading function
r	Radial coordinate
S	Stress

t	Plate thickness
u	Radial displacement of a circular plate
W	Weight of ribbed plate
w	Normal displacement
Y	Plate deflection
$\eta = Y/h$	Dimensionless displacement (see Equation 22.30)
λ	Weight density
ν	Poisson's ratio
σ	Stress

REFERENCES

1. S. Timoshenko and S. Woinowsky-Kreiger, *Theory of Plates and Shells*, McGraw Hill, New York, 1959, p. 402.
2. A. Nadai, *Theory of Flow and Fracture of Solids*, McGraw Hill, New York, 1963.
3. J. Presscott, *Applied Elasticity*, Longeans, Green, 1924.
4. W. C. Young and R. G. Budynas, *Roark's Formulas for Stress and Strain*, 17th ed., McGraw Hill, New York, 2002, pp. 451–452.
5. T. Slot and W. J. O'Donnell, Effective elastic constants for thick perforated plates with square and triangular penetration patterns, *Journal of Engineering for Industry*, 93, 1971, pp. 935–972.
6. *ASME Boiler and Pressure Vessel Code*, Section VIII, American Society of Mechanical Engineers, New York, 1971.
7. J. F. Harvey, *Theory and Design of Modern Pressure Vessels*, Van Nostrand Reinhold, New York, 1974.
8. W. A. Nash, Effect of a concentric reinforcing ring of stiffness and strength of a circular plate, *Journal of applied Mechanics*, 14, Paper 47-A-15, 1947.
9. J. Harvey and J. P. Duncan, The rigidity of rib-reinforced cover plates, *Proceedings of the Institute of Mechanical Engineers*, 177(5), 1963, 115–123.
10. M. M. Frocht and H. N. Hill, Stress concentration factors around a central circular hole in a plate loaded through pin in the hole, *Journal of Applied Mechanics*, 7, 1940.
11. J. O. Almen and A. Laszlo, The uniform-section disk spring, *Transactions of the American Society of Mechanical Engineers*, RP-58-10, 1936.
12. G. Ashworth, The disk spring or Belleville washer, *Proceedings of the Institute of Mechanical Engineers*, 155, London, 1946, pp. 93–100.
13. Associated Spring Corporation, *Handbook of Mechanical Spring Design*, Associated Spring Corporation, Bristol, CT, 1964.
14. E. T. Fortini, Conical-disk springs, *Machine Design*, 29, 1958.

PART V

Dynamic Loadings, Fatigue, and Fracture

The vast majority of structural engineering designs are based on static analyses. This is usually satisfactory since most structures and structural components experience primarily static loads, or if there are dynamic loads, the static loads greatly dominate them. Moreover, for many structures if dynamic loads exist, they are often small, short-lived, and only occasional. Therefore, for most structural component designs based on static analyses, a safety factor of 2 generally ensures that stress limits will not be approached.

There are nevertheless many occasions where dynamic loadings are not small, where they are frequent, unexpected, and/or sudden. In these instances, failure can be sudden and dramatic. But failure can also gradually occur through fatigue, leading to fracture. Alternatively, dynamic loadings can produce large unintended displacements destroying the efficacy of a structure. This can occur, for example, during resonance, during earthquakes, and during blast (explosion) loadings.

In this part, we consider structural designs for accommodating dynamic loadings, fatigue, and seismic effects. We investigate both short-term and long-term phenomena and we propose designs and design methodology to accommodate them and to prevent failure.

We begin with a review of simple dynamic loadings and the corresponding structural responses. We then consider seismic loadings and design countermeasures. Finally, we investigate fatigue, fracture and we present preventative designs.

23 Dynamic Behavior of Structures: A Conceptual Review

23.1 INTRODUCTION

In this chapter we review a few elementary dynamics concepts which are pertinent in the analysis and design of structures with dynamic loadings. These concepts include sudden loadings, natural frequency, and free-fall/impact. We attempt to illustrate the concepts with a few simple examples.

23.2 VIBRATION AND NATURAL FREQUENCY

Consider the elementary mass–spring system of Figure 23.1, where a body B with mass m can move left and right over a smooth horizontal surface. The movement of B is restricted by a linear spring with stiffness modulus k. The displacement (or distance) of B away from a neutral (static equilibrium) position is x. An elementary dynamic analysis shows that the differential equation describing the movement of B is [1]:

$$m\mathrm{d}^2x/\mathrm{d}t^2 + kx = 0 \qquad (23.1)$$

The general solution of Equation 23.1 is

$$x = A_1 \cos \omega t + A_2 \sin \omega t \qquad (23.2)$$

where
  A_1 and A_2 are constants
  ω called the "circular frequency" is given by

$$\omega^2 = k/m \qquad (23.3)$$

The integration constraints A_1 and A_2 are determined by initial conditions. For example, suppose that at time $t = 0$, B is displaced to the right by an amount $x(0) = \delta$, with an initial speed $\dot{x}(0) = \delta$. Then A_1 and A_2 are

$$A_1 = \delta \quad \text{and} \quad A_2 = \delta/\omega \qquad (23.4)$$

Then by back substitution from Equation 23.4 into 23.2, we obtain the displacement x as

$$x = \delta \cos \omega t + (\delta/\omega) \sin \omega t \qquad (23.5)$$

Sometimes it is helpful to use trigonometric identities to express Equation 23.2 in the alternative form:

$$x = C \cos (\omega t + \phi) \qquad (23.6)$$

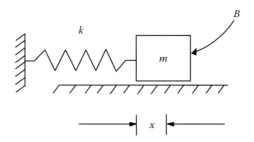

FIGURE 23.1   A simple mass–spring system.

where by comparing Equations 23.2 and 23.6 we see that C and ϕ are

$$C = \left(A_1^2 + A_2^2\right)^{1/2} \quad \text{and} \quad \phi = -\tan^{-1}(A_2/A_1) \tag{23.7}$$

Similarly, in terms of C and ϕ, A_1 and A_2 are

$$A_1 = C\cos\phi \quad \text{and} \quad A_2 = -C\sin\phi \tag{23.8}$$

(Equations 23.7 and 23.8 are immediately obtained by using the identity $\cos(\alpha + \beta) \equiv \cos\alpha \cos\beta - \sin\alpha \sin\beta$.)

Consider the special case in the above example, when δ is zero. That is, body B of Figure 23.1 is displaced to the right by an amount δ and released from rest. Then from Equation 23.5 we see that the subsequent displacement of B is simply

$$x = \delta \cos \omega t \tag{23.9}$$

Equations 23.2 through 23.9 provide insights into the behavior of body B:

1. The movement of B is oscillatory. That is, B vibrates with the phenomenon called "vibration."
2. C (or δ of Equation 23.9) is called the "amplitude" of the vibration.
3. ϕ is called the "phase" of the vibration.
4. As noted earlier, ω is called the "circular frequency."
5. Since the trigonometric functions are periodic with period 2π, the "frequency" f of the oscillation (the number of cycles per unit time) is then

$$f = \omega/2\pi = (1/2\pi)\sqrt{k/m} \tag{23.10}$$

6. The "period" T or the time to complete a cycle is then

$$T = 1/f = 2\pi/\omega = 2\pi\sqrt{m/k} \tag{23.11}$$

7. f is sometimes called the "natural frequency" of the system. It is perhaps the single best parameter describing the nature of the motion of the system. For stiff systems (large k), f is large and the movement is rapid. For heavy or massive systems (large m), f is small and the movement is slow.

23.3 DYNAMIC STRUCTURAL RESPONSE—INTUITIVE DESIGN CRITERIA

Before looking at the details of dynamic structural analysis, it may be helpful to consider a few guidelines or "rules of thumb." First, suppose we know the natural frequency f and consequently the period $T (= 1/f)$ of a structure or a structural component, and assuming we can estimate the time of load application and if the load is slowly applied so that the application time significantly exceeds the period, then we can essentially consider the load as "static."

Specifically, the loading may be regarded as static if

$$\frac{\text{Time of load application}}{\text{Period}} > 3 \text{ (essentially static loading)} \qquad (23.12)$$

Second, suppose the time of loading is somewhat quicker but still greater than the vibration period, then a static analysis may still be applicable but the material properties may need to be adjusted upward. An approximate guide to this condition is

$$1.5 \leq \frac{\text{Time of load application}}{\text{Period}} < 3 \text{ (static loading with increased elastic modulus)} \qquad (23.13)$$

Third, if the time of loading is rapid so that it is only a fraction of the period, a static analysis is usually not very helpful and it could lead to grossly erroneous results. This condition occurs when

$$\frac{\text{Time of load application}}{\text{Period}} < 0.5 \text{ (static loading is not applicable)} \qquad (23.14)$$

Finally, it should be remembered that an analysis of structural response to sudden loading is different from a fatigue analysis. Fatigue occurs with slow loadings applied and released numerous times. We consider fatigue in Chapter 26.

23.4 DYNAMIC STRENGTH

As far as steels and other metals are concerned, those with lower yield strength are usually more ductile than higher strength materials. That is, high yield strength materials tend to be brittle. Ductile (low yield strength) materials are better able to withstand rapid dynamic loading than brittle (high yield strength) materials. Interestingly, during repeated dynamic loadings low yield strength ductile materials tend to increase their yield strength, whereas high yield strength brittle materials tend to fracture and shatter under rapid loading. Figure 23.2 illustrates the strengthening (and lack of strengthening) of steel.

The behavior illustrated in Figure 23.2 is characteristic of other metals as well. Table 23.1 provides some typical data [2–5]. Reference [2] also provides similar data for concrete and brittle materials under compressive loadings.

It should be noted that the behavior of Figure 23.2 and the data listed in Table 23.1 are primarily developed in uniaxial testing, as opposed to combined stress conditions.

23.5 SUDDENLY APPLIED WEIGHT LOADING

Consider a rod R with length ℓ, cross-section area A, and elastic modulus E. Let R be suspended vertically and supported at its upper end as in Figure 23.3. Let there be a small flange at the lower end of R.

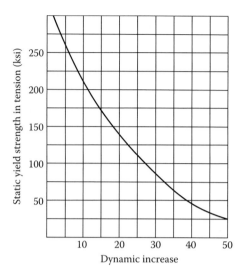

FIGURE 23.2   Effect of rapid loading on yield strength.

Next, let W be a relatively heavy plate or weight W with a slot so that it can be placed around R and supported by the end flange of R as in Figures 23.4 and 23.5. In Figure 23.5, δ_{eq} is the elongation of R due to the weight W. If W is also the weight of the plate, then δ_{eq} is

$$\delta_{eq} = W\ell/AE \tag{23.15}$$

Alternatively, Equation 23.15 may be written as

$$W = k\delta_{eq} \tag{23.16}$$

where by inspection k is simply AE/ℓ. Also if σ_{eq} is the equilibrium stress in R, then σ_{eq} is

$$\sigma_{eq} = W/A \tag{23.17}$$

Next, consider the same rod and weight but let the weight be suddenly applied or dropped onto the rod flange. To visualize this, imagine the weight to be placed just above the flange, resting on removable supports as in Figure 23.6. In this configuration let the supports be removed so that the

TABLE 23.1

Dynamic Strengthening of Metals

Material	Static Strength (psi)	Dynamic Strength (psi)	Impact Speed (ft/s)
2024 Al (annealed)	65,200	68,600	>200
Magnesium alloy	43,800	51,400	>200
Annealed copper	29,900	36,700	>200
302 Stainless steel	93,300	110,800	>200
SAE 4140 steel	134,800	151,000	175
SAE 4130 steel	80,000	440,000	235
Brass	39,000	310,000	216

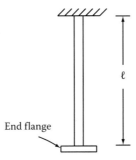

FIGURE 23.3   Suspended rod.

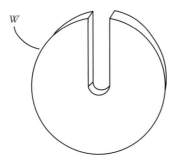

FIGURE 23.4   Plate weight.

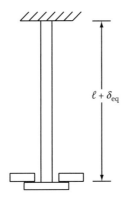

FIGURE 23.5   Rod supporting weight in equilibrium.

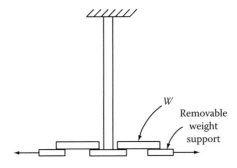

FIGURE 23.6   Weight resting upon removable supports.

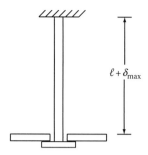

$\ell + \delta_{max}$

FIGURE 23.7 Maximum elongation of a rod due to suddenly applied weight loading.

weight is suddenly resting upon the flange. This creates a sudden loading and stretching of the rod so that in the extreme downward position the rod is elongated by δ_{max} as represented in Figure 23.7.

We can determine δ_{max} by using the work–energy principle of elementary mechanics. Recall that the principle states that for two states, say 1 and 2, of a mechanical system the work done on the system, $_1W_2$ is equal to the change in kinetic energy between the states $K_2 - K_1$. That is,

$$_1W_2 = \Delta K = K_2 - K_1 \tag{23.18}$$

For the rod/weight system of Figures 23.6 and 23.7, let 1 refer to the initial state just as the weight is being released onto the rod as in Figure 23.6, and let 2 refer to the state where the weight has fallen through its greatest drop, δ_{max} or equivalently greatest rod stretch as in Figure 23.7. Interestingly, in each of these states the kinetic energy is zero. Therefore, from Equation 23.18 the work $_1W_2$ is zero.

The work done on the weight is due to two sources: (1) gravity and (2) the "spring" force created by the stretching rod. The work by gravity is simply the weight W multiplied by the drop distance δ_{max}. That is,

$$_1W_2^{gravity} = W\delta_{max} \tag{23.19}$$

The spring force created by the stretched rod is proportional to the stretch δ. The spring force resists the movement of W and acts opposite to the direction of movement. Thus, the work on W due to the rod elasticity is

$$_1W_2^{spring} = -\int_0^{\delta_{max}} k\delta \ d\delta = -(k/2)\delta_{max}^2 = -(AE/2\ell)\delta_{max}^2 \tag{23.20}$$

(The negative sign is due to the force acting opposite to the movement direction.)

Since the total work is zero, we have

$$0 = {}_1W_2 = {}_1W_2^{gravity} + {}_1W_2^{spring} = W\delta_{max} - (AE/2\ell)\delta_{max}^2 \tag{23.21}$$

By solving for δ_{max}, we have

$$\delta_{max} = 0 \quad \text{and} \quad \delta_{max} = 2W\ell/AE \tag{23.22}$$

The first solution of Equation 23.22 is trivial and simply represents the nonapplication or loading of the weight on the rod. The second solution, however, shows that the maximum rod stretch is *twice* that obtained in static equilibrium as documented by Equation 23.15. That is,

$$\delta_{max} = 2\delta_{eq} \qquad (23.23)$$

Consequently, the maximum rod stress due to the dynamic loading is *twice* that found in static equilibrium:

$$\sigma_{max} = 2\sigma_{eq} \qquad (23.24)$$

Equation 23.24 represents a well-known principle of suddenly applied weight loading. Specifically, a suddenly applied loading produces at least twice the static equilibrium stress.

23.6 STRAIN ENERGY—AN ELEMENTARY REVIEW

Consider again the simple example of the foregoing section. Specifically, consider the condition of maximum downward displacement of the weight W as in Figure 23.8, where the level at 0 represents the unloaded position of the end of the rod, and where the elongation δ_{eq} and δ_{max} are the rod extensions under static equilibrium (δ_{eq}) and in the maximum extension due to the suddenly released weight.

From Equations 23.15 and 23.22, δ_{eq} and δ_{max} are

$$\delta_{eq} = W\ell/AE \quad \text{and} \quad \delta_{max} = 2W\ell/AE \qquad (23.25)$$

where, as before,
  W is also the weight of W
  ℓ is the undeformed rod length
  A is the rod cross-section area
  E is the elastic modulus

Since the rod is a linearly elastic body, it may be viewed as being a linear spring with modulus k given by

$$k = AE/\ell \qquad (23.26)$$

When the rod is stretched with its maximum elongation δ_{max}, the rod (viewed as a spring) will have an elastic potential energy (or "strain energy") U given by [1,6]

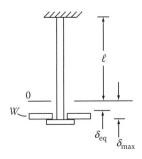

FIGURE 23.8 Dynamically loaded rod by a suddenly applied weight.

$$U = (1/2)k\delta_{max}^2 \qquad (23.27)$$

or by Equation 23.26,

$$U = (1/2)(AE/\ell)\delta_{max}^2 \qquad (23.28)$$

Observe that by comparing Equations 23.20 and 23.28, we see that the strain energy of the rod is exactly the negative of the work done by the rod on the weight as the weight drops. Equivalently, the strain energy of the rod is equal to the work done on the rod by the weight as it drops.

For an arbitrarily shaped and loaded elastic body the strain energy U is defined as [6–8]:

$$U = \int_V \gamma \, dV = \int_V (1/2)\sigma_{ij}\varepsilon_{ij} \, dV \qquad (23.29)$$

where γ, defined by inspection as $1/2\sigma_{ij}\varepsilon_{ij}$ (sums on i and j), is known as the "strain energy density," with σ_{ij} and ε_{ij} being the stress and strain matrix elements as before; and V is the volume of the elastic body.

If we apply Equation 23.29 to stretched hanging rod, with one-dimensional (uniaxial) stress and strain, we have

$$U = \int_V (1/2)\sigma\varepsilon \, dV = (1/2)(W/A)(\delta_{max}/\ell)\int_V dV = (1/2)(W/A)(\delta_{max}/\ell)A\ell$$

$$= (1/2)W\delta_{max} = (1/2)k\delta_{max}^2 = (1/2)(AE/\ell)\delta_{max}^2 \qquad (23.30)$$

This result is identical to the expressions of Equations 23.27 and 23.28.

23.7   LOADING FROM A FALLING WEIGHT

On many occasions, a weight may fall onto a structure from a height, say h, above the structure as represented symbolically in Figure 23.9. By the time the weight reaches the structure, it would have acquired a kinetic energy from the fall. During impact this energy will be transmitted to the structure, creating stress and deformation in the structure. We can study this problem using the work–energy principle as in the hanging rod example of Section 23.5.

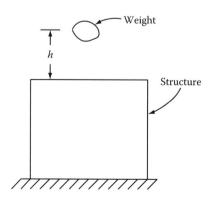

FIGURE 23.9   Weight falling onto a structure.

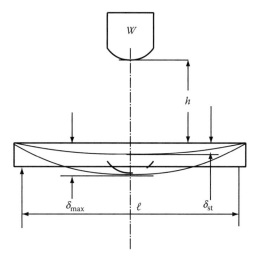

FIGURE 23.10   A weight dropped from a height h onto a simply supported beam.

To illustrate the procedure, consider a body W with weight W dropped onto a simply supported beam as depicted in Figure 23.10, where h is the drop height, ℓ is the beam length, δ_{max} is the maximum beam deflection, and δ_{st} is the static deflection under load W.

The work done on W consists of *positive* work by gravity (the movement of W in the direction of gravity) and *negative* work by the beam as W deflects the beam (the force of the beam on W is upward and thus opposite the downward direction of the movement of W).

The work done by gravity is simply

$$W_{gravity} = W(h + \delta_{max}) \tag{23.31}$$

The work done by the beam on W is

$$W_{beam} = -(1/2)k\delta_{max}^2 \tag{23.32}$$

where k is the beam stiffness under transverse loading by a concentrated center load.

Specifically, recall that if a simply supported beam has a central load as in Figure 23.11 (see Figure 11.12a), then the maximum displacement δ_{max} occurring under the load is $P\ell^3/48EI$ as in Figure 23.12 (see Figure 11.12c), where, as before, E is the elastic modulus and I is the second moment of area of the beam cross section. Thus if the beam force–displacement relationship is expressed as

$$P = k\delta \tag{23.33}$$

Then from Figure 23.12 and from Equation 11.20, k is seen to be

$$k = 48EI/\ell^3 \tag{23.34}$$

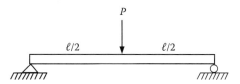

FIGURE 23.11   A centrally loaded, simply supported beam.

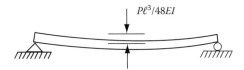

FIGURE 23.12   Displacement of a simply supported beam under a concentrated center load.

Therefore, from Equation 23.32 the work done by the beam on the falling weight W is

$$W_{\text{beam}} = -24EI\delta_{\text{max}}^2/\ell^3 \tag{23.35}$$

(Observe that from the discussion of the foregoing section, $24EI\delta_{\text{max}}^2/\ell^3$ is the strain energy of the beam.)

    The kinetic energy of weight W is zero at both the beginning of the fall and at the instant of maximum beam deflection. The change in kinetic energy is thus zero, and therefore total work done on W is zero. That is,

$$W_{\text{gravity}} + W_{\text{beam}} = 0 \tag{23.36}$$

or from Equations 23.31 and 23.35,

$$W(h + \delta_{\text{max}}) - 24EI\delta_{\text{max}}^2/\ell^3 = 0 \tag{23.37}$$

or

$$\delta_{\text{max}}^2 - (W\ell^3/24EI)\delta_{\text{max}} - (W\ell^3 h/24EI) = 0 \tag{23.38}$$

From Figure 23.12 and Equations 23.33 and 23.34, we see that the static displacement δ_{st} under a load W is

$$\delta_{\text{st}} = W\ell^3/48EI \tag{23.39}$$

Then Equation 23.38 may be written in the simple form:

$$\delta_{\text{max}}^2 - 2\delta_{\text{st}}\delta_{\text{max}} - 2h\delta_{\text{st}} = 0 \tag{23.40}$$

By solving this quadratic equation for δ_{max}, we obtain

$$\delta_{\text{max}} = \delta_{\text{st}} + \left(\delta_{\text{st}}^2 + 2\delta_{\text{st}}h\right)^{1/2} \tag{23.41}$$

or

$$\delta_{\text{max}}/\delta_{\text{st}} = 1 + [1 + 2(h/\delta_{\text{st}})]^{1/2} \tag{23.42}$$

Since the stresses are proportional to the displacements, we can use Equation 24.42 to relate the dynamic stress σ_{dyn} (or σ_{max}) to the static stress σ_{st} (or σ_{eq}) as

$$\sigma_{\text{dyn}} = \sigma_{\text{st}}[1 + (1 + 2h/\delta_{\text{st}})^{1/2}] \tag{23.43}$$

Observe in Equation 24.43 that if the drop height h is zero (as with a suddenly applied load), the dynamic stress is twice the static stress as in Equation 23.24.

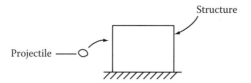

FIGURE 23.13 A projectile directed toward the side of a structure.

23.8 IMPACT FROM A HORIZONTALLY MOVING MASS

Imagine a projectile striking the side of a structure as suggested by Figure 23.13. Since projectiles have constant horizontal velocity, we can study the problem by modeling it as a horizontally moving mass colliding with a structure. We can then use the work–energy principle to estimate the dynamic stress in the structure during the impact. The procedure is similar to that used for the falling mass problem of the foregoing section.

To illustrate the procedure, consider a horizontally moving mass striking the center of a simply supported vertical beam as in Figure 23.14. Let W be the projectile with mass m and let the impact speed be V. Then the kinetic energy K_i of W as it strikes the beam, is

$$K_i = (1/2)mV^2 \tag{23.44}$$

This energy is absorbed by the beam during deformation. Let δ_{max} be the maximum deflection of the beam. Then the work done by the beam on W is

$$W_{beam} = -(1/2)k\delta_{max}^2 \tag{23.45}$$

where, as before, k is the beam stiffness given by (see Equation 23.34):

$$k = 48EI/\ell^3 \tag{23.46}$$

where
  E is the elastic modulus
  I is the second moment of area of the beam cross section
  ℓ is the beam length

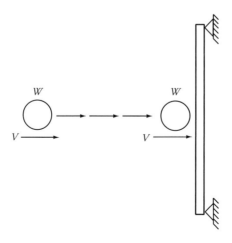

FIGURE 23.14 A horizontally moving mass striking a simply supported vertical beam.

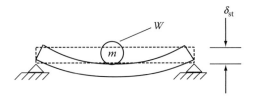

FIGURE 23.15 Simply supported beam deformed by a centrally placed weight.

When the beam is deformed to its maximum deflection δ_{max}, the kinetic energy of W is reduced to zero. The work–energy principle applied to W during impact is then

$$\text{Work} = \Delta K \tag{23.47}$$

or

$$-(1/2)k\delta_{max}^2 = 0 - (1/2)mV^2 \tag{23.48}$$

Hence, δ_{max} is

$$\delta_{max} = V\sqrt{m/k} \tag{23.49}$$

Thus by knowing the mass and speed of the projectile, we can use Equation 23.46 to determine the maximum beam deformation and consequently, the maximum dynamic stress.

To put the dynamic effects into perspective, let δ_{st} be the equilibrium state deflection of a horizontal beam supporting an object with mass m as in Figure 23.15. Then from Equations 23.33 and 23.34, δ_{st} is

$$\delta_{st} = mg/k = mg\ell^3/48EI \tag{23.50}$$

Then by substituting into Equation 23.49, we have

$$\delta_{max} = V\sqrt{\delta_{st}/g} = V\delta_{st}/\sqrt{g\delta_{st}} \tag{23.51}$$

Finally, since the stress is proportional to the displacement, we have

$$\sigma_{dyn} = V\sigma_{st}/\sqrt{g\delta_{st}} \tag{23.52}$$

23.9 ILLUSTRATIVE DESIGN PROBLEMS AND SOLUTIONS

In this section, we briefly consider a few simple examples to illustrate the procedures of the foregoing sections.

23.9.1 CANTILEVER SUBJECTED TO FREE-END-SUDDEN LOADING

A 10 ft long steel cantilever beam has a cross section area I with a second moment of 600 in.[4]. Suppose the beam is designed to carry an end load of 500 lb in a static configuration with a stress factor of safety of 8. Suppose now that the end load is suddenly dropped onto the end of the beam. At what height h above the beam end can the load be dropped so that the stress factor of safety is at least 1.8?

SOLUTION

The end displacement δ_{st} is

$$\delta_{st} = \frac{W\ell^3}{3EI} = \frac{(500)(120)^3}{(3)(30)(10^6)(600)} = 0.016 \text{ in.} \tag{23.53}$$

For a factor of safety of 1.8 for the dynamic loading, and of 8 for the static loading, we have

$$\sigma_{dyn}/\sigma_{st} = 8/1.8 = 4.44 \tag{23.54}$$

From Equation 23.43, we have

$$\sigma_{dyn}/\sigma_{st} = 1 + (1 + 2h/\delta_{st})^{1/2} \tag{23.55}$$

Therefore we have

$$4.44 - 1 = (1 + 2h/0.016)^{1/2}$$

or

$$h = 0.0869 \text{ in.} = 2.2 \text{ mm} \tag{23.56}$$

23.9.2 VEHICLE–BARRIER IMPACT

Vehicle crashworthiness is a principal focus of safety engineers, and the principal measure of crashworthiness is a vehicle's response upon colliding with a "fixed barrier." In reality, the barriers are not rigid. Instead, they deform as any other structure. Thus to estimate the stress on a barrier, it is necessary to know the forces on the barrier during the impact and during the time of maximum deformation.

To explore this, consider an elementary model of a vehicle–barrier impact as in Figure 23.16, where the vehicle having mass m collides with the barrier with speed V. The barrier is modeled as a thick plate supported by a linear spring with modulus k. Assuming that the vehicle deforms plastically upon striking the barrier, develop an expression for the maximum spring force exerted upon the barrier.

SOLUTION

From the work–energy principle, the kinetic energy of the vehicle will be absorbed by the spring deformation and the vehicle deformation. If we assume that the energy absorbed by the vehicle deformation is small compared with the spring deformation energy, we will obtain an upper bound estimation of the spring force. The work–energy principle then leads to the expression

$$(1/2)k\delta_{max}^2 = (1/2)mV^2 \tag{23.57}$$

where δ_{max} is the maximum spring deformation.

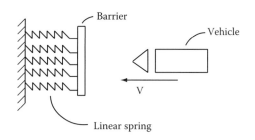

FIGURE 23.16   Vehicle-barrier impact model.

By solving Equation 23.57 for δ_{max}, we obtain

$$\delta_{max} = (\sqrt{m/k})V \tag{23.58}$$

The maximum force F_{max} is then

$$F_{max} = k\delta_{max} = (\sqrt{km})V \tag{23.59}$$

Observe in the result of Equation 23.59 that the force is reduced by decreasing the spring constant k. From Equation 23.58, however, such reduction leads to increased maximum displacement δ_{max} and thus greater "ride-down." But this defeats the purpose of the "fixed" barrier concept. A solution which reduces δ_{max} without increasing the force, is to use a massive thick plate for the barrier. We will explore this in Section 23.12.

23.10   ENERGY LOSS DURING IMPACT

It should be stated in all fairness that the stresses due to impact cannot be determined accurately. The materials involved are never perfectly elastic. Furthermore, when a body strikes another object, simultaneous contact is not realized at all points and the distribution of stresses and strains under impact loading is not the same as that under static loading, particularly at higher velocities of impact. Last but not the least, some kinetic energy of the moving body is dissipated during impact. This loss can be approximated with the aid of Table 23.2 for a number of elementary design cases [9]. The procedure here is to multiply the theoretically calculated energy by the dissipation factor C_e from Table 23.2.

23.11   IMPACT OF FALLING STRUCTURAL COMPONENTS

In Section 23.7, we considered the effect of an object falling onto a structure. Here we consider the reverse problem: a falling structural component. Such an event is usually unintended but

TABLE 23.2

Energy-Loss Factors for Impact Loading

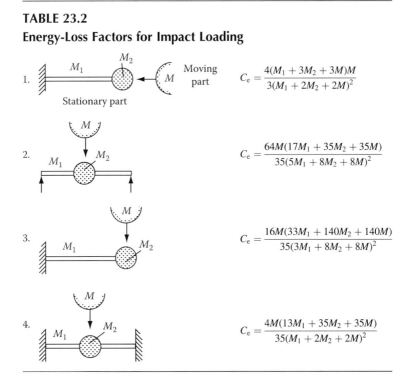

1. $C_e = \dfrac{4(M_1 + 3M_2 + 3M)M}{3(M_1 + 2M_2 + 2M)^2}$

2. $C_e = \dfrac{64M(17M_1 + 35M_2 + 35M)}{35(5M_1 + 8M_2 + 8M)^2}$

3. $C_e = \dfrac{16M(33M_1 + 140M_2 + 140M)}{35(3M_1 + 8M_2 + 8M)^2}$

4. $C_e = \dfrac{4M(13M_1 + 35M_2 + 35M)}{35(M_1 + 2M_2 + 2M)^2}$

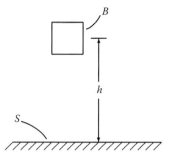

FIGURE 23.17   A falling structural component.

unfortunately, it is not uncommon. Moreover, the damage to the falling/impacting component is often unrepairable.

Companion type problems occur with moving structural components striking fixed surfaces or other structural components. Motor vehicle accidents are a principal source of such collisions. But they can also occur during machine failure with parts breaking off, or during environmental disturbances such as storms and earthquakes.

Knowing that unintended structural collisions can and will occur, the issue for designers is: how should a structural component be designed to minimize the damage from the underlying stress?

To answer this question, it is helpful to understand the mechanism of stress occurrence during impact, and to be able to estimate the magnitude of the stress enhancement. To this end, consider the common problem of a structural component falling onto a fixed surface as represented in Figure 23.17.

Specifically, consider a falling body B with mass m falling onto a fixed surface S from a height h above the surface. Upon impact, B will experience stresses arising from inertia forces due to the immediate deceleration of B. Although these forces are usually quite large, they act for only a relatively short time. Therefore, we can simplify the analysis by using the principles of impulse–momentum analysis [1].

It is well known that when a body falls, its speed upon impact with a fixed surface increases with the square root of the fall height. Specifically, upon falling through a distance h the speed V of a body B is [1]:

$$V = \sqrt{2gh} \tag{23.60}$$

where g is the gravity acceleration (32.2 ft/s^2 or 9.8 m/s^2).

When B impacts the fixed surface, it immediately begins to decelerate generating an inertia force proportional to its mass and the deceleration. That is, the force F on B is (Newton's second law)

$$F = ma \tag{23.61}$$

where
   m is the mass of B
   a is the deceleration

Let T be the time during which an impact force F acts. Then the impulse Imp of F is

$$\text{Imp} = \int_0^T F \, dt \tag{23.62}$$

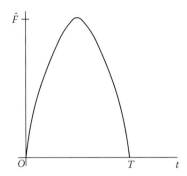

FIGURE 23.18   Sinusoidal forces.

Typically, F will have a triangular or sinusoidal form during time T as in Figure 23.18. By carrying out the indicated integration of Equation 23.62, we have

$$\text{Imp} = \int_0^T F \, dt = \int_0^T \hat{F} \sin\left(\pi t/T\right) dt = \hat{F}(2T/\pi) \tag{23.63}$$

where $\hat{F}$ is the peak value of F as in Figure 23.18.

The impulse–momentum principle [1] states that the impulse is equal to the change of momentum as

$$\text{Imp} = \Delta mV = m\Delta V \tag{23.64}$$

where ΔV is the sudden velocity change. Since B comes to a sudden stop upon striking the surface, we see from Equation 23.60 that ΔV is

$$\Delta V = \sqrt{2gh} \tag{23.65}$$

Therefore from Equations 23.63 and 23.64, we have

$$\hat{F}(2T/\pi) = m\Delta V = m\sqrt{2gh} \tag{23.66}$$

Consequently from Equation 23.61, the maximum or peak deceleration $\hat{a}$ experienced by B is

$$\hat{a} = \hat{F}/m = (\pi/2T)\sqrt{2gh} \tag{23.67}$$

In such analyses, it is helpful to express $\hat{a}$ in terms of multiples of the gravity of acceleration g. That is,

$$\hat{a}/g = (\pi/T)\sqrt{h/2g} \tag{23.68}$$

Equation 23.68 shows that the deceleration of B depends inversely upon the impact time T and upon the square root of the fall height h. Usually h will be expressed in feet, inches, or meters, and T will be expressed in seconds or milliseconds. When T is expressed in seconds, $\hat{a}/g$ is

$$\hat{a}/g = 0.391\sqrt{h}/T \quad (h \text{ in ft}) \tag{23.69}$$

$$\hat{a}/g = 0.113\sqrt{h}/T \quad (h \text{ in in.}) \tag{23.70}$$

$$\hat{a}/g = 0.709\sqrt{h}/T \quad (h \text{ in m}) \tag{23.71}$$

In Equations 23.69 through 23.71, h is likely to be known but the impact time T will probably be less apparent. For relatively hard surfaces, T is often approximated as $1/4f$, where f is the natural frequency of the falling body. We evaluate and estimate natural frequencies in Sections 23.15 and 23.16.

23.12 EXAMPLE—VEHICLE–BARRIER IMPACT

Consider again the vehicle–barrier impact problem and the example discussed in Section 23.9.2. Figure 23.16 (shown again in Figure 23.19) provides a model of the configuration: a vehicle with mass m and speed V collides with a barrier which is supported by a linear spring system as indicated in the figure. In Section 23.9.2, we neglected the mass of the barrier. Here, however, we let the barrier have mass M. The objective, as before, is to develop an expression for the maximum spring force exerted on the barrier.

SOLUTION

As before, we assume that upon impact the vehicle deforms plastically, and subsequent to impact the vehicle and barrier move together deforming the spring. The vehicle–barrier combination has a mass $M + m$. Let $\hat{V}$ be its speed. By the conservation of linear momentum principle [1], we see that $\hat{V}$ is given by

$$(M + m)\hat{V} = mV \quad \text{or} \quad \hat{V} = mV/(M + m) \qquad (23.72)$$

Just after impact the kinetic energy K of the vehicle–barrier combination is

$$K = (1/2)(M + m)\hat{V}^2 = (1/2)m^2 V^2/(M + m) \qquad (23.73)$$

Since the kinetic energy is dissipated by the spring upon maximum deformation δ_{max},

$$1/2 k\delta_{max}^2 = (1/2)m^2 V^2/(M + m) \qquad (23.74)$$

Thus δ_{max} is

$$\delta_{max} = mV/[k(M + m)]^{1/2} \qquad (23.75)$$

Hence the maximum spring force F_{max} is

$$F_{max} = k\delta_{max} = mV[k/(M + m)]^{1/2} \qquad (23.76)$$

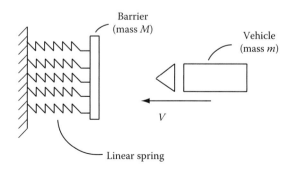

FIGURE 23.19   Vehicle–barrier impact model.

Observe in this result that if the barrier mass M is neglected F_{max} reduces to $\sqrt{km}V$, which is identical to the result of Equation 23.59. Observe further that if the barrier mass M is large the magnitude of F_{max} is reduced. Also if M is large, Equation 23.75 shows that δ_{max} is small thus reducing the "ride down," and the barrier is more like the desired "rigid barrier" intended for crash testing.

23.13   IMPACT MITIGATION

A review of the results of the examples of Sections 23.9.2 and 23.12 shows that the force on the barrier structure is reduced by reducing the stiffness k of the structure. This in turn increases the displacement and consequently, also the "ride-down" distance and time. Similarly, a review of Equations 23.69, 23.70, and 23.71 shows that the peak acceleration upon impact of a falling body, is reduced by increasing the impact time T.

The impact time and "ride-down" may be increased by padding or cushioning upon the impact surfaces as suggested by Figure 23.20. The cushioning medium may be a soft polymer as in an automobile dashboard, or a sand/gravel ramp as for runaway trucks on a steep downhill, or a thick floor carpet.

We illustrate the design concepts by continuing our study of bodies falling onto a surface. The illustrated procedure may also be used with colliding systems as in motor vehicle accidents.

The characteristics of cushioning media depend upon the material of the cushion but generally the cushion exerts a force on the impacting body, proportional to the deformation as represented in Figure 23.21. If k is the proportional coefficient, then the maximum force F_{max} exerted on the body will be proportional to the maximum deformation δ_{max} of the cushioning medium. That is,

$$F_{max} = k\delta_{max} \qquad (23.77)$$

Again, by using the work–energy principle we see that as the falling body or falling structure falls to its deepest penetration as in Figure 23.21, the net work is zero. That is, the kinetic energy K is zero both at the beginning of the fall and during the maximum penetration of the cushion. The work W done on the falling structure is due to (1) gravity and (2) the deforming cushion. That is,

$$\Delta K = 0 = W = mg(h + \delta_{max}) - (1/2)k\delta_{max}^2 \qquad (23.78)$$

By solving Equation 23.78 for $k\delta_{max}$, we obtain

$$k\delta_{max} = \frac{2mg(h + \delta_{max})}{\delta_{max}} = F_{max} = ma_{max} \qquad (23.79)$$

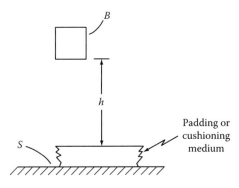

FIGURE 23.20   Cushioning medium for impact.

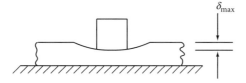

FIGURE 23.21   Deformed cushioning by an impacting body.

Then the maximum acceleration normalized by gravity is

$$(a_{max}/g) = 2(h + \delta_{max})/\delta_{max} \stackrel{D}{=} \bar{a}_{max} \qquad (23.80)$$

where $\bar{a}_{max}$ is defined by inspection.

Equation 23.80 may be used to obtain the maximum dynamic force F_{max} on the falling structure by simply multiplying the weight of the structure by $\bar{a}_{max}$. Observe that if h is zero, $\bar{a}_{max}$ is 2 as in Equation 23.23. Observe further if δ_{max} is large, the effect of the falling height h is diminished, thus demonstrating the effect of the cushioning.

Another way of thinking about cushioning is that the cushioning produces "ride-down time" and "ride-down distance," thus decreasing the forces.

23.14   DESIGN PROBLEM EXAMPLE

A section of steel casing with mean radius r, wall thickness t, and length ℓ falls from a height h into a thick layer of sand. The orientation of the casing is such that the fall is axial as represented in Figure 23.22. If the depth of penetration is δ, determine the approximate axial stress caused by the impact and penetration.

SOLUTION

The weight W of the casing is

$$W = 2\pi rt\ell\gamma \qquad (23.81)$$

where γ is the weight density. The inertia force F^* is then

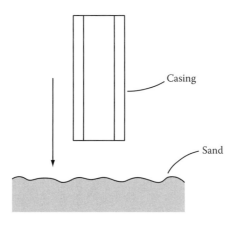

FIGURE 23.22   A casing falling into sand.

$$F^* = -(W/g)a \qquad (23.82)$$

and axial stress (compressive) is

$$\sigma_{dyn} = (2\pi r t \ell \gamma / 2\pi r t)(a/g)$$

or

$$\sigma_{dyn} = \ell \gamma (a/g) \qquad (23.83)$$

Finally, by substituting from Equation 23.80, we have

$$\sigma_{dyn} = 2\ell \gamma (h + \delta)/\delta \qquad (23.84)$$

Observe surprisingly that the stress is independent of the casing cross-section area.

23.15 NATURAL FREQUENCY OF SELECTED STRUCTURAL COMPONENTS

As noted earlier, the natural frequency of a structure or structural component is a good indicator of dynamic response. Thus knowledge of the natural frequency is an aid to design decision when dynamic loading is likely to occur.

The term "natural frequency" may be ambiguous as for structural components there are theoretically an infinite number of natural frequencies. To avoid confusion, we will use the term the "natural frequency" to refer to the lowest or "fundamental" frequency. When a structural component is vibrating at the fundamental frequency, the shape of the deformation (the "mode shape") is the same as the static deformation profile under gravity.

Since beams and plates are the most common structural components, it may be helpful to list the fundamental frequencies of a few common support conditions. Tables 23.3 through 23.6 provide such listings.

TABLE 23.3
Fundamental Natural Frequencies for Longitudinal (Axial) Beam Vibration

Configuration	Frequency
1. Fixed–free ends	$f = (1/4\ell)\sqrt{E/\rho}$
2. Free–free ends	$f = (1/2\ell)\sqrt{E/\rho}$
3. Fixed–fixed ends	$f = (1/2\ell)\sqrt{E/\rho}$

E, elastic modulus; ρ, mass density; ℓ, beam length.

TABLE 23.4
Fundamental Natural Frequency for Torsional Vibration
of Circular Shafts

Configuration	Frequency

1. Fixed–free ends

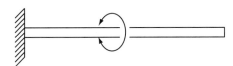

$$f = (1/4\ell)\sqrt{G/\rho}$$

2. Free–free ends

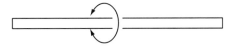

$$f = (1/2\ell)\sqrt{G/\rho}$$

3. Fixed–fixed ends

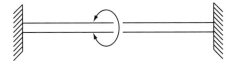

$$f = (1/2\ell)\sqrt{G/\rho}$$

G, shear modulus; ρ, mass density; ℓ, beam length.

TABLE 23.5
Fundamental Natural Frequency for Lateral (Flexural)
Beam Vibration

Configuration	Frequency

1. Simple-supports

$$f = (\pi/2\ell^2)\sqrt{EI/\rho A}$$

2. Fixed–fixed ends

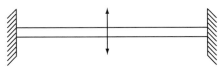

$$f = (3.565/\ell^2)\sqrt{EI/\rho A}$$

3. Fixed–free ends (cantilever)

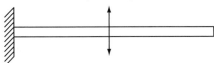

$$f = (0.559/\ell^2)\sqrt{EI/\rho A}$$

(*continued*)

TABLE 23.5 (continued)
Fundamental Natural Frequency for Lateral (Flexural)
Beam Vibration

Configuration	Frequency
4. Free–free ends 	$f = (3.565/\ell^2)\sqrt{EI/\rho A}$

E, elastic modulus; I, second moment of area of the cross-section; A, cross-section area; ρ, mass density; ℓ, beam length.

TABLE 23.6
Fundamental Natural Frequency for Circular
Plates

Configuration	Frequency
1. Simply supported edge	$f = (0.794/r^2)\sqrt{D/\rho}$
2. Fixed-edge	$f = (1.623/r^2)\sqrt{D/\rho}$
3. Free-edge	$f = (0.836/r^2)\sqrt{D/\rho}$

$D = Et^3/12(1 - \nu^2)$; E, elastic modulus; ν, Poisson ratio; t, plate thickness; r, plate radius; ρ, mass density.

23.16   ESTIMATING NATURAL FREQUENCY

Consider again a light rod supporting a weight W as in Sections 23.5 and 23.6, and as represented again with exaggerated displacement in Figure 23.23, where ℓ is the unextended length of the rod and δ_{st} is the static elongation due to the weight W. Then from elementary analysis δ_{st} is

$$\delta_{st} = W\ell/AE \tag{23.85}$$

where, as before,
   W is also the weight of W
   A is the rod cross-section area
   E is the elastic modulus

Alternatively, Equation 23.85 may be written as

$$W = k\delta_{st} \quad \text{where} \quad k = AE/\ell \tag{23.86}$$

More generally the light rod may be viewed as a linear spring. That is, if a force with magnitude F is applied to the end of the rod with the other end supported as in Figure 23.24, then F and the elongation δ of the rod are related as

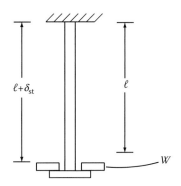

FIGURE 23.23   A rod supporting a weight W.

$$F = k\delta \tag{23.87}$$

Next, refer again to the rod supporting the weight W as in Figure 23.23. Let the weight oscillate vertically as represented in Figure 23.25, where y locates W relative to a reference level. If, as before, δ is the elongation or shortening of the rod, then y is simply

$$y = \ell + \delta \tag{23.88}$$

Consider a free-body diagram of the weight W as in Figure 23.26, where m is the mass of W ($m = W/g$) and $m\ddot{y}$ is the d'Alembert inertia force [1]. Thus by inspection of Figure 23.26, we have

$$m\ddot{y} + k\delta = W \tag{23.89}$$

or by using Equations 23.88 and 23.86

$$m\ddot{\delta} + k\delta = W = k\delta_{\mathrm{st}} \tag{23.90}$$

If in Equation 23.90, we introduce a new dependent variable η as

$$\eta \overset{\mathrm{D}}{=} \delta - \delta_{\mathrm{st}} \tag{23.91}$$

then Equation 23.90 becomes

$$m\ddot{\eta} + k\eta = 0 \tag{23.92}$$

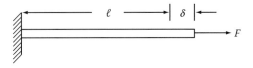

FIGURE 23.24   Rod with applied end loading.

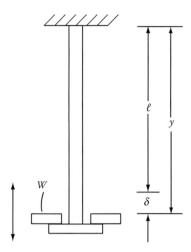

FIGURE 23.25   An oscillating end weight.

Equation 23.92 is of the same form as Equation 23.1. Thus W will oscillate in the same way as the mass–spring system of Figure 23.1 with the oscillation about the static equilibrium position. The frequency f of the oscillation is then

$$f = (1/2\pi)\sqrt{k/m} = (1/2\pi)\sqrt{AE/\ell m}$$ (23.93)

Consider again the configuration of the system where the weight is supported in its undeformed state as in Figure 23.6 and as shown again in Figure 23.27a. If, as before, the supports are removed, W will fall to its maximum low position as in Figure 23.27b. Then from the work–energy principle [1], the total work done on W between the positions of Figure 23.27 is zero (the kinetic energy is zero in both positions). That is,

$$mg\delta_{max} - (1/2)k\delta_{max}^2 = 0$$ (23.94)

or

$$k\delta_{max} = 2mg = 2W$$ (23.95)

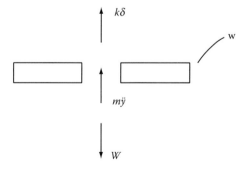

FIGURE 23.26   Free-body diagram of oscillating weight W.

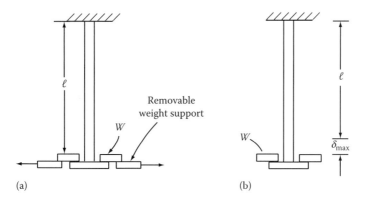

FIGURE 23.27 Undeformed and maxially deformed rod. (a) Undeformed rod with pending weight W. (b) Maximum elongation.

Then from Equation 23.86 δ_{max} is

$$\delta_{max} = 2\delta_{st} \tag{23.96}$$

Thus when the weight W is released it falls through twice the static deformation. Hence, by substituting from Equation 23.94 into Equation 23.95, we have

$$mg\delta_{st} = k\delta_{st}^2 \quad \text{or} \quad k/m = g\delta_{st}/\delta_{st}^2 \tag{23.97}$$

Observe that the behavior of this rod/weight system is the same as the classical spring–mass system of Figure 23.28. Interestingly, the same is the case for a beam oscillating in its fundamental mode as represented for a simple support beam in Figure 23.29. For the oscillating beam in the fundamental mode, the shape of the deformation is the same as the statically deformed beam due to its weight. The behavior is the same for a suddenly applied uniform load. That is, if the beam is held in its undeformed position and then suddenly released in the gravity field, the beam will fall to a maximum displacement which is twice as large as the static equilibrium displacement. The beam will then oscillate about the static equilibrium configuration.

In each of these three cases described above: (1) the vertical rod with the weight; (2) the mass–spring system; and (3) the simply supported horizontal beam, when the systems are at rest in their

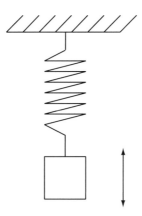

FIGURE 23.28 Spring–mass system.

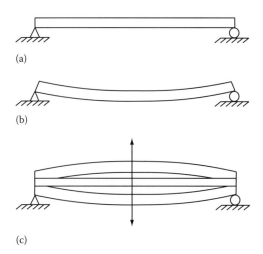

FIGURE 23.29 Simply supported, deformed, and oscillating beam. (a) Undeformed beam. (b) Static deformation from gravity. (c) Oscillating beam (fundamental mode).

uppermost (undeformed) configurations, their kinetic energies are zero. Their potential energies in their uppermost configurations, with the static equilibrium configuration as a reference, are at a maximum. Also, the deformation energies (the "spring energy") in these uppermost undeformed configurations are zero.

    a. Undeformed beam
    b. Static deformation from gravity
    c. Oscillating beam (fundamental mode)

When these systems are released and drop to their lowermost configurations, their kinetic energies are each also zero, their potential energies, due to gravity, are at a minimum, and their deformation energies (spring energies) are at a maximum. The deformation energy at its maximum is twice the potential energy of the equilibrium state. During the oscillation, there is a continuing transfer of energy between the kinetic energy, the potential energy (due to gravity), and the deformation energy. At all times, the sum of the kinetic energy, the potential energy, and the deformation energy is constant.

When the behavior of the systems are viewed from the perspective of the work–energy principle, we see that with the kinetic energies being zero at both the uppermost and lowermost position, the net work done on the systems between these positions is zero. That is, in the movement from the uppermost position to the lowermost position the net work is zero, so that the work by gravity is negative the work due to deformation. For the vertical light rod with the end weight of Figure 23.25 and for the spring–mass system of Figure 23.28, this zero network leads to the expression (see Equations 23.21 and 23.97):

$$\text{Work} = 0 = mg\delta_{\text{max}} - (1/2)k\delta^2_{\text{max}} \tag{23.98}$$

or

$$mg\delta_{\text{st}} = k\delta^2_{\text{st}} \tag{23.99}$$

For the oscillating beam of Figure 23.29, the computation of zero work between the uppermost and lowermost positions leads to the approximate analog of Equation 23.99:

$$mg \int_0^\ell y\, dx = k \int_0^\ell y^2 dx \qquad (23.100)$$

where
  m is the beam mass
  k is the stiffness as in Equation 23.46

The stiffness/mass ratio for the beam is then

$$k/m = \left(g \int_0^\ell y\, dx \right) \Big/ \left(\int_0^\ell y^2 dx \right) = \omega^2 \qquad (23.101)$$

where as before omega is the circular frequency and the last equality follows from Equation 23.3.

Equation 23.101 is a useful expression for estimating the fundamental natural frequency of a beam. By generalization, it may be applied with more complex structures.

Equation 23.91 may be interpreted as follows. The circular frequency squared is well established as the stiffness/mass ratio. This ratio in turn is obtained through a balance between the work of gravity (mass effect) and the work due to deformation (stiffness effect). Next observe that for the beam, viewed as a continuum, the works done by gravity and by deformation are obtained by integration along the beam. Such integration must occur before the ratio is computed. For a concentrated mass as with the light-rod/weight system of Figure 23.23, and as with the spring–mass system of Figure 23.28, we may simply use Equation 23.97 to calculate the stiffness/mass ratio as g/δ_{st}. But for the continuum of the beam, the division k/m for individual particles, or elements, of the beam is premature and inaccurate since those particles (or elements) are not isolated but instead are connected and thus their movements are affected by one another.

To illustrate the use of Equation 23.101, consider estimating the fundamental frequency of a cantilever beam as in Figure 23.30. From Equation 11.8, we see that the downward displacement y of a cantilever beam due to its own weight is

$$y = (w/24EI)(x^4 - 4\ell x^3 + 6\ell^2 x^2) \qquad (23.102)$$

where w is the beam weight per unit length.

By substituting from Equation 23.102 into Equation 23.101 we have the expressions:

$$\int_0^\ell y\, dx = w\ell^5/20EI \quad \text{and} \quad \int_0^\ell y^2 dx = (13/3240)(w^2\ell^9/E^2I^2) \qquad (23.103)$$

FIGURE 23.30 Vibrating cantilever beam.

so that the circular frequency squared (ω^2) is

$$\omega^2 = \left(g \int_0^\ell y^2 dx\right) \Big/ \left(\int_0^\ell y^2 dx\right) = (162/13)gEI/w\ell^4) \tag{23.104}$$

Then the circular frequency and fundamental natural frequency f are

$$\omega = 3.53\sqrt{(gEI/w\ell^4)} \quad \text{and} \quad f = \omega/2\pi = 0.561\sqrt{gEI/w\ell^4} \tag{23.105}$$

(From Table 23.5, the theoretical value of f is $0.559\sqrt{gEI/w\ell^4}$.)

SYMBOLS

A	Cross-section area
A	Acceleration
A_1, A_2	Integration constants
B	Body
C	Amplitude
D	$Et^3/12(1-\nu^2)$
E	Modulus of elasticity
F	Force
f	Frequency (see Equation 23.10)
G	Shear modulus
g	Gravity acceleration
h	Displacement
I	Second moment of area
Imp	Impulse
K	Kinetic energy
k	Spring constant
l	Length
M	Mass
m	Mass
P	Concentrated load
r	Plate radius
S	Surface
T	Period (see Equation 23.11)
t	Time; thickness
U	Potential energy; strain energy
V	Speed
W	Weight; work
w	Weight per unit length
W_{beam}	Work done by a beam
W_{gravity}	Work done by gravity
$_1W_2$	Work done from position 1 to position 2
x	Coordinate; displacement
y	Coordinate
γ	Strain energy density; weight density
ΔV	Velocity change
δ	Initial displacement

$\dot{\delta}$	Initial speed
δ_{eq}	Displacement of equilibrium
δ_{max}	Maximum displacement
δ_{st}	Static displacement
ε	Strain
ν	Poisson's ratio
ρ	Mass density
σ	Stress
σ_{dyn}	Dynamic stress
σ_{st}	Static stress
ϕ	Phase angle
ω	Circular frequency (see Equation 23.3)

REFERENCES

1. H. Josephs and R. L. Huston, *Dynamics of Mechanical Systems*, CRC Press, Boca Raton, FL, 2002 (Chaps. 9, 11, and 13).
2. American Society of Testing and Materials (ASTM), *Symposium on Speed of Testing of Nonmetallic Materials*, ASTM Special Technical Publication No. 185, Philadelphia, PA 1956.
3. J. S. Rinehart and J. Pearson, *Behavior of Metals Under Impulsive Loads*, American Society of Metals (ASM), Cincinnati, OH, 1949.
4. P. G. Skewmon and V. F. Zackay, *Response of Metals to High Velocity Deformation*, Metallurgical Society Conference, Vol. 9, Interscience, New York, 1961.
5. M. Kornhauser, *Structural Effects of Impact*, Spartan Books, New York, 1964.
6. F. P. Beer and E. R. Johnston, Jr., *Mechanics of Materials*, 2nd ed., McGraw Hill, New York, 1992, pp. 575–582.
7. I. S. Sokolnikoff, *Mathematical Theory of Elasticity*, Robert E. Krieger Publishing, Malabar, FL, 1983, pp. 83–86.
8. S. Timoshenko and J. N. Goodier, *Theory of Elasticity*, McGraw Hill, New York, 1951 (Chapter 6).
9. W. C. Young and R. G. Budynas, *Roark's Formulas for Stress & Strain*, 6th ed., McGraw Hill, New York, 1989, pp. 765–768.

24 Elements of Seismic Design

24.1 INTRODUCTION

Earthquakes have stimulated considerable research in the development of shock resistance structures, the so-called "seismic design." Many papers and books document the accomplishments. Seismic design is a rather specialized field and is of considerable importance due to the devastating effect of earthquakes upon the societies and the economies of the regions where they occur.

In this chapter, we present in brief a review of the fundamentals of seismic design. References [1–8] provide the basis for our review. They also provide a starting point for a more in-depth study.

24.2 EARTHQUAKE DESIGN PHILOSOPHIES

In general, there appear to be two schools of thought among engineers concerned with the theory of earthquake-resistant design. One thesis is that a building should be perfectly rigid, so that in the event of seismic motion, the top and bottom positions of this building would move an identical amount during the same time interval. Such a response would, of course, tend to induce rather large stresses in the building structure. The other thesis also goes to an extreme by maintaining that a building should be as flexible as possible in order to literally sway during an earthquake. Then, in an extreme case, one would have to construct a building from vulcanized rubber, which would certainly be resistant to seismic shock. However, such a building would not protect the contents very well because of the possibility of large displacements.

In practice, builders are limited by the availability and cost of construction materials, design codes, and soil conditions that neither of the extreme design criteria outlined above are actually utilized. Good earthquake-resistant design is a compromise in which both stresses and deflections should be evaluated. Furthermore, the criterion of flexibility should not imply flimsy construction. Some designers also suggest that a reasonable compromise could be reached if we were to construct rigid buildings on soft ground and flexible buildings on rock.

Over the years, a rule of thumb developed, which states that a well-designed earthquake-resistant building should be able, at any level, to withstand a horizontal force equal to one-tenth of its weight above that level. However, to the surprise of all concerned, more recent seismological data indicated a maximum acceleration at times exceeding three-tenths of gravity. It also became clear that the damage was not always proportional to the maximum acceleration, and that a more detailed knowledge of the vibrational modes of a structure was important in determining a realistic seismic response.

The general problem of earthquake design has not yet been solved to the satisfaction of seismologists, geologists, and engineers. The principal reason is that it is difficult to predict the character and intensity of an expected earthquake for design purposes. Hence, calculations are necessarily based upon rather crude approximations. These philosophies and insights have led to the development of various design methods discussed in the following sections.

24.3 BUILDING CODE METHOD

A particular difficulty in earthquake design is in predicting structural response to irregular ground motion. Theoretical analyses are difficult and generally impractical. An alternative approach is to

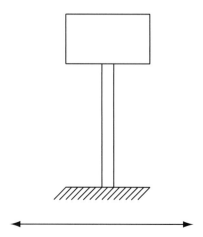

FIGURE 24.1 A simplified seismic structural model.

use regulatory codes based upon past observations of structural behavior in earthquakes. This experience is coupled with some established concepts of dynamic response and vibratory motion using the assumption that only the first vibration mode is important. Thus, the characteristic shape of the deflection curve is taken to be that of the first fundamental vibration mode of a given structure. Equivalent static forces and displacements may then be calculated.

To illustrate the method, consider the simple structural model consisting of a concentrated mass atop a light flexible column as in Figure 24.1. From a commonly accepted California code, the following formulas are used [1]:

$$Q_b = WCK_d \tag{24.1}$$

$$C = 0.05/T^{1/3} \tag{24.2}$$

and

$$\delta_{max} = 0.49/f^{5/3} \tag{24.3}$$

where
   Q_b is the horizontal shear load at the base of the model
   W is the total weight of the structure
   C is an inertia force multiplier, as given by Equation 24.2
   K_d has values between 0.7 and 1.5, with lower values for ductile (or pliable) materials and
      higher values for brittle materials (such as concrete)
   T is the fundamental period of vibration of the structure (in s)
   f is the natural frequency of the fundamental mode (in Hz)
   δ_{max} is the maximum horizontal displacement of the structure

Figure 24.2 illustrates these parameters.

24.4 SPECTRAL VELOCITY METHOD

In this method, the structure is assumed to be flexible and to respond in a fundamental vibration model. Then by having seismic data for a geometric region of interest, we can predict the probable

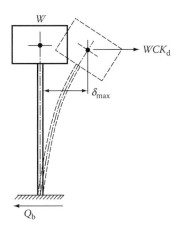

FIGURE 24.2   Seismic model displacement.

acceleration levels (in multiples of gravity) which will impact a structure. This information in turn can be used to calculate stresses. Analogous to the building code method, we can use the concept of an equivalent static force to calculate the stresses. Before illustrating the procedure in more detail, it is helpful to first review the concept of the Richter scale and how it can characterize a seismic event.

24.5   RICHTER SCALE

In most news reports of earthquakes, the Richter scale is mentioned as a measure of the intensity of an earthquake. While the concept of a Richter measure is quite familiar, many readers may not be familiar with its definition. Essentially, the measure of the Richter scale is a correlation between seismic ground motion and the possible energy of the source causing the motion. Thus, the Richter scale represents an attempt to describe the strength of an earthquake by allotting a number Q to it, on a scale from 1 to 10. In mathematical terms, Q is defined as the logarithm (to the base of 10) of the maximum amplitude measured in microns (mm $\times 10^{-3}$) and traced on a standard seismograph at a distance of 100 km from the epicenter. Here, the epicenter is defined as the point on the earth's surface directly above the focus of the earthquake. Recent determinations also indicate that a 1-unit increase in Richter scale Q is equivalent to a 32-fold increase in the energy of the earthquake source. However, the exact calculations of the absolute amount of energy involved remain rather uncertain and can vary by as much as a factor of 10. Practical experience also shows that based on this scale earthquakes of magnitude 5 or greater are potentially destructive to buildings.

Richter scale is useful in classifying the extent of seismic disturbances. However, it should be realized that Q is a magnitude derived from the response of a seismic instrument and, as such, it must be influenced by the sensitivity of the available instrument. Furthermore, because of the nonuniformity of the earth's crust and random orientations of the geological faults, Q cannot be a precise measure of the energy released by an earthquake.

24.6   ILLUSTRATION—SPECTRAL VELOCITY METHOD

For an illustrative procedure, we use spectral velocity data from California for earthquakes near San Fernando, Parkview, and El Centro. These cases have been selected due to the superior quality of the recorded data. The 1940 El Centro movement is indicative of an acceleration of 0.33g, which was one of the larger values known at that times [8].

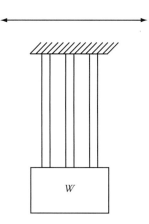

FIGURE 24.3 Structural model.

We illustrate the procedure by considering a structural model consisting of a mass with weight W suspended by a series of light rods as represented in Figure 24.3.

1. We select two structural modes: a single and double cantilever, as in Figure 24.4: (a) single cantilever mode and (b) double cantilever mode.
2. Let the spring constants for the two modes be evaluated separately and have values (a) 60,000 lb/in. and (b) 6600 lb/in.
3. Let the seismic loading for the single and double modes be as represented in Figures 24.5 and 24.6, where n_d is the damping factor.
4. For single degree of freedom spring mass models, the static displacement δ is

$$\delta = W/k \qquad (24.4)$$

where k is the spring constant. For the two extreme modes of movement for an assumed weight W of 7200 lb, we have

$$\delta_{single} = 0.12 \text{ in.} \quad \text{and} \quad \delta_{double} = 1.09 \text{ in.} \qquad (24.5)$$

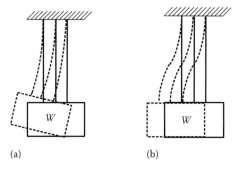

(a)   (b)

FIGURE 24.4 Structural movement modes. (a) Single movement modes. (b) Double cantilever mode.

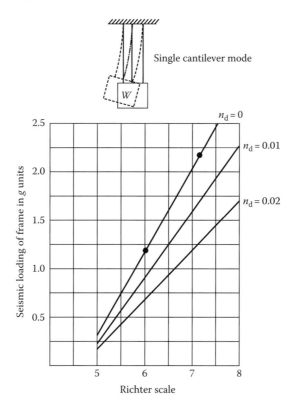

FIGURE 24.5   Seismic loading on the support in the single cantilever mode.

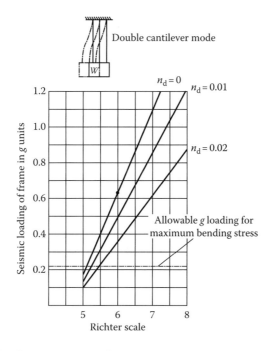

FIGURE 24.6   Seismic loading on the support in the double cantilever mode.

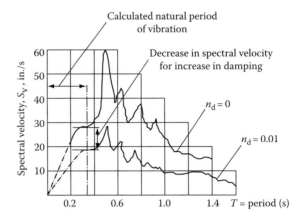

FIGURE 24.7   Spectral velocity chart (El Centro data).

5. The fundamental period T of the movement is (see Equation 23.2.11):

$$T = 2\pi\sqrt{m/k} = 2\pi\sqrt{w/gk} = 2\pi\sqrt{\delta/g} \tag{24.6}$$

where as before m is the mass of the structural model (w/g). We then have

$$T_{\text{single}} = 0.11 \text{ s} \quad \text{and} \quad T_{\text{double}} = 0.33 \text{ s} \tag{24.7}$$

6. We can use a spectral velocity chart such as that of Figure 24.7 to obtain spectral velocities. Specifically for the periods of Equation 24.7 we obtain

$$S_v = 15 \text{ in./s (single)} \quad \text{and} \quad S_v = 25 \text{ in./s (double)} \tag{24.8}$$

Comment: The spectral velocity chart of Figure 24.7 is a rough sketch of the data given here simply for our illustration. In actual design, computations analysts should use more precise seismological data [8].

7. We can now proceed to compute the seismic loading on the models as follows. Let a be an acceleration due to a seismic event evaluated as

$$a = \omega S_v \tag{24.9}$$

where as before ω is the circular frequency given by

$$\omega = 2\pi/T = 2\pi f \tag{24.10}$$

Then by substituting for T from Equation 24.6 we have from Equation 24.9:

$$a = 2\pi S_v/T = S_v\sqrt{g/\delta} \tag{24.11}$$

In g units (multiples of gravity acceleration), the acceleration $\bar{a}$ is

$$\bar{a} = a/g = S_v/\sqrt{g\delta} \tag{24.12}$$

Finally, by substituting the results of Equations 24.5 and 24.6 for δ and S_V, we obtain the seismic accelerations as

$$\bar{a}_{single} = 2.2 \quad \text{and} \quad \bar{a}_{double} = 1.22 \tag{24.13}$$

or

$$a_{single} = 2.2g \quad \text{and} \quad a_{double} = 1.22g \tag{24.14}$$

These results are based upon the El Centro spectrum where the maximum ground acceleration is known to be $0.33g$. Assuming that the El Centro earthquake measured about 7.2 on the Richter scale, we see from Figures 24.5 and 24.6 that the seismic results of Equation 24.14 correspond to the points at 7.2 on the zero damping ($n_d = 0$) lines. Thus, by linear extrapolation, we have the entire characteristics of structural loading as a function of the Richter scale measure.

8. To further illustrate the application, consider the graph of Figure 24.8 where the design curves are marked by solid curves. Based upon the San Fernando and Parkville experiences, we can extrapolate the design curves with dashed curves.

   The structural seismic loadings may now be developed using Figure 24.5 as follows. Suppose that the distance to the fault is about 25 miles. For single damping in a single-cantilever mode, the level of $2.2g$ (Figure 24.5) corresponds to seismic loading on the structure for a $0.33g$ ground acceleration. For an intermediate ground acceleration such as $0.17g$ at the 25-mile distance, and a Richter scale of 6, the structural loading is $(0.17/0.33)$ (2.2) or $1.13g$. (These illustrations are indicated by small circles in Figure 24.5.)

9. The allowable g loading on a structure may be determined on the basis of either strength or stability. In the illustration, the maximum bending stress in a double-cantilever mode was used as a criterion for the design chart of Figure 24.6. The dashed line corresponds to the maximum g loading on the structure.

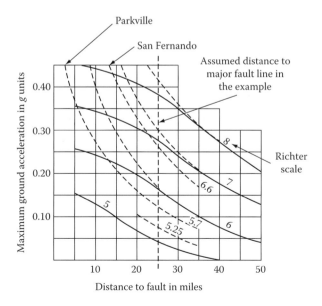

FIGURE 24.8   Maximum ground acceleration as a function of distance to the fault.

TABLE 24.1

Representative Damping Values for Various Types of Structural Components and Induced Stresses

	No More Than One-Half Stress (%)	At or Just Below Yield (%)
Piping	1–2	2–3
Welded steel	2–3	5–7
Prestressed concrete	2–3	5–7[a]
Well-reinforced concrete	2–3[b]	7–10
Bolted or riveted steel	5–7	10–15
Bolted wood	5–7	10–15
Nailed wood	5–7	15–20

[a] Without complete loss of prestress.
[b] Only slight cracking.

10. Interpretation of the design charts may be conducted as follows: Consider, for example, the critical double-cantilever mode shown in Figure 24.6. [The point of intersection between the dashed line and the line denoted by $n_d = 0.01$ yields the maximum allowable earthquake intensity of 5.25 on the Richter scale, assuming the El Centro type of spectral response and about 25 miles distance to the major fault line.] Interpolating in Figure 24.8, we get a corresponding maximum ground acceleration of about $0.075g$.

The procedure described above contains only the elementary concepts of stress analysis. Yet, from a practical point of view, this approach will be found useful because it reduces the nonsteady response of a structure to a statics problem, which can then be handled with ease. Because of the assumptions involving linear elasticity, the results obtained should be sufficiently conservative for most practical needs.

When the seismic input is such that the engineering structure is plastically deformed, it is not particularly appropriate to determine a design stress, because it will always be at the yield level. In such a case the maximum displacement rather than the maximum stress should be taken as a criterion. The practice also indicates that the potential for producing a failure in a ductile structure is relatively low for the majority of seismic spectra available.

24.7 STRUCTURAL DAMPING

Even a small amount of damping can significantly decrease the dynamic response of a structure. For example, as indicated in Figure 24.7, the peak response for a damping factor $n_d = 0.01$ is approximately half that for zero damping. This attenuation exists in all structures and its magnitude varies with the material used and the type of construction. It also depends upon how close to yield are the induced stresses. Table 24.1 provides some representative damping values for various structural components and induced stresses [9,10].

SYMBOLS

a	Acceleration
$\bar{a}$	Normalized acceleration (acceleration expressed in multiples of g)
C	Inertia force multiplier
f	Natural frequency

g	Gravity acceleration
K_d	Ductile to brittle multiplier (0.7–1.5) (see Equation 24.1)
k	Spring constant
m	Mass
n_d	Damping factor
Q	Richter scale number
Q_b	Horizontal shear load
S_v	Spectral velocity
T	Fundamental vibration period
W	Total weight of a structure
δ	Displacement
δ_{double}	Double cantilever displacement
δ_{max}	Maximum displacement
δ_{single}	Single cantilever displacement
ω	Circular frequency

REFERENCES

1. J. M. Biggs, *Introduction to Structural Dynamics*, McGraw Hill, New York, 1964.
2. C. E. Crede, *Vibration and Shock Isolation*, John Wiley & Sons, New York, 1957.
3. R. L. Wiegel, *Earthquake Engineering*, Prentice Hall, Englewood Cliffs, NJ, 1970.
4. N. M. Newmark and E. Rosenblueth, *Fundamentals of Earthquake Engineering*, Prentice Hall, Englewood Cliffs, NJ, 1971.
5. M. A. Biot, Analytical and experimental methods in engineering seismology, *Transactions of the American Society of Civil Engineering*, 42, 1942, pp. 49–69.
6. A. W. Anderson et al., Lateral forces of earthquake and wind, *Transactions of the American Society of Civil Engineers*, 78, 1952, D-66.
7. J. A. Blume, N. M. Newmark, and L. H. Corning, *Design of Multistory Reinforced Concrete buildings for Earthquake Motions*, Portland Cement Association, Chicago, IL, 1961.
8. G. W. Housner, Intensity of earthquake ground shaking near the causative fault, *Proceedings of the Third World Conference of Earthquake Engineering*, Vol. 1, 1965, pp. III-94–III-115.
9. N. M. Newmark and W. J. Hall, Dynamic behavior of reinforced and pre-stressed concrete buildings under horizontal loads and the design of joints (including wind, earthquake, blast effects), *Proceedings, Eighth Congress*, International Association for Bridge and Structural Engineering, New York, 1968.
10. N. M. Newmark and W. J. Hall, Seismic design criteria for nuclear reactor facilities, *Proceedings, Fourth World Conference of Earthquake Engineering*, Santiago, Chile, 1969.

25 Impact Stress Propagation

25.1 INTRODUCTION

Stress propagation is becoming increasingly important in structural design due to increasing occurrences of high-speed impacts with structures. Stress propagation phenomenon has been studied for many years, but due to mathematical complexities, the phenomenon is still not completely understood nor readily accessible to engineers and designers.

In this chapter, we review some elementary concepts of stress propagation as well as a few simple relations and formulas that may be of use. We focus upon simple structures and events where time of load application is less than half the fundamental natural period. In these instances, the propagation phenomenon is greatly dependent upon the properties of the structural materials.

A starting point for stress propagation analyses is often the one-dimensional wave equation describing the displacement of a taut, vibrating string [2–5]:

$$a^2 \frac{\partial^2 u}{\partial x^2} = \frac{\partial^2 u}{\partial t^2} \qquad (25.1)$$

where
  u is the transverse displacement
  x is the geometric position along the string
  t is time
  a^2 is a physical constant (string tension divided by mass density per unit length)

In the solution of this equation, a is identified with the speed of wave propagation along the string.

25.2 A SIMPLE CONCEPTUAL ILLUSTRATION

The elementary illustration of wave propagation phenomenon can be based on the analogy of a locomotive starting to pull a long string of stationary freight cars, or a similar train running into a barrier of fixed buffers. In the first case, each car starts up the one behind it, while the last car is, so to speak, "unaware" of the load applied to the front sections of the train. This analogy applies also to a rod to which an axial force is suddenly applied, sending a tensile stress wave along its axis. Any section of the rod, other than that experiencing a wave propagation phenomenon, remains unstressed. By analogy to a freight train, the particles of the rod at the impacted end are displaced and create a wave, which begins to travel from one end of the rod to the other.

By reference again to the train analogy, we note that each car resists the motion by the inertia of the car behind it, except the last car in the string, which will run after the train faster than the cars in front of it, initiating a compressive type of wave. This wave is expected to proceed until the front of the train is reached. On the other hand, when a freight train runs into a barrier of fixed buffers, each car is brought to rest in turn. However, the last car rebounds and initiates a tensile wave, which begins to travel up to the locomotive.

The foregoing simplified illustration of mechanics of wave propagation leads to the following basic conclusions:

1. A compression wave reaching a free end transforms into a tensile wave and vice versa
2. A wave is reflected from a fixed end without a stress reversal

25.3 STRESS PROPAGATION THEORY

In a one-dimensional impact (such as a rod struck on its end), the stress in the rod begins at the end and propagates along the rod. Viewed as a whole, the stress (in the rod) will be a function of both, the position (x along the rod) and time t. That is,

$$\sigma = \sigma(x, t) \tag{25.2}$$

Wasley [1] shows that the stress may be expressed as

$$\sigma = \rho c V(x, t) \tag{25.3}$$

where
ρ is the mass density
V is the speed of a particle at a point along the rod
c is the material propagation speed, known as "sonic speed," given by

$$c = \sqrt{E/\rho} \tag{25.4}$$

where, as before, E is the modulus of elasticity.

If the stress becomes sufficiently large, the material may experience plastic deformation. With plastic deformation, the wave propagation speed C_p is smaller, or slower, than with elastic deformation. Analogous to Equation 25.3, for plastic deformation, the stress may be expressed as

$$\sigma = \rho C_p V(x, t) \tag{25.5}$$

where C_p is

$$C_p = [(d\sigma/d\varepsilon)/\rho]^{1/2} \tag{25.6}$$

where $d\sigma/d\varepsilon$ is the slope of the stress–strain curve in the elastic region with ε being the strain (see Figure 25.1).

Observe that when $d\sigma/d\varepsilon$ approaches zero, as the stress–strain curve becomes horizontal, the stress is near the ultimate strength of the material. At this point, the stress propagation speed approaches zero and the material begins to rapidly fail.

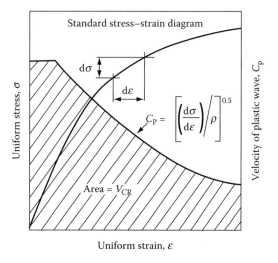

FIGURE 25.1 Graphical interpretation of critical impact velocity.

If we imagine a rod being struck axially on an end, the end speed at impact, which will produce the ultimate stress and material failure, is called the "critical velocity," V_{cr}. We can estimate the critical velocity by using the work–energy principle [6] as follows. Let σ_{ult} be the ultimate strength of the material. Then the work done on the bar, per unit volume, by the impact force through the strain at failure ε_f is approximately

$$\text{Work} \approx (1/2)\int_0^{\varepsilon_f} \sigma_{ult}d\varepsilon = (1/2)\int_0^{\varepsilon_f} \rho C_p V_{cr}d\varepsilon = (1/2)\rho\left(\int_0^{\varepsilon_f} C_p d\varepsilon\right)V_{cr} \tag{25.7}$$

where we have used Equation 25.5. (Recall that in the elastic region, the work done by a force P in elongating a rod by an amount δ is $(1/2)P\delta^2$.)

The kinetic energy per unit volume of a rod struck with the critical velocity is approximately

$$KE \approx (1/2)\rho V_{cr}^2 \tag{25.8}$$

By equating the work and kinetic energy of Equations 25.7 and 25.8 we have

$$(1/2)\rho\left(\int_0^{\varepsilon_f} C_p d\varepsilon\right)V_{cr} = (1/2)\rho V_{cr}^2$$

or

$$V_{cr} = \int_0^{\varepsilon_f} C_p d\varepsilon = \int_0^{\varepsilon_f} [(d\sigma/d\varepsilon)/\rho]^{1/2}d\varepsilon \tag{25.9}$$

The critical velocity may be interpreted as the area under the C_p curve as in Figure 25.1.

25.4 ELASTIC IMPACT

Consider further the phenomena of elastic impact. To develop the concepts, consider a rod with uniform cross section being subjected to a tensile impact as represented in Figure 25.2. Let the left end of the rod suddenly have a speed V_O to the left as represented in the figure. Then as the rod is strained, a tensile stress wave will move along the rod as in the figure. The stress wave moves with acoustic speed C.

Let t be the time it takes for the wave front to move a distance L. Then L is simply

$$L = Ct \tag{25.10}$$

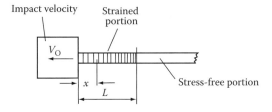

FIGURE 25.2 Model of a uniform rod under tensile impact.

Let δ be the displacement, or elongation, of the left end of the rod. Then in time t, δ is

$$\delta = \varepsilon L = V_O t \tag{25.11}$$

where as before, ε is the strain in the rod. Therefore, from Hooke's law (see Equation 5.1.3), the stress σ is

$$\sigma = \varepsilon E = E V_O t / L = E V_O / C \tag{25.12}$$

where the final equality follows from Equation 25.10.

If the acoustic speed C is $\sqrt{E/\rho}$ as in Equation 25.4 then the elastic modulus E is

$$E = \rho C^2 \tag{25.13}$$

Consequently from Equation 25.12 the stress σ is

$$\sigma = \rho C V_O \tag{25.14}$$

Observe that Equation 25.12 is the same as Equation 25.3.

25.5 ACOUSTIC (SONIC) SPEED AND CRITICAL SPEED

In Equation 25.14, we see that the stress is proportional to the impact speed V_O. Thus, for very large impact speeds, the stress could well exceed the yield stress of the material. As noted in Section 25.3, the speed at which this occurs is called the "critical speed." Equation 25.14 also shows that the stress depends upon the physical properties of the material and specifically upon the sonic speed C. The sonic speed depends upon the geometry of the structure. Table 25.1 provides a listing of sonic speed formulas for a number of common structural components.

As an illustration of the source of the formulas of Table 25.1, consider the last entry, for an elastic continuum. Recall from Equation 7.67 the equilibrium equations for an elastic continuum may be written in index notation as

$$\sigma_{ij,j} + \rho a_i = 0 \tag{25.15}$$

where a_i is the acceleration of a point within an element of the continuum.

From Equation 7.67 the stress–strain equations may be written as

$$\sigma_{ij} = \lambda \delta_{ij} \varepsilon_{kk} + 2G \varepsilon_{ij} \tag{25.16}$$

TABLE 25.1

Formulas for Sonic Velocity

Uniform bar of infinite length	$C = \left(\dfrac{E}{\rho}\right)^{1/2}$
Infinite slab or plate	$C = \left[\dfrac{E}{(1 - \nu^2)\rho}\right]^{1/2}$
Cylinder	$C = \left[\dfrac{E}{(1 - \nu^2)\rho}\right]^{1/2}$
Elastic continuum	$C = \left[\dfrac{E(1 - \nu)}{(1 + \nu)(1 - 2\nu)\rho}\right]^{1/2}$

---

where as before

ε_{ij} are the elements of the strain tensor
δ_{ij} is Kronecker's delta symbol
λ and G are Lame coefficients given by (see Equations 7.68 and 7.48)

$$\lambda = \frac{E\nu}{(1 + \nu)(1 - 2\nu)} \quad \text{and} \quad G = \frac{E}{2(1 + \nu)} \tag{25.17}$$

where ν is Poisson's ratio. G will be recognized as the shear modulus.

From Equation 5.20, the strain–displacement equations are

$$\varepsilon_{ij} = (1/2)(u_{i,j} + u_{j,i}) \tag{25.18}$$

By substituting from Equations 25.16, 25.17, and 25.18 into Equation 25.15 we obtain

$$Gu_{i,jj} + (\lambda + G)u_{k,k_i} + \rho\ddot{u}_{i,i} \tag{25.19}$$

where we have replaced the acceleration a_i by $\ddot{u}_i$.

Now, if we differentiate with respect to x_i we have

$$Gu_{i,ijj} + (\lambda + G)u_{k,kii} + \rho\ddot{u}_{i,i} = 0 \tag{25.20}$$

Observe that with repeated indices there is no difference between $u_{k,k}$ and $u_{i,i}$. This quantity is often called the "dilatation" and represented by the symbol Δ. That is,

$$\Delta = u_{k,k} = u_{i,i} \tag{25.21}$$

Equation 25.20 may now be written as

$$(\lambda + 2G)\Delta, jj + \rho\ddot{\Delta} = 0$$

or as

$$[(\lambda + 2G/\rho]\nabla^2\Delta + \partial^2\Delta/\partial t^2 = 0 \tag{25.22}$$

where $\nabla(\ )$ is the vector differentiation operator:

$$\nabla(\ ) = \mathbf{n}_1\partial(\ )/\partial x_1 + \mathbf{n}_2\partial(\ )/\partial x_2 + \mathbf{n}_3\partial(\ )/x_3 \tag{25.23}$$

with the $\mathbf{n}_i$ $(i = 1, 2, 3)$ being mutually perpendicular unit vectors parallel to the coordinate axes.

Equation 25.23 will be recognized as the three-dimensional wave equation:

$$C^2\nabla^2\Delta + \partial^2\Delta/\partial t^2 = 0 \tag{25.24}$$

with C being the acoustic speed. Then C^2 is

$$C^2 = (\lambda + 2G)/\rho \tag{25.25}$$

By substituting for λ and G from Equations 25.17 we have

$$C^2 = \frac{E(1 - \nu)}{(1 + \nu)(1 - 2\nu)\rho} \tag{25.26}$$

(See the final entry in Table 25.1.)

Observe the prominent role of Poisson's ratio ν in Equation 25.26. For a typical metallic value of 0.3, C increases approximately 16% over the value for a uniform bar.

25.6 ILLUSTRATION

Equation 25.14 is useful for estimating the critical speed at which the material will fail due to the stress wave. To illustrate this, consider a straight steel rod with elastic modulus: $E = 30 \times 10^6$ psi; weight density: $\gamma = 0.289$ lb/in.^{3}; and yield strength: $S_y = 80,000$ psi. The objective is to determine the critical speed.

SOLUTION

The mass density ρ may be obtained from the weight density by simply dividing by the gravity acceleration g (32.2 ft/s^2)

$$\rho = \gamma/g = \frac{(0.289)(1728)}{32.2} = 15.5 \, \text{slug/ft}^3 \tag{25.27}$$

where 1728 converts in.^{3} to ft^3.
   In these units, the elastic modulus is

$$E = 30 \times 10^6 \times 144 = 43.2 \times 10^8 \, \text{lb/ft}^2$$

From Table 25.1 the sonic speed C is then

$$C = \sqrt{E/\rho} = [(43.2/15.5) \times 10^8]^{1/2} = 16,910 \, \text{ft/s} \tag{25.28}$$

Similarly, the yield strength S_y is

$$S_y = (80,000) \times (144) = 11.5 \times 10^6 \, \text{lb/ft}^2 \tag{25.29}$$

From Equation 25.14 the critical speed V_O is then

$$V_O = S_y/\rho C = \frac{11.5 \times 10^6}{(15.5)(16,910)} = 43.9 \, \text{ft/s} = 30 \, \text{mph} \tag{25.30}$$

25.7 AXIAL IMPACT ON A STRAIGHT BAR

Consider a rigid body M, with mass M, moving with speed V and colliding with the end of a fixed rod as represented in Figure 25.3.
   When the mass M strikes the end of the rod, speed will be V. Upon striking the end of the rod, the mass M will be resisted and slowed by the rod. Thus the speed v of M after impact will be less than V. The equation of motion for M is then given as

$$\sigma A = -M \, dv/dt \tag{25.31}$$

whereas before
   σ is the stress in the rod
   A is the cross-section area of the rod

From Equation 25.14, we see that the stress in the rod is

$$\sigma = \rho C v \tag{25.32}$$

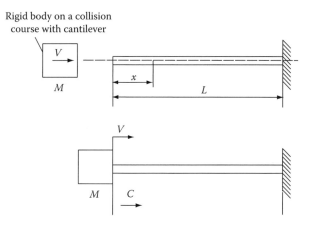

FIGURE 25.3 Uniform rod subjected to compressive impact.

But, from Table 25.1, the sonic speed C is $\sqrt{E/\rho}$. Thus the speed v of a point within the rod is

$$v = \sigma/\sqrt{\rho E} \tag{25.33}$$

In Equation 25.33 if ρE is assumed to be constant, we see that the rate of change of v, (dv/dt), is simply

$$dv/dt = (d\sigma/dt)/\sqrt{\rho E} \tag{25.34}$$

By eliminating dv/dt between Equations 25.31 and 25.34, we have

$$M(d\sigma/dt)/\sqrt{\rho E} + \sigma A = 0 \tag{25.35}$$

By solving for σ, we obtain

$$\sigma = \sigma_O \exp\left[-(A\sqrt{\rho E}/M)\right]t \tag{25.36}$$

where σ_O is the end stress of the rod when $t = 0$.

Equation 25.36 shows that the stress decreases exponentially in time while the stress wave travels toward the fixed end ($x = L$). At this point, the compression is suddenly reflected and doubled. This in turn sends a wave back along the rod. The complete analysis of the impact may thus require several stress wave reflections.

25.8 CONDITIONS OF SPALL

The phenomenon of spall is a direct result of a high-amplitude compressive wave encountering a free surface. The basic features of spall, for instance, may be observed at the ground surface disturbed by an underground explosion or at the ruptured surface opposite the point of impact of a projectile striking a thick plate. A compressive wave reaching a free surface generates a reflected tensile wave of the same amplitude. If this stress amplitude happens to exceed the tensile strength of the material, a layer near the free boundary will spall off. The corresponding plane of failure must be below the surface since the peak tensile stress can occur only after the reflected wave starts to travel back toward the original point of impact. The thickness of the spalled layer is known to depend on the wave amplitude, wave shape, and tensile strength of the material.

The mechanical model of spall is complex and it requires a rather sophisticated analysis of the various phases of stress propagation where the impinging compressive wave interacts with the reflected tensile wave throughout the thickness of the particular medium. The reader interested in theoretical ramifications of the spall problem is advised to consult the specialized literature on the subject [1]. From the practical point of view, it will suffice here to state that spall can occur in materials whose tensile strength is numerically smaller than the compressive strength, or when the amplitude of the impinging stress wave is significantly higher than the tensile strength of the material.

25.9 EXAMPLE DESIGN PROBLEMS

25.9.1 OBJECT FALLING ONTO A COLUMN

A rigid object falls a distance of 10 ft onto the top of a steel column built in at the base. Assuming that the impact produces a purely axial response in the column, calculate the maximum compressive stress at the base if the sonic speed for the column material is 16,820 ft/s, with the weight density γ of steel being 0.289 lb/in.^{3}.

SOLUTION

If the falling object has sufficient mass, the speed of the end of the column at the time of the impact is the same as the speed V_O of the object after it has fallen 10 ft. Specifically, this speed is

$$V_O = \sqrt{2gh} = [(2)(32.2)(10)]^{1/2} = 25.3 \text{ ft/s} \qquad (25.37)$$

where as before
  g is the gravity acceleration
  h is the fall

From Equation 25.14, the propagated compressive stress σ is then

$$\sigma = \rho C V_O \qquad (25.38)$$

where
  ρ is the mass density of the column material
  C is the sonic speed

With the weight density γ being 0.289 lb/in.^{3}, ρ is

$$\rho = \gamma/g = (0.289)(1728)/(32.2) = 15.5 \text{ slug/ft}^3 \qquad (25.39)$$

Then the propagated stress is

$$\sigma = (15.5)(16,820)(25.3) = 6.6 \times 10^6 \text{ lb/ft}^3 = 45,832 \text{ psi} \qquad (25.40)$$

At the fixed end, the stress wave is reflected with no change in sign so that the maximum stress value σ_{max} is double that of the propagated wave. That is,

$$\sigma_{max} = (2)(45,832) = 91,664 \text{ psi} \qquad (25.41)$$

25.9.2 OBJECT IMPACTING A LONG CYLINDER

A rigid object weighing 5000 lb collides with the free end of a long cantilevered metal rod, or cylinder, as in Figure 25.4. Let the rod be 28 ft long with a cross-section area of 8 in.^{2}, the specific weight γ of the rod material be 0.283 lb/in.^{3} and the elastic modulus E be 28×10^6 psi.

28 ft

5000 lb

FIGURE 25.4 Object about to collide with end of a long rod.

   Suppose the impact produces a stress wave with amplitude of 6000 psi. Estimate the compressive stress at the struck end of the rod 0.001 s after impact and determine where the wave front is at that time.

SOLUTION

Equation 25.36 provides an expression for the requested stress and the stress wave will propagate at the sonic speed C given by (Equation 25.4)

$$C = \sqrt{E/\rho} \tag{25.42}$$

where, as before, ρ is the mass density.
   Equation 25.36 is

$$\sigma = \sigma_0 \exp[-(A\sqrt{\rho E}/M)]t \tag{25.43}$$

where
   σ_0 is the initial and stress (6000 psi)
   A is the rod cross section (8 in.^{2})
   M is the mass of the striking object

From the given data ρ and M are

$$\rho = \gamma/g = (0.283)(1728)/(32.2) = 15.19 \text{ slug/ft}^3 \tag{25.44}$$

and

$$M = \text{Weight}/g = 5000/32.2 = 155.28 \text{ slug} \tag{25.45}$$

By substituting the data into Equation 25.43 for $t = 0.001$ we have

$$\sigma = 6000 \exp\{-(8/144)[(15.19)(28)(10^6)(144)]^{1/2}/(155.28)\}(0.001)$$
$$= 6000 \exp\{-90.1\}(0.001)$$

or

$$\sigma = 5483 \text{ psi} \tag{25.46}$$

From Equation 25.42 the sonic speed C is

$$C = [(29)(10^6)(144)/(15.19)]^{1/2} = 16{,}580 \text{ ft/s} \tag{25.47}$$

Then at $t = 0.001$ s the distance d traveled by the wave front is

$$d = Ct = (16580)(0.001) = 16.5 \text{ ft} \tag{25.48}$$

25.10 AXIAL AND RADIAL MODES OF ELEMENTARY STRUCTURES

We can obtain approximate dynamic analyses of structures by modeling them as lumped masses supported by springs. The mass represents the structural inertia and the springs represent the elasticity of the structure. This is often called "mass–spring modeling." It enables a simplified analysis of structural response to impacts and also to continually applied forces and/or enforced displacements.

A lumped-mass model is especially useful for calculating the fundamental frequency of a structure: recall that the fundamental frequency is proportional to the square root of the ratio of stiffness to mass $\sqrt{k/m}$ (see Equation 23.3). From Equation 25.4, we see that this is of the same form as the sonic speed. This analogy is independent of the type of structural response, be it longitudinal or radial. For example, for simple shells we can describe the fundamental vibration as a "breathing mode," a radial expansion/contraction.

Table 25.2 provides expressions for the fundamental frequencies in terms of sonic speeds. The formulas for the cylinder and sphere define the breathing modes [7,8].

TABLE 25.2

Fundamental Frequency Data as a Function of Sonic Velocity (Axial and Radial Modes)

Direction of impulse

L	Free/free	$f = \dfrac{C}{2L}$
L	Fixed/free	$f = \dfrac{C}{4L}$
$a \uparrow b$	Simply supported h = thickness	$f = \dfrac{\pi Ch/a^2}{4[3(1-v^2)]^{1/2}}\left[1 + \left(\dfrac{a}{b}\right)^2\right]$
$a \uparrow b$	Fixed h = thickness	$f = \dfrac{3Ch}{\pi a^2}\left[\dfrac{7}{6(1-v^2)}\right]^{1/2}\left[\left(\dfrac{a}{b}\right)^4 + \dfrac{4}{7}\left(\dfrac{a}{b}\right)^2 + 1\right]^{1/2}$
$\downarrow h$, a_0	Simply supported plate	$f \cong \dfrac{0.3Ch}{a_0^2}$
$\downarrow h$, a_0	Fixed plate	$f \cong \dfrac{0.49Ch}{a_0^2}$
r	Thin ring or cylinder	$f = \dfrac{C}{2\pi r}$
r	Thin sphere	$f = \dfrac{0.27C}{r}$

The expressions of Table 25.2 may be useful in the following configurations:

1. Where an oscillating loading has a frequency nearly coinciding with the fundamental natural frequency of a structural component. Such a loading can lead to increasing vibration amplitudes and, consequently, increasing stresses. A simple check of the fundamental frequency can reveal at a glance if a component should be redesigned to avoid the resonance.
2. Where knowledge of the fundamental natural period of vibration could help select the method of dynamic analysis. This is important if the time of external loading nearly coincides with the fundamental period.
3. In experiments where observance of the fundamental natural frequency can provide information about the sonic speed.

25.11 RESPONSE OF BURIED STRUCTURES

In underground nuclear weapons testing and on other occasions, it is of interest to evaluate the response of an underground structure in the vicinity of the explosion. A detailed study of such configurations requires knowledge of soil material properties and ground motion parameters. However, the inhomogeneity and anisotropy of soil makes a characterization of soil properties difficult. Nevertheless, if a structure is near the point of detonation, the surrounding medium behaves as a fluid subjected to intense stress and velocity fields. The structure then experiences rigid-body displacements.

We can estimate the magnitude of the ground shock if we know the duration of the stress wave and the fundamental natural vibration period. Although a spherical stress wave is generated from the detonation point, the analysis is simplified without losing much accuracy by considering the wave as one-dimensional, that is, neglecting the effect of lateral inertia. With this assumption we can use a one-directional structural frequency response as in Table 25.2.

For example, suppose that a ground shock envelops a buried steel pipe and let the explosion duration be 30 ms. Let the pipe have radius r of 24 in, weight density γ of 0.284 lb/in.^{3}, elastic modulus E of 30×10^6 psi, and Poisson ratio ν of 0.3. Then, from Table 25.1, the sonic speed C is

$$C = [E/(1 - \nu^2)\rho]^{1/2} \tag{25.49}$$

Using the given data C is seen to be

$$C = \left\{ \frac{(30)(10)^6(144)(32.2)}{[1 - (0.3)^2](0.284)(1728)} \right\}^{1/2} = 17649 \text{ ft/s} \tag{25.50}$$

From Table 25.2 the fundamental natural period T is

$$T = 1/f = 2\pi r/C \tag{25.51}$$

Using the given and calculated data T is

$$T = (2\pi)(24/12)/(17649) = 0.71 \text{ ms} \tag{25.52}$$

The time of load application (30 ms) is thus seen to be considerably longer than the natural period of the breathing mode (0.71 ms). Hence, in view of the rule of Chapter 23, with the ratio being much greater than 3, with only minor error the loading of the explosion can be represented by a step pulse of infinite duration. That is, the pipe can be analyzed statically.

25.12 STRESS PROPAGATION IN GRANULAR MEDIA

The radial and tangential stresses in a surrounding medium, such as soil, rock, or stemming materials, must be determined before a buried structure such as a pipe, canister, or underground room can be designed. The fundamentals of stress wave propagation for these media are essentially the same as those utilized in the study of the homogeneous materials such as metals. Unfortunately, the nature of granular, soil-like materials or rock formations represents a multitude of boundaries affecting the propagation and reflections of sonic stress waves. Consequently, only the approximate, gross values of ground motion parameters such as sonic velocity, particle velocity, and density can reasonably be estimated.

The mechanical model for predicting particle motions for ground motion due to underground explosions can be based on the following three assumptions [9,10]:

1. Particle velocity decreases as $1/R^3$ in the inelastic region and as $1/R^2$ in the elastic region, where R is the scaled range, defined as $R_0/W^{1/3}$. In this relation, R_0 denotes the distance from the point of explosion in feet and W is the yield in kilotons.
2. Peak particle velocity depends on the square of the sonic velocity of the media in which the explosion takes place.
3. Propagation velocity of the peak stress depends linearly on the sonic velocity.

Knowledge of wave initiation, propagation, and reflection is closely tied to seismic studies. Out of these three features, the initiation process is probably least understood despite a number of empirical observations and long field experience.

When a stress wave produces impulsive external pressure all around a cylindrical shell, complex vibration modes can be excited involving "breathing" and flexural response [9,10]. The interaction between the purely extensional and flexural modes is found to precipitate permanent wrinkles. The subject of this response is too lengthy for inclusion in this brief review of wave propagation. It may be observed, however, that under dynamic conditions, the buckling modes depend not only on the structure but also on the magnitude of the applied loads. The dynamic buckling often occurs where the impulsive loads are sufficiently high to cause an appreciable plastic flow.

25.13 APPLICATIONS IN MACHINE DESIGN

In many instances in routine machine design, a theoretical analysis of dynamic response may be intractable, too time-consuming, or too costly. Alternative approaches are to use numerical analyses or empirical procedures. Numerical analyses, particularly finite element methods, are especially popular but they are not always available or accessible. Moreover, the analyses are very sensitive to support, loading, and modeling assumptions. Empirical, or experimental, procedures also present difficulties in that they require testing equipment and modeling.

For moderately high speed impact, however, testing machines can provide useful information about sonic speeds for various materials and common structural components. Many high-speed impact machines have been developed using weight and pneumatic forces. The speeds of impact generally vary between 100 and 200 ft/s. Some impact machines have energy capacities exceeding 15,000 ft lb. Force–time and force–speed relations can be obtained for correlating with classical theoretical results. Also, critical impact speeds for proposed structural materials can be obtained.

SYMBOLS

A Cross-section area
a, b Dimensions (see Table 25.2)
c Sonic speed

C_p	Plastic wave propagation
d	Distance
E	Modulus of elasticity
G	Shear modulus (see Equation 25.17)
h	Height; thickness
KE	Kinetic energy per unit volume
L	Length; distance
P	Force
r	Radius
R	Sealed range
S_y	Yield strength
t	Time
u	Transverse displacement
u_i	Displacement components
V	Speed
V_{cr}	Critical velocity
V_O	Impact velocity
v	Speed
x	Axial coordinate
γ	Weight density
Δ	Dilatation
δ	Elongation
δ_{ij}	Kronecker's delta function
λ	Lame coefficient (see Equation 25.17)
ν	Poisson's ratio
ρ	Mass density
ε	Strain
ε_f	Strain at failure
σ	Stress
σ_O	End stress
σ_{ult}	Ultimate material strength

REFERENCES

1. R. J. Wasley, *Stress Wave Propagation in Solids*, Marcel Dekker, New York, 1973, p. 161.
2. W. Kaplan, *Advanced Mathematics for Engineers*, Addison Wesley, Reading, MA, 1981, p. 405.
3. C. R. Wiley, *Advanced Engineering Mathematics*, 4th ed., McGraw Hill, New York, 1975, p. 323.
4. I. S. Sokolnikoff and R. M Redheffer, *Mathematics of Physics and Modern Engineering*, 2nd ed., McGraw Hill, New York, 1966, p. 438.
5. E. Kreyszig, *Advanced Engineering Mathematics*, 2nd ed., John Wiley & Sons, New York, 1967, p. 490.
6. H. Josephs and R. L. Huston, *Dynamics of Mechanical Systems*, CRC Press, Boca Raton, FL, 2002 (Chap. 10).
7. M. Kornhauser, *Structural Effects of Impact*, Spartan Books, New York, 1964.
8. L. S. Jacobsen and R. S. Ayre, *Engineering Vibrations*, McGraw Hill, New York, 1958.
9. V. E. Wheeler and R. G. Preston, Scaled free-field particle motions from underground nuclear explosions, research Laboratory Report 50563, University of California, 1968.
10. H. C. Rodean, *Nuclear-Explosion Seismology*, U.S. Atomic Energy Commission, Oak Ridge, TN, 1972.

26 Fatigue

26.1 INTRODUCTION

As the name implies, "fatigue" is a description of a material getting "tired" or "worn-out." Technically, fatigue refers to material degradation due to repeated loading and unloading. When fatigue becomes excessive, the material will fracture and fail. The familiar example of repeatedly bending and unbending a paper clip until it breaks provides an illustration of fatigue failure.

Various theories and design methods have been proposed concerning fatigue and its induced structural failure [1–8]. For cyclic loading and unloading, it is convenient and also reasonably accurate to model the loading by a sinusoidal function as in Figure 26.1. The figure also shows the maximum, minimum, and mean stress.

Intuitively, the larger the maximum stress the fewer the number of cycles needed for failure to occur. Conversely, for low levels of stress the material may last indefinitely without failure. Figure 26.2 illustrates these concepts. It is of interest to note that when the stress is plotted against the number of cycles, there is a rapid change in the shape of the curve at about 10^6 to 10^7 cycles beyond which a constant stress value is approached. This value is known as the "endurance limit." That is, materials surviving 10^7 cycles are expected to last indefinitely.

Finally, it happens that the endurance limit depends upon the type of loading experienced by a material. Specifically, in tension–compression testing (i.e., "push–pull" tests), the endurance limit is only about 75% of that obtained in bending. Also, there is often a significant scatter in fatigue testing. Therefore for designs with anticipated loads near the elastic limit, a statistical analysis is recommended.

26.2 CUMULATIVE DAMAGE CRITERIA

On occasion, a structural or machine component will experience several different stress amplitudes and periods of operation during the lifetime. When this occurs, the concept of "cumulative damage" may be of use [7]. Although this method should be used with caution (no allowance is made for other variables), a fairly satisfactory approximation can be made using the following criterion:

$$\sum n/N = 1 \tag{26.1}$$

where
   n is the number of cycles at a particular working stress
   N is the number of cycles to failure at the same stress level

The term n/N represents the cycle ratio. The value of N may be obtained from a fatigue diagram as in Figure 26.2.

Illustration: To illustrate the method, consider the following example problem: Suppose that we want to perform a safety check for a structure undergoing stress fatigue. Suppose further that $\sigma - N$ of Figure 26.2 is applicable and that we have the following data:

$$
\begin{aligned}
&80{,}000 \text{ cycles at } \sigma_{max} = 42{,}000 \text{ psi} \\
&50{,}000 \text{ cycles at } \sigma_{max} = 30{,}000 \text{ psi} \\
&100{,}000 \text{ cycles at } \sigma_{max} = 21{,}000 \text{ psi}
\end{aligned} \tag{26.2}
$$

The objective is to use Equation 26.1 to evaluate the safety of the structure.

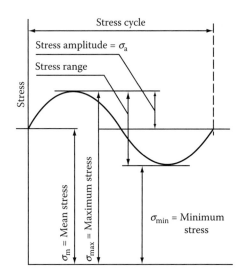

FIGURE 26.1   Cyclic stress notation.

SOLUTION

The required values of the N in Equation 26.1 may be obtained from Figure 26.2 for the assigned stress levels. We have

$$
\begin{aligned}
\sigma_{max} &= 42,000 \text{ psi}, \quad N = 100,000 \\
\text{For } \sigma_{max} &= 30,000 \text{ psi}, \quad N = 1,000,000 \\
\sigma_{max} &= 21,000 \text{ psi}, \quad N = 10,000,000
\end{aligned}
\tag{26.3}
$$

Then by using Equation 26.1 we have

$$
\begin{aligned}
\sum n/N &= \frac{80,000}{100,000} + \frac{50,000}{1,000,000} + \frac{100,000}{10,000,000} \\
&= 0.8 + 0.05 + 0.01 = 0.86
\end{aligned}
\tag{26.4}
$$

Since the sum of the cycle ratios is less than 1, the structure is presumed to be safe. In practice, however, a limiting value smaller than unity is generally used to have a factor of safety. For example, $\sum n/N < 0.8$ may be taken as a suitable design criterion.

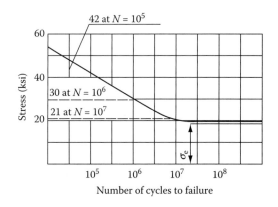

FIGURE 26.2   Illustration of a fatigue/failure ("$\sigma - N$") diagram.

26.3 NEUBER EFFECT

It is known that with ductile materials, stress concentrations may be relieved by plastic yielding. With cyclic loading, however, the full value of theoretical stress concentration factors K must be used in structural design. Numerous fatigue experiments conducted with notched bars and sharp radii led to the establishment of the Neuber effect [9]. The concept is that there is a small limiting value of the notch radius below which no additional stress increase in fatigue is expected.

The ratio between the anticipated stress increase due to the cyclic loading and that predicted by elasticity theory is called the notch sensitivity factor, q, [10,11], given by

$$q = \frac{K_f - 1}{K - 1} \tag{26.5}$$

where K_f denotes a fatigue stress concentration factor derived from tests.

A structural member is considered to have no notch effect when $q = 0$. Alternatively, $q = 1$ defines maximum theoretical notch sensitivity. Design tables [11] show how q varies with the notch or hole radius. Fatigue theory suggests, however, that the actual size of the hole or the depth of the notch is more influential than geometric parameter ratios.

26.4 ELEMENTS OF DESIGN FOR FATIGUE

Design for fatigue is relatively straightforward when a $\sigma - N$ diagram as in Figure 26.2 for the material is known. If such data is not available, however, various theoretical estimates can still be made [6]. For example, Soderberg's law [12] provides a conservative prediction for maximum and minimum for uniaxial loading as

$$\sigma_{max} = \sigma_e + \sigma_m \left(1 - \frac{\sigma_e}{\sigma_y}\right) \tag{26.6}$$

and

$$\sigma_{min} = -\sigma_e + \sigma_m \left(1 + \frac{\sigma_e}{\sigma_y}\right) \tag{26.7}$$

where as before
   σ_e is the endurance limit as in Figure 26.2
   σ_y is the yield stress
   σ_{max}, σ_{min}, and σ_m are the maximum, minimum, and mean stresses as in Figure 26.1

A difficulty with Equations 26.6 and 26.7 is that the endurance limit still needs to be known. Experiments can help: Table 26.1 provides an estimate of the endurance limit in terms of the ultimate tensile strength σ_u.

A difficulty with material property data as in Table 26.1 is that a given structural material may have more internal flaws than ideal materials.

26.5 EFFECT OF SURFACE FINISH

Even though applied stress is the most important single factor in fatigue phenomenon, surface finish effects are also important. Stress risers due to roughness, residual stresses, and nonuniformity of the material properties between the surfaces and the core influence fatigue life.

TABLE 26.1

Preliminary Fatigue Strength Ratios (σ_e/σ_u) for Various Structural Materials

Recommended Fatigue Strength (σ_e/σ_u) Ratios for Preliminary Design

Cast aluminum, 220-T4	0.17
Cast aluminum, 108	0.52
Cast aluminum, F132,-T5	0.38
Cast aluminum, 360-T6	0.40
Wrought aluminum, 2014-T6	0.29
Wrought aluminum, 6061-T6	0.45
Beryllium copper, HT	0.21
Beryllium copper, H	0.34
Beryllium copper, A	0.47
Naval brass	0.35
Phosphor bronze	0.32
Gray cast iron (No. 40)	0.48
Malleable cast iron	0.56
Magnesium, AZ80A-T5	0.29
Titanium alloy, 5A1, 2.5Sn	0.60
Steel, A7-61T	0.50
Steel, A242-63T	0.50
Spring steel, SAE 1095	0.36
Steel, SAE 52100	0.44
Steel, SAE 4140	0.42
Steel, SAE 4340	0.43
Stainless steel, Type 301	0.30
Tool steel, H.11	0.43
Maraging steel, 18 Ni	0.45

Note: T, heat-treated; H, hard; HT, hardened; A, annealed.

In the usual procedures of surface finishing, highly polished surfaces have fatigue life, which is at least 10% longer than that of relatively rough surfaces [13]. The effect of finish on the endurance limit for shorter fatigue appears to be less pronounced [14]. There are also indications that surface sensitivity increases with tensile strength [15]. Indeed, the magnitude of residual stresses in high-strength steels, produced by milling and grinding, can be very high.

In design of machine shafts and other structural components where bending and torsion can combine to create high-stress gradients, hardening of the surface by carburizing, flame hardening, or nitriding produces beneficial effects. In this way, a surface stronger than the core can be created. If, through this treatment, we can obtain a surface state such that the mean stress in fatigue is reduced, we can expect longer component life. Shot peening provides a simple way of reducing surface tensile stresses [16]. With shot peening, a compressive residual stress is induced on the surface, thus increasing the tensile strength on the surface.

The quality of a machined surface can be affected by a number of finishing processes. The surface quality may be characterized by "surface roughness." A definition of roughness alone, however, does not convey the entire picture. The process of material removal is an extremely complex process, which can result in detrimental surface layer alterations. This, in turn, can lead to such problems as grinding burns, stress corrosion, cracks, distortion, and thus increased residual stresses. These features influence the fatigue resistance in a deleterious way. Abrasive grinding can be especially harmful. These kinds of surface degradations can decrease the endurance limit by 10%

or much more. Where high-temperature nickel-base and titanium alloys are involved, the loss in fatigue resistance can become as high as 30%. This is especially noticeable in high-cycle fatigue.

In addition to the purely mechanical means of surface finishing electrical discharge machining (EDM) can seriously affect the fatigue strength, particularly in the case of highly stressed components. In the event of EDM roughing, the outer layer of metal can undergo microstructural changes, local overheating, and cracking, which may extend further into the adjacent layers of the material. Other methods of chemical and electrochemical machining appear to have a relatively smaller effect on the endurance limit. The surface finish can, of course, be improved by mitigating heat-affected zones, cracks, and other detrimental layers in critically stressed components [17].

26.6 EFFECT OF CREEP

For temperatures below 650°F, there is virtually no thermal effect on the endurance limit of a conventional structural material. At higher temperatures, however, creep can affect the endurance limit. Figure 26.3 provides a simple relation between fatigue and creep [18]. In this figure, the fatigue strength at a completely reversed cycle and the static creep strength are represented on the two axes for a given design temperature. Two points provide a straight line approximation, which is considered to be relatively conservative.

26.7 EFFECT OF CORROSION

Corrosion can significantly lower the endurance limit. The combination of fatigue and corrosion, called corrosion fatigue, is a serious problem because the corrosion products can act as a wedge, opening a crack. This can result in as much as a 20% reduction of the endurance limit for a conventional carbon steel. In salt water and other corrosive media, the reduction can be even more pronounced. To counter this effect, alloying elements, as well as protective coatings are used. The rate of cycling can also lower the fatigue life, but it is seldom considered in normal design procedures because such data are poorly defined ahead of time.

26.8 EFFECT OF SIZE

Fatigue failure generally occurs by fracture, which begins as a small or microscopic crack within the material. Such cracks arise due to microscopic defects, dislocations, or irregularities within the material. These irregularities occur randomly, but generally uniformly during material fabrication

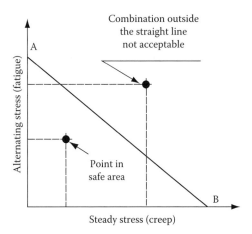

FIGURE 26.3 A simplified method for combining creep and fatigue.

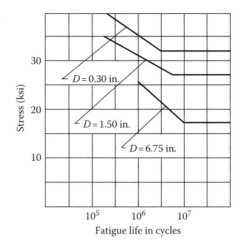

FIGURE 26.4   Effect of bar diameter on fatigue life.

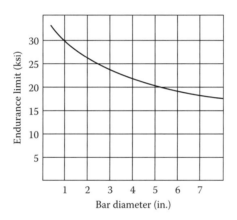

FIGURE 26.5   Effect of bar diameter on endurance limit.

processes. Therefore, if a structure has several components with similar shapes, the smaller components are likely to have longer life, simply because they have fewer material defects owing to their smaller size.

Figure 26.4 provides an illustration of the concept where it is seen that a bar with smaller diameter has a better resistance to fatigue failure due to the fewer sites in the material with excessively high stress.

This size concept can be useful in design where there are similarly shaped members with various thicknesses. Specifically, if the endurance limit is known for at least some of the thicknesses, then predictions of the endurance limit for other thicknesses can be obtained. For example, Figure 26.5 shows the endurance limit as a function of thickness for steel bars as given in Figure 26.4.

26.9   LOW-CYCLE FATIGUE

In applications where fatigue life is an issue, the designer is usually concerned with configurations where there are millions of cycles. On occasion, a component may be loaded to failure at a much lower number of cycles. It is customary to refer to "low-cycle fatigue" as the occasion when the total number of cycles to failure is less than 10,000.

Faupel and Fisher [12] have established an empirical relation predicting low-cycle fatigue as

$$\sigma_a = \sigma_e + \frac{CE}{2N^{1/2}} \tag{26.8}$$

where as before

σ_a and σ_e are the stress amplitude and endurance limit
E is the elastic modulus
N is the number of cycles to failure
C is given by

$$C = (1/2)\ell n \left[\frac{100 - \%AR}{100} \right] \tag{26.9}$$

where %AR is the percentage reduction of area in a conventional tensile test. [Observe that C is negative except for the ideally brittle case where C is zero (zero %AR).] Equations 26.8 and 26.9 agree well with experimental data for most structural metals.

26.10   LOW-CYCLE FATIGUE EXAMPLE

To illustrate the foregoing concepts and procedures, consider the following problem.

Problem: Calculate the mean and maximum fatigue stresses in uniaxial loading for a machine part made of steel with an endurance limit σ_e of 35,000 psi expected to be used for 20,000 cycles. Let the yield strength σ_y be 80,000 psi and the ultimate strength σ_u be 100,000 psi with an area reduction AR of 30%.

SOLUTION

From Figure 26.1 the maximum stress, σ_{max}, the stress amplitude, σ_a, and the mean stress σ_m are related as

$$\sigma_{max} = \sigma_m + \sigma_a \tag{26.10}$$

Then from Equation 26.6 we have

$$\sigma_m + \sigma_a = \sigma_{max} = \sigma_e + \sigma_m \left(1 - \frac{\sigma_e}{\sigma_y} \right)$$

or

$$\sigma_m = \sigma_y \left(1 - \frac{\sigma_a}{\sigma_e} \right) \tag{26.11}$$

From Equation 26.9 the parameter C is

$$C = (1/2)\ell n \left(\frac{100 - 30}{100} \right) = -0.1783 \tag{26.12}$$

Then, from Equation 26.8 we find σ_a to be

$$\sigma_a = 35,000 - \frac{(0.1783)(30)(10)^6}{(2)(20,000)^{1/2}} = 16,000 \text{ psi } (110.3 \text{ N/mm}^2) \tag{26.13}$$

Hence, from Equation 26.11, the mean stress σ_m is

$$\sigma_m = 80,000\left(1 - \frac{16,000}{35,000}\right) = 43,400 \text{ psi } (299.2 \text{ N/mm}^2) \tag{26.14}$$

Finally, from Equation 26.10, the requested maximum stress σ_{max} is

$$\sigma_{max} = 43,400 + 16,000 = 59,400 \text{ psi } (409.6 \text{ N/mm}^2) \tag{26.15}$$

The maximum stress of Equation 26.15 is sometimes called the "pseudoelastic limit," since low-cycle fatigue is expected to extend beyond the elastic limit. When the stress amplitude σ_a is zero, the mean stress σ_m and the yield stress σ_y are equal (see Equation 26.11). Then from Equation 26.8, the number N of cycles to failure is

$$N = \frac{C^2 E^2}{4\sigma_y^2} \tag{26.16}$$

For the numerical data of this problem, the lowest number of cycles to failure at a mean stress of 80,000 psi is 1120. Thus, the lowest number of cycles to failure at yield depends only upon the strain at yield and the area reduction in the standard tensile test.

Equations 26.9 and 26.16 indicate that as the brittleness of the material increases, the number of cycles to failure falls off rather rapidly. The ideally brittle material, based upon Equation 26.9 is that at which C tends to zero.

SYMBOLS

C Coefficient defined by Equation 26.9
E Elastic modulus
K Static stress concentration factor
K_f Fatigue stress concentration factor
N Number of cycles to failure
n Number of cycles
q Notch sensitivity factor (see Equation 26.5)
σ Stress
σ_a Stress amplitude
σ_e Endurance stress
σ_m Mean stress
σ_{max} Maximum stress
σ_{min} Minimum stress
σ_y Yield stress

REFERENCES

1. E. Orowan, Theory of the fatigue of metals, *Proceedings of the London Royal Society, Series A*, 171, 1939.
2. A. M Freudenthal and T. J. Dolan, The character of the fatigue of metals, 4th Progress Report, ONR, Task Order IV, 1948.
3. E. S. Machlin, Dislocation theory of the fatigue of metals, NACA Technical Note 1489, 1948.
4. L. F. Coffin, The resistance of material to cyclic thermal strains, *Transactions, American Society of Mechanical Engineers*, 1957, Paper 57-A-286.
5. L. F. Coffin, The stability of metals under cyclic plastic strain, *Transactions, American Society of Mechanical Engineers*, 1959, Paper 59-A-100.

6. H. J. Grover, S. A. Gordon, and L. R. Jackson, *Fatigue of Metals and Structures*, U.S. Government Printing Office, Washington, D.C., 1954.

7. M. A. Miner, Cumulative damage in fatigue, *Journal of Applied Mechanics*, Vol. 12, 1945.

8. S. S. Manson, Cumulative fatigue damage, *Machine Design*, 1960.

9. H. Neuber, *Theory of Notch Stresses*, J.W. Edwards, Ann Arbor, MI, 1946.

10. R. E. Peterson, *Stress Concentration Design Factors*, John Wiley & Sons, New York, 1953.

11. R. E. Peterson and A. M. Wahl, Two and three-dimensional cases of stress concentration, and comparison with fatigue tests, *Journal of Applied Mechanics*, 3(1), 1936, A-15.

12. J. H. Faupel and F. E. Fisher, *Engineering Design*, 2nd ed., John Wiley & Sons, New York, 1981 (chap. 12).

13. O. J. Horger and H. R. Neifert, Effect of surface condition on fatigue properties, *Surface Treatment Metallurgy*, American Society of Metals Symposium, 1941.

14. A. C. Low, Short endurance fatigue, *International Conference on Fatigue Metallurgy*, New York, 1956.

15. G. C. Noll and C. Lipson, Allowable working stresses, Proceedings, Society of Experimental Stress Analysis, Vol. 3, 1946, p. 11.

16. H. F. Moore, *Shot Peening and the Fatigue of Metals*, American Foundry Equipment Company, Tulsa, OK, 1944.

17. *Machining Data Handbook*, Metcut Research Associates, Inc., Cincinnati, OH, 1972.

18. J. J. Tapsell, Fatigue at high temperatures, *Symposium on High Temperature Steels and Alloys for Gas Turbines*, Iron and Steel Institute, London, 1950.

27 Fracture Mechanics: Design Considerations

27.1 INTRODUCTION

A principal objective of fracture mechanics is the stress analysis of components, which are sensitive to crack propagation and brittle failure. To this end, it may be useful to briefly review concepts of ductile and brittle behavior.

Unfortunately, the distinction between ductile and brittle behavior is not precise. At normal working temperatures, steel is generally considered to be ductile whereas cast iron is considered to be a brittle material. A dividing boundary between the two is often defined as the amount of elongation during a tensile test. If a rod elongates more than 5% under tension, the rod material is said to be "ductile." If the rod fractures before elongating 5%, the material is said to be "brittle."

The 5% limit, however, may be a bit low and it may even be challenged in terms of strength theories. It is generally regarded that brittle failure may be related to maximum principal stresses, whereas ductile failure is often interpreted with the aid of traditional plastic failure criterion [1].

But these rules cannot be applied rigidly since in component design, the component strength involves both geometric and material parameters. A designer is thus often forced to make a final decision with incomplete information. Under these conditions, the margin between failure and success is expanded only by using a high factor of safety.

The complexity of the design process is also increased due to requirements of fracture control. It is generally recognized that the metallurgical phenomenon of a fracture toughness transition with temperature is exhibited by a number of low- and medium-yield-strength steels. This transition results from the interactions among temperature, strain rate, microstructure, and the state of stress. One of the more perplexing aspects of this behavior is that the customary elongation property of the material appears to have virtually no relation to the degree of fracture toughness. For example, a well-known mild steel such as the ASTM Grade A36, having an elongation greater than 20%, exhibits brittle behavior not only at lower, but also at room temperatures.

It appears, therefore, that it may be advisable to characterize the materials with respect to their brittle tendencies before selecting the method of stress analysis. Traditional mechanical properties, in the form of yield point, ultimate strength, elongation, and elastic constants, must be supplemented with the thermomechanical data. The response of a stressed component, particularly at lower working temperatures, may be impossible to predict without a knowledge of fracture mechanics and the material's toughness.

The application aspects of fracture mechanics given in this chapter are treated in an elementary fashion. The aim of the presentation is simply to alert designers to some potential problem areas and to indicate the nature of modern trends in stress analysis and fracture control. It points to the necessity of characterizing the material's behavior under stress in terms of new parameters.

27.2 PRACTICAL ASPECTS OF FRACTURE MECHANICS

Fracture mechanics has been studied and documented as early as 1920 [2]. Since then, it has received considerable attention from many analysts with a focus on high-strength metals. A fracture is seen to occur even due to a low nominal stress setting off a brittle behavior phenomenon. Once initiated, such a brittle process can propagate at a high velocity to the point of complete failure. Thus, failure can occur even if the yield strength is not reached.

As a general guide, steels with yield strength above 180 ksi, titanium alloys above 120 ksi, and aluminum alloys above 60 ksi are in this high-strength but brittle category. They should thus be evaluated on the basis of fracture toughness rather than pure yield strength.

Extensive experimentation and some recent developments in continuum mechanics have been aimed at defining a quantitative relationship between stress, the size of a crack, and the mechanical properties. It is relevant to point out that the advent of fracture mechanics does not negate the traditional concepts of stress analysis, which allow us to design for stresses exceeding the yield strength in the vicinity of such structural discontinuities as holes, threads, or bosses, provided that the material deforms plastically and redistributes the stresses. This concept is still valid unless the material contains critical flaws that produce unstable crack propagation below the design value of the yield strength.

The difficulty with introducing the correction for flaws in design is that, in many cases, flaws cannot be easily detected. It becomes necessary, therefore, to develop a procedure that would define the maximum crack length permissible at a particular level of stress. According to the theory of fracture mechanics, this stress level is inversely proportional to the flaw size. The stress σ at which crack propagation is expected to occur is given by the relation

$$\sigma = \frac{K_{1C}}{(\pi a)^{1/2}} \tag{27.1}$$

where
K_{1C} is called the "plane-stress fracture toughness"
a is the half crack length

Fracture toughness K_{1C}, with units psi (in.)$^{1/2}$ or [N(mm)$^{-3/2}$], is a mechanical property, approaching a limiting minimum value as the specimen thickness increases. Its magnitude in Equation 27.1 constitutes an absolute minimum corresponding to a plane strain condition at which the fractured surface has a brittle appearance. This type of failure is associated with very limited plastic deformation and is typical of fractures in heavy sections.

Equation 27.1 can be utilized in calculations involving through cracks in relatively large containers and pressure vessels. If the calculated stress of fracture is found to be higher than the operating stress based on a minimum acceptable factor of safety, the component should be satisfactory for service.

The stress calculated in Equation 27.1 will be tensile. The expression may be used for both static and dynamic loading conditions. For conventional structural steels having a static yield strength σ_y up to 100,000 psi, the corresponding dynamic yield σ_{yd} can be estimated as

$$\sigma_{yd} = \sigma_y + 30,000 \tag{27.2}$$

According to Lange [3], the upper limit for this prediction can be as high as 140,000 psi while at the same time the numerical term in Equation 27.2 is gradually reduced from 30,000 to zero.

The concept of K_{1C} refers to brittle fracture and plane strain conditions. In mathematical terms, plane strain is defined as the state of zero plastic flow parallel to a crack front. Thick materials normally develop plane strain fracture characteristics and the broken surface is essentially flat. As the material's thickness decreases, the degree of constraint decreases, creating a plane stress condition and a maximum amount of plastic flow associated with the fracture.

Figures 27.1 and 27.2 show the effect of temperature on conventional strength and crack resistance for two high-strength steels. These examples indicate poor resistance to brittle fracture within the specific temperature ranges associated with either a tempering process or testing. It is seen that the conventional tensile property gives little insight into the brittle behavior of the materials.

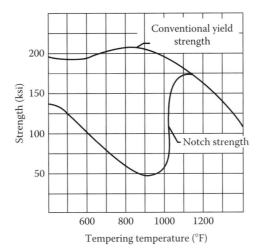

FIGURE 27.1 Effect of tempering temperature on high-strength steel (12Mo-V stainless).

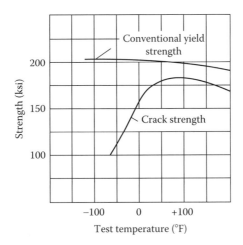

FIGURE 27.2 Effect of test temperature on high-strength steel.

27.3 DESIGN IMPLICATIONS OF THE CRACK SIZE PARAMETERS

Typically in material fabrication, there will be surface defects involving cracks. In such cases, it is prudent to use the principles of fracture mechanics to estimate the maximum stress σ associated with component imperfection. Specifically, σ is estimated as

$$\sigma = \frac{K_{1C}(\psi)^{1/2}}{\left[3.77b + \left(0.21K_{1C}^2/\sigma_y^2\right)\right]^{1/2}} \tag{27.3}$$

where as before

K_{1C} is the plane strain fracture toughness number

σ_y is the yield strength

ψ is the "crack shape parameter," which is a function of depth b and half length a of a crack [4]

Figure 27.3 provides a typical representation of the fracture toughness parameter as a function of tensile strength and Figure 27.4 shows the relation between the crack shape parameter ψ and the crack aspect ratio b/a.

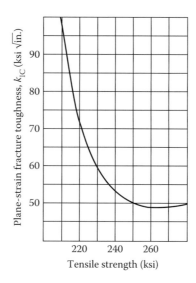

FIGURE 27.3   Fracture toughness for 4340 steel.

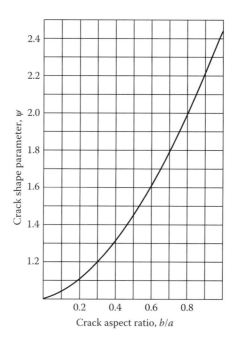

FIGURE 27.4   Crack shape parameter.

27.4   ILLUSTRATIVE DESIGN PROBLEMS AND SOLUTIONS

In the following sections, we present a few example problems intended to illustrate design procedures to guard against fracture failure.

27.4.1   Example Problem 1

Calculate the maximum stress at failure for a large structural component made of a 4340 steel plate having an ultimate tensile strength of 220,000 psi. The surface flaw discovered is 0.2 in. deep and 2 in. long.

SOLUTION

For the specified strength of 220,000 psi, the fracture toughness parameter K_{1C} is seen from Figure 27.3 as

$$K_{1C} = 73,000 \text{ psi (in.)}^{1/2} \qquad (27.4)$$

Since a is 1 in. and b is 0.2 in., the ratio b/a is 0.2. Then Figure 27.4 gives the shape parameter ψ as

$$\psi = 1.09 \qquad (27.5)$$

Assuming the critical fracture stress to be equal to the yield strength, Equation 27.3 may be written as

$$\sigma = \frac{K_{1C}\psi^{1/2}}{\left[3.77b + 0.21\ K_{1C}^2/\sigma^2\right]^{1/2}} \qquad (27.6)$$

By squaring and solving for σ^2 and σ, we have

$$\sigma^2 = \frac{K_{1C}^2\psi}{\left[3.77b + 0.21\ K_{1C}^2/\sigma^2\right]}$$

or

$$3.77b\sigma^2 + 0.21\ K_{1C}^2 = K_{1C}^2\psi$$

or

$$\sigma = \frac{K_{1C}(\psi - 0.21)^{1/2}}{(3.77b)^{1/2}} \qquad (27.7)$$

Finally, by substituting from Equations 27.4 and 27.5, we find the failure stress to be

$$\sigma = \frac{73,000(1.09 - 0.21)^{1/2}}{[(3.77)(0.21)]^{1/2}} = 78,860 \text{ psi } (544 \text{ N/mm}^2) \qquad (27.8)$$

27.4.2 EXAMPLE PROBLEM 2

Select structural material to tolerate the following flaw dimensions: Total crack length: 1 in. and crack depth: 0.1 in.. The maximum applied stress is expected to be equal to the yield strength of the material, approaching 220,000 psi for 4340 grade steel. Assume the ratio of ultimate to yield strength is 1.13.

SOLUTION

With the stress expected to be equal to the yield stress, Equation 27.7 states that

$$\sigma = \sigma_y = \frac{K_{1C}(\psi - 0.21)^{1/2}}{(3.77b)^{1/2}} \qquad (27.9)$$

By solving for the fracture toughness parameter K_{1C}, we have

$$K_{1C} = 1.92\sigma_y \left(\frac{b}{\psi - 0.21} \right)^{1/2} \tag{27.10}$$

Using the specified crack dimensions, we obtain the geometric ratio b/a as

$$b/a = 0.2 \tag{27.11}$$

(Note the crack length is $2a$.)
    From Figure 27.4, the crack shape parameter ψ is approximately

$$\psi = 1.09 \tag{27.12}$$

Therefore, with $b = 0.1$ in. and σ_y being 220,000 psi, Equation 27.10 produces the fracture toughness K_{1C} as

$$K_{1C} = (1.92)(220,000) \left[\frac{0.1}{1.09 - 0.21} \right]^{1/2} = 142,390 \text{ psi (in)}^{1/2} \text{ [4948 N(mm)}^{-3/2}] \tag{27.13}$$

From Figure 27.3, the plane strain fracture toughness is only approximately 73,000 psi (in.)$^{1/2}$. The conclusion, therefore, is that we either have to select material with a higher fracture toughness or decrease the corresponding working stress.

27.4.3 EXAMPLE PROBLEM 3

Provide an estimate of the leak-before-break internal pressure for a high-strength alloy steel cylinder with mean radius r of 36 in. and wall thickness t of 2.5 in. Assume that the minimum plane strain fracture toughness is 93 psi (in.)$^{1/2}$. Neglect end effects and assume that the conventional membrane theory of stress analysis applies.

SOLUTION

Recall from Equation 27.1, from the theory of fracture mechanics, that the stress σ at which fracture leading to crack propagation is likely to occur is

$$\sigma = \frac{K_{1C}}{(\pi a)^{1/2}} \tag{27.14}$$

where a is the half length of the crack.
    For a through-thickness crack for a wall thickness t, Equation 27.14 becomes

$$\sigma = \frac{K_{1C}}{(\pi t)^{1/2}} \tag{27.15}$$

Recall further that hoop stress σ_h is given by the simple expression [5,6]

$$\sigma_h = pr/t \tag{27.16}$$

where p is the internal pressure, r is the hoop radius, and t is the hoop thickness.
    By identifying σ with σ_h between Equations 27.15 and 27.16 and solving for the pressure p, we obtain

$$p = \frac{tK_{1C}}{r(\pi t)^{1/2}} \tag{27.17}$$

Then by substituting the given numerical values, the desired leak-before-break pressure is

$$p = \frac{(2.5)(93)(1000)}{(36)(2.5\pi)^{1/2}} = 2300 \text{ psi} \tag{27.18}$$

27.5 COMMENT

The examples presented in Section 27.4 illustrate the quantitative approach to the prediction of stresses for a given maximum size of flaw. Since the fracture toughness parameter K_{1C} is a compound quantity involving stress and the crack size, it becomes necessary to rely on empirically obtained curves of K_{1C} before calculating the strength of a component for a given size of defect.

When a structural element of a brittle nature is subjected to a combined loading with the lack of symmetry, it is difficult to rely on the size and orientation of the flaw. For example, in the case of glass, it may be necessary to describe the fracture mechanics property in terms of the strain energy release rate G_C. This may be described as the quantity of energy released per unit area of crack surface as the crack extends [7]. The release rate parameter G_C is then

$$G_C = \frac{\sigma^2 \pi a}{E} \tag{27.19}$$

where as before
  a is the half crack length
  E is the elastic modulus

On the premise that a denotes one-half crack length for a through-thickness flaw in a semiinfinite plate in tension and σ is the nominal prefracture stress, Equation 27.1 gives

$$K_{1C} = \sigma(\pi a)^{1/2} \tag{27.20}$$

By eliminating πa between Equations 27.19 and 27.20, we have

$$K_{1C}^2 = EG_C \tag{27.21}$$

The parameter G_C may be defined as a measure of the force driving the crack. Once this quantity is established, it should be possible to obtain a preliminary estimate of the critical crack size.

27.5.1 EXAMPLE PROBLEM

Calculate the approximate critical length of a crack on a glass panel subjected to tension on the premise that the modulus of elasticity E is 10×10^6 psi, the ultimate strength σ_u is 40,000 psi, and the strain energy release rate G_C is 0.08 (in lb/in.^{2}).

SOLUTIONS

From Equation 27.21, the equivalent fracture toughness is

$$K_{1C} = [10 \times 10^6 \times 0.08]^{1/2} = 894 \text{ psi (in.)}^{1/2} \tag{27.22}$$

Next, from Equation 27.20, the half crack length a is

$$a = \frac{K_{1C}^2}{\pi\sigma^2} = \frac{(894)^2}{\pi(40,000)^2} = 0.00016 \text{ in.} \tag{27.23}$$

27.6   IMPLICATIONS OF FRACTURE TOUGHNESS

The plane strain fracture toughness parameter K_{IC} has received rather wide attention because of its role in fracture mechanics and fracture-safe design. As a distinct mechanical property, K_{1C} depends on the temperature and the rate of strain. Experience also shows that in the case of higher-strength materials, which tend to be relatively brittle, the determination of K_{1C} does not present any special problems. However, when using a medium or a lower material strength combined with high toughness, the testing procedure becomes more complex and requires a higher loading capacity for the test equipment. In some situations, the required high specimen thickness for the test becomes almost impractical. This is because, to assure a valid K_{1C} number in a tough material, the test specimen must satisfy the plane strain conditions as noted previously in the section dealing with practical aspects of mechanics.

The characterization of fracture properties for intermediate and lower strength steels is not as precise or convenient as we often need in structural design. We do not have a simple reliable test leading to values of the toughness parameter K_{1C}.

A good approximation, however, may often be obtained by using the popular and relatively simple Charpy V-notch test. It may be recalled that this test is conducted on a specimen containing a centrally placed sharp notch. The specimen is broken by impact in a beamlike configuration with a three-point loading. The amount of impact energy at the temperature of interest, sometimes denoted by "CVN," is usually expressed in ft-lb. Since a K_{1C} value is involved in the fracture mechanics concept described by Equation 27.1, numerous attempts have been made to develop simple correlations between CVN and K_{1C}. These correlations have generally been of the type

$$K_{1C} = A(\text{CVN})^n \tag{27.24}$$

where A and n denote the specific numerical constants for a selected material and design conditions. Although this correlation may be suitable for design analysis in the elastic range, a continuous debate of the merits and limitations of CVN created an impasse in the various committees that deal with specifications and test standards. In the meantime, other test methods have evolved with the specific charter of plane strain fracture characterization, resulting in a standard document known as ASTM-E399 [8]. Whatever means the designer plans to adopt for the decision regarding the value of K_{1C}, the designer would be well advised to remember that no sweeping generalizations are available within the bounds of the current state of the art in this field and that each design case should be treated individually.

Table 27.1 provides a listing of minimum values of the toughness parameter K_{1C} for a few common materials. On average, the maximum values of the K_{1C} are approximately 1.1 to 1.3 times the minimum values quoted in the table. Hence the use of these minimum values should lead to conservative predictions.

The information given in Table 27.1 is intended to provide an illustration of the order of magnitude of K_{1C} values. Precision cannot be implied because of the unavoidable variations in chemical composition, heat treatment, mechanical working of structural materials, as well as fabrication temperatures. Nevertheless, the K_{1C} values are relatively versatile since, when used in design, they will produce a conservative, or relatively safe component geometry.

Knowledge of K_{1C} data should help in material selection and in judging the performance of a component when the extent of a defect can be defined. Minute cracks are inherently present in any structural component and their actual size is often below the limit of the sensitivity of nondestructive test equipment. Clearly, the higher the value of K_{1C}, the greater will be the resistance of the material to brittle failure and the greater stress required to produce such a failure.

TABLE 27.1

Yield and Toughness of Engineering Materials

Material	Condition	Form	Minimum Yield (ksi)	Minimum K_{1C} [ksi(in.)$^{1/2}$]
Alloy steel				
18 Ni maraging (200)	Aged 900°F, 6 h	Plate	210	100
18 Ni maraging (250)	Aged 900°F, 6 h	Plate	259	78
18 Ni maraging (300)	Aged 900°F	Plate	276	44
4330 V	Tempered at 525°F	Forging	203	77
4330 V	Tempered at 800°F	Forging	191	93
4340	Tempered at 400°F	Forging	229	40
4340	Tempered at 800°F	Forging	197	71
Stainless steel				
PH13-8 Mo	H1000	Plate	210	78
PH13-8 Mo	H950	Forging	210	70
Titanium alloys				
Ti-6A1-4V	Annealed	—	120	81
Ti-6A1-6V-2Sn	Annealed	—	144	45
Ti-6A1-6V-2Sn	Solution treated and aged	—	179	29
Aluminum alloys				
2014	T651	Plate	57	22
2021	T81	Plate	61	26
2024	T851	Plate	59	19
2124	T851	Plate	64	22
7049	T73	Forging	61	29
7049	T73	Extrusion	73	28
7075	T651	Plate	70	25
7075	T7351	Plate	53	31

Source: Blake, A. (Ed.), *Handbook of Mechanics, Materials, and Structures*, Wiley, New York, 1985.

27.7 PLANE STRESS PARAMETER

The concept of plane strain fracture toughness, K_{1C}, briefly outlined in the previous sections can be viewed as a measure of a material's ability to arrest a crack. But, when a crack begins to propagate, a brittle fracture can occur. Moreover, the speed of propagation can be several thousand feet per second. The amount of plastic energy consumed in the propagation of a crack is assumed to be rather small, so that the process of crack extension is primarily governed by the release of elastic strain energy.

As a material's thickness decreases, lateral constraint relaxes and the size of the plastic zone around the crack tip grows suddenly. This phenomenon is known as "crack-tip blunting," indicating that a relatively large volume of material has deformed. This process may be compared to the behavior of the neck region in a tensile test specimen. The velocities of crack propagation in this instance drop drastically and the crack is arrested.

If, after a crack has been arrested, the loading on the structure is increased so that the stress again exceeds the yield stress of the material, the crack may propagate again. We are then in a state of what might be called "arrestable instability." In this state, the fracture toughness parameter no longer applies.

In the state of arrestable instability, it is helpful to introduce a new parameter K_C called the "plane stress parameter." K_C may be correlated with the fracture toughness parameter K_{1C}, the yield strength σ_y, and the material thickness B, by the expression

$$K_C = K_{1C}\left[1 + \frac{1.4}{B^2}\left(\frac{K_{1C}}{\sigma_y}\right)^4\right]^{1/2} \tag{27.25}$$

Since the energy required to propagate the fracture under plane stress conditions must be high because of the arrestable crack characteristics, it follows that the critical flaw size under plane stress should be higher than that under plane strain conditions. According to the science of fracture mechanics, the relation between the two crack lengths may be stated as follows:

$$a_C = a_{C1}\left[1 + \frac{1.4}{B^2}\left(\frac{K_{1C}}{\sigma_y}\right)^4\right] \tag{27.26}$$

where a_C and a_{C1} are the critical lengths of the cracks in plane stress and plane strain, respectively.

Equation 27.26 may be obtained directly from Equation 27.25 by making the crack ratio directly proportional to the square of the ratio of the respective stress intensity factors K_C and K_{1C}.

Although the concept of a plane stress arrestable instability factor (or "plane stress fracture toughness parameter") K_C is easy to adapt to the analysis of thin-walled components, the task of experimental determination of the K_C values is not without some serious limitations. The fracture corresponding to the K_C parameter is a mixed-mode type involving large amounts of crack-tip plastic flow. Also, the critical length of the crack is difficult to establish because of the limitations of the instrumentation. Nevertheless, approximate values of K_C can be derived experimentally in such cases as, for example, wide panels and sheets. Such a process may be based on the initial crack lengths and the stresses to failure. In the case of heavier sections in low-strength materials, however, more complex correlation techniques are needed before the results can be considered applicable to design.

27.8   PLANE STRESS CRITERION FOR PRESSURE VESSEL DESIGN

The concept of plane stress fracture toughness K_C, as discussed in Section 27.7, is well suited for application in pressure vessel design. We may wish to predict the burst pressure of the vessel for a given surface defect such as a part-through, longitudinal flaw. The choice of the longitudinal orientation of the flaw relates well to the nature of loading in a pressurized cylinder where the critical membrane stress is likely to be in the hoop direction. The task then is to estimate the internal pressure to failure when the dimensions of the flaw and a specific value of the plane stress parameter K_C are known. The basic question in this type of analysis is concerned with the essential design criterion of leak-before-break. This situation has, over the years, provoked a number of scientific investigations and has always had an important issue of industrial safety attached to it. Ideally, given a through-the-wall crack, we prefer that the vessel would leak rather than fail suddenly due to unstable crack propagation through the plate proper or in the weld region of the vessel. The designer should also know what kind of a specific relation exists between the critical crack length and the stress for a given material characterization and the material's thickness intended in pressure vessel applications.

Assuming that plane stress conditions exist in the wall of the vessel, the relevant fracture toughness parameter for a longitudinal through-the-wall crack in a cylindrical geometry can be expressed by the equation

$$K_C^2 = \frac{\pi(1 + 5\nu)a\sigma_m^2}{2(1 + \nu)\cos(\pi\sigma_m/2\sigma_u)}\left[1 + \frac{1.7a^2}{Rt}(1 - \nu^2)^{1/2}\right] \tag{27.27}$$

where, as before, ν is Poisson's ratio and a is the half length of the crack, or imperfection. The parameter σ_m denotes the nominal circumferential stress in the pressure vessel given by the familiar formula: pR/t where p is the internal pressure, R is the vessel radius, and t is the wall thickness. σ_u denotes the ultimate strength of the material.

Equation 27.27 is based upon classical plate theory and it involves a number of corrections to conform to pressure vessel geometry and the nature of critical stress [10]. The corrections include the effect of curvature in going from a flat plate to a cylinder, the influence of the plastic zone at the tip of the defect, and the allowance for the effect of the biaxial state of stress.

The plane stress parameter K_C is given in psi(in.)$^{1/2}$ when σ_m and σ_u are expressed in psi, and a, R, and t are measured in inches.

Suppose that we have a cylindrical vessel of thickness t, which contains a part-through longitudinal flaw having a maximum depth d. The flaw is assumed to be symmetrical and in the form of a hacksaw slot. The presence of this flaw reduces the wall thickness locally and without full penetration. Hence, by this definition, t/d must be greater than unity. If we denote the actual area of the part-through flaw by A_f, its corresponding equivalent length can be defined as A_f/d. The original half-length of the crack entering Equation 27.27 under conditions of full penetration is, of course, a. Hence, the approximate nominal stress σ_m for the part-through flaw can be estimated from Equation 27.27, provided that we introduce $a = A_f/2d$. On substituting this equivalent quantity into Equation 27.27, we obtain

$$K_C^2 = \frac{\pi(1+5\nu)A_f\sigma_m^2}{4d(1+\nu)\cos(\pi\sigma_m/2\sigma_u)}\left[1 + \frac{0.425A_f^2}{Rtd^2}(1-\nu^2)^{1/2}\right] \tag{27.28}$$

The membrane stress to failure σ_f in a cylindrical vessel containing a part-through flaw can be expressed empirically [10] as

$$\sigma_f = \frac{(t-d)\sigma_u^2}{t\sigma_u - d\sigma_m} \tag{27.29}$$

Equations 27.28 and 27.29 may be used to predict the failure stress for a thin vessel with a part-through, longitudinal flaw. If we know the area A_f of the part-through flaw and the maximum depth d, we can iteratively calculate σ_m from Equation 27.28. To do this, we need to know K_C, ν, σ_u, R, and t. If necessary, K_C may be approximated using Equation 27.25 provided that the plane strain fracture toughness K_{1C}, material thickness B, and yield strength σ_y, are known.

In summary, the leak-before-break criterion can be verified analytically by comparing the two critical stress levels, σ_f and σ_m. If the stress to failure σ_f, obtained from Equation 27.29 and intended for a given surface defect exceeds σ_m, calculated from Equation 27.28 for an equivalent length of through-the-wall crack, we can expect the flaw to propagate, leading to a structural failure of the vessel. However, if the failure stress σ_f, calculated for the surface defect such as a part-through flaw proves to be lower than σ_m, the corresponding vessel may be expected to leak before a catastrophic break.

27.9 REMARKS

It should be emphasized that although quite useful, the presented method of fracture mechanics does not provide complete answers for all design situations. For instance, the effect of nonapplied loads such as residual stresses induced by welding is very difficult to interpret. The design parameters, such as K_C or K_{1C}, have to be determined with reference to the direction of the working stresses during the manufacture as well as service. Other special considerations, such as neutron radiation, may also enter the picture. Fracture toughness of ferritic steels is known to be reduced by such a process. Furthermore, the design criteria may be based on the choice of the flaw size, its geometry, and its orientation with respect to a working stress field different from that required for a particular design case. Highly stressed regions such as nozzle junctions and similar transitions pose separate problems of interpretations of test results, inspection techniques, stress analysis methods, and fracture mechanics criteria, which are certainly beyond the scope of this introductory treatment of fracture analysis. Last, but not least, the design factors of safety will be affected by all the technical issues noted above, together with considerations of production economics and the potential consequences of failure.

SYMBOLS

A	Numerical constant
A	Half crack length
a_C, a_{C1}	Critical crack lengths
B	Material thickness
B	Depth of crack
b/a	Aspect ratio
CVN	Impact energy of Charpy V-notch test
E	Modulus of elasticity
G_C	Release rate parameter
K_C	Plane stress parameter
K_{1C}	Plane stress fracture toughness
N	Numerical constant
p	Pressure
R	Vessel radius
R	Hoop radius
T	Wall thickness
v	Poisson's ratio
σ	Stress
σ_f	Membrane stress
σ_h	Hoop stress
σ_m	Nominal circumferential stress
σ_u	Ultimate strength of material
σ_y	Yield stress
σ_{yd}	Dynamic yield stress
ψ	Crack shape parameter

REFERENCES

1. S. Timoshenko, *Strength of Materials, Part II, Advanced Theory and Problems*, D. Van Nostrand Co., New York, 1941 (chap. IX).
2. A. A. Griffith, The phenomena of rupture and flow in solids, *Transactions, Royal Society of London*, 221, 1920.
3. E. A. Lange, Fracture toughness measurements and analysis for steel castings, *AFS Transactions*, 1978.
4. G. R. Sippel, Processing affects fracture toughness, *Met. Prog.*, November 1967.
5. F. L. Singer, *Strength of Materials*, 2nd ed., Harper and Row, New York, 1962, p. 19.
6. F. P. Beer and E. R. Johnston, Jr., *Mechanics of Materials*, 2nd ed., McGraw Hill, New York, 1992, p. 377.
7. G. C. Sih, The role of fracture mechanics in design technology, *ASME Journal of Engineering for Industry*, 98, 1976, pp. 1223–1249.
8. American Society for Testing and Materials, Plane strain crack toughness testing of high strength metallic materials, ASTM 410, Philadelphia, 1967.
9. A. Blake (Ed.), *Handbook of Mechanics, Materials, and Structures*, Wiley, New York, 1985.
10. R. L. Lake, F. W. DeMoney, and R. J. Eiber, Burst tests of pre-flawed welded aluminum alloy pressure vessels at −220°F. In *Advances in Cryogenic Engineering*, 13, Plenum Press, New York, 1968.

28 Fracture Control

28.1 INTRODUCTION

The recognition of the basic problem of "ductile-to-brittle transition" in metallic materials dates back to the time when the welded fabrication of World War II ships was plagued by catastrophic failures. The incidents were characterized by almost instantaneous fractures of entire ships. Although Navy records indicate that World War II ship steel exhibited some 40% elongation, there was obviously no beneficial effect of this property on the structural integrity of the ship plate in the particular environment. This experience has not been limited to the Navy materials and structures. At times, other costly failures were observed, such as a sudden burst of a multimillion-gallon storage tank or unexpected break of a main aircraft spar in flight. The problems were very serious, of course, and, over the years, many large-scale investigations have been sponsored by the U.S. government and private industry to develop remedial measures. In particular, the Naval Research Laboratory has been very active in the studies of fracture phenomena. This work has provided an excellent theoretical and experimental background for fracture-safe design [1–4], with special regard to low- and medium-strength materials. Normally, this implies "fracture control" in design utilizing steels with the yield strength lower than about 120,000 psi. In this category, we encounter the majority of quenched and tempered steels used in modern applications involving rolling stock, shipping, bridges, lifting gear, storage tanks, and automotive components.

High-alloy steels having high resistance to fracture at room temperature include HY-80, HY-100, HY-130, and A-543 ASTM grades and may be acceptable up to thicknesses of about 3 in. In the low-alloy class, A-514 and A-517 ASTM grades serve as examples. For thicker sections, in the range of 6–12 in., the alloying elements have to be increased. However, experience with thick pressure vessels indicates that even good-quality quenched and tempered steels may become highly brittle. One of the best fracture-tough materials has been HY-80 steel, specified by the Navy, which even in welded regions performed successfully in hull structures of both surface ships and submarines since the mid-1950s.

Current manufacturing specifications are designed to give full assurance that the HY-80 weldment system is a well-proven structural material. However, even in this and similar cases, special attention must be paid to any changes in the fabrication processes that may significantly affect the quality of the material, as well as the meaning of a standard test procedure such as Charpy V notch. This is especially important during the process of quality control and the correlation of fracture-tough parameters for design purposes.

28.2 BASIC CONCEPTS AND DEFINITIONS

Fracture control is based upon the basic concepts of fracture mechanics as outlined in Chapter 27. Fracture control utilizes the plane stress fracture toughness parameter K_{1C}, which is a measure of the fracture resistance in the brittle state of a material. The parameter, however, does not represent the full range of fracture toughness between brittle and plastic behavior. Thus it is helpful to have correlation between K_{1C} values and other test parameters.

The dynamic tear (DT) test can provide such correlations. In DT testing, a specimen with a deep sharp crack is broken by a pendulum machine. The upswing of the pendulum following the break is a measure of the energy needed for the fracture. References [2–4] detail methods for relating DT data to K_{1C} values.

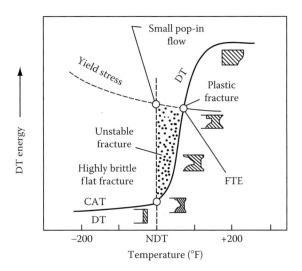

FIGURE 28.1 Transition characteristics for low-strength steel with high shelf fracture toughness. (Adapted from Pellini, W.S., Evolution of engineering principles for fracture-safe design of steel structures, NRL Report 6957, Naval Research Laboratory, 1969.)

The DT test is a relatively inexpensive method for determining fracture toughness as a function of temperature. Figure 28.1 illustrates the kind of information that can be obtained. In this figure (adapted from Ref. [2]), the solid line is known as the crack arrest temperature (CAT) curve or DT curve. The point on the curve denoted by NDT is known as the nil ductility transition temperature. As the temperature is decreased to the NDT point, brittle fracture will occur from a dynamically loaded small crack. As the temperature is increased above the NDT point, we reach a point called the fracture transition elastic (FTE) point. At the FTE point, we have the highest temperature where unstable crack propagation will be driven by elastic stress.

The shaded portion of the diagram of Figure 28.1 thus represents unstable fracture conditions. To the right of the CAT (or DT) curve, the material is stable and will not propagate a crack. To the left of the dashed vertical line, the material produces highly brittle flat fracture. The diagram of Figure 28.1 also shows a superimposed yield stress curve, indicating a small downward trend with an increase in temperature.

The CAT or DT curves have a distinct S-shape. The horizontal branches of the curves care called the lower and upper shelves of fracture toughness. Brittle failure is manifested by a flat fracture, indicating a condition of plane strain. Alternatively, a plastic shear-type failure is characteristic of plane stress. With plane strain (brittle), the elastic portion of the structure envelopes a smaller volume of plastic flow than with plane stress (ductile). That is, with plane strain the elastic past constrains the plastic flow.

This can also be explained in terms of the degree of through-toughness lateral contraction developed during testing of a DT sample. For brittle materials, this contraction is very small. But with ductile materials, the plastic flow is less restricted and considerable notch blunting occurs. Ductile fracture, therefore, is associated with high-energy absorption due to a relatively large plastic zone continuously being formed ahead of the propagating crack.

28.3 CORRELATION OF FRACTURE PROPERTIES

Fracture control involves specification and procurement decisions regarding material toughness. Since specifications, codes, and fabrication techniques frequently change, it is advantageous to obtain the most reliable and current information.

Figure 28.1 shows transition characteristics for a low-strength steel derived from measured values of DT energy. In a similar manner, a transition curve may be constructed from Charpy V-notch (CVN) energy tests. Both these methods are popular in industry despite the finding that the two approaches can produce different results along the energy and temperature axes [5]. The primary intent in developing the DT tests is to obtain a sensitive fracture resistance criterion for the elastic–plastic regions for a broad range of materials.

The use of these tests has led to the development of correlation techniques. In the case of a high-strength steel casing used in oil field explorations, the correlation along the energy axis between the DT and CVN data is

$$CVN = 0.12\,DT + 15 \tag{28.1}$$

The discrepancy along the temperature axis for the casing material is approximately 60°F with the CVN results being less conservative.

Similarly, with conventional bridge and other steels, the shift in the DT energy curve is toward the higher temperature. Barsom [6] suggests the expression

$$T_S = 215 - 1.5S_y \tag{28.2}$$

where
T_S is expressed in °F
the yield strength at room temperature is expressed in ksi (kilopounds per square inch)

This equation should be valid for yield strength values between 36 and 140 ksi.

There is evidence that CVN energy increases with tensile strength of steels for various components such as plates, forgings, and welded parts [7]. The correlation between the plane strain fracture toughness parameter K_{1C} and the CVN values has been empirically established as

$$\left(\frac{K_{1C}}{S_y}\right)^2 = 5\left[\left(\frac{CVN}{S_y}\right) - 0.05\right] \tag{28.3}$$

where
K_{1C} is expressed in ksi (in.)$^{1/2}$
S_y is the 2% yield strength in ksi
CVN is the Charpy V-notch energy in ft lb

28.4 PRACTICAL USE OF CRACK ARREST DIAGRAMS

We can reduce the analysis of fracture safety to examination of a single curve, relating the stress at crack arrest and the temperature of the material. This relationship is known as the CAT curve (see Figure 28.1). The curve represents a practical, conservative criterion upon which a design may be based. In the literature, the CAT curve is generally shown as part of a more complex picture of material behavior known as a generalized fracture analysis diagram [1].

A complete fracture analysis diagram has three basic variables: (1) nominal stress, (2) crack size, and (3) temperature. By adopting a conservative view, only stress and temperature need to be considered. We assume that our particular structure is flaw-free and we further stipulate that it is not necessary to know the actual fracture property of the material, such as the plain strain fracture toughness, K_{1C}. This is a very convenient assumption for the entire process of fracture-safe design.

Figure 28.2 illustrates the principal features of the design procedure. The curve *FABC* represents a portion of the CAT curve designated as CAT, and it is the most important element of the fracture analysis diagram compatible with our conservative philosophy. This line divides the two main regions of the material's behavior. The shaded area under the curve represents, so to speak, a "safe" region

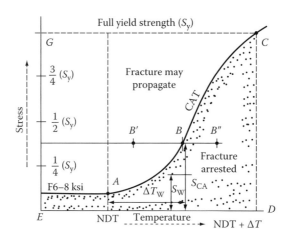

FIGURE 28.2   Crack arrest diagram (CAT).

with respect to fracture initiation and propagation. Suppose that ΔT_W will denote the temperature increment above the NDT at which we expect our structure to be. We assume that the *FABC* curve has been developed experimentally for the selected material and the temperature range containing the NDT. The limiting stress corresponding to ΔT_W is S_{CA}. Point *B* denotes this in Figure 28.2.

We also assume that the working stress is S_W. Hence, we have a positive margin of safety when $S_{CA} > S_W$. When the actual working temperature is lower than (NDT $+ \Delta T_W$), such as that corresponding to, for example, *B′*, a fracture may or may not propagate at a working stress level of S_{CA}. The region above the curve *FABC* is always difficult to analyze without knowledge of the actual crack size. It can only be stated in general that for a given temperature of NDT $+ \Delta T$, the crack propagation stress should increase with the decrease in crack size. Figure 28.2 also suggests that any increase in ΔT_W at a constant stress level shifts our working point to *B″*, progressing further into the "safe" operational region *FABCDE*. From the point of view of fracture safety, however, the upper region *FGCBA* should be considered as more "uncertain" for most applications.

The fracture arrest relationship, expressed in the form of the *FABC* curve, relates to the brittle behavior of the material for various levels of the applied stress. The *FA* portion of this curve represents the lower stress limit in fracture propagation criteria. Numerically, it corresponds to a level of 5 to 8 ksi fracture extension stress in the plane strain region.

A combination of crack-tip blunting and low nominal stresses should prevent rapid crack propagation above NDT, regardless of crack size. Experience with ship structures, involving weld residual stresses, indicates that when the crack moves out of the welded region into a stress field of about 5 to 8 ksi, there is insufficient elastic energy to propagate brittle fracture.

In 1953, Robertson [8] introduces the concept of a lower-bound stress, as seen in Figure 28.2. This concept provides a practical and conservative approach to fracture-safe design particularly in those areas of material control where the existing structures cannot be certified as fracture-resistant. This area then includes off-the-shelf items without prior history of satisfactory fracture toughness as well as new designs in structural steel, which exhibits K_{1C} values of not more than about 25 ksi (in.)$^{1/2}$. This level of plane strain fracture toughness represents a practical lower limit of K_{1C} for what may be termed as "garden variety steel." The corresponding limit for aluminum is taken here as 15 ksi(in.)$^{1/2}$.

When edge-notch criteria of linear elastic fracture mechanics are used, the lower-bound nominal stress becomes a function of the material's thickness and fracture toughness. This observation enables the development of useful design charts. Figure 28.3 illustrates such a chart for steel and aluminum. If a designer limits a working stress to the area below the appropriate curve, any existing crack should not propagate catastrophically under usual conditions of loading and geometry.

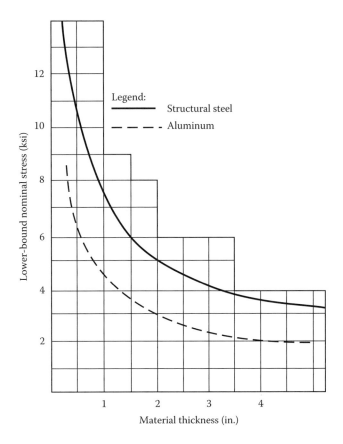

FIGURE 28.3 Lower-bound design curves for steel and aluminum structures.

The lower-bound regions of toughness for steel and aluminum, which form the basis of design limits of Figure 28.3, are consistent with fracture control studies of load-carrying members including welds and heat-affected zones [9]. For other materials, the lower-bound fracture toughness should be obtained from empirical sources.

As we progress along the CAT curve through points *A*, *B*, and *C* in Figure 28.2, the highest temperature of fracture propagation for purely elastic stresses is reached at point *C*. In the literature of fracture-safe design, this point is referred to as "fracture transition elastic." The corresponding stress level reaches the yield strength of the material. The generalized fracture analysis diagram developed over the years indicates that point *C* coincides with NDT + 60°F, which is the more frequently quoted temperature consistent with the last, purely elastic point on the CAT curve. This correlation has been developed primarily for use with steels. If we restrict our use to temperatures above, NDT + 60°F, the problems of crack size becomes unimportant, provided that the level of stress does not exceed the yield strength of the selected steel. Under these conditions the fractures can neither initiate nor propagate.

28.5 THICKNESS CRITERIA

The plane strain fracture toughness K_{1C} is perhaps the most important parameter in fracture mechanics. However, K_{1C} determines only the degree of brittleness. The complete range of fracture toughness between plane strain and plane stress conditions may be characterized by also considering the DT parameter. The yield strength S_y is yet another important parameter, particularly, its magnitude relative to K_{1C}. The ratio K_{1C}/S_y and its relation to the component thickness B is a useful design criterion [10].

The temperature, flaw size, and geometry are also important design parameters. But if we consider only the fracture toughness K_{1C}, yield strength S_y, and thickness B, we can develop relatively simple key recommendations for fracture-safe design,

$$B \leq (K_{1C}/S_y)^2 \quad \text{and} \quad B \geq 2.5(K_{1C}/S_y)^2 \tag{28.4}$$

Design thickness B, meeting or exceeding these recommendations, provides an assurance that through-thickness flaws are unlikely to propagate unless the nominal stresses exceed the yield strength S_y of the material.

The first equation of Equations 28.4 defines the yield criterion and the second corresponds to the theoretical plane strain limit. In general, high K_{1C}/S_y ratios require large section size and large flaws for plane strain fracture initiation. Alternatively, low K_{1C}/S_y ratios correspond to small section size and small cracks for plane strain fracture.

Theoretically, K_{1C}/S_y ratios can vary between 0.5 and 2. Values exceeding 2 are generally unattainable irrespective of flaw size because ductility becomes too high to allow the state of plane strain.

In thicker sections, the level of the mechanical constraint is indicated by the extent of through-thickness contraction adjoining the fracture surface. In thinner sections, the imposed mechanical constraint must be small. These considerations have been explored and reported extensively [4].

For practical design needs, Equations 28.4 provide a satisfactory guide for the development of safe fracture control plans. The only difficulty encountered throughout the various phases of application of fracture mechanics and fracture-safe design principles lies in the basic stress analysis used to determine the nominal stresses. It is important to emphasize that a fracture critical component is defined as a tension-loaded member, which in the presence of geometrical discontinuities, may have significant stress gradients. The accuracy of the analysis of such gradients and local stress concentrations should determine the final factors of safety based on the thickness criteria given by Equation 28.4.

28.6   SIGNIFICANCE OF STRESS AND STRENGTH

Consider a structural component with a flaw or crack. If the components are loaded so that there is stress in the interior of the component, and at the crack, then it is the component of the stress normal to the crack, which tends to open and propagate the crack. Therefore, it is this "normal to the crack" stress, which should be used in fracture control analyses and with design charts such as in Figure 28.2.

According to the basic criteria involving the transition temperature, the designer can aim at either fracture-tough or fracture-safe design. From the stress point of view, both approaches constitute the major elements of the fracture control process. The object of fracture-tough design would be to select a material that could prove to be insensitive to crack propagation and brittle failure throughout the working range of temperatures, material thicknesses, and design stresses.

In fracture-safe design with a known CAT curve, as in Figure 28.2, the working stress S_W should be less than the limiting stress S_{CA} at a selected temperature. For example, at temperature $NDT + \Delta T_W$ (nil ductility transition temperature plus temperature margins above NDT corresponding to the working stress) corresponding to point B, the maximum stress that can be arrested is approximately 37.5% of the yield strength S_y of the material, on the vertical axis in Figure 28.2. When the operating temperature is less than NDT, the rate of 5000 to 8000 psi applies.

The problem with such a low shelf value is that a weight penalty may occur in developing the low-stress system. Also, unless the stress is low in the design region, left of the dashed vertical line in Figure 28.2, the material is essentially brittle and plane strain failure can be initiated in the presence of a very small flaw.

In many cases, structural members are designed to perform in an as-welded but not stress-relieved condition. The residual stresses in this instance can be as high as the yield strength, and they are oriented in the direction parallel to the weld. Such stresses result from the longitudinal shrinkage during cooling of the weld area, which is restrained by the adjacent colder metal. In practice, these peak stresses extend to about one to two weld widths and therefore they can contain only relatively small cracks. We therefore have a highly localized residual stress region in the weld or heat-affected zone, which may initiate fracture due to a small crack for a wide range of stresses and when the service temperature falls below the NDT.

The brittle condition in the heat-affected zone is further aggravated by the formation of a coarse grain structure, which is inherently more brittle than the finer grain size of the parent metal. Additional aggravation is encountered if the rapid cooling of the weld results in the formation of martensite, a hard, brittle microconstituent. However, at temperatures above the NDT, such a fracture will not occur because of the requirement for extensive plastic deformation. This theory has been verified experimentally [11].

When the transition curve corresponds to a low-strength steel, it normally exhibits a high shelf fracture toughness. On the other hand, a flat S curve, indicating no significant transition temperature, is characteristic of high-strength steels. This is an important practical consideration, because increasing service temperature for a high-strength material may not necessarily assure that the new system will become much more fracture-safe.

28.7 CONCLUDING REMARKS

In this chapter, we have only presented some basics of fracture control. For composites, nonhomogeneous materials and geometrically complex components, the designer may want to consult the ever growing literature on fracture mechanics and also employ the increasingly sophisticated numerical procedures which are becoming available.

We list here a few basic concepts for reference:

1. Yield strength alone is not an indicator of fracture resistance.
2. Ductile materials can fail in a brittle manner when they are below the NDT.
3. Fracture mechanics and fracture-safe design relate only to failure in tension.
4. When small flaws are present, the flaw and not the section size control the initiation and progress of the fracture.
5. For large flaws, the section size is important. It shifts the CAT-DT curve toward higher temperature.
6. Cracks will not propagate under normal conditions when the stress is less than lower bound design values as in Figure 28.3.
7. A crack will be arrested when the crack enters either a lower-stressed region, a region of higher fracture toughness, or both.
8. A safe metal thickness can be determined when both the plane strain fracture toughness parameter K_{1C} and the yield strength S_y are known.
9. Increasing the section thickness may not provide additional safety margin unless the K_{1C}/S_y ratio indicates an improvement.
10. Design modification lowering the NDT parameter is not likely to be cost-effective.

Whatever method is adopted for the development of a fracture control plan, it is safe to assume that real engineering materials must contain some manufacturing imperfections. We have, therefore, the task of designing around a potential problem area using both linear elastic fracture mechanics and experimental data on crack arrest characteristics.

SYMBOLS

A Cross-section area
B Material thickness
CAT   Crack arrest temperature
CVN   Charpy V-notch energy
DT   Dynamic tear
FTE   Fracture transition elastic/plastic
K Stress concentration factor
K_{1C} Plane stress fracture toughness
NDT   Nil ductility transition temperature
P Nominal axial force
S_{CA} Limiting stress
S_W Working stress
S_y Yield strength
T_s Temperature shift
ΔT Temperature margin above NDT
ΔT_W Temperature margin above NDT, corresponding to working stress

REFERENCES

1. W. S. Pellini and P. P. Puzak, Fracture analysis diagram procedures for the fracture-safe engineering design of steel structures, NRL Report 5920, Naval Research Laboratory, 1963.
2. W. S. Pellini, Evolution of engineering principles for fracture-safe design of steel structures, NRL Report 6957, Naval Research Laboratory, 1969.
3. W. S. Pellini, Integration of analytical procedures for fracture-safe design of metal structures, NRL Report 7251, Naval Research Laboratory, 1971.
4. W. S. Pellini, Principles of fracture-safe design, *Welding Journal* (Suppl.), 50, 1971, pp. 147–162.
5. E. A. Lange and L. A. Cooley, Fracture control plans for critical structural materials used in deep-hole experiments, NRL Memorandum Report 2497, Naval Research Laboratory, 1972.
6. J. M Barsom, Development of the ASSHTO fracture toughness requirements for bridge steels, *Engineering Fracture Mechanics*, Vol. 7, Permagon Press, London, 1975.
7. P. P. Puzak and E. A. Lange, Significance of Charpy-V test parameters as criteria for quenched and tempered steels, NRL Report 7483, Naval Research Laboratory, 1972.
8. T. S. Robertson, Propagation of brittle fracture in steel, *Journal of the Iron and Steel Institute*, 175, 1953.
9. T. S. Rolfe and J. M. Barsom, *Fracture and Fatigue Control in Structures*, Prentice-Hall, Englewood Cliffs, NJ, 1977.
10. American Society for Testing and Materials (ASTM), Plane strain crack toughness testing of high strength metallic materials, ASTM 410, Philadelphia, PA, 1967.
11. A. A. Wells, Brittle fracture strength of welded steel plates, *British Welding Journal*, 1961, pp. 259–274.

Part VI

Piping and Pressure Vessels

Pipes and pressure vessels have been used extensively for hundreds of years to transport and store liquids and gasses. When viewed as structural components, pipes and vessels present unique and special problems for designers and engineers. Undoubtedly, the most important consideration is safety: preventing leaks and explosions.

In Part IV, we considered plates and panels and their structural characteristics. Plate theory is considerably more complex than beam theory. Plates, however, with plane surfaces are considerably simpler than shells with their curved surfaces. Nevertheless, shells are ideally suited for modeling pipes and pressure vessels.

The theory and analysis of shells is an extensive field of study. Shells with simple geometry have been investigated for many years and the literature on shell theory is extensive. No single book part or even an entire book can comprehensively document the many advances. Here, we will simply summarize the results and procedures useful in pipe and vessel design.

The simplest loading of pipes and pressure vessels is via internal pressure. External pressure and other loadings may also be of interest to structural analysts and designers. Accordingly, we devote a rather lengthy chapter to externally pressurized cylindrical structures and another short chapter to externally pressurized spherical structures.

We begin this part with a study of internally pressured structures. Then, following a chapter on externally loaded members, we look at axial and bending response and some special problems in cylinder structures.

29 Vessels with Internal Pressure

29.1 INTRODUCTION

The more common forms of pressure vessels involve spheres, cylinders, and ellipsoids, although conical and toroidal configurations are also found. When such components have small thickness compared with the other dimensions and offer a limited resistance to bending, the stress can be calculated with the aid of the membrane theory. Such stresses, taken as average tension or compression over the thickness of the vessel wall, act in the direction tangential to the surface. Since the middle surface of the wall extends in two dimensions, the analysis can become complicated where more than one expression for the curvature is required to describe the displacement of a particular point. In a more rigorous sense, it would be necessary to define a normal force, two transverse shearing forces, two bending moments, and a torque in order to describe the entire state of stress. Fortunately, membrane theory allows us to neglect the bending, shearing, and twisting effects. In a number of elementary but practical cases, the simple equations of equilibrium of forces are sufficient for deriving the necessary design formulas.

29.2 THIN CYLINDERS

Consider a thin-walled circular cylinder as in Figure 29.1. Let the cylinder have end caps as shown and let there be an internal pressure P. This pressure induces both, longitudinal and circumferential stresses.

Analogous to a circular ring under internal pressure, the resultant membrane force F_t, shown in Figure 29.2, may be obtained by adding force components along the X-axis: That is

$$2F_t = \int_0^{\pi/2} PR_i \sin\theta \, d\theta \quad \text{or} \quad F_t = PR_i \tag{29.1}$$

where R_i is the inner radius of the cylinder.

In the development of Equation 29.1, the longitudinal length cancels and thus F_t may be assumed to act on a unit length. By dividing by the cylinder thickness T in Equation 29.1, we obtain the familiar pressure vessel formula:

$$S_t = \frac{PR_i}{T} \tag{29.2}$$

In spite of its simplicity, Equation 29.2 has wide application with many structures of practical importance such as boiler drums, accumulators, piping, casing, chemical processing vessels, and nuclear pressure vessels.

Equation 29.2 is often referred to as the "hoop stress formula." It provides the maximum hoop (tangential) stress in a vessel wall assuming that the end closures provide no support as in long cylinders. It is also evident that the inner radius R_i may be replaced by the mean radius R without any significant error if the cylinder wall is sufficiently thin.

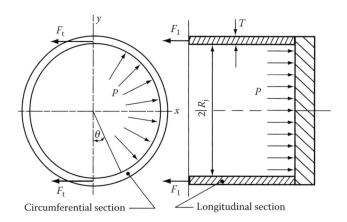

FIGURE 29.1 A thin-walled cylinder with sealed ends and internal pressure.

Since the ratio R/T appears consistently in the analysis, it is convenient to denote it by the parameter m. Equation 29.2 then becomes

$$S_t = mP \quad (m = R/T) \tag{29.3}$$

By considering equilibrium in the longitudinal direction we see that the longitudinal membrane force F_ℓ (see Figure 29.1) is simply

$$F_\ell = P\pi R^2 \tag{29.4}$$

The corresponding axial stress S_ℓ is then:

$$S_\ell = mP/2 \tag{29.5}$$

In the analysis of a general shell of revolution, the term "meridional" is sometimes used instead of "axial" or "longitudinal." It is now evident from Equations 29.3 and 29.5 that the efficiency of the circumferential joints need only be half that of the longitudinal joints. It is also clear from the equilibrium of forces in a spherical shell that the relevant maximum stress in the shell is represented by Equation 29.5. This simple deduction is of great importance in the design of pressure vessels

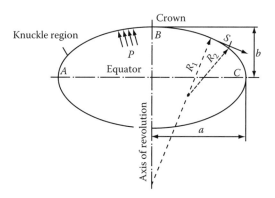

FIGURE 29.2 Ellipsoidal geometry.

because the thickness requirement for a spherical vessel, for the same material strength and parameter m, is only one half of that necessary for a cylinder. Hence, the sphere is the most efficient configuration for a pressure vessel.

29.3 RADIAL GROWTH (DILATION)

An important criterion in pipe and vessel analysis is "radial growth," or dilation, under internal pressure. For a long cylindrical vessel this dimensional change is given by

$$\delta = \frac{(2 - \nu)PR^2}{2ET} \tag{29.6}$$

where, as before,
  ν is Poisson's ratio
  E is the modulus of elasticity

The corresponding dilation of a spherical vessel is

$$\delta = \frac{(1 - \nu)PR^2}{2ET_s} \tag{29.7}$$

where T_s is the thickness of the spherical shell.

When a vessel is made out of a cylindrical portion and hemispherical heads, we can equate Equations 29.6 and 29.7 to find a relation between the thicknesses so that there is no displacement difference at the seam. The result is

$$T_s = \frac{T(1 - \nu)}{2 - \nu} \tag{29.8}$$

When $\nu = 0.3$, Equation 29.8 gives $T_s = 0.41T$. For the complete range of ν between 0.0 and 0.5, the thickness ratio T_s/T is between 0.5 and 0.33.

The analysis of a general shell of revolution leads to other formulas, such as those required in the calculation of the conical and ellipsoidal vessels. For the cone configurations subtending angle 2α, the hoop and meridional stresses are

$$S_t = \frac{mP}{\cos \alpha} \tag{29.9}$$

and

$$S_\ell = \frac{mP}{2 \cos \alpha} \tag{29.10}$$

As α approaches zero, the hoop stress for a conical shell approaches that for a cylinder. However, when the cone begins to flatten and α approaches 90°, the stress becomes unreasonably large, indicating that a flat membrane cannot support loads perpendicular to its plane. Similar reasoning can be applied to Equation 29.10.

Analogous to a cylindrical vessel, the radial growth delta of a conical vessel is

$$\delta = \frac{(2 - \nu)PR^2}{2ET \cos \alpha} \tag{29.11}$$

29.4 ELLIPSOIDAL SHELLS

Ellipsoidal shells are frequently used as closures for pressure vessels and storage tanks. Stress analyses of ellipsoid shells are complicated because the curvature varies from point to point. Figure 29.2 illustrates the geometry.

In Figure 29.2, the portion ABC is frequently used as a vessel head with its axis of revolution coinciding with the cylinder axis. The meridional stress S_m is then the same as the longitudinal stress, S_ℓ, in the adjoining cylinder. That is (see Equation 29.5)

$$S_\ell = \frac{Pa}{2T} = \frac{PR}{2T} \tag{29.12}$$

The hoop stress, acting in a plane perpendicular to the plane of the figure, is

$$S_t = \frac{Pa(2b^2 - a^2)}{2Tb^2} \tag{29.13}$$

It is of some interest to note that the hoop stress can change its sign and become compressive with $a/b > \sqrt{2}$, despite the fact that the vessel is loaded internally. For this condition also, the maximum shear stress τ found at the equator (plane AC) becomes

$$\tau = \frac{Pa(a^2 - b^2)}{4Tb^2} \tag{29.14}$$

The stress of Equation 29.14 is important for vessels made of ductile materials. Experiments indicate that a buckle case envelops in the knuckle region under internal pressure due to the change from tensile to compressive hoop stress for certain values of a/b [1].

The equatorial dilation of an ellipsoidal head is not always positive and may be expressed by the formula

$$\delta = \frac{PR^2}{ET} \left(1 - \frac{a^2}{2b^2} - \frac{\nu}{2} \right) \tag{29.15}$$

When $\nu = 0.3$ and $a/b > 1.3$, the equatorial dilation can become negative, and it can cause an increase in the discontinuity stresses when a purely ellipsoidal head is used with a cylindrical shell of the same thickness.

29.5 TOROIDAL VESSELS

In many cases, a torus or a doughnut-shaped pressure vessel is used in construction. This could be a steam generator, a bent tube, or a containment vessel in a nuclear reactor system. The stress analysis formulas can be developed from the equilibrium of forces with reference to the sketch shown in Figure 29.3. For instance, the vertical load on a shaded portion of the torus is

$$V = \pi(x^2 - \zeta^2)P \tag{29.16}$$

Also, from the component of hoop stress, this load is

$$V = 2\pi x T \cos \theta \, S_t \tag{29.17}$$

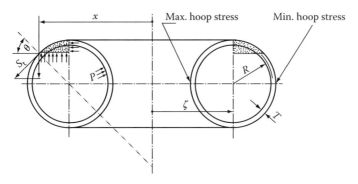

FIGURE 29.3 Circular torus geometry and notation.

Equating Equations 29.16 and 29.17 yields

$$S_t = \frac{P(x^2 - \zeta^2)}{2xT\cos\theta} \tag{29.18}$$

where, from Figure 29.3, x is

$$x = \zeta + R\cos\theta \tag{29.19}$$

Then, the general design formula for hoop stress is

$$S_t = \frac{PR}{2T}\left(\frac{2\zeta + R\cos\theta}{\zeta + R\cos\theta}\right) \tag{29.20}$$

When θ is zero, Equation 29.20 gives the minimum possible value of S_t as

$$S_t = \frac{PR}{2T}\left(\frac{2\zeta + R}{\zeta + R}\right) \tag{29.21}$$

For the case of the torus centerline, that is, $\theta = \pi/2$, we obtain the standard hoop formula of Equation 29.3. However, when the θ line rotates in such a manner that $\theta = \pi$ and $x = \zeta - R$, the hoop stress in the torus becomes

$$S_t = \frac{PR}{2T}\left(\frac{2\zeta - R}{\zeta - R}\right) \tag{29.22}$$

The meridional stress is given by Equation 29.5 as

$$S_\ell = mP/2 \quad m = R/T \tag{29.23}$$

The meridional stress value is independent of location on the torus.

The same formulas can also be obtained from the Laplace equation for a thin axisymmetric shell [2] given by

$$\frac{S_\ell}{R_1} + \frac{S_t}{R_2} = \frac{P}{T} \tag{29.24}$$

where R_1 and R_2 are the relevant radii of curvature.

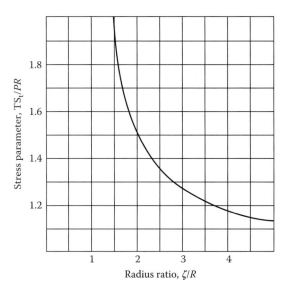

FIGURE 29.4   Variation of maximum hoop stress with torus geometry.

Figure 29.4 provides a variation of the maximum hoop stress with the ratio of the bend radius ζ to the cross-section radius R. The vertical axis is a dimensionless parameter obtained by dividing the stress in torus by the hoop stress in the corresponding straight pipe. Theoretically, hoop stress becomes infinitely large for very small bend radii, that is, when the doughnut hole is essentially closed. Experience indicates that caution must be exercised when applying torus theory to the process of tube bending to a small radius. Strain hardening of the material and wall thickening on the inside of the sharp bend may cause the pipe to fail on the centerline of the bend instead of the place where the stress is essentially the same as that in a straight cylinder. In such a case, Equation 29.22 would not be recommended.

The majority of pipe and vessel configurations currently used can be classified as relatively thin, and the dividing line between the thin and thick vessels may be set at an m value of about 10. The lower practical limit of m is about 2.5 for cylinders and 3.5 for spheres. With these types of m ranges, it is customary to use the membrane stress criteria for thin shells and Lamé theory for thick shells. Use of these theories, however, rests entirely on the application of the elastic principles, neglecting the criteria of materials failure.

29.6   THICK CYLINDER THEORY

Although it is beyond our scope to discuss details of the mathematical theory of elasticity, it may be helpful to review some concepts of strain in cylindrical coordinates and in spherical coordinates (see Section 6.9). This will enable us to develop expressions for stresses and displacements in thick-walled vessels.

Figure 29.5 shows an element undergoing displacement and deformation in a polar, or cylindrical, coordinate system. From the figure and also from Equations 6.133, 6.134, and 6.136, we see that the radial, tangential, and shear strains are

$$\varepsilon_r = \frac{\partial u}{\partial r} \tag{29.25}$$

$$\varepsilon_t = (1/r)\frac{\partial v}{\partial \theta} + u/r \tag{29.26}$$

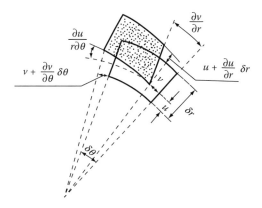

FIGURE 29.5   Displacement and deformation of a two-dimensional element in cylindrical polar coordinates.

$$\varepsilon_{rt} = (1/2)\left[(1/r)\frac{\partial u}{\partial \theta} + \frac{\partial v}{\partial r} - (1/r)v\right] \tag{29.27}$$

For axisymmetric problems, the tangential displacement v and the shear strain ε_{rt} are zero so that the above strains reduce to

$$\varepsilon_r = \frac{du}{dr}, \quad \varepsilon_t = \frac{u}{r}, \quad \text{and} \quad \varepsilon_{rt} = 0 \tag{29.28}$$

Thick cylinder theory is applicable in gun barrels, hydraulic ram cylinders, heavy piping, and similar configurations. Figure 29.6 provides a view of stresses on an element. In a general case, the wall can be subjected to an internal pressure P_i and external pressure P_o, corresponding to the radii, R_i and R_o, respectively. Furthermore, because of the symmetry of loading, it is sufficient to analyze one element of the cylinder subtended by a small angle $d\theta$ and a small radial thickness dr. Summing up all the forces in the direction of the bisector of the angle $d\theta$ and making $\sin d\theta = d\theta$ gives

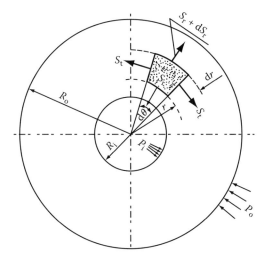

FIGURE 29.6   Stresses on an element of a thick cylinder.

$$S_r r \, d\theta + S_t \, dr \, d\theta - \left(S_r + \frac{dS_r}{dr} dr \right)(r + dr) \, d\theta = 0 \tag{29.29}$$

Neglecting small quantities in Equation 29.29 leads to the equation

$$S_t - S_r - r \frac{dS_r}{dr} = 0 \tag{29.30}$$

Using Hooke's law (see Section 7.7, Equations 7.55 through 7.60), we can express S_r and S_t in terms of ε_r and ε_t as

$$S_r = E \frac{\nu \varepsilon_t + (1 - \nu)\varepsilon_r}{(1 + \nu)(1 - 2\nu)} \tag{29.31}$$

and

$$S_t = E \frac{\nu \varepsilon_r + (1 - \nu)\varepsilon_t}{(1 + \nu)(1 - 2\nu)} \tag{29.32}$$

Using Equations 29.28, we can express the stresses in terms of the displacement u as

$$S_r = E \frac{\nu \frac{u}{r} + (1 - \nu)\frac{du}{dr}}{(1 + \nu)(1 - 2\nu)} \tag{29.33}$$

and

$$S_t = E \frac{\nu \frac{du}{dr} + (1 - \nu)\frac{u}{r}}{(1 + \nu)(1 - 2\nu)} \tag{29.34}$$

Finally, by substituting from Equations 29.33 and 29.34 into Equation 29.30 and simplifying we obtain

$$\frac{d^2 u}{dr^2} + \frac{1}{r} \frac{du}{dr} - \frac{u}{r^2} = 0 \tag{29.35}$$

The general solution of this equation is

$$u = C_1 r + C_2/r \tag{29.36}$$

where C_1 and C_2 are integration constants to be evaluated by the boundary condition. Thus, du/dr is

$$\frac{du}{dr} = C_1 - C_2/r^2 \tag{29.37}$$

The radial and tangential stresses are then (see Equations 29.33 and 29.34)

$$S_r = E \frac{\nu(C_1 + C_2/r^2) + (1 - \nu)(C_1 - C_2/r^2)}{(1 + \nu)(1 - 2\nu)} \tag{29.38}$$

$$S_t = E \frac{\nu(C_1 - C_2/r^2) + (1 - \nu)(C_1 + C_2/r^2)}{(1 + \nu)(1 - 2\nu)} \tag{29.39}$$

Equations 29.38 and 29.39 form a pair of simultaneous linear equations for C_1 and C_2, which may be obtained by enforcing the conditions that S_r is P_i when $r = R_i$ and P_o when $r = R_o$. Then, by substituting these values for C_1 and C_2 back into Equations 29.38 and 29.39, we obtain the stresses as

$$S_r = \frac{R_i^2 P_i - R_o^2 P_o}{R_o^2 - R_i^2} - \frac{R_i^2 R_o^2 (P_i - P_o)}{r^2 (R_o^2 - R_i^2)} \tag{29.40}$$

and

$$S_t = \frac{R_i^2 P_i - R_o^2 P_o}{R_o^2 - R_i^2} + \frac{R_i^2 R_o^2 (P_i - P_o)}{r^2 (R_o^2 - R_i^2)} \tag{29.41}$$

Equations 29.40 and 29.41 are known as Lamé formulas, named after a French engineer who obtained this solution in 1833. This theory has withstood the test of time, advancing technology, and modern trends of numerical analysis in a remarkable manner. The maximum shearing stress at any point of a thick cylinder follows from Equations 29.40 and 29.41 and is equal to

$$\tau = \frac{(P_i - P_o) R_o^2 R_i^2}{(R_o^2 - R_i^2) r^2} \tag{29.42}$$

Finally, from Equation 29.36, the radial displacement u is

$$u = \frac{1 - \nu}{E} \frac{R_i^2 P_i - R_o^2 P_o}{R_o^2 - R_i^2} r + \frac{1 + \nu}{E} \frac{R_i^2 R_o^2 (P_i - P_o)}{(R_o^2 - R_i^2) r} \tag{29.43}$$

When the cylinder is subjected only to internal pressure, we have

$$S_r = \frac{R_i^2 P_i}{(R_o^2 - R_i^2)} \left(\frac{r^2 - R_o^2}{r^2} \right) \tag{29.44}$$

$$S_t = \frac{R_i^2 P_i}{(R_o^2 - R_i^2)} \left(\frac{r^2 - R_o^2}{r^2} \right) \tag{29.45}$$

$$\tau = \frac{P_i R_o^2 R_i^2}{(R_o^2 - R_i^2) r^2} \tag{29.46}$$

and

$$u = \frac{1 - \nu}{E} \frac{R_i^2 P_i r}{R_o^2 - R_i^2} + \left(\frac{1 + \nu}{E} \right) \frac{R_i^2 R_o^2 P_i}{(R_o^2 - R_i^2) r} \tag{29.47}$$

It follows from the Lamé theory that the sum of the two stresses given by Equations 29.40 and 29.41 remains constant, suggesting that the deformation of all the elements in the axial direction is the same and the cylinder cross sections remain plane after the deformation. Equations 29.44 and 29.45 also indicate that both stresses reach maximum values at $r = R_i$ and that S_r is always compressive and smaller than S_t. The minimum value of the tangential stress is found at $r = R_o$, which is smaller than that at the inner surface.

So far, the equations considered have been applicable to an infinitely long cylinder, in which no axial stress was present. However, when the cylinder contains rigid closures and no change in length is possible, the axial stress under internal pressure is

$$S_\ell = \frac{2\nu P_i R_i^2}{R_o^2 - R_i^2} \tag{29.48}$$

When the vessel contains closures but is free to change its length under strain, the axial stress is

$$S_\ell = \frac{P_i R_i^2}{R_o^2 - R_i^2} \tag{29.49}$$

The corresponding axial strain is

$$\varepsilon_\ell = \frac{(1 - 2\nu)P_i R_i^2}{E\left(R_o^2 - R_i^2\right)} \tag{29.50}$$

Finally, when the pressure is maintained by a piston-type closure at each end of the cylinder and there is no connection between the cylinder and the piston, the axial stress is zero and the strain is

$$\varepsilon_\ell = -\frac{2\nu P_i R_i^2}{E\left(R_o^2 - R_i^2\right)} \tag{29.51}$$

29.7 THICK-WALLED SPHERE

We can develop governing equations for a thick-walled spherical container by considering an element of the sphere as in Figure 29.7. Due to the symmetry, there are no shearing stresses on the faces of the element. Adding forces on a radial plane through the center of the element produces the expression [3]

$$\left(S_r + \frac{dS_r}{dr}dr\right)(r + dr)^2(d\theta)^2 - S_r r^2 (d\theta)^2 - 4 S_t r dr \, d\theta \sin\left(\frac{d\theta}{2}\right) = 0 \tag{29.52}$$

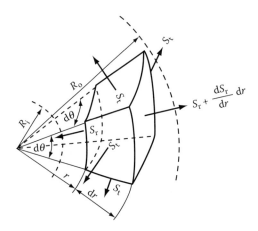

FIGURE 29.7 Equilibrium of an element of a spherical shell.

Simplifying this expression and neglecting higher-order terms yields

$$\frac{dS_r}{dr} + \frac{2}{r}(S_r - S_t) = 0 \tag{29.53}$$

Since the transverse stress S_t is the same in all directions, as indicated in Figure 29.7, the radial and tangential strains can be expressed as

$$\varepsilon_r = \frac{S_r}{E} - \frac{2\nu S_t}{D} \tag{29.54}$$

and

$$\varepsilon_t = \frac{(1-\nu)S_t}{E} - \frac{\nu S_r}{E} \tag{29.55}$$

Due to the symmetry, the radial and transverse strains may be expressed in terms of the radial displacement u as

$$\varepsilon_r = du/dr \quad \text{and} \quad \varepsilon_t = u/r \tag{29.56}$$

By using Hooke's law and substituting these expressions into the equilibrium expression of Equation 29.53 and simplifying, we obtain

$$\frac{d^2u}{dr^2} + \frac{2du}{rdr} - \frac{2u}{r^2} = 0 \tag{29.57}$$

The general solution of Equation 29.57 may be written in the form

$$u = \frac{(1-2\nu)Ar}{E} + \frac{(1+\nu)B}{Er^2} \tag{29.58}$$

where A and B are integration constants. If we solve Equations 29.54 and 29.55 for the stresses and use the strain–displacement relations of Equations 29.56 we obtain

$$S_r = E\frac{2\nu\frac{u}{r} + (1-\nu)\frac{du}{dr}}{1 - \nu - 2\nu^2} \tag{29.59}$$

and

$$S_t = E\frac{\frac{u}{r} + \nu\frac{du}{dr}}{1 - \nu - 2\nu^2} \tag{29.60}$$

By substituting the displacement u given by Equation 29.58 into Equations 29.59 and 29.60 and by imposing the boundary pressures P_i and P_o on the interior and exterior of the vessel, we can solve for A and B. The results are

$$A = \frac{P_iR_i^3 - P_oR_o^3}{R_o^3 - R_i^3} \tag{29.61}$$

and

$$B = \frac{R_i^3R_o^3(P_i - P_o)}{2(R_o^3 - R_i^3)} \tag{29.62}$$

Finally, by substituting back into Equation 29.59 and 29.60, the radial and transverse stresses take the form

$$S_r = \frac{P_i R_i^3 - P_o R_P^3}{R_o^3 - R_i^3} - \frac{R_i^3 R_o^3 (P_i - P_o)}{r^3 (R_o^3 - R_i^3)}$$

(29.63)

and

$$S_t = \frac{P_i R_i^3 - P_o R_o^3}{R_o^3 - R_i^3} + \frac{R_i^3 R_o^3 (P_i - P_o)}{2r^3 (R_o^3 - R_i^3)}$$

(29.64)

The corresponding expression for the radial displacement is

$$u = \frac{(1 - 2\nu)r \left(P_i R_i^3 - P_o R_o^3\right)}{E \left(R_o^3 - R_i^3\right)} + \frac{(1 + \nu)R_i^3 R_o^3 (P_i - P_o)}{2Er^2 \left(R_o^3 - R_i^3\right)}$$

(29.65)

For the case of internal pressure alone the stresses and displacement of Equations 29.63, 29.64, and 29.65, become

$$S_r = \frac{P_i R_i^3}{R_o^3 - R_i^3} \left(1 - \frac{R_o^3}{r^3}\right)$$

(29.66)

$$S_t = \frac{P_i R_i^3}{R_o^3 - R_i^3} \left(1 + \frac{R_o^3}{2r^3}\right)$$

(29.67)

and

$$u = \frac{P_i R_i^3}{E \left(R_o^3 - R_i^3\right)} \left[(1 - 2\nu)r + \frac{(1 + \nu)R_o^3}{2r^2}\right]$$

(29.68)

29.8 DESIGN CHARTS FOR THICK CYLINDERS

A common design problem is to estimate stresses and displacements at the surfaces of thick cylinders under internal pressure. The maximum tangential stress S_t at the inner surface is

$$S_t = K_1 P_i$$

(29.69)

where as before

P_i is the internal pressure

K_1 is a dimensionless design factor which depends upon the thickness of the cylinder or specifically the radius/thickness ratio

Figure 29.8 provides a representation of K_1 for various ratios. Observe that K_1 is always larger than 1, indicating that the tangential stress exceeds the applied internal pressure.

Correspondingly, the displacements u_i and u_o at the inner and outer surfaces, respectively are

$$u_i = K_2 P_i (R_o - R_i)/E \quad \text{and} \quad u_o = K_3 P_i (R_o - R_i)/E$$

(29.70)

where the factors K_2 and K_3 are also given by Figure 29.8.

The factors given by Figure 29.8 have been calculated for metals, such as steel, with a Poisson's ratio ν of 0.3. It happens, however, that ν has a relatively small effect on the magnitude of the displacement, therefore, the values of Figure 29.8 should be reasonable for nonmetallic elastic materials as well.

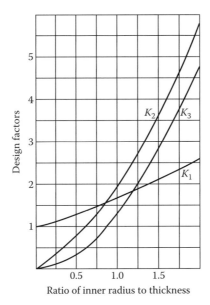

FIGURE 29.8 Stress and displacement factors for thick cylinders under internal pressure.

29.9 ULTIMATE STRENGTH CRITERIA [4]

So far, various formulas have been discussed that allow reasonably accurate design estimates of the elastic stresses and deflections in thin and thick pressure vessels or piping. The complete problem, however, may also require the prediction of the ultimate pressures in cylinders and spheres at which these components can fail by bursting.

The general criteria of shell failure under internal pressure have been developed from the classical theory of plasticity, which has been shown to be suitable to explain burst characteristics of thick-walled vessels made of ductile materials. The essential difference between the elastic and plastic response of a thick shell can be illustrated as follows. In a purely elastic response of a thick shell, the maximum stress under internal pressure develops at the bore. On further load increase at this point, the material reaches the yield but, contrary to what one might expect, failure of ductile fibers does not start at the bore. The reason for this is that the strain at the inner zone of the wall is held at a constant level by the restraint of the outer fibers until the region of plastic flow moves radially outward. The motion of the elastic–plastic interface causes the tangential stresses in the outer fibers to increase until a complete state of plastic strain is established throughout the wall and the fibers along the outer surface of the shell begin to fail. Researchers in the field of plasticity have defined the bursting pressure as the internal pressure required to move the elastic–plastic interface into the outer radius of the vessel.

The state of plastic stress defined above as a criterion of bursting pressure has several important implications. First of all, a vessel made of a ductile material has a considerable amount of strength beyond the onset of yielding at the inner fiber and should retain its usefulness up to the very point of fracture. This reserve of strength, however, must be dependent on the entire history of stress and temperature at a particular critical location. It is therefore necessary to assign a correct level of design stress in predicting the relevant bursting pressures. The formulas for this purpose contain either yield or ultimate strength terms of the material involved. In the case of the majority of vessel applications, work-hardening materials are used, which require knowledge of the stress–strain curves. This condition introduces additional functional relationships in the development of pressure–deformation equations for cylinders and spheres. It has been customary to employ a Ludwik type of a stress–strain curve in which the stress is represented by an exponential function of strain.

The work-hardening capacity of the material is said to increase when the exponent of the stress–strain curve increases. The bursting pressure, in turn, decreases with the increase in work-hardening.

The technical literature contains a multitude of design formulas for the calculation of bursting strength of thick- and thin-walled vessels, derived from the assumed theories of failure and containing various limitations. One development in this area comes from England [5]. This approach is based on the pressure–deformation response of a vessel and on an idealized stress–strain curve of the Ludwik type. The formulas resulting from this work are intended for work-hardening materials, provided that the strain-hardening exponent for the particular material can be established. In the case of commonly employed low- and intermediate-strength low-alloy and carbon steels, a simple relation can be obtained between the strain-hardening exponents and the materials strength ratio β. This ratio is calculated by dividing the yield strength of the material by its ultimate strength. It is useful to recall here that the definition of the yield point on a particular stress–strain plot can lead to some difficulties. To circumvent this problem, it has been a custom in industry to accept a 0.20% offset yield strength for most engineering calculations. This offset strength value is usually higher than the elastic limit of the material by about 10%, but the exact spread between the two values will depend on the type of material involved.

29.10   BURST PRESSURE OF CYLINDERS AND SPHERES

As before, let β be the material strength ratio: S_y/S_u where S_y is the yield stress and S_u is the ultimate strength. Also let m be the radius/thickness ratio R_i/T. Then we can estimate burst pressure P_C and P_S for a cylinder and a sphere by using Svensson's equations [5] and a geometry factor X as

$$P_C = S_y \psi B_1 \text{ (cylinder)} \tag{29.71}$$

and

$$P_S = S_y \psi B_2 \text{ (sphere)} \tag{29.72}$$

where B_1 and B_2 are burst factors that depend upon the strength ratio β as in Figure 29.9, and 29.10 provides the geometry factor.

Equations 29.71 and 29.72 are convenient in design due to the usual availability of material properties necessary for calculating the dimensionless ratio β. The formula selection in this chapter is essentially based on the assumption that the burst pressure/strength ratio relation is more likely to be nonlinear over the major part of the β range. As reported by various investigators, both Svensson's and Faupel's formulas have been found to be useful in correlating experimental data.

When the cylindrical and spherical vessels can be classed as relatively thin, Svensson's theory leads to the working equations

$$P_C = \frac{S_y}{m} B_3 \text{ (cylinder)} \tag{29.73}$$

and

$$P_S = \frac{S_y}{m} B_4 \text{ (sphere)} \tag{29.74}$$

where Figure 29.11 provides values for the burst factors B_3 and B_4 in terms of the strength ratio β. The formulas given by Equations 29.73 and 29.74 are intended for all values of m greater than 10.

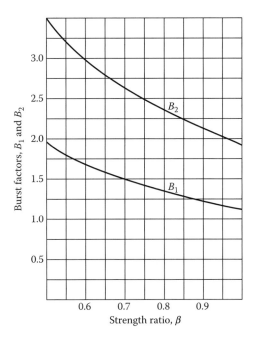

FIGURE 29.9 Burst factors for thick cylinders and spheres.

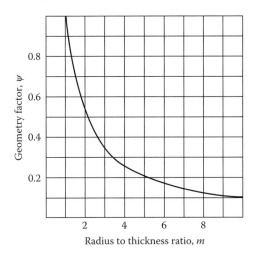

FIGURE 29.10 Geometry factor for thick cylinders and spheres.

29.11 SHRINK-FIT DESIGN

The fundamental objective of a shrink-fit construction is to introduce residual or initial stresses into the material in order to control the critical features of the stress field. This process can, for instance, increase the elastic resistance of a multiwall pressure vessel, strengthen the extrusion die, or enhance the fatigue life of a wheel mounted on the shaft. Such a shaft can be either hollow or solid. Also a shrunk-on shell, applied to the liner of a pressure vessel, should help to retard crack propagation. The shrink-fit design then can mitigate the peak stresses and enhance utilization of the materials.

In this section, we consider a two-shell shrink-fit construction as illustrated in Figure 29.12. In this configuration, the inner cylinder with radii R_s and R_i may represent a pressure vessel or a

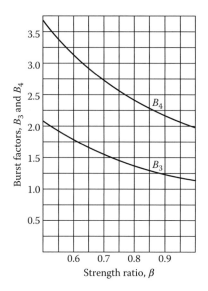

FIGURE 29.11   Burst factors for thin cylinders and spheres.

hollow shaft. The outer cylinder bounded by R_i and R_o may act as a pressure vessel liner or a component in a double-wall cylinder construction. In either case, the system is characterized by the amount of radial interference. When the inner radius of the outer cylinder is made smaller than the outer radius of the inner part, the system can be assembled either by heating the outer cylinder or cooling the inner component. This process results in a contact pressure P often described as the interference or shrink-fit pressure. The amount of radial interference (or lack of it) δ at the common boundary defined by R_i is equal to the sum of the decrease of the outer radius of the inner part and the increase of the inner radius of the outer cylinder.

Although this is somewhat complex and detailed, the two cylinder surfaces simply seek a common boundary depending upon their individual radii and stiffnesses. The resulting design formula is [6]

$$P = \frac{E\delta}{2R_i^3} \frac{\left(R_i^2 - R_s^2\right)\left(R_o^2 - R_i^2\right)}{\left(R_o^2 - R_s^2\right)} \tag{29.75}$$

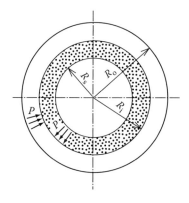

FIGURE 29.12   Two-shell shrink-fit assembly.

The previous expression applies to same materials. When the inner radius R_s is reduced to zero, we obtain the case of a solid shaft held by a hub. The relevant formula is

$$P = \frac{E\delta\left(R_o^2 - R_i^2\right)}{2R_i R_o^2}$$

(29.76)

When the two shrink-fitted cylinders have different mechanical properties such as E_o and ν_o for the outer cylinder and E_i and ν_i for the inner member, the interference pressure can be calculated as

$$P = \frac{\delta}{\frac{R_i}{E_o}\left(\frac{R_i^2 + R_o^2}{R_o^2 - R_i^2} + \nu_o\right) + \frac{R_i}{E_i}\left(\frac{R_s + R_i^2}{R_i^2 - R_s^2} - \nu_i\right)}$$

(29.77)

When R_s tends to zero, we have the case of a solid shaft defined by a simpler formula:

$$P = \frac{\delta}{R_i\left[\frac{1}{E_o}\left(\frac{R_o^2 + R_i^2}{R_o^2 - R_i^2} + \nu_o\right) + \frac{1 - \nu_i}{E_i}\right]}$$

(29.78)

SYMBOLS

A	Constant of integration
a	Major half-axis
B	Constant of integration
B_1 through B_4	Burst factors
b	Minor half-axis
C_1, C_2	Integration constants
E	Modulus of elasticity
E_o	Modulus of elasticity of sleeve
E_i	Modulus of elasticity of shaft
F_1	Longitudinal force
F_t	Tangential force
K_1, K_2, K_3	Thick-cylinder factors
$m = R/T$	Radius-to-thickness ratio
P	General symbol for pressure
P_c	Burst pressure of cylinders
P_i	Internal pressure
P_o	External pressure
R	Mean radius
R_i	Inner radius
R_o	Outer radius
R_s	Inner radius of shaft
R_1, R_2	Radii of curvature
r	Arbitrary radius
S_1	Longitudinal or axial stress
S_r	Radial stress
S_t	Hoop or tangential stress
S_u	Ultimate strength
S_y	Yield strength
T	Thickness of cylinder wall

T_s	Thickness of spherical wall
u	Radial displacement
u_i	Radial displacement of inner surface
u_o	Radial displacement of outer surface
V	Vertical load
v	Tangential displacement
x	Arbitrary distance
X, Y, Z	Rectangular (Cartesian) axis system
x, y, z	Coordinates relative to X, Y, Z axes
α	Cone half-angle
$\beta = S_y/S_u$	Strength ratio
δ	Dilation, also radial interference
ε_1	Longitudinal strain
ε_r	Radial strain
ε_t	Tangential strain
ζ	Major radius of torus
θ	Arbitrary angle
λ_{rt}	Shear strain
ν	Poisson's ratio
ν_i	Poisson's ratio of shaft
ν_o	Poisson's ratio of sleeve
τ	Shear stress, psi
ψ	Geometry factor

REFERENCES

1. J. Adachi and M. Benicek, Buckling of tori-spherical shells under internal pressure, *Experimental Mechanics*, 4, 1964, pp. 217–232.
2. S. Timoshenko and S. Woinosky-Krieger, *Theory of Plates and Shells*, McGraw Hill, New York, 1959.
3. F. V. Warnock and P. P. Benham, *Mechanics of Solids and Strength of Materials*, Isaac Pitman, London, 1965.
4. A. Blake, *Practical Stress Analysis in Engineering Design*, 2nd ed., Marcel Dekker, New York, 1990, pp. 506–507.
5. N. L. Svennson, The bursting pressure of cylindrical and spherical vessels, *Journal of Applied Mechanics*, Vol. 24, 1957.
6. J. H. Faupel, *Engineering Design*, Wiley, New York, 1964.

30 Externally Pressured Cylindrical Vessels and Structures

30.1 INTRODUCTION

Vessels subjected to external pressure occur in many applications (for example, naval and marine vessels). Nevertheless, many texts and treatises devote relatively limited attention to this important topic.

Unfortunately, design errors can occur when stress criteria intended for internally pressured vessels are extended for use with externally pressured containers and structures. Such errors are due to the fact that the externally pressurized vessels can fail by elastic instability long before the relevant compressive stresses can reach a critical magnitude. In addition, the effect of shape imperfections on the externally loaded vessels results in a number of analytical and computational difficulties. Often the problem is accentuated by the presence of manufacturing imperfections and the variations in material properties. These constraints make it mandatory to pursue the development of idealized models and the approximate design formulas based on the empirical data.

We devote this chapter to simplified design solutions and practical considerations of cylindrical vessel sizing criteria. We consider spherical vessels in Chapter 31.

30.2 THINNESS FACTOR

Intuitively, we expect a different response between a short, thick-walled vessel and a long, thin-walled vessel when both are subjected to external pressure. Attempts to quantify shell response to external pressure dates back to over 100 years. Montague [1] provides a review of the question of relating vessel geometry to failure under external pressure. The "thinness ratio" [2] provides a key measure determining vessel response.

The thinness ratio or "thinness factor" is defined as

$$\lambda = 1.2(m)^{1/4}/(k\phi)^{1/2} \tag{30.1}$$

where m, k, and ϕ are dimensionless parameters defined as

$$m = R/T, \quad k = T/L, \quad \phi = E/S_y \tag{30.2}$$

where
  R is the mean radius of the vessel
  T is the wall thickness
  L is the vessel length
  E is the elastic modulus
  S_y is the yield strength of the material

The collapse mechanism under external pressure appears to fall into two distinct patterns. In one type of response, we can observe circumferential lobes and localized buckles as soon as the material

477

begins to yield. In other cases, the mechanical response may be characterized by the development of an hourglass shape, sometimes described as a waisted configuration.

30.3 STRESS RESPONSE

The thinness parameter λ may also be used as a boundary parameter between the lobing (local buckling) and nonlobing (global buckling). A typical boundary value of λ is 0.35.

Low values of λ occur when the radius/thickness ratio m is small and $k\phi$ (product of thickness/length and elastic modulus/strength) is large. The value of λ, at 0.35, is characteristic of a short, thick cylinder made of low-strength material. Hence for λ values smaller than 0.35, single hoop stress should govern the design, and it is not necessary to calculate collapse pressures by the stability formulas. The compressive stress reaches the yield point midway between discontinuities such as vessel heads or stiffening rings. Local yielding is also possible in the vicinity of the stiffness and similar transitions. But, high longitudinal discontinuity effects in these areas are not expected to precipitate overall collapse.

In 1920, von Sanden and Günther [3] developed a simple design formula, based upon the concept of midbay collapse, as

$$P_y = 0.9S_y/m \tag{30.3}$$

where
P_y is the external pressure of yield
S_y is the yield strength of the material
m is the radius/thickness ratio (R/T)

The formula indicates that the collapse pressure corresponding to the yield strength is only about 10% different than that obtained from elementary membrane analyses.

Equation 30.3 is useful in preliminary design for ring-stiffened cylindrical vessels having m values 10 or higher. The parameter m is based upon the mean cylinder radius but the pressure is applied at the outer surface. In most practical cases, this difference is likely to be negligible.

30.4 STABILITY RESPONSE

When the response of a cylindrical vessel under external pressure is purely in a stability mode, preferential yielding at some region of the wall away from stiffeners is initiated, which in turn leads to unstable geometry and sudden formation of a lobe. Such a pattern of deformation is nonaxisymmetric and the parameter λ corresponding to this condition should exceed 2.5. It may be said then that above the limit $\lambda = 2.5$, elastic stability should govern the design. In this category of calculations, the classic long cylinder formula is often used. For a Poisson's ratio of 0.3, the long cylinder formula becomes

$$P_C = 0.275 E/m^3 \tag{30.4}$$

where
P_C is the critical buckling pressure
E is the elastic modulus
m is the mean radius to thickness ratio (R/T)

Equation 30.4 is intended for round cylinders in which collapse occurs at stresses significantly lower than the elastic limit of the material. Such a mode of failure is due to the insufficient flexural rigidity of the cylinder, and the critical buckling pressure P_C is defined as the pressure at which the circular form becomes unstable.

The circular cylinder is assumed to buckle into an elliptical shape. When the theory of a buckled ring is extended to a long vessel, by considering a ring element of the cylinder, Equation 30.4 is obtained.

Equation 30.4 can be modified for use with materials having a pronounced yield point as [4]

$$P_C = \phi S_y / m(\phi + 3.64 \ m^2) \tag{30.5}$$

where ϕ is the inverse strain parameter E/S_y with S_y being the material yield strength. As the thickness ratio m increases, Equation 30.5 approaches the form of Equation 30.4. For other values of m, the critical pressure calculated from Equation 30.5 is less than that calculated from Equation 30.4.

30.5 ILLUSTRATIVE DESIGN PROBLEM

Evaluate the extent of the error involved in estimating the external collapse pressure by means of a membrane stress criterion instead of the stability formulas on the assumption that the material's ratio ϕ corresponds to that of a mild steel known as A-36, having a yield strength of $S_y = 36,000$ psi. Assume the minimum and maximum m values to be 10 and 50, respectively.

SOLUTION

By the definition of the inverse strain parameter ϕ as E/S_y, we have

$$\phi = \frac{30 \times 10^6}{36 \times 10^3} = 833 \tag{30.6}$$

The membrane stress criterion is:

$$S_y = P_m R / T = P_m m \quad \text{or} \quad P_m = S_y / m \tag{30.7}$$

By dividing Equation 30.7 by Equation 30.4 and using the result of Equation 30.6 we obtain

$$P_m / P_C = 3.64 m^2 / \phi = 0.00436 m^2 \tag{30.8}$$

Table 30.1 provides values of the P_m/P_C ratio for m ranging from 10 to 50.

Comment

The results of this design problem show that the use of the elementary membrane stress criterion for predicting collapse pressures can be misleading, particularly for relatively thin vessels or piping. Alternatively, when the thinness parameter λ (see Section 30.2) is 0.35 or less, the use of the membrane stress criterion appears to be appropriate.

TABLE 30.1

Membrane/Critical Pressure Ratio as a Function of the Radius/Thickness Ratio m

m	10	20	30	40	50
P_m/P_C	0.436	1.744	3.924	6.976	10.900

30.6 MIXED MODE RESPONSE

The thinness factor λ, defined by Equation 30.1, provides a good predictor of vessel response. With λ values below 0.35, unstable collapse is unlikely. With λ values above 2.5, stability is the principal design consideration. The λ values of 0.35 and 2.5, however, are not exact, and in either event, there is a large "gray" area between these limits requiring careful design considerations.

Experience indicates that for values of λ up to 1.0, the stress criterion is still reasonably accurate. The hoop stress criteria in general should give conservative results even without due corrections for small initial imperfections and manufacturing tolerances. The reader should be cautioned, however, that gross initial imperfections can be very detrimental and may precipitate the overall cylinder collapse. This aspect of the analysis therefore requires engineering judgment, since the precise definitions of what may be considered as "small" or "gross" imperfections have not been fully established.

The intermediate region bracketed by the values of λ equal to 1.0 and 2.5 represents a complicated picture of cylinder behavior under external pressure where some combination of stress and stability effects is involved. Relevant tests show that under these conditions, some local yielding may take place before the critical elastic instability load is reached, precipitating the onset of ultimate collapse. Furthermore, with the simultaneous involvement of stress and stability response, the stresses arising from the imperfections may prove to be rather significant.

The material parameter ϕ numerically representing the inverse of the elastic strain varies normally between about 100 and 1000 for the great majority of metallic engineering materials. The medium range of yield strength is about 60,000 to 150,000 psi, with the corresponding ϕ values of 500 and 200, respectively. In this range, we may find such steels as A537, HY80, or 4330 series, which are recognized for their superior fracture toughness characteristics. The practical ranges of the dimensionless parameters k and m can be assumed to be as follows:

$$10 \leq m \leq 100 \quad \text{and} \quad 0.001 \leq k \leq 0.200 \tag{30.9}$$

30.7 CLASSICAL FORMULA FOR SHORT CYLINDERS

Design considerations based upon elastic behavior may be regarded as upper-bound estimates [4]. A well-known example of this type of solution is based on the short-cylinder theory describing the buckling response of the cylinder wall midway between the stiffeners. A simplified version of the relevant formula is

$$P_C = \frac{0.87\ Ek}{m^{3/2}} \tag{30.10}$$

where
   E is the elastic modulus
   k is the thickness/length ratio
   m is the radius thickness ratio

Equation 30.10 applies to vessel material that has a Poisson's ratio equal to about 0.3. This theory does not include the effect of the circumferential stiffener on the strength of the shell, and it is assumed that the initial out-of-roundness of the shell is zero. In general, the design of hardware for the various applications of cylindrical vessels and piping, involving external pressure loading, is complicated despite the availability of a variety of design formulas and their respective ranges of practical use. This situation is not surprising when one considers the modes of failure in the elastic or elastoplastic range, in addition to the geometrical and manufacturing features of pressure vessels.

For all values of $\lambda > 1.0$, stability-oriented design formulas should be used. Here the length of the cylinder in relation to the cylinder radius determines the type of design model that is likely to be applicable. Because of the stability criteria affecting the design, the analysis should consider possible effects of manufacturing out-of-roundness on the collapse resistance of a particular cylindrical vessel.

30.8 MODIFIED FORMULA FOR SHORT CYLINDERS

The short cylinder formula of Equation 30.10 may be modified to provide an expression for calculating the ultimate collapse pressure P_u. Specifically, P_u may be expressed as

$$P_u = S_y F_1(k) F_2(m) \tag{30.11}$$

where
S_y is the yield strength
$F_1(k)$ and $F_2(m)$ are

$$F_1(k) = K_1' - \left(K_1'^2 - K_2' \right)^{1/2} \tag{30.12}$$

and

$$F_2(m) = 1/2m^{3/2} \tag{30.13}$$

where K_1' and K_2' in turn are

$$K_1' = m^{1/2} + 0.87k\phi(1 + 6n) \quad \text{and} \quad K_2' = 3.48k\phi m^{1/2} \tag{30.14}$$

where
k is the thickness/length ratio T/L
m is the mean radius/thickness ratio R/T
ϕ is the inverse strain parameter E/S_y
n is the out-of-roundness parameter defined as the ratio of the radial deviation e from the perfect circular shape to the thickness of the vessel e/T

Equation 30.11 is based upon Timoshenko's theory of ellipticity [4] applicable in an intermediate region of a cylindrical vessel where the thinness ratio λ (see Equation 30.1) has values between 1.0 and 2.5. Recall that this is a region of mixed mode possible failure, that is, from excessive stress and/or instability.

The range of applicability of Equation 30.11 may be approximated using the following general criteria, which gives the appropriate ratios of cylinder length to radius. The length can be measured between the stiffeners or cylinder heads

$$\left(\frac{L}{R} \right)_{min} = 0.63(m)^{-1/2} \left(\frac{1.44\phi}{m} + 1 \right) \tag{30.15}$$

and

$$\left(\frac{L}{R} \right)_{max} = 3.1(m)^{1/2} \tag{30.16}$$

When the L/R ratio is between the values computed from Equations 30.15 and 30.16, the modified formula of Equation 30.11 is deemed to apply. The designer should be cautioned, however, that in some circumstances Equations 30.15 and 30.16 may indicate that the range of applicability of

Equation 30.11 is nonexistent. This should not be surprising when we consider the complex nature of the functions and the number of parameters involved.

The range of applicability of the formula of Equation 30.11 decreases substantially with the simultaneous decrease in the R/T ratio and the yield strength S_y of the material. In the majority of design configurations, however, R/T ratios are generally higher than 20 and ϕ values are lower than 500. This allows for a relatively large range of applicability of Equation 30.11.

30.9 SIMPLIFIED CRITERION FOR OUT-OF-ROUNDNESS

The out-of-roundness parameter n can be defined in terms of the extreme diametral measurements D_{max} and D_{min} and the wall thickness T as

$$n = \frac{D_{max} - D_{min}}{4T} = \frac{e}{T} \tag{30.17}$$

where e is the radial deviation from an exact circle.

It should be noted that Equation 30.17 applies only to cases where radial deviation can be related to the even number of circumferential lobes of the out-of-roundness pattern. For a conservative estimate based on Equations 30.11 and 30.17, a purely elliptical mode of cylinder collapse may be recommended for design calculations. Unfortunately, the actual out-of-roundness parameter n will seldom be known before manufacture. Nevertheless, as the first rational step in design, the extent of the anticipated maximum out-of-roundness can be deduced from knowledge of the customary manufacturing tolerances. For instance, when the tolerance on the diameter of a particular cylindrical canister is given, say, as ±0.05 in., the corresponding value of n can be taken as $n = 0.025/T$.

For a long time now, the effect of out-of-roundness on the collapse strength of vessels and piping under external pressure has been duly recognized by the ASME code. For the purpose of eliminating the potential of any gross out-of-roundness, the ASME code recommends that the ratio $e/2R$ not exceed 1%, where R is the nominal radius. This conservative rule is at times used as the upper limit in establishing the design criteria. The actual manufacturing and field experience tends to indicate, however, that 1% grossly overestimates the extent or radial deviation from the perfect circularity found by the measurements. Advances in mechanical technology result in relatively small increases in values or n, even for large-diameter vessels.

30.10 LONG CYLINDER WITH OUT-OF-ROUNDNESS

When a pressure vessel is relatively long and its characteristics fall outside the range of short-cylinder geometry, as required for use of Equation 30.11, we can use the following conservative formula for the collapse strength under external pressure:

$$P_{u\ell} = S_y A_m A_n \tag{30.18}$$

where $P_{u\ell}$ is the long cylinder collapse pressure and S_y is the yield strength. The parameters A_m and A_n are

$$A_m = 1/2m^3 \quad \text{and} \quad A_n = A_1 - (A_1^2 - A_2)^{1/2} \tag{30.19}$$

where A_1 and A_2 in turn are

$$A_1 = m^2 + 0.275\phi(1 + 6n) \quad \text{and} \quad A_2 = 1.1\, m^2\phi \tag{30.20}$$

where

   m is the mean radius/thickness ratio

   ϕ is the elastic modulus/strength ratio, E/S_y, also known as the "inverse strain parameter"

When the effect of out-of-roundness is neglected, by setting $n = 0$, Equation 30.18 is to be replaced by the classical formula for elastic buckling of a long cylinder given by Equation 30.4.

30.11 EFFECTIVE OUT-OF-ROUNDNESS

Defects in materials are virtually impossible to eliminate and difficult to define or identify ahead of time. Nevertheless it is still possible to make reasonable assumptions about out-of-roundness of a cylindrical vessel for design purposes. It is important to consider this feature because it can become significant in relation to other effects, such as variation in wall thickness or the residual stress patterns. It should be added here that in the case of well-known structural members such as columns, the collapse is essentially of the bending type, leading to the formation of a plastic hinge. Unfortunately, as far as the vessels and piping subjected to external pressures are concerned, the collapse mechanism is much more sensitive because even small local deformations can give rise to significant bending moments, because of the presence of large compressive forces. For example, when conducting a typical structural test of a pressure vessel, it is easy to note that the relevant load–deflection curve develops smoothly almost up to the level of external collapse pressure, at which point a rather violent failure suddenly occurs. This type of structural behavior is known as an "implosion," in contrast to the "explosion" caused by internal pressure.

Comprehensive theoretical analyses, which can account for all factors leading to vessel collapse, are difficult to obtain even incorporating numerical and computer analyses. Therefore it is prudent to include out-of-roundness corrections in prototype design and then to test the integrity of the pressure vessel by experiment.

To date, correlations of measured and calculated collapse pressures appear to indicate a trend of improved agreement with decreasing radius to thickness m, or R/T ratio. Differences in measured and calculated collapse pressures for thick-walled piping and vessels may be caused mainly by the variation in material properties and the initial imperfections. It is less severe for thick-walled tubes and many test results for thick-walled piping should indicate collapse pressures that exceed the calculated values. While thick-walled vessels and tubes need perhaps less restrictive criteria, the results on thin-walled tubes may still require somewhat larger factors of safety.

The foregoing discussion indicates that the collapse of thick-walled and thin-walled vessels is based on different criteria, and it is often difficult to determine the exact boundary between the response of thick and thin vessels. Short- and thick-walled vessels fail at the yield point of the material, while long and thin-walled vessels tend to become elastically unstable at wall stresses far below the yield strength.

The thinness factor (see Section 30.2) is useful in developing a method for assessing out-of-roundness. This is motivated by the fact that the effect of out-of-roundness on the collapse of long, thin-walled vessels is more significant than for short, thick-walled vessels.

Recall in Section 30.8 that the out-of-roundness parameter n was defined as the ratio of radial deviation to thickness, e/T. Extensive analyses and experiments suggest that for design purposes an effective out-of-roundness parameter n_e may be defined as

$$n_e = n \sin^2 (36\lambda) \tag{30.21}$$

Assuming an elliptical shape of a vessel as the most likely mode of failure under external pressure, the procedure is to estimate the out-of-roundness n by Equation 30.17:

$$n = \frac{D_{max} - D_{min}}{4T} = \frac{e}{T} \tag{30.22}$$

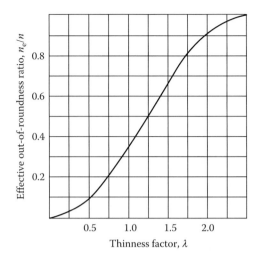

FIGURE 30.1   Graph for effective out-of-roundness.

Next, the thinness factor λ may be calculated using Equation 30.1 as

$$\lambda = 1.2(m)^{1/4}/(k\phi)^{1/2} \qquad (30.23)$$

Then the ratio n_e/n follows directly from Equation 30.21. Figure 30.1 provides a graphical representation of Equation 30.21.

30.12   ILLUSTRATIVE DESIGN EXAMPLE

To illustrate the foregoing procedures, consider the following design problem: determine the maximum allowable out-of-roundness for a large cylindrical canister at a design pressure of 750 psi, assuming a clearance distance between the stiffeners equal to 60 in., a mean radius of 35 in., and a wall thickness of 0.875 in. The material is HY80 with a minimum expected yield strength of 80,000 psi and the modulus of elasticity of 30×10^6 psi. Plot the variation of the collapse pressure with radial deviation from a perfect circular shape, and calculate the corrected design pressure based on the theory of effective out-of-roundness.

SOLUTION

The relevant dimensionless ratios k (thickness/length, T/L), m (mean radius/thickness, R/T), and ϕ (inverse strain parameter E/S_y) are

$$k = \frac{0.875}{60} = 0.0146, \quad m = \frac{35}{0.875} = 40, \quad \text{and} \quad \phi = \frac{30 \times 10^6}{80 \times 10^3} = 375 \qquad (30.24)$$

By performing a "λ check" using Equation 30.1 we have

$$\lambda = 1.2m^{1/4}/(k\phi)^{1/2} = 1.2(40)^{1/4}/(0.0146 \times 375) = 1.29 \qquad (30.25)$$

This result indicates that the canister is likely to behave as a short cylinder where a mixed mode of failure involving stress and stability occurs. Then from Equation 30.15, we can determine the minimum critical length:

$$\left(\frac{L}{R}\right)_{min} = 0.63(m)^{-1/2}\left(\frac{1.44\phi}{m} + 1\right) = 0.63(40)^{-1/2}\left(\frac{1.44 \times 375}{40} + 1\right) = 1.44 \qquad (30.26)$$

That is, the minimum length L_{min} is

$$L_{min} = 1.44 \times 35 = 50.4 \text{ in.} \tag{30.27}$$

Since L_{min} is less than 60, the theoretical short-cylinder formula of Equation 30.11 is applicable. Specifically, the collapse pressure P_u is

$$P_u = S_y F_1(k) F_2(m) \tag{30.28}$$

A glance at Equation 30.16 shows that the critical maximum length will be considerably higher than 60 in. That is,

$$\left(\frac{L}{R}\right)_{max} = 3.1 m^{1/2} = 3.1 \times (4D)^{1/2} = 19.6 \tag{30.29}$$

or

$$L_{max} = 19.6 \times 35 = 686 \text{ in.} \tag{30.30}$$

Therefore, we can compute the K_1' and K_2' parameters from Equation 30.14. That is

$$K_1' = m^{1/2} + 0.87 k\phi(1 + 6n) = (40)^{1/2} + (0.87)(0.0146)(375)\left(1 + \frac{6e}{0.875}\right)$$

$$= 11.09 + 32.66e \tag{30.31}$$

and

$$K_2' = 3.48 k\phi m^{1/2} = (3.48)(0.0146)(375)(40)^{1/2} = 120.5 \tag{30.32}$$

Using these results in Equation 30.12, $F_1(k)$ becomes

$$F_1(k) = K_1' - \left(K_1'^2 - K_1'\right)^{1/2} = 11.09 + 32.66e - 1.58(1 + 291e + 429.8e^2)^{1/2} \tag{30.33}$$

Similarly, from Equation 30.13 $F_2(m)$ becomes

$$F_2(m) = 1/2 m^{3/2} = 1/(2)(40)^{3/2} = 0.00198 \tag{30.34}$$

Finally, using these results in Equation 30.11, we obtain the expression for the collapse pressure as

$$P_u = S_y F_1(k) F_2(m) = 1756.7 + 5173.3e - 250.3(1 + 291e + 429.8e^2)^{1/2} \tag{30.35}$$

Figure 30.2 provides a graphical representation of Equation 30.35. From the figure, we see that the maximum allowable deviation from a perfect radius for the design pressure of 750 psi is 0.123 in. At $\lambda = 1.29$, however, we see from Equation 30.21 that the effective out-of-roundness n_e is

$$n_e = 0.52n = (0.52)(0.123)/T = 0.064/T \tag{30.36}$$

The corrected design pressure of Figure 30.1 depends upon the assumed value of cylinder thickness $T = 0.875$ in.

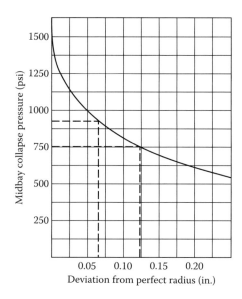

FIGURE 30.2   Deviation from perfect radius (in.).

Comment

The method illustrated by this example is relatively simple, especially in view of the complexity of pressure vessel technology. The procedure effectively relaxes the conservatism inherent in Equation 30.11. This is due to the assumption of an elliptical mode of failure. This is manifested in the example through an increase in the estimated design pressure above 750 psi.

The formulas of this chapter for critical pressure, together with the design example, are offered as an illustration of the process of application of relatively simple solid mechanics principle to complex design problems.

The formulas given by Equations 30.11, 30.18, and 30.21 may be used in preliminary design, development, and testing, but only within the limits defined by Equations 30.1, 30.15, and 30.16. The method offered here is an alternative but not a substitute for established and proven code practices such as those recommended by the ASME Boiler and Pressure Vessel Code.

30.13   EMPIRICAL DEVELOPMENTS

There are considerable theoretical difficulties in obtaining a comprehensive analysis of ring-stiffened cylinders subjected to external pressure. As an alternative considerable effort has been expended by pressure vessel specialists to gather well-documented experimental data [5]. Figure 30.3 provides a summary of these results using dimensionless values for ordinate and abscissa. Specifically, the parameter $(1.05 m P_e/S_y)$ is shown as a function of $(0.92 k\phi/m^{1/2})$ where P_e denotes the experimentally determined interstiffener collapse pressures for data having $m(R/T)$ values between 6 and 250 and the wide range of L/R ratios between 0.04 and 50. (As before, k is T/L, S_y is the yield strength, and ϕ is the inverse strain parameter E/S_y.)

In the majority of test cylinders reported in the literature, the out-of-roundness was much less than 1% of the radius. In some cases, however, the values of out-of-roundness were found to be in the order of 1%, representing what can be judged to be relatively poor manufacturing practices. Because of this series of relatively low experimental collapse pressure, however, a lower-bound curve was selected here for illustration.

It should be recognized that the lower-bound curve represents a wide range of geometrical proportions and that the correlation must break down for cylinders with oversized and very closely

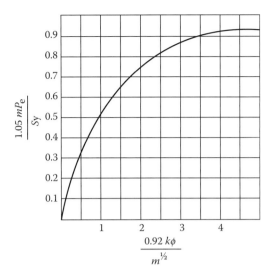

FIGURE 30.3 Lower-bound empirical curve for ring-stiffened cylinders under external pressure.

spaced stiffeners. Naturally, the critical buckling pressures for such structures would become much higher than those with the simple hoop stress criterion at midbay would tend to indicate.

Figure 30.3 is representative of the maximum hoop stress at the midpoint between the stiffeners. It represents the ratio of the critical pressure sought and that expressed by Equation 30.3.

The parameter on the abscissa of Figure 30.3 ($0.92k\phi/m^{1/2}$) is obtained by Equations 30.10 and 30.3, and by simplifying the expression in the usual range of the geometries where m is much larger than 1. For example, in many practical design applications, the numerical value of m is seldom found to be lower than 20. Under such circumstances, the error introduced by the foregoing simplification should be less than 3%.

The advantage of referring to an empirical curve is that the results cover a relatively wide range of L/R ratios without drawing any particular distinction among short, intermediate, or long cylinders. The disadvantages of the empirical curve, however, are (1) lack of exact details of the out-of-roundness corresponding to the chart and (2) the practical necessity of relying on the lower-bound curve, resulting in a conservative prediction of the ultimate collapse pressure.

Nevertheless, we can make computational use of the empirical data for externally pressurized cylindrical vessels using the following approximate formula for the critical buckling pressure:

$$P_{CR} = S_y Z_1 (Z_2 - nZ_3) \tag{30.37}$$

where Z_1, Z_2, and Z_3 are

$$Z_1 = \exp\left[-0.815m^{1/2}/k\phi\right] \tag{30.38}$$

$$Z_2 = 1/[(m)^{0.95}(k\phi)^{0.10}] \tag{30.39}$$

$$Z_3 = [50/(m)^{1.95}(k\phi)^{0.10}] - 33/m^2 \tag{30.40}$$

where the essential variables such as S_y, k, m, n, and ϕ quoted here make the comparison with other existing formulas relatively straightforward. However, the main reason for developing a general formula of the type given by Equation 30.37 is to have a continuous mathematical model, which could be used with short, intermediate, and long cylinders alike, thereby eliminating the problems

TABLE 30.2

Examples of Collapse Pressures Calculated for Cylindrical Vessels

Type of Vessel	R (in.)	T (in.)	L (in.)	e (in.)	S_y (ksi)	E (psi)	P_u (psi)[a]	P_{CR} (psi)[b]	P_e (psi)[c]
Experimental canister	35.88	1.125	82.5	0.0625	80	30×10^6	1260	1000	1000
Experimental canister	35.75	0.500	72	0.0625	36	30×10^6	180	152	160
Line-of-sight pipe	8.81	0.375	96	0.0450	35	30×10^6	500	430	470
Well casing	24.38	0.750	96	0.1220	36	30×10^6	410	490	530
Well casing	27.00	0.625	28	0.0400	100	30×10^6	1190	1010	1060
Experimental canister	40.50	1.500	272	0.1000	36	30×10^6	340	510	540
Experimental canister	27.75	0.750	48	0.1425	36	30×10^6	600	570	690
Experimental canister	3.1	0.4	42	0.0050	152	26×10^6	8550	4980	5230
Concrete room	91	27	240	0.4550	12	4×10^6	2920	2640[d]	3120

[a] See Equation 36.6.
[b] See Equation 36.20.
[c] See Figure 36.3.
[d] See Equation 36.2.

associated with the transitions and gray areas. The formula represented by Equation 30.37 is intended for the following ranges of the parameters:

$$0 \leq n \leq m/100, \quad 10 \leq m \leq 100, \quad 0.001 \leq k \leq 0.200, \quad 100 \leq \phi \leq 1000 \qquad (30.41)$$

To obtain a brief comparison between theoretical and experimental results, various data have been randomly selected for the comparison. The data include both typical and some extreme configurations, providing a general assessment of the potential differences. Table 30.2 presents the results. The last three columns of the table indicate reasonable agreement among the various estimates despite the marked variations in cylinder proportions. The actual discrepancies often appear to be on the conservative side, and in general the differences are not too serious when one considers the inherent complexity of the collapse problem.

The approximate formula (Equation 30.37) is easy to use but requires further refinements. It is hoped that its practical use will provide, in due course, the necessary background for improving the proposed model. It is offered here as an example of a plausible starting point in the search for a general formula applicable to all proportions of cylindrical vessels.

In the region of relatively low values of k and m, the minimum collapse pressure of casing, tubing, or drill piping can be estimated on the basis of specifications recommended by the American Petroleum Institute [6]. The formulas suggested by the institute include experimental corrections and apply to yield strengths between 40,000 and 150,000 psi. They are basically of the Lamé [7] or Stewart [2] type.

The classical buckling response of a thin and long cylinder depends primarily on the modulus of elasticity of the material and the cube of the ratio of mean radius to thickness, as shown by Equation 30.4. In other cases, the pressure vessels, well casing, or piping may fall into the category of intermediate or short cylinders. In this range, governed by Equation 30.4, the collapse strength may increase or decrease as a function of raising or lowering the yield strength of the material. Also, when an externally pressurized and relatively thick pipe is subjected to axial tension, such as may be the case with a long string of casing emplaced vertically in the ground, the question may be raised as to the effect of tension on collapse. This effect is briefly analyzed in the next section.

30.14   EFFECT OF AXIAL STRESSES ON COLLAPSE

From a simple geometric prospective, we can obtain insight regarding the effect of axial loading on collapse. Specifically to the extent that axial compression of a cylinder tends to increase wall thickness through the Poisson's ratio effect, axial tension does the opposite. Hence, "wall thinning" should contribute to lowering of the classical buckling pressure, which is directly proportional to the cube of thickness. Naturally, such geometrical effects must be relatively small, and one has to examine additional aspects of the problem such as the role of wall thinning in accentuating manufacturing imperfections as well as the influence of biaxial loading on the yield strength of the material.

Unfortunately, theoretical work on externally pressurized cylinders has not been fully successful in providing rigorous solutions to stability problems in the presence of manufacturing imperfections, superimposed axial stresses, or local plastic deformations. However, the need for practical solutions to some of these problems has prompted extensive experimental studies of the effect of the combined longitudinal loading and the external pressure [8]. This particular work was conducted in support of the requirements of the oil-drilling industry. It involved more than 200 tests on seamless tubing loaded simultaneously by external pressure and longitudinal tension. The ratios of tube radius to wall thickness for this experiment ranged from 5 to 11. The yield strength of the tube material varied from about 30,000 to 80,000 psi. Unlike some earlier speculations, this study has clearly established that the effect of combined loading can substantially reduce both the collapse strength and the tensile strength of the tubing. The results of this study have since been utilized in the development of practical handbook data for the soil drilling industry [9].

Conventional theories of the strength of materials indicate that in a biaxial state of stress, where tension and compression act at right angles to each other, the effective yield strength of the material appears to be lowered. The particular case of the maximum strain–energy theory gives a good approximation to the experimental data where ductile materials are involved. This strength theory is based on the assumption that the quantity of strain energy stored in a unit volume of an elastic material attains a maximum value at the instant of the material's failure. Therefore, knowing the energy required to case the failure in a simple tensile test specimen, the approximate limiting stresses for the combined loading can probably be estimated.

Figure 30.4 presents an ellipse of biaxial yield stress based upon the concept of maximum strain energy. The effect of biaxial loading in the case under consideration corresponds, therefore, to

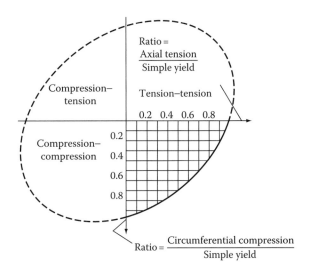

FIGURE 30.4   Diagram for maximum strain energy theory.

the functional relationship depicted by the right lower quadrant of the ellipse. In this region, the longitudinal tensile stress is represented by the horizontal axis set out to the right of the zero point, and the compressive hoop stress, resulting from the external pressure, conforms to the lower portion of the vertical line. The ellipse diagram shown in Figure 30.4 is expressed in terms of the nondimensional ratios to ensure its generality. The experimental points derived from tests on steel tubing were found to correlate quite well with the elliptical curve drawn in the compression–tension quadrant [8,9]. This rather remarkable agreement between the maximum strain energy formulation and experimental evidence provides a firm basis for design of well casing employed in the oil industry and in similar engineering ventures.

While investigations involving piping are likely to be successful, the problem of sizing a large-diameter vessel may present a different task because the intermediate range of mean radius to thickness ratio is higher and varies between about 10 and 40. There seems to be no available experimental data in this range of the geometry, which could be directly used in evaluating the effect of the tensile stresses on collapse, although it is known that the yield strength in circumferential compression is definitely one of the principal factors in controlling the collapse resistance of oil well casing, characterized by relatively low radius to thickness ratios. Our intuition and experience may suggest that thin-walled, high-strength casing should be less affected by changes in the yield strength due to the biaxiality of low-strength casing. However, the response of a large-diameter casing, characterized by an intermediate ratio of radius to thickness and the average yield strength, can only be inferred from the oil industry experimental data discussed above and a suitable theoretical model, such as that defined by the theory of the maximum strain energy.

It may be instructive at this point to briefly analyze the combined effect of thickness and material strength parameters, which should have a finite influence on the susceptibility of the externally loaded pressure vessels to biaxial effects. This can be approximated with the aid of the thinness factor λ, expressed in terms of the geometry and materials parameters k, m, and ϕ, as before.

Equation 30.1 is useful in determining the failure characteristics for a cylindrical vessel or piping subjected to external pressure. Recall that when the thinness factor λ exceeds 2.5, stability criteria are predominant and the effect of biaxial loading is less important. Alternatively, when λ is small, say less than 0.35, the circumferential stress is likely to govern the collapse. In this case, the vessel is more susceptible to biaxial effects.

A relatively long cylinder is one whose length equals or exceeds the lengths obtained from Equation 30.16. That is

$$(L/R)_{max} = 3.1m^{1/2} \quad \text{or} \quad L = 3.1Rm^{1/2} \tag{30.42}$$

where as before
　R is the mean radius
　m is the radius/thickness ratio R/T

Combining Equations 30.1 and 30.42 yields the thinness factor for a relatively long pipe or casing as

$$\lambda = 2.11m/\phi^{1/2} \tag{30.43}$$

Equation 30.43 provides limiting values of m for the limiting λ values of 0.35 and 2.5. Figure 30.5 provides a chart showing the relative sensitivity of the collapse pressure to superimposed axial tensile stresses.

Consider, for example, two pressure vessel designs using (1) a high-strength steel, thin-walled cylinder and (2) a low-yield-strength vessel with a thicker wall. Specifically, let ϕ and m be

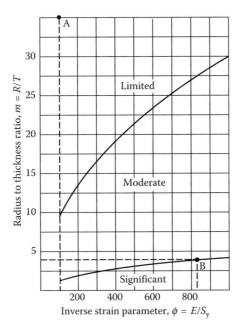

FIGURE 30.5 Limiting values for the effect of tension on collapse pressure.

1. $\phi = 100$ and $m = 35$ (Point A, Figure 30.5)
   and
2. $\phi = 833$ and $m = 4$ (Point B, Figure 30.5)

In the first case, the effect of the tensile stresses is likely to be small whereas in the second case the tension is likely to be important.

This example and the chart given in Figure 30.5 are intended only as a general guide for designing relatively long cylindrical components. Similar interpretation of the sensitivity of the collapse resistance to tension can be developed for intermediate and short cylinders using the relevant limits of applicability of specific formulas.

The theoretical correction factor for the effect of tension can be derived from analytical expressions for the maximum strain energy. The work done on the elastic material in the state of biaxial tension may be described as

$$U = \frac{S_1^2 + S_2^2}{2E} - \frac{\nu S_1 S_2}{E} \qquad (30.44)$$

where
   U is the strain energy
   S_1 and S_2 are principal stresses
   E and ν are the elastic modulus and Poisson's ratio

The maximum amount of the elastic work that can be done in direct tension is obtained when the corresponding stress approaches the yield strength for a given material. The formula representing this amount of work is

$$U = \frac{S_y^2}{2E} \qquad (30.45)$$

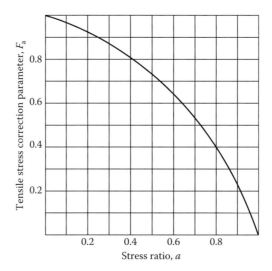

FIGURE 30.6 Design chart for the effect of tension on collapse.

The state of midbay hoop stress and axial wall stress in a casing subjected to external pressure and longitudinal tension simultaneously can be defined by $S_2 = -S_c$ and $S_1 = S_t$, respectively. Here S_c denotes hoop compressive stress due to external pressure and S_t is the average axial tensile stress. Substituting these terms and combining Equations 30.44 and 30.45 gives

$$S_y^2 = S_c^2 + S_t^2 + 2\nu S_c S_t \tag{30.46}$$

For the majority of metallic materials, we have $\nu = 0.3$. Using this value and putting $a = S_t/S_y$, Equation 30.46 yields

$$\frac{S_c}{S_y} = F_a \tag{30.47}$$

where F_a is

$$F_a = (1 - 0.91a^2)^{1/2} - 0.3a \tag{30.48}$$

Since the absolute value of the compressive wall stress is proportional to S_y and F_a, as given by Equation 30.47, the correction for tension may be obtained by multiplying the conventionally calculated collapse pressure by F_a. Figure 30.6 provides a design chart based upon Equation 30.48.

In field applications involving piping or casing emplaced vertically underground and backfilled, the ratio a is expected to be relatively small, so that the correction factor F_a does not differ significantly from unity. However, the chart given in Figure 30.6 indicates that the influence of tension can be significant. This effect manifests itself as that which would pertain if the pipe or tubing were made of a lower-strength material. We can make a comparison between values obtained using Equation 30.48 and available experimental data [8]. Table 30.3 provides the comparison.

30.15 STRENGTH OF THICK CYLINDERS

For thick and moderately thick vessels and piping, the stability criteria of the previous sections may not be important or applicable. In this case, Lamé's theory provides a convenient approach for

TABLE 30.3

Effect of Axial Stress on Collapse by Test and by Theory

m	a	Yield strength	Experimental Collapse Pressure	Calculated Collapse Pressure
9.1	0.138	500	44.4	45.2
9.1	0.276	500	43.1	41.9
9.1	0.404	500	35.8	38.1
6.6	0.107	290	43.4	43.8
6.6	0.317	290	35.8	37.0
10.9	0.135	560	28.9	29.1
10.9	0.385	560	29.3	24.9
5.6	0.289	565	96.5	96.4
5.6	0.474	565	75.8	82.7

Note: All values in megapascals: 1 megapascal $= 1$ N/mm$^2 = 145$ psi $= 1$ MPa.

design. For example, the maximum tangential stress S_t at the inner surface of the cylinder may be expressed as

$$S_t = P_o K_4 \qquad (30.49)$$

where P_o is the external pressure and K_4 is a design factor depending upon the ratio of the inner radius to thickness, as provided by Figure 30.7. This stress is compressive. It may be of interest to note that when the cylinder is very thick to the point of becoming a solid shaft, the stress tends to a value equal to twice the externally applied pressure.

The displacement u_i of the inner surface of the cylinder toward the central axis may be calculated from the expression:

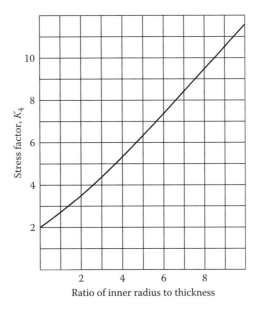

FIGURE 30.7 Stress factor for a thick cylinder under external pressure (see Equation 30.49).

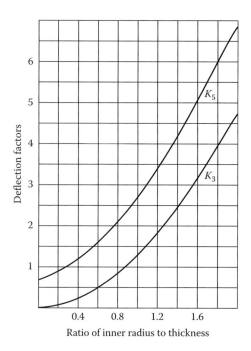

FIGURE 30.8 Displacement factors for a thick cylinder under external pressure (see Equations 30.50 and 30.51).

$$u_i = \frac{P_o T}{E} K_3 \qquad (30.50)$$

where
  T is the cylinder thickness
  E is the elastic modulus

where K_3 is a displacement factor depending upon the ratio of the inner radius to thickness, as provided by Figure 30.8.
  Similarly, the radial displacement u_o of the outer surface of the cylinder is given by

$$u_o = \frac{P_o T}{E} K_5 \qquad (30.51)$$

where K_5 is a displacement factor depending upon the ratio of the inner radius to thickness, as provided by Figure 30.8.
  Observe in Figure 30.8 that the factor K_5 does not vanish when the inner radius tends to zero. Then for a solid shaft, the displacement of the outer surface u_{osolid} is

$$u_{osolid} = 0.7 P_o T / E \qquad (30.52)$$

30.16 ILLUSTRATIVE DESIGN PROBLEM

Figure 30.9 depicts a straight cylindrical hub and a solid shaft assembly. Assuming that both components are made of steel, calculate the shrinkage allowance δ to develop a shrink-fit pressure P of 10,000 psi.

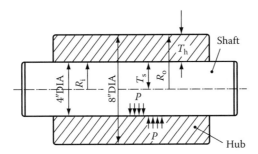

FIGURE 30.9 Shrink-fit assembly.

SOLUTION

The hub is subjected to internal pressure whereas the shaft experiences external pressure. Equation 29.12 provides the radial deformation u_i of the hub as

$$u_i = \frac{PT_h K_2}{E} \tag{30.53}$$

where T_h is the hub thickness: $R_o - R_i$.

Correspondingly, Equation 30.51 provides the radial compression u_o of the shaft as

$$u_o = \frac{PT_s K_5}{E} \tag{30.54}$$

The sum of the displacements of Equations 30.53 and 30.54 must be the interference δ. That is

$$\delta = u_i + u_o = (P/E)(T_h K_2 + T_s K_5) \tag{30.55}$$

For the hub $R_i/T_h = 1$ and for the shaft $R_i/T_s = 0$. Then Figures 29.8 and 30.8 give K_2 and K_5 as

$$K_2 = 1.97 \quad \text{and} \quad K_5 = 0.7 \tag{30.56}$$

Therefore δ is

$$\delta = \frac{10^4(2 \times 1.97 + 2 \times 0.7)}{30 \times 10^6} = 0.00178 \text{ in.} = 0.045 \text{ mm} \tag{30.57}$$

30.17 OUT-OF-ROUNDNESS CORRECTION FOR STRESS

The maximum wall stress in an out-of-round cylinder can be obtained through superposition of the tangential stress calculated for a perfectly circular cylinder and the offset bending stress caused by the out-of-roundness. The offset stress follows from the bending moment, which may be determined by multiplying the average hoop load times the amount of radial deviation. For a relatively thin cylinder where m is greater than 10, we have

$$S = Pm(1 + 6n) \tag{30.58}$$

where
  S is the stress
  m is the radius/thickness ratio (R/T)
  n is the deviation to thickness (out-of-roundness) ratio (e/T)

Usually, the magnitude of n defining the particular out-of-roundness is markedly less than unity. Nevertheless, Equation 30.58 shows how quickly the total stress can increase as a function of n. It is also relevant to note that in the case of internal pressure, the hoop load tends to make the cylinder more circular. The actual amount of this straightening effect, however, is not easy to estimate because of the nonlinear relation between the hoop load and radial deformation. Hence, Equation 30.58 in its present form is likely to be conservative. Also, the model implied by Equation 30.58 may be related to the stress theory of oval tubes subjected to internal pressure [10], with the relevant agreement appearing to be good for values of n smaller than 1. This bracketing condition should certainly cover the range of radial deviations most likely to be encountered in pressure vessel design.

Since Equation 30.58 is based on the membrane stress theory, there is no need to differentiate between the internal and external loading on a cylindrical shell. When the cylinder ratio m is less than 10, thick-shell theory dictates the choice of the design formulas. In such a case, the total tangential stress should be calculated as the sum of the maximum Lamé stress and the offset bending stress. The latter may be obtained from the hoop load, which can be taken as the product of the average stress and the longitudinal cross-sectional area of the cylinder. In this manner, the corrected stress for the case of internal pressure can be described as

$$S = P\left[\frac{4m^2 + 1}{4m} + \frac{3n(4m^2 - 2m + 1)}{2m}\right]$$
(30.59)

Similarly, for the condition of the external pressure, we obtain

$$S = P\left[\frac{(2m + 1)^2}{4m} + \frac{6n(m^2 + 0.5m + 0.25)}{m}\right]$$
(30.60)

Observe in these last two expressions that when m is large (as with a thin cylinder), both expressions reduce to the thin-cylinder formula of Equation 30.58.

30.18 DESIGN CRITERION FOR THICK CYLINDERS

In Section 30.17, we considered various formulas for calculating and estimating collapse pressures for cylindrical containers subjected to uniform external loading. The formulas are developed using the concepts of elastic stability and an elliptical mode of cylinder failure.

This implies that the majority of containers found in industry can be characterized by relatively thin walls and predominantly elastic response. Although this statement is largely true, there are exceptions where the vessels may have thicker walls made of a lower-strength material and where stress rather than buckling may govern the design. Hence, substituting $\lambda \cong 0.35$ in Equation 30.1 results in the following approximate criterion:

$$\frac{100m}{k^0\phi^2} \leq 0.75$$
(30.61)

where
   m is the radius/thickness ratio (R/T)
   k is the thickness/length ratio
   ϕ is the modulus/strength ratio (E/S_y)

For example, if we take values like $m = 10$, $k = 0.1$, and $\phi = 1,000$, Equation 30.61 gives 0.1, which is less than 0.75. According to this rule, the selected cylindrical geometry and the material are such that stress criteria should govern the design. Hence, any combination of k, m, and ϕ, which gives a numerical quantity smaller than 0.75, determines the condition of the governing stress for the cylinders loaded externally. However, when the vessels are loaded internally, the design is always governed by stress. The only differentiation needed in the latter type of loading concerns the relative thickness of the vessel wall, which determines the choice of the analytical approach.

SYMBOLS

A_1, A_2	Long cylinder factors
A_m, A_n	Long cylinder parameters
A_n	Long cylinder parameter
$a = \frac{S_t}{S_y}$	Stress ratio
a_0	Projected radius of spherical cap
E	Elastic modulus
E_s	Secant modulus
E_t	Tangent modulus
e	Deviation from perfect radius
F_a	Axial load correction factor
$F(k)$, $F(m)$	Short cylinder parameters
$F(m)$	Short cylinder parameter
H	Depth of shallow cap
h	Reduced thickness of shell
K	Buckling coefficient
K_1', K_2', K_3'	Short cylinder factors
K_1 through K_5	Thick cylinder design factors
$k = T/L$	Dimensionless ratio
L	Length of cylinder
L_c	Critical arc length
$m = R/T$	Ratio of mean radius to thickness
$m_i = R_i/h$	Ratio of mean radius to local thickness
$n = e/T$	Ratio of radial deviation to thickness (out-of-roundness)
n_e	Effective out-of-roundness
$n_0 = a_0/T$	Shallow cap ratio
P_c	Classical buckling pressure
P_{CR}	General symbol for buckling pressure
P_e	Experimental collapse pressure
P_m	Pressure to cause membrane yield stress
P_u	Short cylinder collapse pressure
P_{ul}	Long cylinder collapse pressure
P_y	External pressure at yield
P_0	External pressure
R	Mean radius
R_i	Inner radius
R_o	Outer radius
S	General symbol for stress
S_1, S_2	Principal stresses
S_c	Compressive stress
S_t	Tensile stress
S_y	Yield strength
T	Thickness of wall
T_h	Thickness of hub

T_s	Thickness of hollow shaft
U	Elastic strain energy
u_i	Displacement of inner surface
u_o	Displacement of outer surface
Z_1, Z_2, Z_3	Collapse pressure formula parameters
Δ	Shrinkage allowance
θ	Central half-angle of spherical cap
λ	Thinness factor
λ_0	Shallow cap parameter
ν	Poisson's ratio
$\phi = E/S_y$	Inverse strain parameter

REFERENCES

1. Q. Montague, Experimental behavior of thin-walled cylindrical shells subjected to external pressure, *Journal of Mechanical Engineering Science*, 11(1), 1969, 40–56.
2. D. F. Windenburg and C. Trilling, Collapse by instability of thin cylindrical shells under external pressure, *Transactions, American Society of Mechanical Engineers (ASME)*, Vol. 56, 1934, pp. 819–825.
3. K. von Sanden and K. Günther, Über das Festigkeitsproblem quer versteiffer Hohlzylinder unter Allseitig Gleichmässigem Aussendruck, *Werft Recderei*, 1, 1920; 2, 1921.
4. S. Timoshenko, *Theory of Elastic Stability*, Engineering Societies Monograph, McGraw Hill, New York, 1936.
5. S. S. Gill, *The Stress Analysis of Pressure Vessels and Pressure Vessel Components*, Permagon Press, Oxford, 1970.
6. American Petroleum Institute, Performance Properties of Casing and Tubing, API Bulletin, 5C2, Dallas, TX, 1970.
7. S. Timoshenko, *Strength of Materials*, D. Van Nostrand, New York, 1956.
8. S. H. Edwards and C. P. Miller, Discussion on the effect of combined longitudinal loading and external pressure on the strength of oil-well casing, *Drilling and Production Practice*, American Petroleum Institute, Dallas, TX, 1939.
9. Armco Steel Corporation, *Oil Country Tubular Products Engineering Data*, Armco Steel Corporation Middletown, OH, 1966.
10. J. C. Weydert, Stresses in oval tubes under internal pressure, *Society of Experimental Stress Analysis*, 12(1), 1954.

31 Buckling of Spherical Shells

31.1 INTRODUCTION

By "spherical shell," we mean complete spherical configurations, hemispherical heads (such as pressure vessel heads), and shallow spherical caps. In analyses, a spherical cap may be used to model the behavior of a complete spherical vessel with thickness discontinuities, reinforcements, and penetrations.

Although the response of a spherical shell to external pressure has received considerable attention from analysts, the calculation of collapse pressure still presents substantial difficulties in the presence of geometrical discontinuities and manufacturing imperfections. The bulk of the theoretical work carried out so far has had a rather limited effect on the method of engineering design, and therefore much experimental support is still needed. At the same time, the application of spherical geometry to the optimum vessel design has continued to be attractive in many branches of industry dealing with submersibles, satellite probes, storage tanks, pressure domes, diaphragms, and similar systems. This chapter deals with the mechanical response and working formulas for spherical shell design in the elastic and plastic ranges of collapse, which could be used for underground and aboveground applications. The material presented is based on state-of-the-art knowledge in pressure vessel design and analysis.

31.2 ZOELLY–VAN DER NEUT FORMULA

R. Zoelly and A. Van der Neut conducted significant original theoretical work on the buckling of spherical shells [1]. They used the classical theory of small deflections and the solution of linear differential equations. Based upon this work, the elastic buckling pressure P_{CR} for complete, thin spherical shell was found to be

$$P_{CR} = 2E/m^2 \sqrt{3(1 - \nu^2)} \tag{31.1}$$

where
 E is the elastic modulus
 ν is Poisson's ratio
 m is the radius/thickness ratio (R/T)

For a typical Poisson's ratio ν of 0.3, Equation 31.1 becomes simply

$$P_{CR} = 1.21E/m^2 \tag{31.2}$$

31.3 CORRECTED FORMULA FOR SPHERICAL SHELLS

At the time of the development of the classical theory, which led to Equation 31.1, no systematic experimental work was done. Several years later, however, some tests reported at the California Institute of Technology [2] showed that the experimental buckling pressure could be as low as 25% of the theoretical value given by Equation 31.1. The value derived by means of Equation 31.1 was then considered as the upper limit of the classical elastic buckling, while several investigators embarked on special studies with the aim of explaining these rather drastic differences between the theory and experiment. There was no reason to doubt the classical theory of elasticity, which worked well for flat plates, and it was soon suspected that the effect of curvature and spherical shape imperfections could have been responsible for the discrepancies.

This thesis led to the realization that the classical theory must have failed to reveal the fact that for a vessel configuration, not far away but somewhat different from the perfect geometry, lower total potential energy was involved, and therefore a lower value of buckling load could be expected, such as that indicated by tests. The theoretical challenge then became to formulate a solution compatible with such a lower boundary of collapse pressure at which the spherical shell could undergo the "oil canning" or "Durchschlag" process.

After making a number of necessary simplifying assumptions, von Kármán and Tsien [2] developed a formula for the lower elastic buckling limit for collapse pressure, which for $\nu = 0.3$ was found to be

$$P_{CR} = 0.37E/m^2 \qquad (31.3)$$

This level of collapse pressure may be said to correspond to the minimum theoretical load necessary to keep the buckled shape of the shell with finite deformations in equilibrium. The lower limit defined by Equation 31.3 appeared to compare favorably with experimental results, also given in the literature [2]. On the other hand, the upper buckling pressure given by Equation 31.1 could be approached only if extreme manufacturing and experimental precautions were taken. In practice, the buckling pressure is found to be closer to the value obtained from Equation 31.3 and therefore this formula is often recommended for design.

The exact calculation of the load–deflection curve for a spherical segment subjected to uniform external pressure is known to involve nonlinear terms in the equations of equilibrium, which cause substantial mathematical difficulties [3].

31.4 PLASTIC STRENGTH OF SPHERICAL SHELLS

Equations 31.2 and 31.3 may be regarded as design formulas based upon results using elasticity theory. Bijlaard [4], Gerard [5], and Krenzke [6] conducted subsequent studies to determine the effect of including plasticity upon the classical linear theory. To this end, Krenzke [6] conducted a series of experiments on 26 hemispheres bounded by stiffened cylinders. The materials were 6061-T6 and 7075-T6 aluminum alloys, and all the test pieces were machined with great care at the inside and outside contours. The junctions between the hemispherical shells and the cylindrical portions of the model provided good natural boundaries for the problem. The relevant physical properties for the study were obtained experimentally. The best correlation was arrived at with the aid of the following expression:

$$P_{CR} = \frac{0.84(E_s E_t)^{1/2}}{m^2} \qquad (31.4)$$

where E_s and E_t are the secant and tangent moduli, respectively, at the specific stress levels. These values can be determined from the experimental stress–strain curves in standard tension tests. The relevant test ratios of radius to thickness in Krenzke's work varied between 10 and 100 with a Poisson's ratio of 0.3. The correlation based on Equation 36.4 gave the agreement between experimental data and the predictions within +2% and −12%.

The extension of the Krenzke results to other hemispherical vessels should be qualified. Although his test models were prepared under controlled laboratory conditions, the following detrimental effects should be considered in a real environment:

Local and/or overall out-of-roundness
Thickness variation
Residual stresses
Penetration and edge boundaries

These effects are likely to be more significant when spherical shells are formed by spinning or pressing rather than by careful machining.

31.5 EFFECT OF INITIAL IMPERFECTIONS

In a subsequent series of collapse tests, Krenzke and Charles [7] aimed at evaluating the potential applications of manufactured spherical glass shells for deep submersibles. Because of the anticipated elastic behavior of glass vessels, the emphasis was placed on verifying the linear theory that resulted in Equation 31.2. Prior to this series of tests, very limited experimental data existed, which could be used to support a rational, elastic design with special regard to the influence of initial imperfections.

The formula for the collapse pressure of an imperfect spherical shell can be expressed in terms of a buckling coefficient K and a modified ratio m_i as

$$P_{CR} = \frac{KE}{m_i^2} \quad (K \approx 0.84) \tag{31.5}$$

where, based upon the work of Krenzke and Charles [7], the modified radius/thickness ratio m_i may be approximated as

$$m_i = R_i/h \tag{31.6}$$

where Figure 31.1 illustrates the modified radius R_i and thickness h.

According to the results obtained by Krenzke and Charles on glass spheres, the buckling coefficient K in Equation 31.5 was about 0.84. Their study showed that the elastic buckling strength of initially imperfect spherical shells must depend on the local curvature and the thickness of a segment of a critical arc length, L_c. For a Poisson's ratio of 0.3, this critical length can be estimated as

$$L_c = 2.42h(m_i)^{1/2} \tag{31.7}$$

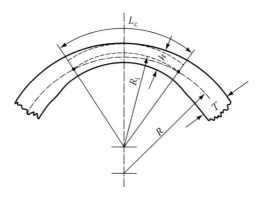

FIGURE 31.1 Notation for defining a local change in wall thickness.

In a related study conducted at the David Taylor Model Basin Laboratory, for the Department of Navy, the effect of clamped edges on the response of a hemispherical shell was evaluated. The relevant collapse pressure was found to be about 20% lower than that for a complete spherical shell having the same value of the parameter m and the elastic modulus E. Although these tests on accurately made glass spheres tended to support the validity of the small-deflection theory of buckling, there appeared to be little hope that metallic shells would yield a similar degree of correlation even under controlled conditions.

The investigations reviewed above may be of particular interest to designers dealing with complete spherical vessels as well as domed-end configurations. From a practical point of view, the most satisfactory method of predicting the collapse pressure would be to use a plot of experimental data as a function of the following well-defined dimensional quantities:

Experimental collapse pressure, P_e
Pressure to cause membrane yield stress, P_m
Classical linear buckling pressure, P_{CR}

31.6　EXPERIMENTS WITH HEMISPHERICAL VESSELS

Using experimental data for collapse of hemispherical vessels subjected to external pressure, Gill [8] provides information for a nondimensional plot suitable for preliminary design purposes. Figure 31.2 shows this plot for the following dimensionless ratios:

$$\frac{0.83P_e m^2}{E} = \frac{P_e}{P_{CR}} \quad \text{and} \quad \frac{0.61E}{mS_y} = \frac{P_{CR}}{P_m} \tag{31.8}$$

where
　　P_e is the experimental collapse pressure
　　P_{CR} is the classical linear buckling pressure

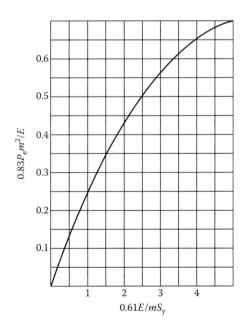

FIGURE 31.2　Lower-bound curve for hemispherical vessels under external pressure.

m is the radius/thickness ratio (R/T)
E is the elastic modulus
S_y is the yield stress

The accuracy with which the collapse pressure can be predicted on the basis of experimental data must be influenced by the maximum scatter band involved. Since this scatter is sensitive to material and geometry imperfections, their probable extent should be known before a more reliable, lower-bound curve can be developed. The results given in Figure 31.2 include hemispherical vessels in the stress-relieved and as-welded condition without, however, specifying the extent of geometrical imperfections, which, in this particular case, were known to be less pronounced. It follows that Figure 31.2 is applicable only to the design of hemispherical vessels, where good manufacturing practice can be assured. Further research work is recommended to narrow the scatter band to assure better correlation for the lower bound.

The dimensionless plot given in Figure 31.2 is sufficiently general for practical design purposes. For example, consider a titanium alloy hemisphere with $m = 60$, $E = 117,200$ N/mm^2, and the compressive yield strength, $S_y = 760$ N/mm^2. From Equation 31.8, we get $0.61E/mS_y = 1.57$. Hence, Figure 31.2 yields $0.83P_e m^2/E = 0.36$, from which $P_e = 14.1$ N/mm^2.

It may now be instructive to look briefly at the empirical result in relation to the theoretical limits defined by Equations 31.2 and 31.3 for the complete spherical vessels.

Making $P_e = P_{CR} = 14.1$ N/mm^2 and solving Equation 31.5 for the magnitude of the buckling coefficient gives $K = 0.43$. This value is close to the theoretical lower limit of 0.37 given by Equation 31.3 for a complete spherical vessel, and it appears to suggest that certain portions of such a vessel under uniform external pressure may behave in a manner similar to that of a complete vessel. This observation may be of special importance in dealing with the spherical shells containing local reinforcements and penetrations. It is also generally consistent with the elastic theory of shells, according to which the influence of geometrical discontinuities is local and does not extend significantly beyond the range determined by the value of the parameter $T(m)^{1/2}$.

31.7   RESPONSE OF SHALLOW SPHERICAL CAPS

Consider a relatively thin and shallow spherical cap fully clamped at its edge and subjected to uniform external pressure as represented in Figure 31.3 [9]. A key parameter characterizing a spherical cap is λ_o, defined as

$$\lambda_o = \frac{1,82 a_o}{T(m)^{1/2}} \quad \text{or} \quad \lambda_o = 2.57(H/T)^{1/2} \tag{31.9}$$

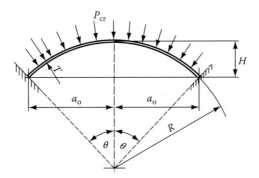

FIGURE 31.3   A spherical cap and notation.

where

a_o is the support radius
T is the shell thickness
m is the radius/thickness ratio (R/T)
R is the shell radius
H is the shell height above its support (see Figure 31.3)

The structural response of the cap for a typical Poisson ratio ν of 0.3 may be described as

$$\lambda_o < 2.08 \quad \text{continuous deformation with buckling}$$
$$\lambda_o > 2.08 \quad \text{axisymmetric snap-through}$$
$$4\lambda_o > 6 \quad \text{local buckling}$$

From Figure 31.3, the half-central angle θ is related to a_o, R, and H as

$$a_o = R\sin\theta \quad \text{and} \quad H = R(1 - \cos\theta) \tag{31.10}$$

By squaring and adding these expressions we obtain, after simplification,

$$H^2 - 2HR + a_o^2 = 0 \tag{31.11}$$

Assuming that H is small, H^2 is considerably smaller than $2HR$. Then by neglecting H^2 in Equation 31.11, the equation may be written as

$$H = \frac{a_o^2}{2R} \tag{31.12}$$

By substituting this expression for H into the second expression of Equation 31.9, we obtain the first expression of Equation 31.9. Thus the two expressions of Equation 31.9 are equivalent for shallow caps (that is, H considerably smaller than R).

As a guide, a spherical cap may be regarded as thin when $m > 10$. Shallow geometry is then approximately defined as $a_o/H \geq 8$. Once the spherical cap parameter λ_o is calculated by either of the equations in (Equation 31.9), we can estimate the critical buckling pressure by using the curve of Figure 31.4. This curve is based upon numerical data quoted by Flügge [9].

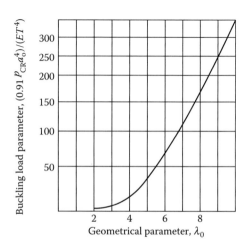

FIGURE 31.4　Design chart for a shallow spherical cap under external pressure.

The curve of Figure 31.4 is smoothed out somewhat in the midregion of the parameter λ_o, which involves a transition between the theoretical and experimental data in simplifying the curve fitting process. By using the curve of Figure 31.4, the following expression for the critical buckling pressure can be developed:

$$P_{CR} = 0.075\ En_0^{-4}\lambda_0^{4.15}e^{-0.095\lambda_0} \tag{31.13}$$

where n_0 is the dimensionless ratio a_o/T.

As an example application of Equation 31.13 let $R = 127$ mm, $a_o = 31.8$ mm, $T = 2.1$ mm, and $E = 117,200$ N/mm². From this data, we obtain

$$m = R/T = 60.5 \quad \text{and} \quad n_0 = a_o/T = 15.1 \tag{31.14}$$

Then from the first equation of Equation 31.9 we obtain λ_o as

$$\lambda_o = 3.53 \tag{31.15}$$

Finally, by substituting the data and results into Equation 31.13, we obtain

$$P_{CR} = 22.7 \text{ N/mm}^2 \tag{31.16}$$

In a special situation where a spherical cap is very thin, with a range of m values between 400 and 2000, the following empirical formula has been suggested for the relevant buckling pressure [10]:

$$P_{CR} = \frac{(0.25 - 0.0026\theta)(1 - 0.000175m)E}{m^2} \tag{31.17}$$

where θ is the half central angle of Figure 31.3 in degrees. In Equation 31.17, θ is intended to have values between 20° and 50°.

Although Equation 31.17 is useful within the indicated brackets of m, it may not be quite suitable for bridging the boundaries between the shallow caps and hemispherical shells without a careful study. Ideally, the formula for the collapse pressure of a spherical shell should be reduced to the form of Equation 31.5 with the K value representing a continuous function of the shell geometry and manufacturing imperfections. For inelastic behavior, the parameter $(E_s E_t)^{1/2}$ appears to have the best chance of success for a meaningful correlation of theory and experiment. In the interim, however, the formulas given in this chapter are recommended for the preliminary design and experimentation.

31.8 STRENGTH OF THICK SPHERES

When a thick-walled spherical vessel is subjected to an external pressure P_0, the maximum stress S occurs at the inner surface as

$$S = \frac{3P_0R_o^3}{2\left(R_o^3 - R_i^3\right)} \tag{31.17}$$

where R_i and R_o are the inner and outer sphere radii.

The displacement of the inner surface toward the center of the vessel is

$$u_i = \frac{3P_0R_iR_o^3(1 - \nu)}{2E\left(R_o^3 - R_i^3\right)} \tag{31.18}$$

where

  E is the elastic modulus
  ν is Poisson's ratio

The corresponding displacement of the outer surface is

$$u_\text{o} = \frac{P_0 R_\text{o}}{2E\left(R_\text{o}^3 - R_\text{i}^3\right)} \left[(1 - \nu)\left(2R_\text{o}^3 - R_\text{i}^3\right) - 2\nu\left(R_\text{o}^3 - R_\text{i}^3\right)\right] \tag{31.19}$$

For a solid sphere subjected to external pressure, the amount of radial compression in the elastic range becomes

$$u_\text{o} = \frac{P_0 R_\text{o}(1 - 2\nu)}{E} \tag{31.20}$$

SYMBOLS

a_o	Support radius
E	Elastic modulus
E_s	Secant modulus of elasticity
E_t	Tangent modulus of elasticity
H	Depth of spherical cap
h	Reduced thickness of shell (see Figure 31.1)
K	Buckling coefficient
L_c	Critical arc length (see Figure 31.1)
m	Radius/thickness (R/T) ratio
m_i	Mean radius/local thickness ratio
P_CR	Elastic buckling pressure
P_e	Experimental collapse pressure
P_m	Membrane yield stress
P_o	External pressure
R	Shell radius
R_i	Inner radius
R_o	Outer radius
S	Stress
S_y	Yield strength
T	Shell thickness
u_i	Inner surface displacement
u_o	Outer surface displacement
λ_o	Shallow cap parameter
ν	Poisson's ratio

REFERENCES

1. S. P. Timoshenko and J. M Gere, *Theory of Elastic Stability*, 2nd ed., McGraw Hill, New York, 1961, pp. 512–519.
2. T. von Kármán and H. S. Tsien, The buckling of thin cylindrical shells under axial compression, *Journal of Aeronautical Sciences*, 8, 1941, pp. 303–312.
3. C. B. Biezeno, Über die Bestimmung der Durchschlagkraft einer schmach gekrümmten kreisförmigen Platte, AAMM, Vol. 19, 1938.

4. P. P. Bijlaard, Theory and tests on the plastic stability of plates and shells, *Journal of the Aeronautical Sciences*, 16(9), 1949, pp. 529–541.

5. G. Gerard, Plastic stability of thin shells, *Journal of the Aeronautical Sciences*, 24(4), 1957, pp. 269–274.

6. M. A. Krenzke, Tests of Machined Deep Spherical Shells Under External Hydrostatic Pressure, Report 1601, David Taylor Model Basin, Department of the Navy, 1962.

7. M. A. Krenzke and R. M. Charles, The Elastic Buckling Strength of Spherical Glass Shells, Report 1759, David Taylor Model Basin, Department of the Navy, 1963.

8. S. S. Gill, *The Stress Analysis of Pressure Vessels and Pressure Vessel Components*, Permagon Press, Oxford, 1970.

9. W. Flügge, *Handbook of Engineering Mechanics*, McGraw Hill, New York, 1962.

10. K. Kloppel and O. Jungbluth, Beitrag zum Durchschlagproblem dünnwandiger Kugelschalen, *Stahlbau*, 1953.

32 Axial and Bending Response

32.1 INTRODUCTION

In Chapter 15, we considered the general problem of stability and buckling resistance of various structural components subjected to axial and compressive loads. In this chapter, we extend these concepts to axial loading of piping and pressure vessels. Specifically, we examine a number of theories and formulas related to the axial response of cylindrical components. We also consider some special topics such as the axial response of a pipe constrained in the transverse direction and a rolling diaphragm theory. These topics involve some mathematical models useful in many practical applications.

32.2 APPROXIMATION OF CROSS-SECTION PROPERTIES

Recall that in beam theory the maximum bending stress σ_{max} in the beam cross section occurs at a distance c away from the neutral axis expressed as (see Equation 8.2)

$$\sigma_{max} = Mc/I \tag{32.1}$$

where
   M is the bending moment at the cross section
   I is the second moment of area about the neutral axis

For a rectangular cross section with base b and height h, as in Figure 32.1, I is (see Section 8.6; Equation 8.12):

$$I = bh^3/12 \tag{32.2}$$

For a circular cross section with radius R as in Figure 32.2, I is

$$I = (\pi/4)R^4 \tag{32.3}$$

Finally, for a hollow circular cross section, as with a pipe, as in Figure 32.3, I is

$$I = (\pi/4)(R_o^4 - R_i^4) \tag{32.4}$$

Let T be the pipe thickness defined as

$$T = R_o - R_i \tag{32.5}$$

Then

$$R_o = R_i + T \tag{32.6}$$

By substituting for R_o in Equation 32.4 we obtain

$$I = (\pi/4)\left[(R_i + T)^4 - R_i^4\right] \approx \pi R_i^3 T \tag{32.7}$$

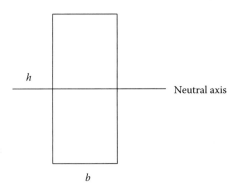

FIGURE 32.1 A rectangular cross section.

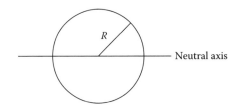

FIGURE 32.2 A circular cross section.

where the last term is obtained through the binomial expansion of $(R_i + T)^4$ and assuming T is small. Referring to Equation 32.1, the section modulus Z is often defined as

$$Z \overset{D}{=} I/C \tag{32.8}$$

Then for a thin, hollow pipe Z is approximately

$$Z = \pi R_i^2 T \tag{32.9}$$

32.3 COLUMN BEHAVIOR OF PIPE

When a relatively long and flexible pipe, of uniform cross section, is loaded as a column with pinned ends, the buckling load P_{CR} and the corresponding critical buckling stress S_{CR} are seen to be (see Equation 15.8)

$$P_{CR} = \frac{\pi^3 E R^3 T}{L^2} \tag{32.10}$$

and

$$S_{CR} = \frac{\pi^2 E R^2}{2L^2} \tag{32.11}$$

where we assume the pipe is hollow and thin with thickness T, and as before, L is the length, R is the inner radius, and E is the elastic modulus.

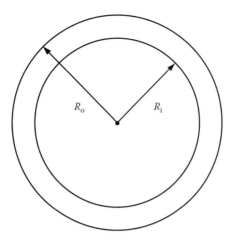

FIGURE 32.3 A hollow circular cross section (as with a pipe).

The results of Equations 32.10 and 32.11 correspond to the first fundamental mode of buckling requiring the smallest value of axial load to produce instability. However, where transverse constraint is present preventing development of the fundamental buckling mode, the issue of higher-order buckling modes arises [1]. In this case, the general expression for the buckling load P becomes

$$P = \frac{\pi^3 E R^3 T (1 + 2\alpha)^2}{L^2} \tag{32.12}$$

where α is the number of buckling mode shapes.

Figure 32.4 provides illustrations of higher-order buckling of a long pin-jointed column. Observe how rapidly the buckling load P increases with α. (The modes are assumed to be symmetrical about the midpoint of the pipe.) The response of the pipe corresponding to $\alpha = 0$ leads to Equation 32.10, which also represents the condition of minimum elastic energy.

Equation 32.12 leads to an interesting statement concerning higher-mode equilibrium consistent with the yield strength of the pipe. Specifically, the yield load P_y may be expressed in terms of the yield strength S_y as

$$P_y = AS_y = 2\pi RTS_y \tag{32.13}$$

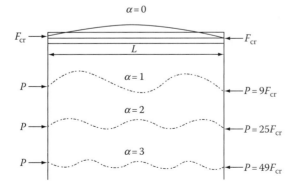

FIGURE 32.4 Higher-mode buckling of a long compressed simple pipe.

where A is the pipe cross section area. By equating the expressions of Equations 32.12 and 32.13 and solving for α, we obtain

$$\alpha = 0.225 \left(\frac{L}{R}\right) \left(\frac{S_y}{E}\right)^{1/2} - 0.5 \tag{32.14}$$

For example, if $L/R = 100$ and $E/S_y = 300$, Equation 32.14 yields $\alpha = 0.8$. The nearest integer is 1, which according to Figure 32.4, corresponds to

$$P = 9P_{CR} \tag{32.15}$$

where P_{CR} is given by Equation 32.10.

32.4 PIPE ON AN ELASTIC FOUNDATION

When a pipe is continuously supported along its length, we can model the pipe as a bar on an elastic foundation [1]. The energy method provides a convenient method of analysis and solution. In this method, the lateral bar deflection is represented by a trigonometric series. Then the work done in compressing the bar axially is made equal to the strain energy of bending the bar and that of deformation of the medium, supporting the bar. The critical (buckling) value of the compressive axial force P_{CR} is then found by solving the energy balance equation by minimizing the energy of deformation.

The result of this analysis in terms of the usual parameters is

$$P_{CR} = \frac{\pi^3 E R^3 T}{L_e^2} \tag{32.16}$$

where L_e is an "equivalent length" of an axially loaded bar without lateral support (see Equation 32.10). That is, Equation 32.16 is the same as Equation 32.12 with L replaced by L_e.

The selection of the reduced length L_e depends on a parameter Z_0 given by

$$Z_0 = 0.02 \frac{\beta L^4}{E R^3 T} \tag{32.17}$$

where

L denotes the actual length of the pipe resting against the elastic foundation
the parameter β defines the modulus of the foundation in psi

The dimensional check of Equation 32.17 shows that the result is compatible with the definition of β. The ratio of L_e/L depends on the parameter Z_0, and this relation can be conveniently represented in a graphical form or in algebraic terms. For example, when the parameter Z_0 varies between 0 and 30, the relevant magnitudes of the ratio L_e/L follow directly as in Figure 32.5.

For the higher values of Z_0, the required length ratios can be calculated from the following approximate relations:

$$\text{When } 30 \leq Z_0 \leq 1000, \quad \frac{L_e}{L} = 0.6675 - 0.1575 \log Z_0$$

and

$$\text{When } 1000 \leq Z_0 \leq 10{,}000, \quad \frac{L_e}{L} = 0.450 - 0.085 \log Z_0 \tag{32.18}$$

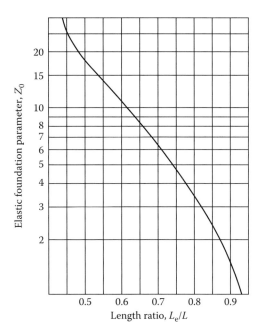

FIGURE 32.5 Elastic foundation parameter Z_0 as a function of the length ratio L_e/L.

The theory of column buckling at higher modes may be related to the theory of a column of finite length supported by an elastic medium. The maximum axial stress at the point of instability must increase with the increase in the number of half waves into which the originally straight column transforms. Similarly, the axial stress in a finite-length column, supported by a continuous elastic medium, must increase with the increase in the spring constant of the elastic foundation. Hence, a definite correlation exists between the buckling mode and the foundation modulus.

Assuming the end support conditions to be the same for both cases and combining Equations 32.22 and 32.16 leads to the simple expression:

$$\alpha = \frac{L - L_e}{2L_e} \qquad (32.19)$$

When $L = L_e$, elastic parameter Z_0 is zero. For a finite pipe geometry this can be true only if $\beta = 0$, or in other words, when the surrounding medium offers no resistance to the transverse pipe bending. Equation 32.19 shows that the mode factor α must also be zero, which corresponds to the fundamental buckling mode of a column with pin-jointed ends as shown in Figure 32.4.

Alternatively, when L_e tends to a very small value, the column offers progressively more resistance to the buckling deformation and, at least theoretically, the number of half-waves should become very high. This process has, of course, a natural boundary condition depending on the yield strength of the material or some other mode of failure, such as that of a cylinder in axial compression. Before the latter boundary is reached, however, Equation 32.19 may be used to approximate the buckling characteristics of the pipe by calculating the mode factor α for a given ratio of L_e/L.

32.5 ONE-WAY BUCKLING

In majority of the problems involving concepts of elastic stability and buckling of axially loaded members, the assumption is made that the bar is perfectly straight before the end load is applied.

The equilibrium based on a direct axial compression is considered stable until the critical load is reached, at which time even the slightest lateral force can immediately produce a lateral deflection, which will not vanish on the removal of the lateral force. In other words, a new state of equilibrium is attained in a slightly bent configuration, and the critical load is just sufficient to maintain such equilibrium.

Since in various practical situations we may have to consider pipe sections that are not perfectly straight to start with, and since such components as the utility pipe or a well casing can be supported by the surrounding medium in the transverse direction, the mathematical model of the pipe equilibrium should include the effects of eccentricity and the transverse resistance to buckling.

Standard solutions to column problems seldom include all such effects simultaneously. For example, the analysis of buckling of a bar on an elastic foundation usually does not account for any initial curvature of the bar before the axial load is applied. In another instance, the effect of the initial curvature on the transverse deflection of the column may be calculated without any allowance of the transverse loading distributed along the axis of the column. Such effects from the two separate solutions are not directly additive, and therefore the problem of a buckling column, with some residual curvature and transverse loading occurring simultaneously, should be formulated and developed from first principles. This form of the pipe behavior will be referred to in this discussion as "one-way buckling."

Allen [2] discussed the problem of buckling of this type. His analysis is based on the assumption that the transverse resistance of medium is constant along the pipe length and that the pipe behaves as a beam with built-in ends at which the deflection and the slope are zero. Furthermore, in addition to the shear reaction and axial forces, the fixing moments are considered at the supports.

Specifically, the assumptions are

1. The bending moment equation includes the change of the lever arm of the axial compressive load.
2. The tangential slope is relatively small, so that the pipe curvature may be represented by the strength-of-materials formula involving second derivative of the displacement.

To relate this analysis to a standard treatment of a beam-column problem, it will be noted that the equilibrium of forces in a one-way buckling type of solution includes the transverse loading and the original shape of the bar simultaneously. The procedure of setting up the differential equation of equilibrium for the portion of the pipe behaving as a pre-bent beam with the transverse restraint follows the general rule of the second-order theory of structural analysis. This rule simply states that the equations of equilibrium are written for the geometry of the deformed structure [3].

Figure 32.6 shows a simplified sketch of the pipe, where L_O defines the half-length of a symmetrically deformed portion of the pipe. There are thus three locations of zero slope with two of these shown in the figure.

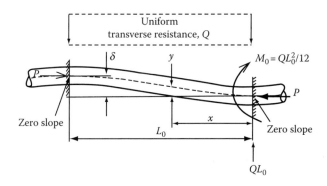

FIGURE 32.6   Half-length of pipe for analysis of one-way buckling.

Let Q represent the uniform resistance of the surrounding medium. End-fixing moment and shear reactions are denoted by M_0 and QL_0, respectively. The compressive forces P are shown to be offset by the amount δ. The deformed pipe is considered to remain in equilibrium under the assumed system of forces as long as the value of P is unchanged. Hence, the basic analytical problem in this instance is to develop a working formula for the limiting compressive force in terms of the transverse resistance of the medium, flexural rigidity of the pipe, and the maximum eccentricity δ. No stipulation is made here as to how this value of δ was originally established, and the basic reason for including this parameter is to indicate the extent of the existing eccentricity for the end load. Hence, from the practical point of view, δ may be looked upon as the manufacturing tolerance, the original bending deflection due to the lack of straightness, or a combination of these effects.

Considering bending moment equilibrium at a typical point x, measured from the right-hand side support as in Figure 32.6, leads to the expression

$$EI\frac{d^2y}{dx^2} = QL_0x - \frac{Qx^2}{2} + \frac{QL_0^2}{12} - Py \tag{32.20}$$

or simply

$$\frac{d^2y}{dx^2} + K^2y + F(x) = 0 \tag{32.21}$$

where by inspection of Equation 32.20 K^2 and $F(x)$ are

$$K^2 = P/EI \quad \text{and} \quad F(x) = \frac{QK^2}{12P}\left(6x^2 - 12L_0x - L_0^2\right) \tag{32.22}$$

When $F(x)$ is a polynomial of not more than fifth degree, the general solution of Equation 32.21 may be obtained directly with the aid of the relation

$$y = A\sin Kx + B\cos Kx - \frac{F(x)}{K^2} + \frac{d^2F(x)}{K^4dx^2} - \frac{d^4F(x)}{K^6dx^4} \tag{32.23}$$

From Equation 32.21, we have

$$\frac{d^4F(x)}{K^6dx^4} = 0 \quad \text{and} \quad \frac{d^2F(x)}{K^4dx^2} = \frac{Q}{PK^2} \tag{32.24}$$

Finally, by substituting from Equations 32.22 and 32.24 in Equation 32.23, we obtain

$$y = A\sin Kx + B\cos Kx - \frac{Q\left(6x^2 - 12L_0x - L_0^2\right)}{12P} + \frac{Q}{PK^2} \tag{32.25}$$

The constants of integration A and B in Equation 32.25 can be determined from the support conditions. Figure 32.6, shows that the displacement y and its slope dy/dx are zero at $x = 0$. This leads to

$$A = -\frac{QL_0}{PK} \quad \text{and} \quad B = -Q\left(\frac{L_0^2}{12} - \frac{1}{K^2}\right) \tag{32.26}$$

By substituting these results into Equation 32.25, we have

$$y = \frac{QEI}{P^2}\left[1 - KL_0 \sin Kx + \frac{-K^2 L_0^2 - 12}{12}\cos Kx - \frac{K^2}{12}\left(6x^2 - 12L_0 x - L_0^2\right)\right] \qquad (32.27)$$

By knowing this displacement, we can proceed to evaluate stresses and critical loads. For example, for the model of Figure 32.6, Blake, in his second edition of this book [4], using this procedure, develops the expression for the critical column stress S_{CR} as

$$S_{CR} = 1.27\left(\frac{QER}{\delta T}\right)^{1/2} \qquad (32.28)$$

where
   Q is the transverse resistance (lb/in.)
   δ is the maximum initial offset

32.6 AXIAL RESPONSE OF CYLINDERS

Instability of thin pipes and containers subjected to end compression is still one of the most challenging problems in stress analysis for design. Wide disparity between theoretical and experimental results often occur. This is due to the complexity of the mathematical formalism and the modeling assumptions. Also, experimental techniques intended to test the theory involve difficulties and practical limitations. Finally, the influence of manufacturing imperfections on a design is difficult to estimate. In this section, therefore, we consider a few proven rules and formulas directly applicable in design.

Essentially all typical responses of cylindrical components under axial compression due to static and pseudostatic loading can be put into three categories:

1. Diamond-shaped buckles of local character
2. Bellows-type wall deformation
3. Direct yield stress criterion

Experiments indicate that thin pipes, vessels, and cans develop isolated buckles of the diamond pattern. This is largely due to the fact that the local bending rigidity of a shell is proportional to T^3, while the resistance to membrane tension depends on the first power of thickness. The thinner the cylindrical surface, therefore, the higher the tendency to develop a diamond-shaped pattern. Also a random appearance of diamond buckles suggests that initial imperfections must be responsible for this type of local buckling. Such defects may not be visible, yet they can cause sufficient variation in the compressive stress distribution around the cylinder to trigger the onset of instability. Furthermore, experiments show that diamond-shaped buckles can proceed with great rapidity, resulting often in a "snap-through" type of response.

The classical theory of elastic stability for a cylindrical component predicts a buckling stress equal to [1]

$$S_{CR} = \frac{0.605E}{m} \qquad (32.29)$$

where
   E is the elastic modulus
   m is the mean radius to thickness ratio (R/T)

The corresponding critical load P_{CR} is

$$P_{CR} = 3.8ET^2 \tag{32.30}$$

Wide discrepancies between theory and experiments have led to an empirical formula attributed to Donnell [1]:

$$S_{CR} = E\frac{0.605 - 10^{-7}m^2}{m(1 + 0.004\phi)} \tag{32.31}$$

where ϕ is the inverse strain parameter E/S_y with S_y being the yield stress. This expression appears to correlate well with experimental data, particularly in the range of higher values of m.

The classical theory of symmetrical buckling of a cylindrical shell under the action of uniform axial compression applies essentially to perfect thin shells of revolution, for which the critical buckling stress does not exceed the proportional limit of the material. In this regard, a cylindrical component may be regarded to be thin when its ratio of mean radius to thickness exceeds 20. The theoretical response in axial compression also indicates that the critical buckling load is independent of the pipe length. However, the number of half-ways into which the cylinder may buckle can be expressed as

$$n_e = \frac{0.58L}{(RT)^{1/2}} \tag{32.32}$$

The corresponding elastic buckling stress is still given by Equation 32.29.

Equations 32.29 and 32.32 are both intended for isotropic materials with Poisson's ratio 0.3 and only for the case of a purely elastic response.

32.7 PLASTIC BUCKLING IN AN AXIAL MODE

When the pipe ratio R/T is decreased, axial buckling stress can approach and exceed the proportional limit of the material. Under these conditions, the length of the half waves into which the pipe buckles becomes shorter. When short pieces of pipe are joined by means of couplings thicker than the pipe itself, lateral expansion of the pipe at the joints becomes restricted. This restriction causes local bending of the pipe wall which, when combined with the direct axial compression, gives rise to the formation of the first axisymmetric half-way buckle. With further increase in axial compression of the pipe, the buckle splits open. Tests also confirm that the formation of the buckle and the onset of split is likely to be accentuated by the degree of load eccentricity.

For the type of deformation experienced with relatively thicker piping such as that found in the field, the theory indicates that buckling should occur beyond the proportional limit of the material. The relevant calculation of the number of the half waves and the corresponding plastic buckling load require the introduction of a reduced modulus concept [5]. This leads to the expressions

$$n_p = \frac{0.58L(E)^{1/4}}{(RT)^{1/2}(E_r)^{1/4}} \quad \text{and} \quad S_p = \frac{0.605E_rT}{R} \tag{32.33}$$

where
n_p is the number of plastic waves
S_p is the plastic stress
L is the pipe length
R is the mean radius
T is the thickness
E is the elastic modulus

E_r is the reduced modulus given by

$$E_r = \frac{4EE_t}{\left(E^{1/2} + E_t^{1/2}\right)^2} \tag{32.34}$$

where E_t is the tangent modulus of elasticity. (Recall that the tangent modulus of elasticity is the slope of the stress–strain curve at any working stress, beyond the yield stress.)

It should be noted that, depending on the shape of the actual stress–strain curve of the material, the reduced modulus E_r can be either constant or variable. It follows that for a constant value of E_t and fixed pipe dimensions R and T, the length of a half-wave remains constant for all the strains beyond the proportional limit of the material. The corresponding critical buckling stress also remains constant. These features are evident from a review of Equations 32.33 and 32.34.

32.8 ANALYSIS OF BELLOWS-TYPE BUCKLE

Based on the analysis of experimental data and the concept of the reduced modulus of elasticity, the design estimate of the ultimate pipe capacity for a bellows type of buckling may be accomplished according to the following four-point procedure:

1. Create a bilinear approximation to the stress–strain curve of the pipe material
2. Obtain the tangent modulus of elasticity from the slope of the upper bilinear stress–strain approximation
3. Calculate the reduced modulus E_r
4. Calculate the ultimate axial load W and the corresponding length L_W of a half-wave buckle using the following expressions:

$$W = 3.80E_r T^2 \quad \text{and} \quad L_W = 1.72(RT)^{1/2}\left(\frac{E_r}{E}\right)^{1/4} \tag{32.35}$$

The minimum value of $E_r(E_{r\,min})$ for which Equation 32.35 is applicable may be determined by expressing the load W in terms of the compressive yield stress S_y. That is,

$$W = 2\pi RTS_y = 3.80E_{r\,min}T^2 \tag{32.36}$$

Solving for $E_{r\,min}$, we obtain

$$E_{r\,min} = 1.65RS_y/T \tag{32.37}$$

The maximum value of $E_r(E_{r\,max})$ for which Equation 32.35 is applicable may be taken as

$$E_{r\,max} = E_{r\,min} \tag{32.38}$$

on the assumption that the ultimate stress is twice the yield stress (that is, $S_u = 2S_y$). The above four-step procedure is not recommended outside the range limited by $E_{r\,min}$ and $E_{r\,max}$.

It should be noted that the absolute size and strength of the standard pipe coupling or a similar local reinforcement does not appear to enter the calculations. Its presence in the pipe string, subjected to axial compression, only serves as the local constraint and the origin of the first perturbation in the continuous process of bellows-type buckling.

Finally, the simplest estimate of the yield force P_y is determined from the yield stress S_y as

$$P_y = 2\pi RTS_y \tag{32.39}$$

32.9 EXAMPLE OF LOAD ECCENTRICITY

Under special conditions of underground explorations and tests, a long string of piping can be subjected to the effect of load eccentricity. Figure 32.7 illustrates this concept where we assume a superposition of tensile and bending stresses in the wall.

Let e be the lateral misalignment of the centerlines as shown in Figure 32.7. Let the eccentricity ratio n be defined as e/T and let m be the mean radius to thickness ratio R/T. That is,

$$n = e/T \quad \text{and} \quad m = R/T \tag{32.40}$$

The combined stress criterion is then

$$S = \frac{W}{T^2}\left(\frac{0.16}{m} + 0.32\frac{n}{m^2}\right) \tag{32.41}$$

The variation of the stress given by Equation 32.41 indicates that the effect of load eccentricity on the combined stress in the pipe wall decreases rather rapidly with an increase in the radius to thickness ratio. This is not surprising when we consider the effect of the increase of pipe radius on the numerical values of the wall bending stress. It is noted that the maximum combined stress is assumed to be tensile. For a relatively high eccentricity, however, it is possible to visualize another condition under which the resultant stresses are changed to compressive. When this happens, an additional bending criterion and its effect on the possibility of a local pipe buckling should be examined.

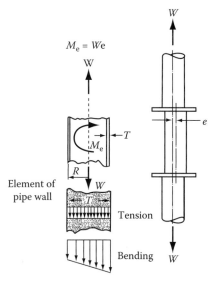

FIGURE 32.7 Combined stress in an offset pipe.

We provide a brief description of this type of analysis here for a general illustration. It is unlikely, however, that this mode of failure will commonly occur. But there may be special applications where this consideration could be of value.

The eccentricity of axial loading at which local buckling of pipe is likely to develop may be estimated from the expression:

$$e = 1.1 \frac{E}{W} mT^3 + \frac{mT}{2} - 1.57 \frac{P_e}{W} m^3 T^3 \tag{32.42}$$

where
W is the applied axial load
P_e is the external pressure on the pipe

In deriving Equation 32.42, the assumptions were made that the theoretical local bending stress in the pipe wall producing elastic instability is equal to the algebraic sum of the following stresses:

1. Membrane compressive stress due to external pressure
2. Compressive bending stress due to eccentric loading
3. Tensile stress due to axial loading

Since the external pressure P_e is often relatively low, the last term of Equation 32.42 is generally negligible. Equation 32.42 then becomes

$$e = 1.1 \frac{E}{W} mT^3 + \frac{mT}{2} \tag{32.43}$$

Equations 32.42 and 32.43 show that the extent of eccentricity and the proportions of the canisters or emplacement piping may be such that the elastic instability due to bending would not take place. This may be due to the effect of the direct tension on the stress distribution across the pipe wall, as well as the influence of a relatively high ratio of E/W.

32.10 THEORY OF A ROLLING DIAPHRAGM

One of the more interesting phenomena of axial response of a cylindrical component is concerned with the formation of a convolution. Such a mechanism is found in rolling cylindrical diaphragms, inverted tube shock absorbers, and positive expulsion devices, to mention a few. A simplified model of this mechanism can be established on the assumption that the neutral axis of the sheet material coincides with its centerline and that the strain energy due to axial and shear forces can be neglected as being small in relation to bending and hoop extension energy. Furthermore, it is assumed that the strain energy of bending and unbending during the process of diaphragm inversion is the same and that the material is bent continuously through a full angle of 180°. The strain energy of bending with a constant bending moment is obtained as the product of the plastic moment and the full angle of bend are expressed in radians. The unit energy due to circumferential strain is taken as the product of the plastic hoop stress and the corresponding total hoop strain.

Figure 32.8 illustrates the basic geometry of the cylindrical diaphragm. The diaphragm is considered to be held rigidly in the plane of circumference *A-A*. The mean radii of diaphragm and bend are *R* and *r*, respectively. Assuming a continuous plastic deformation under pressure differential ΔP, the diaphragm end moves from plane *B–B* to *C–C*, as shown. During this process, each element of the diaphragm undergoes longitudinal strains due to pure bending and hoop strains as the result of the overall increase in radius as seen in Figure 32.9. At the same time the plane O'' $O''G''G'''$ rotates through the angle $\theta = \pi$ into a new position $O'OGG'$.

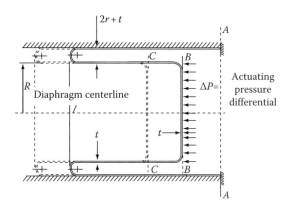

FIGURE 32.8 Notation for a cylindrical diaphragm.

For a small bend element of unit width cut out of the diaphragm, the cross-sectional geometry is approximately rectangular and the corresponding section modulus for a fully developed plastic condition is given by the elementary theory of plasticity as

$$Z_p = \frac{t^2}{4} \tag{32.44}$$

By the rules of strength-of-materials theory, the ultimate bending moment for a relatively wide beam is

$$M_p = \frac{S_y Z_p}{1 - \nu^2} \tag{32.45}$$

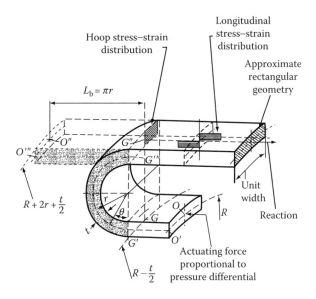

FIGURE 32.9 Geometry of a bent element.

where ν is Poisson's ratio. By substituting from Equation 32.44 into 32.45 and setting ν equal to 0.3 we obtain

$$M_p = 0.275 S_y t^2 \tag{32.46}$$

The longitudinal strain ε_b for pure bending may be expressed in terms of the mean radius of curvature r of the bend and the wall thickness t as

$$\varepsilon_b = t/2r \tag{32.47}$$

From the illustration of Figure 32.10, we see that the length L_b of the bend, which undergoes complete plastic deformation, is

$$L_b = \pi r \tag{32.48}$$

The energy U_b due to the bending deformation for length L_b is πM_p. Then by substituting from Equation 32.46, we obtain

$$U_b = 0.87\, S_y t^2 \tag{32.49}$$

The tensile hoop strain E_h developed during the straightening of the bend element may be developed and expressed as

$$
\varepsilon_h = \frac{\text{Final circumference less original circumference}}{\text{Original circumference}}
$$
$$
= \frac{2\pi\left(R - \frac{t}{2} + 2r + t\right) - 2\pi\left(R - \frac{t}{2}\right)}{2\pi\left(R - \frac{t}{2}\right)} \tag{32.50}
$$

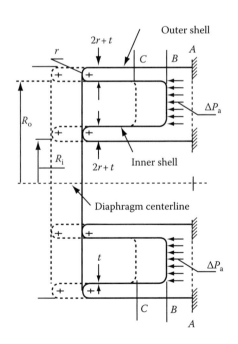

FIGURE 32.10   Annular rolling diaphragm.

or

$$\varepsilon_h = \frac{4r + 2t}{2R - t} \tag{32.51}$$

Since the unit energy due to hoop strain is $S_y\varepsilon_h$, the corresponding strain energy per unit width of the diaphragm becomes

$$U_h = S_y\varepsilon_h L_b t \tag{32.52}$$

Combining Equations 32.47, 32.48, and 32.52, U_h may be expressed as

$$U_h = 1.57 S_y t^2 \left(\frac{\varepsilon_h}{\varepsilon_b}\right) \tag{32.53}$$

By introducing nondimensional parameters k and m as

$$k = R/t \quad \text{and} \quad m = r/t \tag{32.54}$$

and by substituting into Equations 32.47 and 32.51, and in turn into Equation 32.53, we see that the hoop energy per unit width of the diaphragm is

$$U_h = \frac{\pi m(4m + 2)}{2K - 1} S_y t^2 \tag{32.55}$$

From Equations 32.49 and 32.55, we see that the total deformation energy U for a bend element per unit width is

$$U = \left[0.87 + \frac{\pi m(4m + 2)}{2k - 1}\right] S_y t^2 \tag{32.56}$$

The external work W_e done on the diaphragm by the actuating pressure ΔP for one length of bend is simply the total actuating force multiplied by the distance traveled. This gives

$$W_e = \pi R^2 \Delta P L_b \tag{32.57}$$

(The pressure acting on the side of the diaphragm is omitted since it acts perpendicular to the direction of the actuating pressure.)

By substituting from Equation 32.48 into Equation 32.57, we have

$$W_e = \pi^2 \Delta P R^2 r \tag{32.58}$$

By multiplying the unit width deformation energy U for a bend element, in Equation 32.56, by the average length of the circumference involved, gives

$$U_t = 2\pi(R + r)U \tag{32.59}$$

Finally, by setting the external work done W_e to the total internal energy of deformation U_T, we obtain (from Equations 32.58 and 32.59):

$$\Delta P = \frac{2(R + r)U}{\pi R^2 r} \tag{32.60}$$

Then by substituting for U from Equation 32.56, we have an approximate design formula for actuating on the cylindrical diaphragm as

$$\Delta P = \frac{(k + m)(0.55k + 4m^2 + 2m)S_y}{mk^3} \tag{32.61}$$

In many practical cases, m is small compared with k. Thus Equation 32.61 may be expressed in the reduced form:

$$\Delta P = \frac{(0.55k + 4m^2 + 2m)S_y}{mk^2} \tag{32.62}$$

To increase the capacity of the diaphragm we can form an annular geometry, as represented in Figure 32.10. Using the same type of analysis as above, the approximate design formula for the actuating pressure ΔP_a becomes

$$\Delta P_a = \frac{2S_y}{\pi m(k_0^2 - n_0^2)} \left\{ \frac{k_0 + m}{2k_0 - 1} [0.87(2k_0 - 1) + \pi m(4m + 2)] \right.$$
$$\left. + \frac{n_0 + m}{(2n_0 - 1)} [0.87(2n_0 - 1) + \pi m(4m + 2)] \right\} \tag{32.63}$$

When parameter m is relatively small and when $(2k_0 - 1)$ and $(2n_0 - 1)$ are approximated by $2k_0$ and $2n_0$, respectively, Equation 32.63 can be simplified to

$$\Delta P_a = \frac{[0.55(k_0 + n_0) + 8m^2 + 4m]S_y}{m(k_0^2 - n_0^2)} \tag{32.64}$$

Although the inherent complexity of the mechanics of a convolution presents a limited opportunity for a more detailed and fully representative theoretical analysis, the study indicates at least that the materials selected for the rolling diaphragms should exhibit good elongation and ductility. However, since the characteristics of uniform elongation in a metallic material are closely associated with the work-hardening properties and are subject to local variations resulting from heat treatment and alloy content, it is desirable, whenever possible, to select nonheat-treatable alloys. On this basis, aluminum alloys with the limited alloy content appear to be most promising, provided that they are compatible with other metals in the systems.

The equations derived indicate that the actuating pressure depend on geometric parameters and are directly proportional to the yield strength of the material. Hence, for the same thickness of a diaphragm wall, the higher the yield strength, the higher the pressure required to produce the rolling action. This simple relation was deduced on the assumption of the idealized stress–strain characteristics of material undergoing continuous plastic deformation. Under such conditions, the working plastic stress remains sensibly constant while the material is subjected to gradually increasing strains. To fulfill such stress–strain requirements, the relevant material must have good ductility and must be able to sustain appreciable elongation without rupture. From the point of view of practical design, it is better to keep in mind the fact that because of large strains associated with the process of atoroidal inversion of the cylinder wall, thin-gage materials and large diameters may be required for minimizing the working stresses. However, large ratios of diaphragm radius to wall thickness imply a decrease in the resistance of the wall to local buckling. Hence, a suitable design compromise may well be required.

SYMBOLS

A	Constant of integration
B	Constant of integration
b	Rectangular beam width (see Figure 32.1)
c	Half beam depth
E	Modulus of elasticity
E_r	Reduced modulus
$E_{r\,min}$	Minimum reduced modulus
$E_{r\,max}$	Maximum reduced modulus
E_t	Tangent modulus
E	Load eccentricity
$F(k)$	Auxiliary function
H	Rectangular beam height (see Figure 32.2)
I	Second moment of area
K	Curvature parameter
$k = R/t$	Cylindrical diaphragm ratio
$k_0 = R_o/t$	Annular diaphragm ratio
L	Length
L_b	Length of half-wave buckle
L_e	Reduced column length
L_0	Half-length of pipe
L_w	Length of half-wave buckle
M_b	Bending moment to cause buckling
M_e	Moment due to load offset
M_0	End moment
M_p	Plastic moment
m	Mean radius to thickness ratio
$n = e/T$	Eccentricity ratio
$n_0 = R_i/T$	Annular diaphragm ratio
n_p	Number of plastic half-waves
P	Axial load on pipe
P_{CR}	Critical axial load
P_e	External pressure
P_y	Axial load at yield
ΔP	Pressure difference across cylindrical diaphragm
ΔP_a	Pressure difference across annular diaphragm
Q	Transverse resistance
R	Mean radius of cylindrical component; cylinder radius
R_i	Inner radius
R_o	Outer radius
r	Mean radius of convolution
S	General symbol for stress
S_c	Stress in direct compression
S_{CR}	Critical column stress

S_p	Plastic stress
S_u	Ultimate stress
S_y	Yield stress
T	Thickness of pipe or vessel
t	Thickness of diaphragm wall
U	Total elastic energy per inch of convolution
U_b	Unit energy due to bending
U_h	Unit energy due to hoop strain
U_t	Total elastic energy
W	Ultimate axial load
W_e	External work done
x	Arbitrary distance
y	Initial offset at any point
z	Elastic section modulus
Z_0	Elastic foundation parameter
Z_p	Plastic section modulus
α	Buckling mode number
β	Modulus of foundation
δ	Maximum initial offset
ε_b	Bending strain
ε_h	Hoop strain
ν	Poisson's ratio
σ_{max}	Maximum bending stress
$\phi = E/S_y$	Inverse strain parameter

REFERENCES

1. S. P. Timoshenko and J. M. Gere, *Theory of Elastic Stability*, McGraw Hill, New York, 1961.
2. T. Allen, Experimental and analytical investigation of the behavior of cylindrical tubes subject to axial compressive forces, *Journal of Mechanical Engineering Science*, 10, 1968.
3. W. Flügge, *Handbook of Engineering Mechanics*, McGraw Hill, New York, 1962.
4. A. Blake, *Practical Stress Analysis in Engineering Design*, 2nd ed., Marcel Dekker, New York, 1990.
5. F. Bleich, *Buckling Strength of Metal Structures*, McGraw Hill, New York, 1952.

Part VII

Advanced and Specialized Problems

In this final part of our treatise we consider some advanced and specialized problems.

We begin with a brief chapter on stress concentration. We then consider thermal stresses. In the next three chapters we consider axial loadings on bars; rings and arches; and links and eyebars. We conclude with a couple of chapters on mechanical springs.

There is, of course, no limit to the number of special problems we could consider. The above problems were selected due to their general interest and utility.

33 Special Cylinder Problems

33.1 INTRODUCTION

In addition to issues involving stresses in pressurized components, there are a few specialized topics of related interest. These include such matters as dilation of cylinders, nested cylinder effects, and the effectiveness of circumferential stiffeners. In some of these cases, even well-established formulas and practices have speculative features despite the continuing progress in design technology. In this chapter, we briefly consider these problems, beginning with the dilation of closed cylinders.

33.2 DILATION OF CLOSED CYLINDERS

Dilation is the radial growth of a closed vessel (in this case a cylinder) subjected to internal pressure. The magnitude of the dimensional change may be of interest to designers concerned with the limited assembly tolerances in a given mechanical system. Additionally, one of the more important applications in this area involves the development of pressure transducers in the field of instrumentation. Transducer manufacturers are concerned with the optimum use of the strain gages to assure an electrical signal proportional to the internal pressure in a closed-end tube. It is clear, therefore, that some detailed knowledge of the dilation characteristics as a function of tube geometry is essential. Unfortunately, only limited information on this topic is available in the open literature [1].

If the outer and inner tube diameters are denoted by D and d, respectively, the amount of radial growth δ for a thin closed-end tube is

$$\delta = \frac{PD(2 - \nu)(1 + k)^2}{16E(1 - k)} \tag{33.1}$$

where
   P is the pressure
   D is the outer diameter
   k is the ratio of the outer diameter to the inner diameter D/d
   ν and E are Poisson's ratio and the elastic constant, respectively

Equation 33.1 is useful in the region of the cylinder not affected by the end closures. Figure 33.1 shows a transition region marked by a perturbation observed during tests [1].

The region of uniform radial growth, defined by x in Figure 33.1, depends on two dimensionless parameters: L/D and d/D, with L being the cylinder length. Figure 33.2 is a design chart providing the variation of the dimensionless span coordinate x/L in terms of the parameters d/D and L/D.

Experimental evidence also suggests that the ratio of the maximum dilation to the dilatation δ predicted by Equation 33.1 should not exceed 1.10. This ratio, however, will depend upon the basic shell parameters and the accuracy of manufacture. The effects of geometric irregularities such as out-of-roundness, bore eccentricity, and thickness variation has not yet been fully accounted for in experimental and theoretical studies.

33.3 NESTED CYLINDERS

In most design situations, the calculation of wall thickness for a cylindrical canister subjected to external pressure is performed well ahead of hardware development and manufacture. There may be

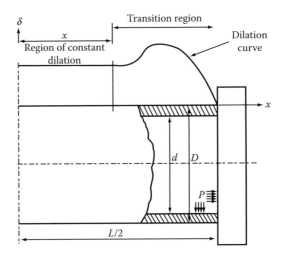

FIGURE 33.1 Partial representation of a closed-end cylinder under dilatation.

special cases, however, where an existing canister has to be modified to meet particular load specifications of increased safety considerations. Under these conditions, the following options are available:

1. Providing a new canister design
2. Providing circumferential stiffeners
3. Providing a continuous structural sleeve

If the time is limited, solution (1) may not be acceptable. If manufacturing and metallurgical problems arise, solution (2) may have to be excluded. This leaves the alternative (3).

We have to assume here that sleeve reinforcement can be provided without undue fabrication difficulties. For the purpose of this analysis, we assume that the composite canister, consisting of the

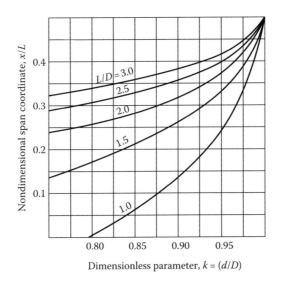

FIGURE 33.2 Design chart for the region of uniform dilation.

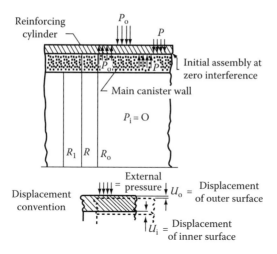

FIGURE 33.3   Notation for reinforcement analysis.

main shell and the reinforcing sleeve, can be assembled with zero or minimal initial interference. Thus, the practical question to be answered is: What degree of pressure attenuation can one reasonably expect in a reinforced system? Although the elastic theory suitable for defining the attenuation factor is rather elementary, few practical design formulas are readily available in engineering handbooks.

Let us consider, for example, a double-wall system such as that sketched in Figure 33.3. The displacement of the inner surface of the reinforcing cylindrical sleeve, in terms of a given external pressure P_o and the unknown contact pressure P, follows directly from Equation 29.33:

$$U_{io} = \frac{(1 - \nu)\left(P_o R_o^2 - PR^2\right)R + (1 + \nu)(P_o - P)R_o^2 R}{E\left(R_o^2 - R^2\right)} \tag{33.2}$$

where ν and E are Poisson's ratio and the elastic modulus, respectively. Figure 33.3 illustrates the other symbols and parameters.

The displacement of the outer surface of the canister may be expressed as

$$U_{oi} = \frac{(1 - \nu)\left(PR^2 - P_i R_i^2\right)R + (1 + \nu)(P - P_i)RR_i^2}{E\left(R^2 - R_i^2\right)} \tag{33.3}$$

For the assumed condition of zero radial interference between the sleeve and the canister, we have

$$U_{io} = U_{oi} \tag{33.4}$$

By substituting from Equations 33.2 and 33.3 in 33.4 and by setting $\nu = 0$ and $P_i = 0$, we obtain the following expression for the attenuation ratio ψ as

$$\psi = P/P_o = \frac{2\left(1 - k_1^2\right)}{\left(1.3 + 0.7k_2^2\right)\left(1 - k_1^2\right) + \left(0.7 + 1.3k_1^2\right)\left(1 - k_2^2\right)} \tag{33.5}$$

where k_1 and k_2 are defined as

$$k_1 = R_i/R \quad \text{and} \quad k_2 = R/R_0 \tag{33.6}$$

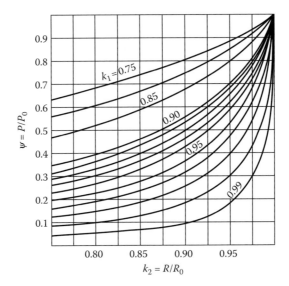

FIGURE 33.4   Pressure attenuation chart for nested cylinders.

When $k_1 = 1$, the thickness of the canister wall vanishes and the interference pressure becomes zero. When $k_2 = 1$, that is, the sleeve is assumed to be infinitely thin, Equation 33.5 yields $\psi = 1$, indicating no pressure attenuation. When $k_1 = 0$ or $k_2 = 0$, real solutions can still be obtained from Equation 33.5, although practical applications in cases reflecting such bracketing values are not very likely to exist. Figure 33.4 provides a series of curves yielding the attenuation ratio for various values of k_1 and k_2.

33.4 DESIGN OF RING STIFFENERS

We can reinforce cylindrical vessels and large pipes using circumferential stiffeners. The issues with this procedure are the determination of the spacing and the cross section dimensions of the stiffeners to prevent collapse of the vessel under external pressure. Figure 33.5 shows the relevant notation for our analysis.

Figure 15.25 provides a formula for the unit external load q_{CR} sufficient to elastically buckle a ring. Specifically, q_{CR} is

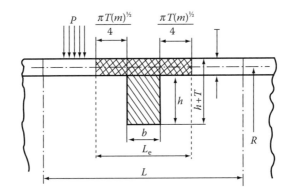

FIGURE 33.5   Notation for a ring stiffener.

$$q_{CR} = 3EI/R^2 \qquad (33.7)$$

where

E is the elastic modulus
I is the second moment of area of the ring
R is the ring radius

Let the length of the cylinder per stiffener be L. Then the external load q_{CR} per unit circumferential length is

$$q_{CR} = PL \qquad (33.8)$$

From Equation 33.7, the second moment of area I for the onset of elastic buckling is

$$I = \frac{PLR^3}{3E} \qquad (33.9)$$

In Figure 33.5, the effective length L_e of the stiffener may be approximated by the rule [2]:

$$L_e = b + 1.57(RT)^{1/2} \qquad (33.10)$$

Consider now the following expression for finding the second moment of area:

$$I = I_b + I_g - \frac{J^2}{A} \qquad (33.11)$$

where I_b, I_g, J, and A are

$$I_b = \frac{b(h+T)^3}{4} + \frac{1.57T^2(m)^{1/2}(2h+T)^2}{4} \qquad (33.12)$$

$$I_g \cong \frac{b(h+T)^3}{12} \qquad (33.13)$$

$$J = \frac{b(h+T)^2}{2} + \frac{1.57T^2(m)^{1/2}(2h+T)}{2} \qquad (33.14)$$

and

$$A = (h+T)b + 1.57T^2(m)^{1/2} \qquad (33.15)$$

Let λ and ε be defined as

$$\lambda = b/T \quad \text{and} \quad \varepsilon = h/T \qquad (33.16)$$

Then by combining Equations 33.10 through 33.16, we can develop an approximate design formula for ring stiffener sizing as

$$\frac{PL}{ET} = \eta = \frac{0.24\lambda^2(\varepsilon+1)^4 + 0.39\lambda(m)^{1/2}(\varepsilon+1)(4e^2 + 2\varepsilon + 1)}{m^3\left[\lambda(\varepsilon+1) + 1.57(m)^{1/2}\right]} \qquad (33.17)$$

Ring-stiffened vessels are generally most efficient. Although the behavior of stiffened cylinders has been studied by numerous investigators, the question of L_e given by Equation 33.10 has not yet been fully resolved, and it remains an interesting area for research.

As indicated previously in connection with the concept of the effective out-of-roundness, this influence becomes less important as the spacing of ring stiffeners decreases. Hence, the stiffening rings not only make a lighter design but also allow some relaxation of the fabrication tolerances. The experiments on ring-stiffened vessels show that the entire shell and stiffeners work together until local buckling of the panel takes place. Such buckling is often followed by a tear at the junction of the shell and the stiffener. The stiffener, however, seldom fails unless the panels on both sides of the ring become unstable and buckle prior to shell tear.

SYMBOLS

A	Cross-sectional area
b	Width of stiffener cross section
D	Outer diameter of cylinder
d	Inner cylinder diameter; bolt diameter
E	Modulus of elasticity
h	Depth of ring stiffener
I	Second moment of area
I_b, I_g	Second moments of area
J	Static moment
$k = d/D$	Diameter ratio
k_1, k_2	Radius ratios (see Equation 33.6)
L	Length of cylinder
L_e	Effective length
$m = R/T$	Radius-to-thickness ratio
P	Symbol for uniform pressure
P_o	External pressure
P_i	Internal pressure
q_{CR}	Critical buckling load
R	Interface or average radius
R_o	Outer radius
R_i	Inner radius
T	Thickness of wall
U_{io} (or U_i)	Displacement of inner surface
U_{oi} (or U_o)	Displacement of outer surface
x	Arbitrary distance
δ	Dilation
$\varepsilon = h/T$	Depth ratio
η	Dimensionless parameter
$\lambda = b/T$	Width ratio
ν	Poisson's ratio
$\psi = P/P_o$	Attenuation ratio

REFERENCES

1. J. S. Dean, Approximate calculation for the radial expansion of the thin tubes used in pressure transducers, *Journal of Strain Analysis*, 8(4), 1973.
2. E. W. Kiesling, R. C. DeHart, and R. K. Jain, Testing of ring-stiffened cylindrical shell encased in concrete—instrumentation and procedures, *Experimental Mechanics*, 10, 1970, pp. 251–256.

34 Stress Concentration

34.1 INTRODUCTION

"Stress concentration," as it is commonly used in design, refers to a high local stress relative to the overall macroscopic stress. Stress concentration has unique meaning for plane problems in terms of average stress. If, for example, a small hole is drilled in an end-loaded plate as in Figure 34.1, the stress is essentially unchanged in regions away from the hole. But at the edge of the hole, the tangential stress is increased dramatically.

By the concept of a macroscopic stress, we understand the average calculated stress related to the material's volume, characterized by a very fine structure. In terms of the practical requirements, this assumption is sufficiently accurate for the great majority of design situations.

The notion of a stress concentration factor K is that it is the multiple of the average stress, S'', which produces the high, or maximum, state of stress, S_{max}. That is,

$$S_{max} = KS'' \tag{34.1}$$

34.2 ELASTIC STRESS FACTORS

As a rule, stress concentrations arise due to the various local changes in shape, such as sharp corners, screw threads, abrupt changes in thickness, and even curved members of sharp curvature. This phenomenon is characteristic of elastic behavior. On the other hand, plastic yielding accompanies high stresses and tends to mitigate stress concentrations even in relatively brittle materials. This is a very important practical rule to keep in mind in developing rational designs. Particularly in the case of ductile response under static conditions, such as rivet holes in structural steel members, high local stresses based on the elastic theory can, indeed, be tolerated.

Under the conditions of static loading applied to the parts made of brittle materials, stress raisers cannot be ignored. This is also true in the case of some inherently ductile materials, which, at lower temperatures, fail due to the acquired brittle characteristics.

Stress concentrations of any kind of cyclic loading should be avoided or at least mitigated. Furthermore, tests show that a single isolated hole or a notch appears to have a worse effect than that due to a number of similar stress raisers placed relatively close together.

We can obtain elastic stress concentration factors either analytically or experimentally. The published literature contains extensive design tables for stress concentration factors as well as design procedures for guarding against fatigue failure in the presence of stress raisers [1–13].

34.3 COMMON TYPES OF STRESS RAISERS

Design experience indicates that there are at least two groups of questions which frequently come up during structural reviews. One concerns the effect of holes in plate and shell members. The other involves stress concentration due to the fillets and grooves under various conditions of loading.

We can illustrate the first group by results listed in Table 34.1 for plates and rectangular bars. These results are based upon long-established rules [13–16]. The results are also applicable with curved surfaces, provided the local curvature is not too sharp.

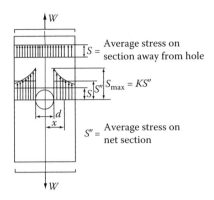

FIGURE 34.1   Example of stress concentration.

In the second group of stress raiser problems, a question frequently encountered concerns the difference in the type of loading on shafts and bars with transversely drilled circular holes. Figure 34.2 illustrates stress concentrations in round and square bars with holes. The upper curve of the figure is based on the case of uniaxial tension from Table 34.1, using however a more exact empirical formula [17,18]. Observe that both curves start at $K = 3$. Thus, as a conservative guide, a factor of 3 can be used in many circumstances.

In the case of a hole drilled near the free edge, however, as shown in Table 34.1, good practice is to make the dimension e equal to at least two hole diameters. Figure 34.3 shows another comparison between rectangular and round bars in this case with surface grooves. Round bars, with the dashed curves, are seen to be less susceptible to the effect of stress raisers. Again the ultimate values of the concentration factor K is less than 3, that is, $K \leq 3$ [19]. Figure 34.4 provides a comparison of stress concentration in bending and torsion.

Where sharp grooves and notches are involved, the theoretical values of stress concentration can be very high, and for this reason, the theory should be corrected for small radii of curvature. Under

TABLE 34.1

Effect of Circular Hole on Direct Stress for Flat Plates and Rectangular Bars

	Uniaxial tension of central hole	$K = \dfrac{3b}{b+d}$ (approximate formula)
	Uniaxial tension of center hole	$\begin{array}{c\|cccccc} e/d & 0.67 & 0.77 & 0.91 & 1.07 & 1.39 & 1.56 \\ \hline K & 4.37 & 3.93 & 3.61 & 3.40 & 3.25 & 3.16 \end{array}$
	Biaxial tension (d/b small)	$K = 2$
	Biaxial tension and compression (d/b small)	$K = 4$

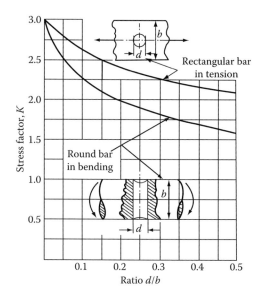

FIGURE 34.2 Stress concentration in round and square bars.

repeated loading, sharp notches can be especially detrimental. The highest stress concentration will develop when the notch depth is large while the notch radius and the angle are small.

Stress concentration in the presence of a groove produces the effect of a combined stress pattern, decreasing the shear stress, for instance, in the middle of a grooved cylindrical specimen. This effect results in a cup-and-cone type of failure of a tensile specimen, so that the ductile material appears to have the characteristics of brittle failure on the inside.

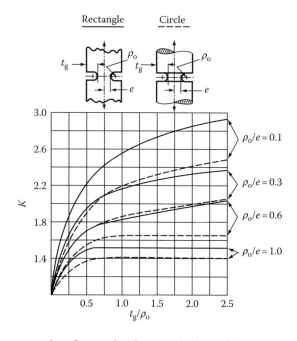

FIGURE 34.3 Stress concentrations for round and rectangular bars with grooves.

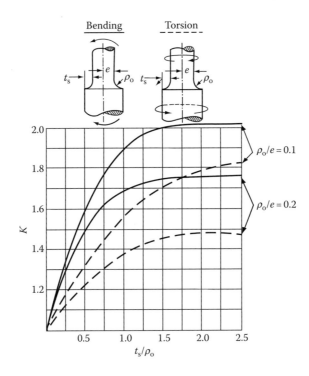

FIGURE 34.4 Comparison of stress concentration between bending and torsion.

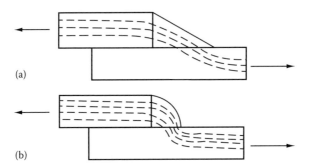

(a)

(b)

FIGURE 34.5 Approximate stress transfer in a welded joint.

In the design of steel structures [20], it should be noted that, in addition to problems of grooves and sharp corners, nonuniform stress distributions can occur in welded joints as illustrated in Figure 34.5. An elongated transverse weld, as in Figure 34.5a, produces a more uniform stress gradient than a shorter transverse weld, as in Figure 34.5b.

34.4 STRESS DISTRIBUTION

When the structural geometry of sharp transitions, worms, notches, or holes creates a stress concentration, a question which often arises is: What is the stress distribution around the concentration?

To discuss this, consider for example, the stretched plate of Figure 34.1, illustrating stress concentration in tension. Theoretical and experimental studies suggest that in this case the stress distribution can be represented as

$$S_{max} = S'' \left[1 + \frac{1}{8} \left(\frac{d}{x} \right)^2 + \frac{3}{32} \left(\frac{d}{x} \right)^4 \right] \tag{34.2}$$

where S'' is the general, overall, or average stress, and S_{max} represents the stress enhancement as a result of the hole, as a function of distance x from the hole center. When x is $d/2$, at the edge of the hole, $S_{max} = 3S''$. That is, the maximum theoretical stress occurs at the hole edge and is three times the average stress S''. Alternatively, when x becomes large in comparison with the hole diameter, S_{max} approaches the average stress S''.

Equation 34.2 shows that stress disturbance is highly localized. Practical rules often state that the maximum theoretical stress concentration for a plate in tension is encountered when the width of the plate is more than about four times the diameter of the hole. Putting $b = 4d$ into an approximate formula, given in Table 34.1, we get $K = 2.4$. Only when d becomes very small, the theoretical value of 3 attained.

In general, the effect of open holes in beams is not easy to evaluate despite the various theoretical and experimental tools available. For example, when holes are present in the flange, the problem of location of the neutral axis can lead to many interesting speculations. Furthermore, the effect of a hole in the tension flange of a beam is difficult to assess if the beam does not fracture and the compression flange carries the significant share of the load. On this basis, it would seem that the effect of holes in flanges can often be ignored, particularly when rivets are used. Under these circumstances, the American Institute of Steel Construction allows us to neglect the reduction of beam area and girder flanges of up to 15% of the gross area.

34.5   PLASTIC REDUCTION OF STRESS FACTORS

So far in this chapter, we have considered stress concentration from an elastic perspective. While this approach has been adopted by designers for many years, economic and environmental considerations require a closer look at the stress concentration factors to see if they can be reduced. More practical models are continually being sought.

In 1913, Inglis [21] proposed a simple formula for estimating the increase in stress due to a finite discontinuity such as an elliptical opening in a plate, porthole, or hatchway. Figure 34.6 shows the Inglis model. The proposed design formula for this model is

$$S_{max} = S \left[1 + 2 \left(\frac{L}{r} \right)^{1/2} \right] \tag{34.3}$$

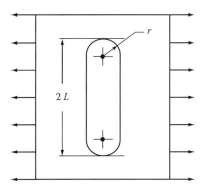

FIGURE 34.6   Inglis model. (From Inglis, C.E., *Proc. Inst. Naval Architec.*, 60, 1913.)

The application of this expression can be extended to the geometry of cracks, notches, scratches, and similar stress raisers as the L/R ratio is increased. The corresponding S_{max}/S ratio becomes the conventional stress concentration factor where the symbol S denotes the nominal stress existing at a point away from the discontinuity. Observe in Figure 34.6 that when $L = r$, the hole becomes circular, and the structure reduces to that of a symmetrically placed hole in a plate in uniaxial tension with $K = 3$.

While the Inglis formula planted some interesting ideas in the minds of engineers, it became necessary to invoke the principles of ductile response in the face of disturbing theoretical results derived from purely elastic considerations. This process has led to the development of a plastic reduction procedure [22,23] for the elastic stress factors. It became obvious that in the case of a truly ductile material under static loading, the conventional elastic factor should be modified with the aid of the appropriate stress–strain diagram of the material [24]. One of the simplest approaches to the correction of any type of elastic stress concentration factor [25] is

$$K_p = 1 + (K - 1)\left(\frac{E_s}{E}\right) \tag{34.4}$$

where E_s is the secant modulus (the slope of the stress–strain curve beyond the yield point) and E is the elastic modulus.

For the case of a circular hole in a wide plate, Equation 34.4 becomes

$$K_p = 1 + 2\left(\frac{E_s}{E}\right) \tag{34.5}$$

This method provides the opportunity for rounding off the calculated higher peaks of the elastic stresses and in this manner assuring a more reasonable value of the design stress factor. The magnitude of the plastic stress concentration factor depends then on the shape of the stress–strain curve while the conventional K factor is a function of the geometry of the part alone.

It should be stated in closing that stress concentrations in general are virtually inevitable in real structures and machines due to the presence of grooves, fillets, holes, threads, and similar discontinuities. The worst situations, of course, include machine errors, gravel nicks, nonmetallic inclusions, and microvoids, which may be difficult or even impossible to detect.

Stress intensities due to cracks in inherently brittle materials and in some ductile materials displaying brittle behavior under specific environmental conditions may lead to fracture. We dealt with the concepts and methods of fracture control in Chapters 27 and 28.

SYMBOLS

b	Width of rectangular bar and diameter of round bar
d	Diameter of bolt or rivet hole
e	Hole-to-edge distance, groove-to-bar center distance
E	Modulus of elasticity
E_s	Secant modulus
K	Stress concentration factor
K_p	Plastic stress concentration factor
L	Length
r	Radius
S	Average tensile stress
S''	Average stress on net cross section
S_{max}	Maximum stress

t_g Depth of groove
t_s Depth of shoulder
W Concentrated load
x Distance from bar center
ρ_0 Fillet or groove radius

REFERENCES

1. H. Neuber, *Theory of Notch Stresses*, N. W. Edwards Publishers, Ann Arbor, MI, 1946.
2. M. Hetenyi, *Handbook of Experimental Stress Analysis*, Wiley, New York, 1950.
3. R. E. Peterson, *Stress Concentration Design Factors*, Wiley, New York, 1953.
4. W. C. Young and R. C. Budynas, *Roarke's Formulas for Stress and Strain*, 7th ed., McGraw Hill, New York, 2002.
5. W. D. Pilkey, *Formulas for Stress, Strain, and Structural Matrices*, John Wiley & Sons, New York, 1994.
6. Battelle Memorial Institute, *Prevention of Fatigue of Metals*, Wiley, New York, 1941.
7. R. B. Haywood, *Designing Against Fatigue of Metals*, Van Nostrand Reinhold, New York, 1962.
8. P. G. Forest, *Fatigue of Metals*, Permagon Press, Elmsford, NY/Addison Wesley, Reading, MA, 1962.
9. G. Sines and J. L. Waismau (Eds.), *Metal Fatigue*, McGraw Hill, New York, 1959.
10. R. C. Juvinall and K. M. Marshek, *Fundamentals of Machine Component Design*, 2nd ed., John Wiley & Sons, New York, 1991.
11. J. E. Shigley and C. R. Mischke, *Mechanical Engineering Design*, 6th ed., McGraw Hill, New York, 2001.
12. B. J. Hamrock, B. O. Jacobson, and S. R. Schmid, *Fundamentals of Machine Elements*, McGraw-Hill, New York, 1999.
13. J. H. Faupel, *Engineering Design*, Wiley, New York, 1964.
14. E. G. Cokes and L. N. G. Filon, *Treatise on Photoelasticity*, Cambridge University Press, London, 1931.
15. J. B. Jeffrey, Plane Stress and Plane Strain in Bipolar Coordinates, *Philosophical Proceedings, Royal Society of London, Series A*, 221, 265, 1921.
16. C. Dumont, Stress concentration around an open circular hole in a plate subjected to bending normal to the plane of the plate, NACA, Technical Note 740, 1939.
17. M. Frocht, Factors of Stress Concentration Photoelastically Determined, *Journal of Applied Mechanics*, 2(2), 1935.
18. R. E. Peterson and A. M. Wahl, Two and three-dimensional cases of stress concentration, and comparison with fatigue tests, *Journal of Applied Mechanics*, 5(1), A-15, 1936.
19. F. Sass, C. Bounche, and A. Leitner, *Dubbels, Taschenbuck für den Maschinenbau*, Springer-Verlag, Berlin, 1966.
20. E. H. Gaylord and C. N. Gaylord, *Design of Steel Structures*, McGraw-Hill, New York, 1957.
21. C. E. Inglis, Stress in a plate due to the presence of cracks and sharp corners, *Proceedings of the Institute of Naval Architecture*, 60, 1913.
22. H. F. Hardrath and L. Ohman, A study of elastic and plastic stress concentration factors due to notches and fillets in flat plates, NACA Technical Note 2566, 1951.
23. A. Blake, Ed., *Handbook of Mechanics, Materials, and Structures*, Wiley, New York, 1985.
24. F. R. Shanley, *Strength of Materials*, McGraw Hill, New York, 1957.
25. J. A. Collins, *Failure of Materials in Mechanical Design: Analysis, Prediction, Prevention*, Wiley, New York, 1981.

35 Thermal Considerations

35.1 INTRODUCTION

The response of a structural member subjected to heating (or cooling) is an expansion (or contraction). The expansion, or contraction, can in turn produce stresses if

1. External forces constrain the expansion, or contraction
2. The shape of the structural member is incompatible with the tendency to expand or contract

The thermal behavior under the first condition may be represented, for example, by a uniform, straight bar held at the ends and subjected to a constant temperature gradient. The case corresponding to the second condition can be best illustrated with reference to a cylinder having a temperature gradient across its wall. Although in this case no external forces of constraint are applied, thermal stresses are produced because the strains are incompatible with the free thermal deformation. The two cases outlined above can also be characterized as those structural systems, which are governed by external or internal constraints.

35.2 BASIC STRESS FORMULA

The simplest expression for calculating the thermal stresses is based on the treatment of an elastic, uniformly heated or cooled bar, restrained firmly at the ends. That is,

$$\sigma = E\alpha\Delta T \tag{35.1}$$

where
  σ is the thermal stress
  E is the elastic modulus
  α is the coefficient of lines thermal expansion
  ΔT is the temperature change

When the temperature of the bar is decreased, the bar develops tension. Compressive stress, on the other hand, is caused by an increase in temperature. Since only the elastic stresses are considered, the total strain is the sum of the stress- and temperature-dependent strains [1].

In Equation 35.1, α denotes the coefficient of linear, one-directional, thermal expansion. For the majority of structural materials, α varies between 5 and 15×10^{-6} per °F. The corresponding volumetric strain due to the temperature change is

$$\left(\frac{\Delta V}{V}\right)_T = 3\alpha\Delta T \tag{35.2}$$

Table 35.1 provides values of thermal constants that are useful in stress calculations [2,3].

In planning the design of equipment for low-temperature use, one is confronted with the basic material decision influenced by various compilations of properties available in industrial literature. Where the thermal and mechanical properties are not readily available, their preliminary estimates can be made using simplified analytical methods [2].

TABLE 35.1

Typical Thermal Constants at Moderate Temperatures

Material	α (°F^{-1} × 10^{-6})	C (Btu/lb °F)	k (Btu/h ft °F)	λ (ft^2/h)
Pure aluminum	14.0	0.22	128.0	3.50
Aluminum alloy	13.0	0.22	91.0	2.30
Pure copper	9.5	0.09	228.0	4.50
Brass (60/40)	10.5	0.09	54.0	1.10
Bronze (90/10)	10.0	0.09	24.0	0.49
Gold	7.8	0.03	180.0	5.10
Silver	11.0	0.06	240.0	6.10
Carbon steel	6.7	0.11	26.4	0.49
Alloy steel	6.7	0.11	13.2	0.25
Lead	16.0	0.03	20.4	0.94
Magnesium alloy	14.0	0.25	47.0	1.73
Pure nickel	7.2	0.11	33.6	0.56
Iron–nickel (64/36)	1.1	0.11	7.2	0.12
Platinum	5.0	0.03	41.0	1.00
Tin	15.0	0.05	37.0	1.60
Zinc	14.5	0.09	67.0	1.67
Ceramics	1.7	0.20	0.72	0.021
Concrete	6.7	0.21	0.60	0.020
Typical glass	4.5	0.18	0.48	0.017
Ice		0.50	0.13	0.045
Typical plastics	11.0	0.37	0.24	0.006
Sandstone	4.5	0.19	0.96	0.029
Granite	4.5	0.19	0.30	0.057

Note: These values are good up to 400°F. *Symbols:* α, thermal coefficient of linear expansion; C, specific heat; k, thermal conductivity; λ, thermal diffusion.

35.3   THERMAL EFFECT ON STRENGTH

It should be recalled that in majority of the cases of solid materials, which do not undergo transitions, their strength in tension, hardness, and resistance to fatigue increase with a decrease in service temperature. This rule applies without reservation to such metals as aluminum, copper, nickel alloys, and austenitic stainless steels. In the case of ordinary carbon steels, however, the advantages of improved properties are seriously compromised by the materials' tendency to become brittle.

Figure 35.1 shows how the ultimate strength of several common metals is dependent upon temperature. The curves represent upper and lower boundaries corresponding to the cold-worked and annealed test specimens [4].

Compared with metals there is relatively little data available for the low-temperature properties of polymers (or plastics). It is known, however, that when a large piece of unreinforced plastic is suddenly cooled, the resulting thermal shock can cause cracking. Many polymers with the exception of Teflon can become relatively brittle at lower temperatures, although these materials tend to resist thermal shock better than glass.

It may be of interest to note that there is a certain similarity between the effects of low temperature and high strain rate. Experiments indicate that both effects tend to increase the yield and ultimate strength of steel while reducing the ductility. The observed influence is greater on the yield than on the ultimate strength, so that the relevant margin between the two properties is reduced. The usual stress–strain curve of the material under a high strain rate degenerates into a stress line parallel to the strain axis as if the material suddenly became completely plastic [5], with the ultimate strength markedly higher than that under static conditions.

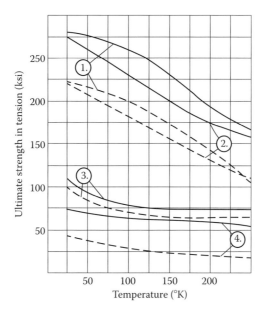

FIGURE 35.1 Low-temperature ultimate strength of metals: (1) 304 stainless; (2) 347 stainless; (3) 2024 aluminum; and (4) 6061 aluminum. Solid curves, cold-worked; dashed curves, annealed.

35.4 MATERIALS FOR SPECIAL APPLICATIONS

In special applications involving nuclear propulsion systems, very high-temperature strength is often required. These may include structural components for fission reactors used as direct heat exchangers, nuclear propulsion systems, and aerospace reentry vehicles, to mention a few. Because of the extreme temperature environment, the choice of materials is very limited. In this class, graphite, tungsten, and rhenium can be considered, although some of the carbides are also most valuable. Table 35.2 provides examples of properties for refractory alloys.

Table 35.3 lists some unusual properties for selected carbides at room temperature. Although these compounds are intended for highly specialized applications [6], it may be of interest to compare their mechanical properties with the typical data known for structural materials in general use.

Figure 35.2 shows the effect of elevated temperatures on the percentage of retained strength for a variety of engineering materials. The diagram illustrates knowledge of materials for high-temperature applications [7].

TABLE 35.2
Selected Properties of Refractory Metals

Metal	Melting Point (°F)	Density (lb/in.³ at 75°F)	Modulus of Elasticity (psi)	NDT (°F)	Tensile Strength at 2200°F (psi)
Chromium	3450	0.76	42×10^6	625	8,000
Columbium (niobium)	4470	0.31	16×10^6	−185	10,000
Molybdenum	4730	0.37	47×10^6	85	22,000
Tantalum	5430	0.60	27×10^6	−320	15,000
Rhenium	5460	0.76	68×10^6	75	60,000
Tungsten	6170	0.70	58×10^6	645	32,000

TABLE 35.3

Room-Temperature Properties for Selected Carbides

Compound	Density (g/cm^3)	Coefficient of Thermal Expansion (10^{-6}/°C)	Modulus of Elasticity (10^{-6} psi)	Compressive Strength (10^3 psi)
Hafnium carbide (HfC)	12.7	6.0	61	—
Tantalum carbide (TaC)	14.5	6.5	55	—
Zirconium carbide (ZrC)	6.7	6.7	69	235
Niobium carbide (NbC)	7.8	6.5	49	—
Titanium carbide (TiC)	4.9	7.7	65	196
Tungsten carbide (WC)	15.8	5.0	102	900
Silicon carbide (SiC)	3.2	3.9	69	200
Boron carbide (B$_4$C)	2.5	4.5	42	420

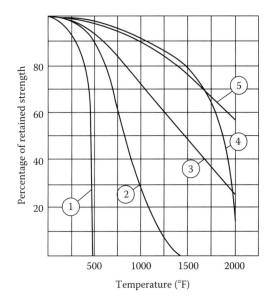

FIGURE 35.2 Strength retention for special materials: (1) Nylon; (2) "E" glass; (3) boron; (4) super alloys; and (5) refractory metal alloys.

35.5 THERMAL STRESS INDEX

The thermal stress index (TSI) is a measure of a material's ability to resist thermal gradients and consequently thermally induced stresses. The index is defined as

$$\text{TSI} = \frac{\sigma_u k}{\alpha E} \tag{35.3}$$

where
  σ_u is the ultimate tensile strength of the material
  E is the elastic modulus
  k is the thermal conductivity
  α is the coefficient of linear expansion

TABLE 35.4

Comparative Values of Thermal Stress Index for Typical Materials

Material	Ultimate Tensile Strength (psi)	Coefficient of Thermal Expansion (in./in. °F)	Thermal Conductivity (Btu/h ft °F)	Modulus of Elasticity (psi)	Thermal Stress Index (Btu/ft h)
Aluminum alloy, 24 ST	68,000	13×10^{-6}	75	10.4×10^{-6}	37,800
Magnesium alloy	34,000	15×10^{-6}	45	6.5×10	15,300
Structural steel	60,000	7×10^{-6}	22	29×10^{-6}	6,500
Nickel steel, A8	100,000	8×10^{-6}	15	30×10^{-6}	6,250
Titanium	80,000	5.5×10^{-6}	10	16×10^{-6}	9,100

In the English system, TSI has the units BTU/ft h. The higher the value of TSI the greater the material's ability to resist thermal gradients. Table 35.4 provides a list of TSI values for a few common structural materials.

35.6 THERMAL SHOCK

The index, given by Equation 35.3, is particularly useful in correlating the material properties with reference to the resistance to fracture by thermal shock [7]. The maximum temperature, T_{max}, that the material can withstand under the thermal shock conditions is dependent on the mechanical properties of the material but independent of thermal conductivity. That is

$$T_{max} = \frac{\sigma_u(1 - \nu)}{\alpha E} \tag{35.4}$$

The TSI and maximum thermal shock temperature, T_{max}, provide useful parameters for determining thermal shock resistance of a given material. In this context, "thermal shock" is the maximum sudden change in temperature that can be withstood by an infinite plate without fracture.

35.7 THERMAL CONDITIONS IN PIPING

In many practical cases of a long tube carrying a hot liquid and being cooled at the outer periphery, the temperature distribution can be established by evaluating heat balance across all the elemental rings of the cylinder.

When the heat flow is specified, the temperature differential ΔT across the tube wall is

$$\Delta T = \frac{Q \ln (R_o/R_i)}{2\pi k} \tag{35.5}$$

where
R_o and R_i denote the outer and inner tube radii, respectively
Q is the quantity of heat that must be conducted per unit length of tube

In terms of the stress parameters applicable to tubular members, the TSI can be stated as follows:

$$\text{TSI} = \frac{Q}{4\pi(1 - \nu)} \left(1 - \frac{2R_i^2}{R_o^2 - R_i^2} \ln \frac{R_o}{R_i} \right) \tag{35.6}$$

Equation 35.6 is useful in comparing brittle materials. It shows that the material with the highest value of the TSI will be able to withstand the highest amount of heat flow.

35.8 THERMAL STRESS FATIGUE

Fatigue from changes in thermal stresses is similar to mechanical fatigue. But there are several differences:

1. Plastic thermal strains tend to concentrate in the warmest regions of the body.
2. Accumulation of strain from thermal stress fatigue is localized.
3. Temperature cycling can adversely affect a material even in the absence of fatigue loading.
4. Thermal and mechanical strains, occurring together, have superimposed effects. This makes it difficult to use low-cycle mechanical fatigue tests to interpret thermal stress fatigue.
5. Tests indicate that for the same magnitude of total strain the life of a material is considerably shorter for thermal fatigue than for mechanical fatigue.

35.9 PRELIMINARY THERMAL DESIGN

To design for thermal stresses, we need to know the expected temperature distribution of a proposed structural component. This in turn requires knowledge of the fundamentals of heat transfer in conduction, convection, and radiation [8,9]. It is usually assumed that the elastic constant E, and the coefficient of thermal expansion α are approximately constant within a reasonable temperature range. Tables 35.5 and 35.6 [10–12] illustrate and provide solutions for thermally stressed components with external and internal constraints.

TABLE 35.5

Thermal Stresses due to External Constraint

1. Uniform bar, both ends fixed $\qquad \sigma = \alpha E \Delta T$

2. Uniform plate, edges fixed $\qquad \sigma = \dfrac{\alpha E \Delta T}{(1 - \nu)}$

3. 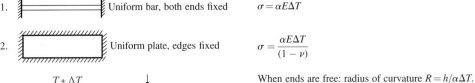 Bar of rectangular cross section

   When ends are free: radius of curvature $R = h/\alpha \Delta T$.
   When ends are fixed, the end couples $M_0 = \alpha \Delta T E I/h$.
   The maximum bending stress $\sigma = \alpha\, E \Delta T/2$.

4. When instead of a bar of a rectangular cross section, a plate of thickness h is used, the radius of curvature is the same as that for case 3. The plate adopts a spherical curvature. The maximum bending stress $\sigma = \alpha E \Delta T/2(1 - \nu)$.

5. 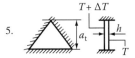 Equilateral triangle, plate fixed at the edges

   Temperature: $T + \Delta T$ on hot side;
   T on cool side
   Uniform edge pressure against hot edge: $q = \alpha E \Delta T h^2/8 a_t$.
   Concentrated pressure at corners against cool face:
   $q = 0.14 \alpha E \Delta T h^2$.
   The maximum bending stress at corners: $\sigma = 0.75 \alpha E \Delta T$.

6. Square plated fixed at the edges.
   Thermal gradient as above.

   Bending stress at the edge: $\sigma = \alpha E \Delta T/2$ (approx.).

TABLE 35.6

Thermal Stresses due to Internal Constraint

1. Solid body of arbitrary shape

Local gradient applied suddenly to a surface

Compressive stress in surface layer:

$\sigma = \alpha E \, \Delta T / (1 - \nu).$

2.

Thin circular disk with heated central portion of radius a_0

Maximum stress within heated zone: $\sigma = \alpha E \Delta T / 2.$
Radial stress outside heated zone:

$\sigma_r = \alpha E \Delta T \, a_0^2 / S r^2$ (compression).

Tangential stress outside heated zone:

$\sigma_b = -\sigma_r$ (tension).

Maximum shear stress at $r = a_0$: $\sigma_s = \alpha E \Delta T / 2.$

3.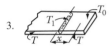

Uniform heating across thickness and width at $x = 0$

Tension along the edges at x: $\sigma_x = E\alpha(T - T_0).$
Maximum tensile stress: $\sigma = E\alpha(T_1 - T_0).$
Maximum compression stress at half-width of the plate: $\sigma_c = -E\alpha(T_1 - T_0).$

4. The plate is heated as above, except that hotter surface has temp. T_2 and lower surface has temp. T_1. Min. temp. T_0.

Maximum tensile stress at the edges, where $x = 0$:

$$\sigma_x = \frac{\alpha E}{2} \left[T_1 + T_2 - 2 T_{0000} \frac{(1 - \nu)}{(3 + \nu)} (T_1 - T_2) \right]$$

5.

Thin-walled tube linear gradient

Maximum hoop stress: $\sigma = \alpha E \Delta T / 2(1 - \nu)$
(inner surface in compression, tension outside).
Maximum longitudinal stress: $\sigma = \alpha E \Delta \, T / 2$
$(1 - \nu)$ (compression inside, tension outside).

6.

Rate of surface temp. increase, m in deg/hour
Hollow sphere

$$\sigma_r = \frac{\alpha m E}{15\lambda(1 - \nu)} \left(\phi - \psi - r^2 - \frac{5a^3}{r} \right)$$

$$\sigma_t = \frac{\alpha m E}{15\lambda(1 - \nu)} \left(\phi - 0.5\psi - 2r^2 - \frac{5a^3}{2r} \right)$$

$$\phi = \frac{b^5 + 5b^2a^3 - 6a^5}{b^3 - a^3}$$

$$\psi = \frac{b^5a^3 - 6b^3a^5 + 5b^2a^6}{r^3(b^3 - a^3)}$$

These tables show that only relatively simple structures and configurations are considered. For more complex geometry, the reader is referred to specialized literature and the use of numerical methods (finite element and boundary element methods). Nevertheless, in many practical situations, the formulas given in Tables 35.5 and 35.6 are sufficient for obtaining results to bracket a proposed design. It is useful to have bracketing values for a problem before embarking on a more complex and time consuming investigation.

SYMBOLS

a	Inner radius of sphere
a_0	Radius of heated area
a_t	Height of triangular plate
b	Outer radius of sphere

C	Specific heat
E	Modulus of elasticity
h	Thickness of plate or bar
I	Second moment of area
k	Thermal conductivity
m	Temperature increase
M_0	Fixing couple
Q	Heat transfer per unit length of tube
q	Edge pressure; corner load
R	Radius of curvature
R_i	Inner radius of tube
R_o	Outer radius of tube
r	Arbitrary radius
T_0, T_1, T_2	Temperatures
T_{max}	Maximum temperature under thermal shock
TSI	Thermal stress index
ΔT	Temperature gradient
V	Volume
ΔV	Volume change
x	Arbitrary distance
α	Linear coefficient of thermal expansion
λ	Thermal diffusivity
ν	Poisson's ratio
σ	General symbol for stress
σ_c	Compressive stress
σ_r	Radial stress
σ_s	Shear stress
σ_t	Tangential stress
σ_u	Ultimate strength
σ_x	Stress at any distance x
ϕ'	Auxiliary constant for sphere
ψ	Auxiliary constant for sphere

REFERENCES

1. A. M. Freudenthal, *Introduction to the Mechanics of Solids*, Wiley, London, 1966.
2. R. J. Corrucini, *Chemical Engineering Progress*, 53, 1957, 252, 342, 397.
3. R. B. Scott, *Cryogenic Engineering*, D. Van Nostrand, New York, 1959.
4. R. F. Barron, Low-temperature properties of engineering materials, *Machine Design*, 1960.
5. L. Tall, L.S. Beedle, and T.V. Galambos, *Structural Steel Design*, Ronald Press, New York, 1964.
6. N. G. Ramke and J. D. Latva, Refractory ceramics and intermetallic compounds, *Aerospace Engineering*, 1963.
7. J. M. Kelble and J. E. Bernados, High-temperature nonmetallic materials, *Aerospace Engineering*, 1963.
8. B. E. Gatewood, *Thermal Stresses*, McGraw Hill, New York, 1957.
9. B. A. Boley and J. H. Weiner, *Theory of Thermal Stresses*, John Wiley & Sons, New York, 1960.
10. J. N. Goodier, Thermal stress, *Journal of Applied Mechanics*, 1937.
11. J. L. Maulbetsch, Thermal stresses in plates, *Journal of Applied Mechanics*, 1935.
12. C. H. Kent, Thermal stresses in spheres and cylinders produced by temperatures varying with time, *Transactions, American Society of Mechanical Engineers*, 54(18), 1932.

36 Axial Response of Straight and Tapered Bars

36.1 INTRODUCTION

The simplest components of structural systems are straight rods or bars, with uniform cross section. Typically, these components are subjected to collinear loads along the axis of the bar, producing tension or compression in the bar.

To review the mechanics of axially loaded bars, consider a bar with a round cross section (a rod) with length L and load W as in Figure 36.1.

Recall from Chapters 2 and 3 that the stress S, strain ε, and elongation ΔL of the rod are then

$$S = W/A, \quad \varepsilon = \Delta L/L, \quad \Delta L = WL/AE \tag{36.1}$$

where, as before,
E is the elastic modulus of the rod material
A is the cross-section area of the rod

If we define the resilience or elastic strain energy, U of the rod as

$$U = (1/2)S\varepsilon \tag{36.2}$$

then from Equation 36.1 we see that the elongation ΔL, the stress S, and the load W may be expressed in terms of U as

$$\Delta L = L\left(\frac{2U}{E}\right)^{1/2}, \quad S = (2UE)^{1/2}, \quad W = A(2UE)^{1/2} \tag{36.3}$$

The energy U of Equation 36.3 may be used as a measure for comparing the material properties. For example, U for a typical structural steel can be 30 times larger than that of soft copper.

36.2 TAPERED AND STEPPED BARS

Figure 36.2 shows a tapered bar supporting an end load W. In the following paragraphs, we develop an expression for the elongation of the bar. For simplicity, we neglect the weight of the bar and we let the cross section be circular.

Let $d(\Delta L)$ represent the elongation of the disk like element having thickness dy. The mean radius for this element follows from the considerations of similar triangles CFH and ABH, so that

$$\frac{CF}{y} = \frac{AB}{L} \tag{36.4}$$

Also, as $CF = r - d/2$, and $AB = (D - d)/2$, solving the foregoing proportion for the mean radius r gives

$$r = \frac{d}{2} + \frac{y(D-d)}{2L} \tag{36.5}$$

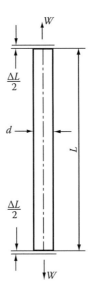

FIGURE 36.1 A rod in tension.

For average cross sectional area πr^2, the elongation of the single element becomes

$$d(\Delta L) = \frac{W dy}{\pi \left[\dfrac{d}{2} + \dfrac{y(D-d)}{2L} \right]^2 E} \qquad (36.6)$$

Integrating this expression between the limits of zero and L with respect to the only variable y, gives the general formula

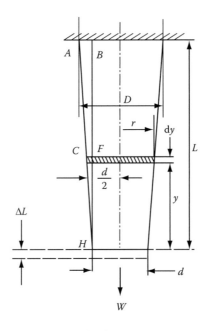

FIGURE 36.2 Tapered round bar (rod) with end load.

$$\Delta L = \frac{4WL}{\pi DdE} \tag{36.7}$$

Observe in Equation 36.7 that when the bar has a uniform cross section (i.e., $D = d$), the expression reduces to the third of Equation 36.1.

36.3 EXAMPLE PROBLEM FOR A STEPPED BAR

Figure 36.3 illustrates a stepped steel bar subjected to a 100,000 lb axial load. The bar segments have circular cross sections with diameters as shown. The segment lengths are also shown. Neglecting the weight of the bar, the objective is to determine the elastic strain energy created in the bar by the load.

SOLUTION

Let S_1 and S_2 be the stresses in the upper and lower segments. Then S_1 and S_2 are

$$S_1 = 100,000/\left[\pi(3)^2/4\right] = 14,150 \text{ psi } (97.6 \text{ N/mm}^2) \tag{36.8}$$

$$S_2 = 100,000/\left[\pi(1)^2/4\right] = 127,350 \text{ psi } (878.3 \text{ N/mm}^2) \tag{36.9}$$

The total strain energy of the stepped bar U_t is the sum of the strain energies U_1 and U_2 of the segments. That is

$$U_t = U_1 + U_2 = \frac{S_1^2 L_1 A_1}{2E} + \frac{S_2^2 L_2 A_2}{2E} \tag{36.10}$$

where L_1, L_2 and A_1, A_2 are the segment lengths and cross-section areas, and as before, E is the elastic modulus (approximately 10×10^6 psi for steel). For the given data, U_t then becomes

$$U_t = 1227 \text{ in lb} = 102.25 \text{ ft lb } (138 \text{ Nm}) \tag{36.11}$$

From Equations 36.8 and 36.9 we see that the step down in diameter shown in Figure 36.3 causes a ninefold increase in the tensile stress. If the bar has a uniform diameter of 3 in., the strain energy would be 282 in. lb. For a bar with 1 in. diameter and the same load, the strain energy is 2547 in. lb.

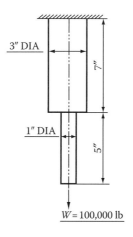

FIGURE 36.3 An axially loaded stepped bar.

36.4 TAPERED BAR UNDER ITS OWN WEIGHT

Figure 36.4 shows a tapered conical bar supported from above and hanging under its own weight. The figure provides the geometric notations for the structure. The objective is to determine the elongation of the bar.

Utilizing the standard expression for the volume of a truncated cone, we see that the weight Q_y of the cone under an elevation y is

$$Q_y = \frac{\pi \gamma y}{3} \left\{ \left[\frac{Ld + y(D-d)}{2L} \right]^2 + \frac{d}{2} \left[\frac{Ld + y(D-d)}{2L} \right] + \frac{d^2}{4} \right\} \tag{36.12}$$

where γ is the weight density.

The tensile stress S at any elevation y is

$$S = \frac{Q_y}{\pi r^2} \tag{36.13}$$

By substituting from Equation 36.12 into 36.13, we can obtain the stress. For example, if $r = D/2$ we see that the stress at the upper support $(y = L)$ is

$$S = \frac{\gamma L}{3D^2} (Dd + D^2 + d^2) \tag{36.14}$$

When $d = D$, the conical bar becomes a bar with a uniform cross section. In this case, the expression of Equation 36.14 for the stress at the support reduces to

$$S = \gamma L \tag{36.15}$$

When d becomes negligibly small (i.e., $d \to 0$) Equation 36.14 yields the support stress as

$$S = \frac{\gamma L}{3} \tag{36.16}$$

Hence, we find another important practical result: the stress for the cone is equal to one third of that in a circular bar, which is evident from Equations 36.13 and 36.14. The same comments also apply to a pyramid configuration.

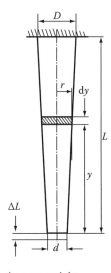

FIGURE 36.4 Conical bar hanging under its own weight.

If the disk like element shown in Figure 36.4 has the thickness dy and average cross-sectional area of πr^2, the elongation of a single element becomes

$$d(\Delta L) = \frac{\gamma}{3} \left\{ \frac{3yL^2d^2 + 3y^2Ld(D-d) + y^3(D-d)^2}{[Ld + y(D-d)]^2E} \right\} dy \tag{36.17}$$

We can add all the elementary elongations by integrating with respect to y from 0 to L. The result is

$$\Delta L = \frac{\gamma L^2}{6D(D-d)^2E}(D^3 + 2d^3 - 3Dd^2) \tag{36.18}$$

This expression is sometimes known as the "tapered bar formula."

36.5 DISCUSSION ABOUT THE TAPERED BAR FORMULA

It is often helpful and reassuring to check if complex expressions such as Equation 36.18 are consistent with simpler and better known results [1]. For example, by letting $d = 0$ in Equation 36.18, we have the standard formula for the elongation of a sharp-pointed conical bar:

$$\Delta L = \frac{\gamma L^2}{6E} \tag{36.19}$$

Next, consider a bar with uniform cross section (with $d = D$). If we equate d and D in Equation 36.18, however, we obtain an indeterminant form: 0/0. But by using L'Hospital's rule [2], we readily obtain

$$\Delta L = \frac{\gamma L^2}{2E} \tag{36.20}$$

This result is the same as that of an elastic bar with the entire weight replaced by a load at the mass center.

36.6 EXAMPLE PROBLEM FOR A LONG HANGING CABLE

Figure 36.5 depicts (not to scale) steel piping consisting of a 200 ft long lower section with a 4.5 in. mean radius and a 0.45 in. wall thickness, and a 400 ft long upper section with a 5.7 in. mean radius and a 0.60 in. wall thickness. The piping is supported at its upper end (ground level) and it in turn supports a 100,000 lb load at its lower end. The objective is to calculate the total elongation of the structure and the maximum stresses in the lower and upper sections.

SOLUTION

A formula for the weight of piping with a uniform cross section is

$$W = 2\pi RTL\gamma \tag{36.21}$$

where
  R is the mean radius of the pipe
  T is the wall thickness
  L is the length
  γ is the weight density

For steel, γ is approximately 0.283 lb/in.^{3}.

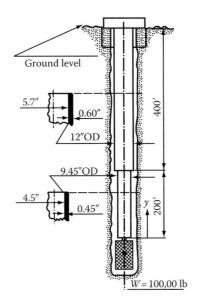

FIGURE 36.5 A downhole piping structure.

Using Equation 36.21, the weight W_1 of the lower section of the piping is

$$W_1 = (2\pi)(4.5)(0.45)(200)(12)(0.283) = 8640 \text{ lb} \tag{36.22}$$

Similarly, the weight W_2 of the upper section is

$$W_2 = (2\pi)(5.7)(0.6)(400)(12)(0.283) = 29{,}190 \text{ lb} \tag{36.23}$$

Hence, the total weight W of the structure (piping and load) is

$$W = W_1 + W_2 + 100{,}000 = 137{,}830 \text{ lb} \tag{36.24}$$

The area A_1 and A_2 of the lower and upper sections of the pipe are

$$A_1 = (2\pi)(4.5)(0.45) = 12.7 \text{ in.}^2 \quad \text{and} \quad A_2 = (2\pi)(5.7)(0.60) = 21.4 \text{ in.}^2 \tag{36.25}$$

The stress S_1 at the top of the lower section is then

$$S_1 = (W_1 + 100{,}000)/A_1 = 8538 \text{ psi} \tag{36.26}$$

Similarly, the stress S_2 at the top of the upper section is

$$S_2 = (W_1 + W_2 + 100{,}000)/A_2 = 6414 \text{ psi} \tag{36.27}$$

By superposition the total elongation ΔL of the structure may be computed from the expression

$$\Delta L = \frac{PL_1}{A_1 E} + \frac{PL_2}{A_2 E} + \frac{W_1 L_2}{A_2 E} + \frac{W_1 (L_1/2)}{A_1 E} + \frac{W_2 (L_2/2)}{A_2 E} \tag{36.28}$$

where P is the load at the bottom (100,000 lb), L_1 and L_2 are lengths of the lower and upper sections (200 and 400 ft) and, as before, E is the elastic modulus (for steel $E = 30 \times 10^6$ psi). By substituting the appropriate values into Equation 36.28, we obtain the structural elongation as

$$\Delta L = 1.58 \text{ in. (40.13 mm)} \tag{36.29}$$

The foregoing method of analysis can be extended to any number of steps. Hence, in the more general case where we have to deal with a rod of variable cross section for which a proper variation of the cross section with the length is difficult to define, the rod shape may be approximated by a finite number of elements. The elongation can then be computed for all the individual elements and added directly to obtain the complete elongation.

36.7 HEAVY HANGING CABLE WITH UNIFORM STRESS ALONG THE LENGTH

Consider the classical case of a cable hanging under its own weight but with a varying cross-section area so that the stress at each elevation is constant. Specifically, let a long hanging cable be designed with an increasing cross-section area, from bottom to top so that even though the weight load increases, the stress remains constant along the length. Figure 36.6 illustrates the concept. In the figure, Y is the vertical coordinate with origin at the bottom of the cable; $A(y)$ is the variable cross-section area, with the bottom end area $A(O) = A_O$; and P is a load at the lower end.

Given P and A_O, the stress σ_O at the bottom and hence uniform along the length, is

$$\sigma_O = P/A_O \tag{36.30}$$

The design objective is to determine $A(y)$.

To this end, consider a small element, or disk of the cable, with thickness Δy as in Figure 36.7. As in the figure, the areas of the bottom and top faces are $A(y)$ and $A(y + \Delta y)$. We can relate these areas using Taylor's series [2] as

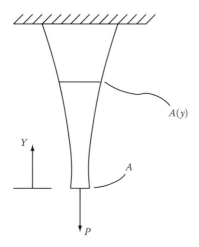

FIGURE 36.6 A long, heavy hanging cable with uniform stress along the length.

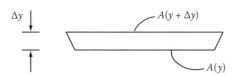

FIGURE 36.7 A disk element of the cable.

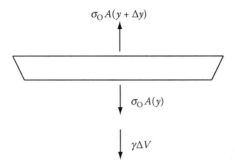

FIGURE 36.8   Free-body diagram of the disk element.

$$A(y + \Delta y) = A(y) + \frac{dA}{dy}\Delta y + \cdots \tag{36.31}$$

where the unwritten terms in the series are of the orders $(\Delta y)^2$ and higher.

Figure 36.8 shows a free-body diagram of the disk element, where γ is the weight density of the cable material and ΔV is the volume of the disk element.

From Figure 36.8, we see that the equilibrium of forces leads to the expression

$$\sigma_O A(y + \Delta y) = \sigma_O A(y) + \gamma \Delta V \tag{36.32}$$

For a relatively slowly varying area, we see from Figure 36.7 that ΔV is approximately

$$\Delta V = \left[\frac{A(y + \Delta y) + A(y)}{2}\right]\Delta y \tag{36.33}$$

By using Equation 36.31 and by neglecting terms of order $(\Delta y)^2$ and higher, ΔV becomes

$$\Delta V = \left[A + \frac{1}{2}\frac{dA}{dy}\Delta y\right]\Delta y = A\Delta y \tag{36.34}$$

Finally, by substituting from Equations 36.31 and 36.34 into the equilibrium expression of Equation 36.32 and by neglecting higher order small terms we obtain

$$\sigma_O\left(A + \frac{dA}{dy}\Delta y\right) = \sigma_O A + \gamma A \Delta y$$

or

$$\sigma_O \frac{dA}{dy} = \gamma A \tag{36.35}$$

Equation 36.35 is a linear first-order ordinary differential equation whose solution is

$$A = A_0 e^{(\gamma/\sigma_O)y} \tag{36.36}$$

Equation 36.36 provides the desired design area. A similar result is obtained for the problem of constructing a tall tower with uniform stress at each elevation.

36.8 EXAMPLE PROBLEM OF AN AXIALLY COMPRESSED TUBE

Figure 36.9 depicts an aluminum tube with a 1 in. mean radius and a 0.25 in. wall thickness compressed between two rigid blocks by a steel bolt with a 0.75 in. outer diameter. The tensile stress in the tightened bolt is 10,000 psi. The objective is to determine the stress and shortening of the aluminum tube.

SOLUTION

The cross-section areas of the bolt and tube are

$$A_b = \pi(0.75)^2/4 = 0.44 \text{ in.}^2 \quad \text{and} \quad A_t = 2\pi(1)(0.25) = 1.57 \text{ in.}^2 \tag{36.37}$$

The bolt load P is

$$P = S_b A_b = (10,000)(0.44) = 4400 \text{ lb} \tag{36.38}$$

The tube stress is then

$$S_t = P/A_t = 4400/1.57 = 2802.5 \text{ psi} \tag{36.39}$$

The tube shortening ΔL_t is

$$\Delta L_t = PL_t/A_t E_t = (4400)(16)/(1.57)(10)(10)^6 = 0.0045 \text{ in.} \tag{36.40}$$

36.9 KERN LIMIT

The structures discussed in the foregoing sections of this chapter were relatively short and rigid. Thus an applied axial compressive loading is unlikely to produce buckling. Therefore if a loading is applied eccentrically, the resulting stresses may be obtained by superposition of direct compression and flexural stresses.

In this section, we briefly consider the concept of kern sometimes referred to as "kernel" or "core of the cross section." Kern can be defined as the area in the plane of the section through which the line of action of the external force should pass to assure the same kind of normal stress at all points of the cross section. Consequently, the kern limit may be defined as a characteristic dimension of the central portion of the cross section, or the locus of points within which the line of action of the external force should fall. The kern limit concept is useful in designing short

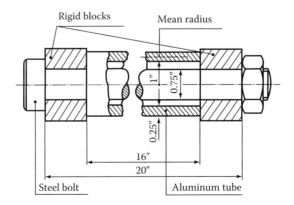

FIGURE 36.9 A tube compressed between rigid blocks.

prisms, columns, and piers subjected to eccentric thrust. Such a thrust causes a direct axial stress and a bending stress, which can be superimposed as long as the response is purely elastic. As a rule, such stresses are not critical in the customary design of machine elements and steel members of various cross-sectional geometry, provided that local buckling is not a problem. However, if a structural member is made of a material that is good in compression but poor in tension, as such is the case with masonry columns, the analysis based on the concept of kern limit should be made [3].

Figure 36.10 demonstrates the principle of kern limit design. Suppose that the external force P is applied at a distance a measured from the centroid of the prism having a symmetrical cross-section, as indicated in Figure 36.10. For simplicity of the derivation, the thrust loading is offset with respect to one axis of the cross section only. The uniform compressive stress is

$$S_c = \frac{P}{bh} \tag{36.41}$$

The tensile stress component due to offset bending is

$$S_t = \frac{6Pa}{bh^2} \tag{36.42}$$

In general, the actual stress distribution can be of the type shown in Figure 36.10. It means that for a certain value of a, a portion of the cross section can be in a state of tension when the direct and bending stresses are superimposed. However, to assure that the combined stress is compressive at all points of the cross section, such as the one indicated in Figure 36.10 as "desired stress distribution," it is necessary to satisfy the following condition

$$S_t - S_c = 0 \tag{36.43}$$

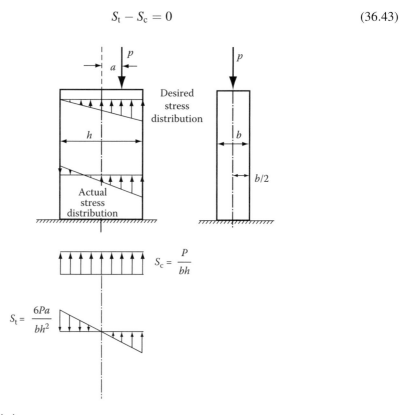

FIGURE 36.10 Kern limit concept.

Then by setting a equal to e, we have

$$\frac{6Pe}{bh^2} = \frac{P}{bh} \tag{36.44}$$

Thus e is

$$e = \frac{h}{6} \tag{36.45}$$

Table 36.1 shows this value in the first case, as one of the dimensions defining a diamond-shaped area characteristic of kern for a rectangular cross section of the structural member. If the line of

---

TABLE 36.1

Kern Limits for Compression Members

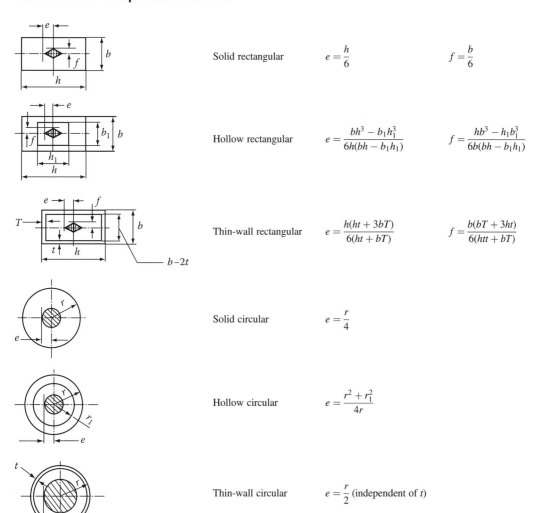

	Solid rectangular	$e = \dfrac{h}{6}$	$f = \dfrac{b}{6}$
	Hollow rectangular	$e = \dfrac{bh^3 - b_1 h_1^3}{6h(bh - b_1 h_1)}$	$f = \dfrac{hb^3 - h_1 b_1^3}{6b(bh - b_1 h_1)}$
	Thin-wall rectangular	$e = \dfrac{h(ht + 3bT)}{6(ht + bT)}$	$f = \dfrac{b(bT + 3ht)}{6(htt + bT)}$
	Solid circular	$e = \dfrac{r}{4}$	
	Hollow circular	$e = \dfrac{r^2 + r_1^2}{4r}$	
	Thin-wall circular	$e = \dfrac{r}{2}$ (independent of t)	

Note: Kern limit areas are defined by shaded portions of sections.

---

thrust of the force P is offset with respect to both axes of symmetry of the cross section, the theoretical basis for the calculation follows directly from the previous derivation. The geometries selected for Table 36.1 cover many a great practical applications, giving relatively simple expressions for defining kern limits. For the case of a prism or a column of arbitrary cross section, the analysis of kern limits is much more involved and requires a semigraphical procedure with successive approximations.

SYMBOLS

A, A_1, A_2	Cross-sectional areas
A_b	Bolt cross-sectional area
A_o	End area
A_t	Tube cross-sectional area
a	Offset of end load
b, b_1	Widths of rectangular cross sections
D	Maximum diameter of bar
d	Minimum diameter of bar
E, E_1, E_2	Elastic moduli
e	Kern limit
F	Bolt load
F_i	Initial bolt load
$F^{/\!/}$	Load on rigid block
f	Kern limit
h, h_1	Depths of rectangular cross sections
k	Spring constant of assembly
k_b	Spring constant of bolt
k_t	Spring constant of tube
L, L_1, L_2	Length dimensions
ΔL	Extension
P	Axial thrust
Q_y	Partial weight of conical bar
R	Mean radius of tube
r, r_1	Tube radii
S	General symbol for stress
S_1, S_2	Tensile stresses
S_c	Compressive stress
S_t	Tensile stress due to bending
T, t	Wall thicknesses
U	Resilience; strain energy
U_t	Total strain energy
ΔV	Volume of disk element
W, W_1, W_2	Downward loads
y	Arbitrary distance
Δy	Disk element thickness
Z	Auxiliary parameter
γ	Specific weight
$\delta, \delta^{/}$	Deflections
σ_o	Uniform stress

REFERENCES

1. J. H. Faupel, *Engineering Design*, Wiley, New York, 1981.
2. L. L. Smail, *Calculus*, Appleton-Century-Crofts, New York, 1949.
3. G. E. Large, *Basic Reinforced Concrete Design: Elastic and Creep*, Ronald Press, New York, 1957.

37 Thin Rings and Arches

37.1 INTRODUCTION

There are numerous references discussing various problems of circular rings and arches [1–14]. These relate to both in-plane and out-of-plane loadings. In this chapter, we review some of these results. For simplicity, we restrict our analysis to in-plane loadings on structures with uniform cross-sections having large radii of curvature compared to cross-section thicknesses.

In evaluating the deflection, it is convenient to use the concept of elastic strain energy and Castigliano's theorem. The deflections are taken to be relatively small, and the strain energy due to bending alone is used in the analysis.

37.2 REVIEW OF STRAIN ENERGY AND CASTIGLIANO'S THEOREM

Consider an elastic body B subjected to various forces as represented in Figure 37.1, where P is a typical force load. These forces cause stresses and strains within B. Let σ_{ij} and ε_{ij} be the elements of the stress and strain tensors (see Chapters 4 and 5). Then the strain energy U is defined as

$$U = \int_V \frac{1}{2}\sigma_{ij}\varepsilon_{ij}\mathrm{d}V \tag{37.1}$$

where V is the volume of B and as before, the repeated index designates a sum.

Consider the typical force P of Figure 37.1. As with all the forces acting on B, P contributes to the stress and strain and also to the strain energy U. That is, U is a function of P as

$$U = U(P) \tag{37.2}$$

Castigliano's theorem [14] states that the deflection δ of B where P is applied, and in the direction of P is

$$\delta = \partial U/\partial P \tag{37.3}$$

Analogously, if a moment or torque M is applied at a point of an elastic body, the resulting rotation θ in the direction of the applied moment, is

$$\theta = \partial U/\partial M \tag{37.4}$$

For most cases in component design, the geometry is relatively simple. In these cases, the form of the strain energy is also relatively simple. For example, consider an axially loaded rod as in Figure 37.2. If P is the load, E the elastic constant, A the cross-section area, ℓ the bar length, σ the stress, ε the strain, and δ the elongation, we have the familiar relations

$$\sigma = P/A, \quad \varepsilon = \delta/\ell, \quad \sigma = E\varepsilon, \quad \delta = P\ell/A\varepsilon \tag{37.5}$$

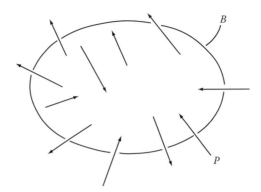

FIGURE 37.1   An elastic body subjected to a set of forces.

For a uniform bar, the stress and strain are constant along the length. The strain energy U is then

$$U = \int_0^\ell (1/2)\sigma\varepsilon dV = \int_0^\ell (1/2)(P/A)(\delta/\ell)Adx$$

$$= \int_0^\ell (1/2)(P^2/AE)dx = P^2\ell/2AE \qquad (37.6)$$

where x is a coordinate along the length. By substitution into Equation 37.3, we have

$$\delta = \partial U/\partial P = P\ell/AE \qquad (37.7)$$

which is consistent with the last expression of Equation 37.4.

For bending, U is [14,15]

$$U = \int_0^\ell (M^2/2EI)dx \qquad (37.8)$$

where, as before,

M is the moment along the beam
E is the elastic modulus
I is the second moment of area of the beam cross section
ℓ is the beam length

To illustrate the use of Equation 37.8, consider first an elastic cantilever beam with a concentrated end moment M_O at the unsupported end as in Figure 37.3. With this simple geometry, the bending moment M along the beam is constant. That is

$$M = M_O \qquad (37.9)$$

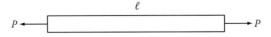

FIGURE 37.2   An axially loaded bar.

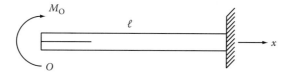

FIGURE 37.3   Cantilever beam with an end moment at the unsupported end.

From Equation 37.8, the strain energy U is

$$U = \int_0^\ell (M_O^2/2EI)\,dx = M_O^2\ell/2EI \tag{37.10}$$

Consequently, from Equation 37.4 the rotation θ at the left end of the beam is

$$\theta = \frac{\partial U}{\partial M_O} = M_O\ell/EI \tag{37.11}$$

Next, consider the same beam but with a force P applied at the unsupported end as in Figure 37.4. Here the bending moment M along the beam is simply

$$M = P_x \tag{37.12}$$

Hence M^2 and the strain energy U are

$$M^2 = P^2 x^2 \tag{37.13}$$

and

$$U = \int_0^\ell (M^2/2EI)\,dx = \int_0^\ell (P^2 x^2/2EI)\,dx$$

or

$$U = P^2\ell^3/6EI \tag{37.14}$$

Therefore, the end deflection δ due to the load P is

$$\delta = \partial U/\partial P = P\ell^3/3EI \tag{37.15}$$

FIGURE 37.4   Cantilever beam with a concentrated force at the unsupported end.

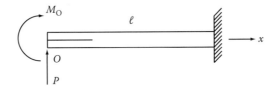

FIGURE 37.5 Cantilever beam with an end moment and concentrated force at the unsupported end.

Finally, consider the same beam but with both a force P and a concentrated moment M_O applied at the unsupported end as in Figure 37.5.

In this case, the bending moment M along the beam is

$$M = M_O + Px \tag{37.16}$$

Hence M^2 is

$$M^2 = M_O^2 + 2M_O Px + P^2 x^2 \tag{37.17}$$

and the strain energy U becomes

$$U = \int_0^\ell (M^2/2EI)\,dx = (1/2EI)\int_0^\ell \left(M_O^2 + 2M_O Px + P^2 x^2\right)dx$$

or

$$U = (1/2EI)\left[M_O^2 \ell + M_O P\ell^2 + P^2 \ell^3/3\right] \tag{37.18}$$

From this expression, we see that the deflection δ at the left, unsupported end of the beam is

$$\delta = \frac{\partial U}{\partial P} = (1/2EI)\left(M_O \ell^2 + \partial P\ell^3/3\right) \tag{37.19}$$

Similarly, the rotation θ at the end is

$$\theta = \frac{\partial U}{\partial M_O} = (1/2EI)\left(2M_O \ell + P\ell^2\right) \tag{37.20}$$

Observe that in the result of Equation 37.19, if we set $P=0$ we have the deflection due to the moment M_O alone. Similarly in Equation 37.20, if we set $M_O = 0$ we have the end rotation due to P alone. In these instances, P and M_O are sometimes regarded as "dummy variables" which vanish at the end of the analysis.

Finally, it is seen that the results of Equations 37.7, 37.11, 37.15, 37.19, and 37.20 are consistent with those of Table 11.1 and with those in handbooks [2,8].

37.3 DIAMETRICALLY LOADED ELASTIC RING

The classical case of a thin elastic ring diametrically loaded in the plane of curvature, is a statically indeterminant problem with accompanying difficulties in analysis. This is somewhat surprising given the geometric symmetry and the usual loading symmetry.

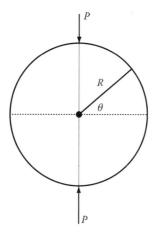

FIGURE 37.6 Diametral loading on a thin elastic ring.

Figure 37.6 shows the simple case of a ring subjected to diametral loading. Due to the symmetry, we can simplify the analysis by considering a single quadrant. Figure 37.7 depicts a free-body diagram of the upper right quadrant, $0 \leq \theta \leq \pi/2$, where θ is the angular coordinate, R is the ring radius, P is the load, and M_O and $M_{\pi/2}$ are the cross-section moments at $\theta = 0$ and $\theta = \pi/2$. Due to the symmetry, there are no rotations of the cross section at $\theta = 0$ and $\theta = \pi/2$.

Consider the bending moment M at an interior point of the ring quadrant as in Figure 37.8.

From equilibrium considerations we see that M is

$$M = M_{\pi/2} - (PR/2)\cos\theta \qquad (37.21)$$

From Equation 37.8, the strain energy U for bending is

$$U = 4 \int_0^{\pi/2} (M^2/2EI)R \, d\theta \qquad (37.22)$$

where the factor 4 is due to our integration (from 0 to $\pi/2$) over only one of the four similar quadrants. By substituting from Equation 37.21, we have

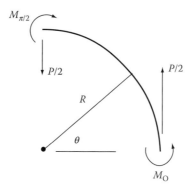

FIGURE 37.7 Free-body diagram of a quadrant of the loaded ring of Figure 37.6.

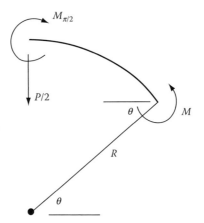

FIGURE 37.8   Bending moment M at an interior location of the ring quadrant of Figure 37.7.

$$U = (2R/EI) \int_0^{\pi/2} \left[M_{\pi/2}^2 - 2M_{\pi/2}(PR/2)\cos\theta + (PR/2)^2 \cos^2\theta \right] d\theta \qquad (37.23)$$

By completing the indicated integration, U becomes

$$U = (2R/EI) \left[M_{\pi/2}^2(\pi/2) - PRM_{\pi/2} + (PR/2)^2(\pi/4) \right] \qquad (37.24)$$

The cross-section rotation ϕ at $\theta = \pi/2$ is then

$$\phi = \partial U/\partial M_{\pi/2} = 0 \qquad (37.25)$$

By substituting from Equation 37.24, we obtain

$$M_{\pi/2} = PR/\pi \qquad (37.26)$$

Thus from Equation 37.21, the bending moment M at any angle θ of the quadrant is

$$M = \frac{PR}{2\pi}(2 - \pi\cos\theta) \qquad (37.27)$$

Observe in Equation 37.27 that when $\theta = \pi/2$, M is

$$M = M_{\pi/2} = PR/\pi \qquad (37.28)$$

and when $\theta = 0$, M is

$$M = M_O = \frac{PR}{2\pi}(2 - \pi) \qquad (37.29)$$

Thus M_O and $M_{\pi/2}$ are different both in magnitude and sign. Also note that

$$M_O + \frac{PR}{2} = M_{\pi/2} \qquad (37.30)$$

which is consistent with Figure. 37.7.

Next, by substituting for $M_{\pi/2}$ from Equation 37.28 into the strain energy result of Equation 37.24, we obtain the strain energy U as a function of the load P. The result is

$$U = \frac{P^2 R^3}{8\pi EI}(\pi^2 - 8) \tag{37.31}$$

Using Costigliano's theorem (see Section 37.2), the deflection Y under the load P is simply

$$Y = \partial U/\partial P = \frac{PR^3(\pi^2 - 8)}{4\pi EI} \tag{37.32}$$

By a similar analysis, the corresponding increase X in the horizontal diameter is

$$X = \frac{PR^3(4 - \pi)}{2\pi EI} \tag{37.33}$$

By setting the moment M of Equation 37.27 equal to zero yields the angle

$$\theta = 50.5° \tag{37.34}$$

At this location the bending moment changes its sign. It may be of interest to note that the ratio Y/X of the vertical to horizontal deflection is

$$Y/X = 1.089 \tag{37.35}$$

37.4 MORE EXACT RESULTS FOR THE DIAMETRICALLY LOADED ELASTIC RING

The results of Equations 37.32, 37.33, and 37.34 are based upon the assumption of a thin ring, neglecting the effects of shear and direct stresses. That is, only bending is considered. The error introduced by ignoring shear and direct stress, is likely to be small in most cases.

For a ring of rectangular cross section, more exact formulas for the deflections are

$$Y = \frac{P\chi}{bE}(1.7856\chi^2 + 0.7854 + 2.0453) \tag{37.36}$$

and

$$X = \frac{P\chi}{bE}(1.6392\chi^2 - 0.5000 + 1.3020) \tag{37.37}$$

where b is the width of the cross section and χ is the ratio of the radius of curvature R to the depth h of the cross section.

To illustrate the magnitudes of the bending, direct, and shear stresses on the ring deflections, let χ $(= R/h)$ have the value 10. Then the terms of Equations 37.36 and 37.37 provide the various contributions. Table 37.1 provides the results.

37.5 DESIGN CHARTS FOR CIRCULAR RINGS

It is evident from Table 37.1 that the simplified analysis of rings, involving bending stresses alone, is acceptable in most design situations. A considerable amount of mathematical work

TABLE 37.1

Percentage Stress Contribution to Maximum Deflection for Thin Rings ($\chi = R/h = 10$)

	Bending Stress	Direct Stress	Shear Stress
Vertical deflection	+98.44	+0.43	+1.13
Horizontal deflection	+99.51	−0.30	+0.79

associated with the derivation of formulas including the effects of direct and shear stresses can, therefore, be avoided.

Usually, a designer is interested in the radial deflection, slope, and bending moment as a function of the angular position θ. The references at the end of the chapter provide the results of numerous closed-form solutions [1,2,8]. Table 37.2 lists the results for common loading configurations. The factors K_u, K_ψ, and K_M are for the deflection, slope, and moment, respectively.

37.6 ESTIMATES VIA SUPERPOSITION

Several important cases summarized in Table 37.2 can be used to answer the majority of practical design questions related to the response of closed thin rings. Should a question regarding a more complex type of ring loading that is not covered specifically by Table 37.2 arise, the designer may wish to solve the problem using the principle of superposition. It is recalled that this principle allows us to add algebraically the stresses or deflections at a point of a structure, caused by two or more independent loads. This principle is applicable as long as the resultant stresses and strains remain elastic.

To illustrate some basic steps in applying the principle of superposition to ring design, consider the response of a thin elastic ring subjected to a four-way tension as shown at the left in Figure 37.9. Symbolically, the principle of superposition is illustrated in Figure 37.9 in terms of a diagrammatic summary of the two effects. The basic component, in this study, is the ring subjected to diametral tension as shown at the top of Table 37.2. In its vertical orientation, it represents a twoway tension along the vertical axis indicated in Figure 37.9, as the first component. The same ring oriented horizontally represents the effect of a two-way tension along the horizontal axis as shown symbolically in Figure 37.9.

It is assumed that in a vertical orientation the angle θ is measured from the vertical axis counterclockwise. For the case of $\theta = 0$, $\sin \theta = 0$, and $\cos \theta = 1$, so that the deflection factor from Table 37.2 is

$$K_U = 0.3927 - 0.3183 = 0.0744 \tag{37.38}$$

The effect of the vertically oriented loading on the horizontal displacement is obtained when $\theta = \pi/2$. Since $\sin (\pi/2) = 1$ and $\cos (\pi/2) = 0$, we get

$$K_U = 0.2500 - 0.3183 = -0.0683 \tag{37.39}$$

The sign here is negative because the ring attains an oval shape, with the vertical diameter increasing and the horizontal diameter decreasing.

Next, if we consider the basic ring to be oriented horizontally as shown in Figure 37.9, and if we still use the same expression for K_u, the question arises as to the appropriate angle convention. Clearly, using the same convention for the two orientations is not satisfactory if we wish to derive

TABLE 37.2

Deflections, Slopes, and Bending Moments for Circular Rings Loaded in a Plane

Formula	Symbol	Function	Range of Application
$u = \dfrac{PR^3}{EI}K_u$	K_u	$0.2500 \sin\theta + (0.3927 - 0.2500\,\theta)\cos\theta - 0.3183$	$0-\pi$
$\psi = \dfrac{PR^2}{EI}K_\psi$	K_ψ	$(0.2500\,\theta - 0.3927)\sin\theta$	$0-\pi$
$M = PRK_M$	K_M	$(0.5000 \sin\theta - 0.3183)$	$0-\pi$
$u = \dfrac{M_oR^2}{EI}K_u$	K_u	$0.5000 - 0.3183\,\theta\sin\theta - 0.4775\cos\theta$	$0-\pi/2$
$\psi = \dfrac{M_oR}{EI}K_\psi$	K_ψ	$0.1592\sin\theta - 0.3183\,\theta\cos\theta$	$0-\pi/2$
$M = M_o K_M$	K_M	$0.5000 - 0.6366\cos\theta$	$0-\pi/2$
		$-0.5000 - 0.6366\cos\theta$	$\pi/2-\pi$
$u = \dfrac{PR^3}{EI}K_u$	K_u	$0.1989\,\theta\sin\theta + (0.4081 - 0.0796\,\theta^2)\cos\theta - 0.3618$	$0-\pi/2$
$\psi = \dfrac{PR^2}{EI}K_\psi$	K_ψ	$(0.3750 - 0.0398\,\theta)\sin\theta + (0.3658 - 0.2500\,\theta)\cos\theta - 0.3618$	$\pi/2-\pi$
		$(0.0796\,\theta^2 - 0.2092)\sin\theta + 0.0398\,\theta\cos\theta$	$0-\pi/2$
		$(0.2500\,\theta - 0.4055)\sin\theta + (0.1250 - 0.0398\,\theta)\cos\theta$	$\pi/2-\pi$
$M = PRK_M$	K_M	$0.3183\,\theta\sin\theta + 0.2387\cos\theta - 0.3618$	$0-\pi/2$
		$0.5000\sin\theta - 0.0796\cos\theta - 0.3618$	$\pi/2-\pi$
$u = \dfrac{PR^3}{EI}K_u$	K_u	$0.5570\,\theta\sin\theta + (0.9382 - 0.1592\,\theta^2)\cos\theta - 0.9053$	$0-\pi/2$
$\psi = \dfrac{PR^2}{EI}K_\psi$	K_ψ	$(0.0796\,\theta - 0.2500)\sin\theta + 0.0681\cos\theta + 0.0947$	$\pi/2-\pi$
		$(0.1592\,\theta^2 - 0.3812)\sin\theta + 0.2387\,\theta\cos\theta$	$0-\pi/2$
		$0.0115\sin\theta + (0.0796\,\theta - 0.2500)\cos\theta$	$\pi/2-\pi$
$M = PRK_M$	K_M	$0.6366\,\theta\sin\theta + 0.7958\cos\theta - 0.9053$	$0-\pi/2$
		$0.1592\cos\theta + 0.0947$	$\pi/2-\pi$

Diagram labels (third loading case): $Q_m = \dfrac{2P}{\pi R}$, $Q = Q_m\cos\theta$

Diagram labels (fourth loading case): $Q_m = \dfrac{4P}{\pi R}$, $Q = Q_m\cos\theta$

a general formula. Let us see what kind of a substitution we would have to make in order to have the two K_u factors consistent with the two orientations of loading. Suppose that in a vertical orientation, angle θ defines a point on the ring at which the displacement is considered. The

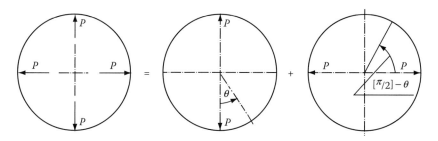

FIGURE 37.9   Symbolic representation of superposition of ring deflection.

same point on the ring when viewed from the horizontal line of loading, can be reached at an angle that is complementary to 90°, resulting in $(\pi/2) - \theta$. This assumption implies, for instance, that with $\theta = \pi/2$ the vertically oriented ring load (center of Figure 37.9) produces a horizontal displacement. For the horizontally oriented load, $\theta = \pi/2$ gives the angle equal to zero and the displacement along the horizontal line of action. This statement follows from the symbolic diagram in Figure 37.9.

For a vertically oriented ring loading, Table 37.2 gives

$$K_u = 0.2500 \sin \theta + (0.3927 - 0.2500 \ \theta) \cos \theta - 0.3183 \tag{37.40}$$

For a horizontally oriented pair of forces P, angle θ must be replaced by $(\pi/2) - \theta$, which gives

$$K_u = 0.2500 \cos \theta + \left[0.3927 - 0.2500 \left(\frac{\pi}{2} - \theta \right) \right] \sin \theta - 0.3183 \tag{37.41}$$

Adding the preceding two expressions, yields

$$K_u' = 0.2500(1 + \theta) \sin \theta + (0.6427 - 0.2500\theta) \cos \theta - 0.6366 \tag{37.42}$$

Hence, the radial deflection for four-way tension becomes

$$u = \frac{PR^3}{EI} K_u' \tag{37.43}$$

Note that for $\theta = 0$ and $\theta = (\pi/2)$, the expression for the deflection factor K_u' in a four-way tension gives two identical results, indicating that the deformation pattern of this ring is radially symmetric. In other words, radial displacement under each of the four loads is

$$u = 0.0061 \frac{PR^3}{EI} \tag{37.44}$$

The same result can be obtained if the deflection factor K_u' is calculated using previously quoted numerical values for $\theta = 0$ and $\theta = (\pi/2)$. That is

$$K_u' = 0.0744 - 0.0683 = 0.0061 \tag{37.45}$$

37.7 RING WITH CONSTRAINT

Consider again the diametrically loaded ring of Figure 37.6 and as shown again in Figure 37.10. We observed in our analysis that this poses a statistically indeterminant problem. As such, the analysis requires knowledge of the deformation to determine the bending moment around the ring. Indeterminacy arises due to the continuity of the ring. That is, the ring may be viewed as a closed curved beam.

The complexity of ring analysis increases rapidly when additional constraints are imposed. Consider for example, a thin elastic ring with a rigid diametral constraint as in Figure 37.11. The equilibrium of one quadrant of the ring can be maintained if we add two statically indeterminate forces, H and M_f. Because of the symmetry of loading and support, all quadrants of the ring must deform in an identical manner. The bending moment M at an arbitrary section defined by θ is

$$M = HR \sin \theta - \frac{PR}{2} (1 - \cos \theta) - M_f \tag{37.46}$$

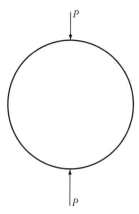

FIGURE 37.10 Diametrically loaded elastic ring.

where H is a horizontal restraining force and M_f is the restraining moment at $\theta = 0$. As before, P is the magnitude of the applied load and R is the ring radius.

Since the horizontal displacement and slope at $\theta = 0$ must be equal to zero, because of the rigid connection between the ring and the bar, the following conditions apply:

$$\int_0^{\pi/2} M \frac{\partial M}{\partial H} d\theta = 0 \quad \text{and} \quad \int_0^{\pi/2} M \frac{\partial M}{\partial M_f} d\theta = 0 \qquad (37.47)$$

By substituting from Equation 37.46 into 37.47 and integrating, we can solve the resulting expressions for H and M_f

$$H = P\left(\frac{4 - \pi}{\pi^2 - 8}\right) \quad \text{and} \quad M_f = PR\left(\frac{4 + 2\pi - \pi^2}{2(\pi^2 - 8)}\right) \qquad (37.48)$$

Finally, by substituting these results into Equation 37.46, we obtain

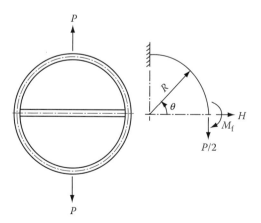

FIGURE 37.11 A thin elastic ring with a rigid restraining bar.

$$M = PR(0.4591 \sin \theta + 0.5000 \cos \theta - 0.6106) \tag{37.49}$$

It may be insightful to observe the effect of the diametral constraint of Figure 37.11. When θ is $\pi/2$ or 90°, Equation 37.49 provides the bending moment M as

$$M = 0.1515PR \tag{37.50}$$

This value is about half or 48% of that obtained from Equation 37.27, for an unconstrained ring. When θ is 0, Equation 37.50 yields

$$M = 0.1106PR \tag{37.51}$$

which is about 61% of the moment for a ring without horizontal restraint.

Hence, as far as the preliminary design is concerned, we have established the important bracketing values of the bending moment. Assuming that the horizontal bar shown in Figure 37.11 is not perfectly rigid, the values of the bending moment for $\theta = 0$ and $\theta = \pi/2$ should increase. The maximum bending stresses are found at the two points of load application. Therefore, when a certain amount of horizontal constraint is present the stresses at the critical locations are reduced.

37.8 A ROTATING RING

Consider a ring rotating in its plane as represented in Figure 37.12. Such a system could simulate the rim of a flywheel.

Let q be the uniform loading per unit length due to the centrifugal inertia forces, and let P be the resulting internal hoop force. Then from equilibrium considerations, P is

$$P = \int_{0}^{\pi/2} qR \sin \theta \, d\theta \tag{37.52}$$

where, as before, R is the ring radius.

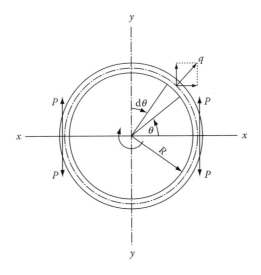

FIGURE 37.12 A ring rotating in its plane of curvature.

If V is the peripheral (tangential) speed of the ring, then the hoop stress σ in the ring is

$$\sigma = \rho V^2 \tag{37.53}$$

where ρ is the mass density of the ring material.

The development of Equation 37.53 follows immediately from the elementary hoop stress formula from the strength of materials [16]:

$$\sigma = \frac{pR}{t} \tag{37.54}$$

where
p is the pressure on the ring
t is the thickness

Consider a small element (e) of the ring as in Figure 37.13, where b is the ring depth (axial length) and $d\theta$ is an incremental angle as in Figure 37.12.

Let F^* be the inertia force on element (e) due to the ring rotation. Then from Newton's second law [17], F^* is

$$F^* = -ma \tag{37.55}$$

where
m is the element mass
a is its acceleration

If dv is the element volume ($b + R\, d\theta$), then m is

$$m = \rho btR\, d\theta \tag{37.56}$$

The acceleration a due to the rotation is simply [17]

$$a = V^2/R \tag{37.57}$$

From Equations 37.54 and 37.55 F^* is then

$$F^* = -\rho btR\, d\theta(V^2/R) \tag{37.58}$$

From Figure 37.13 the pressure p at the base of (e) is

$$P = -F^*/(Rd\theta)b = \rho btd\theta V^2/(Rd\theta)b = \rho tV^2/R \tag{37.59}$$

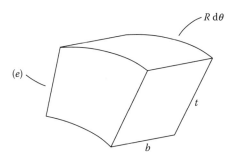

FIGURE 37.13   A ring element.

Finally, by substituting this result into Equation 37.54 we have the result of Equation 37.53,

$$\sigma = \rho V^2 \tag{37.60}$$

We can also obtain this result using Equation 37.52. By integrating we have

$$P = qR \tag{37.61}$$

But q is pb. The hoop stress σ is then

$$\sigma = P/bt = \rho bR/bt$$
$$= \rho t(V^2/R)bR/bt$$

or

$$\sigma = \rho V^2 \tag{37.62}$$

The foregoing analysis assumed that the hoop stress is distributed uniformly over the ring cross section. This approximation is likely to be accurate for many practical applications with thin rings.

37.9 SIMPLY SUPPORTED ARCH

In arch analysis, as with ring analysis, it is convenient to use Castigliano's theorem (see Section 37.2). Perhaps the simplest arch problem is that of a simply supported circular arch under a central load P as in Figure 37.14. If we postulate that there is no friction at the supports and no constraint of any kind, the reactions can be obtained from a simple equation of statics. The bending moment at a section defined by θ is written as follows:

$$M = VR(\cos \alpha - \cos \theta) \tag{37.63}$$

From Equation 37.8, the strain energy in the right side of the arch is

$$U = \int_{\alpha}^{\pi/2} \frac{M^2}{2EI} R \, d\theta \tag{37.64}$$

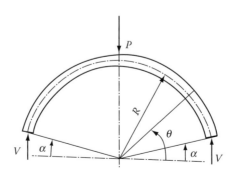

FIGURE 37.14 A simply supported circular arch under a central load.

If the arch at its apex $(\theta = \pi/2)$ is assumed to be fixed, then the displacement Y at the support V is

$$Y = \frac{\partial U}{\partial V} = \int_{\alpha}^{\pi/2} \frac{M \partial M/\partial \theta}{EI} R \, d\theta \tag{37.65}$$

By substituting from Equation 37.63, we find Y to be

$$Y = \frac{VR^3}{EI} \int_{\alpha}^{\pi/2} (\cos \alpha - \cos \theta)^2 d\theta \tag{37.66}$$

By integrating, Y becomes

$$Y = \frac{PR^3}{EI} G_1 \tag{37.67}$$

where G_1 is

$$G_1 = 0.125[(\pi - 2\alpha)(1 + 2\cos^2 \alpha) - 8\cos \alpha + 3\sin 2\alpha] \tag{37.68}$$

and where V is replaced by $P/2$ as seen by equilibrium conditions of Figure 37.14. Figure 37.15 provides a graphical representation of G_1 as a function of angle α.

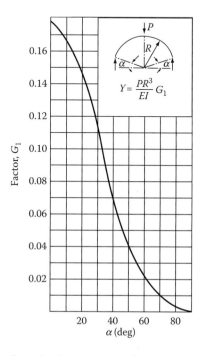

FIGURE 37.15 Deflection factor for a simply supported arch.

37.10 PIN-SUPPORTED ARCH

Figure 37.16 shows a pin-supported arch with a central concentrated load. This is a statically indeterminant system. Static equilibrium shows that the horizontal reactions must be equal and opposite in direction.

The bending moment at a section defined by θ is then

$$M = VR(\cos \alpha - \cos \theta) - H_p R(\sin \theta - \sin \alpha) \tag{37.69}$$

where Figure 37.16 defines the notation.

As before, we can use Castigliano's theorem to determine H_p. Since the displacement of the arch support is zero, the condition determining H_p is

$$\int_{\alpha}^{\pi/2} M \frac{\partial M}{\partial H_p} d\theta = 0 \tag{37.70}$$

From Equation 37.69 $\partial M/\partial H_p$ is

$$\frac{\partial M}{\partial H_p} = -R(\sin \theta - \sin \alpha) \tag{37.71}$$

By substituting from Equations 37.69 and 37.71 into Equation 37.70 and integrating, we obtain

$$H_p = PG_2 \tag{37.72}$$

where G_2 is

$$G_2 = \frac{4 \sin \alpha + 3 \cos 2\alpha - (\pi - 2\alpha) \sin 2\alpha - 1}{2(\pi - 2\alpha)(1 + 2 \sin^2 \alpha) - 6 \sin 2\alpha} \tag{37.73}$$

When α is zero, Equations 37.72 and 37.73 show that H_p is

$$H_p = 0.318P \tag{37.74}$$

which is a standard result. When α is increased, however, G_2 increases dramatically until the arch begins to yield in compression. It is well to keep in mind that relatively flat arches may be subjected to appreciable horizontal forces if the supports are kept apart at a fixed distance.

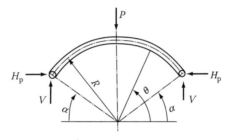

FIGURE 37.16 Pin-supported circular arch under central load.

The maximum bending moment M_{max} occurs when θ is $\pi/2$. From Equation 37.69, M_{max} is

$$M_{max} = PR\left[\frac{\cos\alpha}{2} - (1-\sin\alpha)G_2\right] \tag{37.75}$$

The deflection under load P can be found by Castigliano's theorem, considering one-half of the arch as an arched cantilever subjected to a vertical force V and horizontal thrust H_p acting as two statistically independent forces. The concept of static independence is justified here, as the external work done by H_p in the direction of V is zero. Following the usual procedure, we obtain

$$Y = \frac{PR^3}{EI}G_3 \tag{37.76}$$

where G_3 is

$$G_3 = G_1 - \frac{[4\sin\alpha + 3\cos 2\alpha - (\pi - 2\alpha)\sin 2\alpha - 1]^2}{8(\pi - 2\alpha)(1 + 2\sin^2\alpha) - 24\sin 2\alpha} \tag{37.77}$$

For $\alpha = 0$, Equations 37.76 and 37.77 reduce the well-known formula for a semicircular arch with a horizontal constraint:

$$Y = \frac{0.0189 PR^3}{EI} \tag{37.78}$$

Figure 37.17 provides an illustration of the G_2 and G_3 functions.

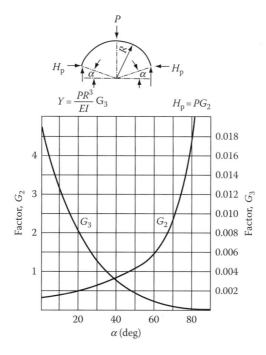

FIGURE 37.17 Force and deflection factors for a pin-supported arch.

37.11 BUILT-IN ARCH

A circular arch built-in at the supports presents a much more difficult analysis, even for the case of a concentrated central load. Figure 37.18 provides a sketch of the configuration. The reason for the complication is that the arch represents a doubly redundant structure with regard to the horizontal thrust H_b and the fixing couple M_f. In terms of these two unknown quantities and the vertical reaction V, which follow from statics, the bending moment at any section defined by θ can be stated as

$$M = VR(\cos \alpha - \cos \theta) - H_b R(\sin \theta - \sin \alpha) - M_f \tag{37.79}$$

The boundary conditions for the calculation of H_b and M_f can be obtained with the aid of the theorem of Castigliano:

$$\int_{\alpha}^{\pi/2} M \frac{\partial M}{\partial H_b} d\theta = 0 \quad \text{and} \quad \int_{\alpha}^{\pi/2} M \frac{\partial M}{\partial H_f} d\theta = 0 \tag{37.80}$$

By substituting for M from Equation 37.79 and then integrating, we obtain

$$2H_b RA_2 - PRA_1 - 8M_f A_3 = 0 \tag{37.81}$$

and

$$2H_b RA_3 - PRA_4 - (\pi - 2\alpha)M_f = 0 \tag{37.82}$$

By solving for H_b and M_f, we have

$$H_b = \frac{P[(\pi - 2\alpha)A_1 - 8A_3 A_4]}{2(\pi - 2\alpha)A_2 - 16A_3^2} \tag{37.83}$$

and

$$M_f = \frac{PR(A_1 A_3 - A_2 A_4)}{(\pi - 2\alpha)A_2 - 8A_3^2} \tag{37.84}$$

where A_1, A_2, A_3, and A_4 are as follows:

$$A_1 = 4\sin \alpha + 3\cos 2\alpha - (\pi - 2\alpha)\sin 2\alpha - 1 \tag{37.85}$$

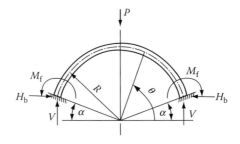

FIGURE 37.18 Built-in circular arch with a central load.

$$A_2 = (\pi - 2\alpha)(1 + 2\sin^2 \alpha) - 3\sin 2\alpha \qquad (37.86)$$

$$A_3 = 0.5(\pi - 2\alpha)\sin \alpha - \cos \alpha \qquad (37.87)$$

$$A_4 = 1 - 0.5(\pi - 2\alpha)\cos \alpha - \sin \alpha \qquad (37.88)$$

The derivation of the design formula for the central deflection follows the same rules as those used in conjunction with other arches, although the amount of algebraic work is substantially increased. The final result in this case is

$$Y = \frac{PR^3}{EI}\left\{ G_1 - \frac{4A_2A_4^2 - 8A_1A_3A_4 + 0.5(\pi - 2\alpha)A_1^2}{4\left[(\pi - 2\alpha)A_2 - 8A_3^2\right]} \right\} \qquad (37.89)$$

All the results for the fixed-end arch can be simplified by introducing additional symbols for the combined trigonometric functions as follows:

$$M_f = PRG_4, \quad H_b = PG_5, \quad \text{and} \quad Y = \frac{PR^3}{EI}G_6 \qquad (37.90)$$

Figure 37.19 provides curves for the force and deflection factors.

The general equation for the bending moment is

$$M = PR[0.5\cos \alpha - G_5(1 - \sin \alpha) - G_4] \qquad (37.91)$$

The foregoing results are simplified by letting α be zero. Specifically, for $\alpha = 0$ we have

$$H_b = \frac{P(4 - \pi)}{\pi^2 - 8} \qquad (37.92)$$

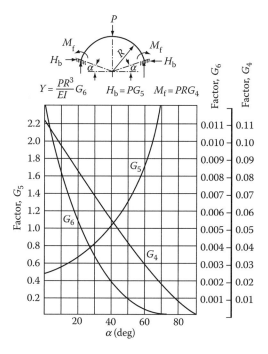

FIGURE 37.19 Force and deflection factors for a built-in circular arch.

$$M_f = \frac{PR(\pi^2 - 2\pi - 4)}{2(\pi^2 - 8)} \tag{37.93}$$

$$M = \frac{PR(2\pi - 6)}{\pi^2 - 8} \tag{37.94}$$

and

$$Y = \frac{PR^3}{EI}\left[\frac{\pi^3 - 20\pi + 32}{8(\pi^2 - 8)}\right] \tag{37.95}$$

37.12   PINNED ARCH UNDER A UNIFORM LOAD

Figure 37.20 shows a pin-supported circular arch subjected to a uniform vertical loading. This type of loading and support is feasible, for instance, in large cylindrical containers or the casing of a compressor of a jet engine subjected to inertia load under a sudden change in the direction of flight. Another possible application is that of a buried, thin-walled cylinder responding to a seismic ground motion or a soil compression wave caused by an underground explosion. Since a long pipe or a cylindrical container can be treated as a number of rings connected together, the model of a pin-jointed arch can be used in the analysis. In the case of a compressor casing, we can have a longitudinal joint holding the two casing halves together, so that the model illustrated in Figure 37.20 is appropriate. When an analysis of the deformation pertains to any point on the arch, such as that defined by θ, the derivation can be rather involved. The relevant statically indeterminate quantity is $qR/2$, where q denotes weight of the arch per inch of circumference.

Accordingly, the bending moment at any section θ is

$$M = 0.5qR^2(\pi - \pi\cos\theta - 3\sin\theta + 2\theta\cos\theta) \tag{37.96}$$

or

$$M = qR^2B \tag{37.97}$$

where B is a "bending moment factor." Figure 37.21 provides values of B in terms of the locating angle θ.

The maximum deflection Y_{max} of the arch, occurring at $\theta = \pi/2$ is

$$Y = \frac{0.0135qR^4}{EI} \tag{37.98}$$

For small angles of θ, on the order of 30°, vertical deflection can undergo a change in sign. At these values, however, the deflection is relatively small.

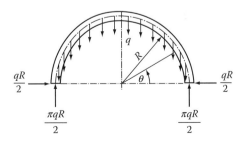

FIGURE 37.20   Pin-supported arch subjected to uniform vertical loading.

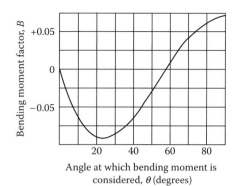

FIGURE 37.21   Chart for bending moment factor B of Equation 37.97.

SYMBOLS

A	Acceleration
A_1, A_2, A_3, A_4	Load factors for built-in arch
B	Bending moment factor; body
b	Width of rectangular section
d	Mean diameter of piston ring before cut
E	Modulus of elasticity
e	Element
F	Pin-joint reaction
F^*	Inertia force
G_1 through G_6	Arch factors
g	Acceleration due to gravity
H	Horizontal load
H_b	Horizontal thrust in built-in arch
H_p	Horizontal thrust in pin-jointed arch
h	Depth of cross section
I	Second moment of area
K_1	Arched cantilever factor
K_M	Factor for bending moment
K_u, K_u'	Factors for radial deflection
K_ψ	Factor for slope
ℓ	Bar or beam length
M	Bending moment
M_f	Fixing moment
M_0	External bending couple
M_q	Bending moment under uniform load
M_1, M_2	Bending moments for various portions
m	Mass
P	Concentrated load
$\bar{P}$	Fictitious load
q	Uniform load
q_m	Maximum load per unit length
R	Mean radius of curvature
S	Stress
S_b	Bending stress

t	Thickness
U	Strain energy
u	Radial displacement
v	Peripheral velocity
V	Vertical reaction; volume; velocity
X	Horizontal deflection
x	Axial coordinate
Y	Vertical deflection
α	Ring or arch angle
γ	Specific weight
δ	Deflection; elongation or shortening
ε	Auxiliary angle; strain
ε_{ij}	Strain matrix elements
θ	Angle at which forces are considered
ρ	Mass density
σ	Stress
σ_{ij}	Stress matrix elements
ϕ	Angle subtended by arched cantilever
$\chi = R/h$	Ratio of radius of curvature to depth of section
ψ	Slope

REFERENCES

1. A. Blake, *Design of Carved Members for Machines*, Robert E. Krieger, Huntington, NY, 1979.
2. W. C. Young and R. G. Budynas, *Roark's Formulas for Stress and Strain*, McGraw Hill, New York, 2002.
3. A. J. S. Pippard, *Studies in Elastic Structures*, Arnold & Co., London, 1952.
4. A. Blake, Circular arches, *Machine Design*, 1958.
5. V. Leontovich, *Frames and Arches*, McGraw Hill, New York, 1959.
6. W. Griffel, *Handbook of Formulas for Stress and Strain*, Frederick Ungar, New York, 1966.
7. F. B. Seely and S. O. Smith, *Advanced Mechanics of Materials*, Wiley, New York, 1966.
8. W. D. Pilkey, *Formulas for Stress, Strain, and Structural Matrices*, Wiley-Interscience, New York, 1994.
9. S. Timoshenko and J. N. Goodier, *Theory of Elasticity*, McGraw Hill, New York, 1951.
10. A. P. Boresi and O. M. Sidebottom, *Advanced Mechanics of Materials*, John Wiley & Sons, New York, 1985.
11. G. P. Fisher, Design charts for symmetrical ring girders, *Journal of Applied Mechanics*, 24(1), 1957.
12. H. R. Meck, Three dimensional deformation and buckling of a circular ring of arbitrary section, *Journal of Engineering for Industry*, 91(1), 1969.
13. J. R. Barber, Force and displacement influence functions for the circular ring, *Journal of Strain Analysis*, 13(2), 1978.
14. R. C. Juvinall and K. M. Marshek, *Fundamentals of Machine Component Design*, 2nd ed., John Wiley & Sons, New York, 1991.
15. B. J. Hamrock, B. Jacobson, and S. R. Schmid, *Fundamentals of Machine Elements*, WCB McGraw Hill, Boston, MA, 1999.
16. F. L. Singer, *Strength of Materials*, 2nd ed., Harper, New York, 1962.
17. H. Josephs and R. L. Huston, *Dynamics of Mechanical Systems*, CRC Press, Boca Raton, FL, 2002.

38 Links and Eyebars

38.1 INTRODUCTION

Rings, coupling links, chains, and eyebars commonly occur in structures and machines. These are closed, multiply connected structural components. In spite of their geometric simplicity, analysis of these members is complicated due to static indeterminacy. Even strain energy methods are tedious and difficult. In many cases, analyses can be simplified by neglecting direct and shear stresses in comparison with those due to bending.

Many of these ringlike components and their loadings may be modeled as thick rings under diametral loading. Therefore, it may be helpful and of interest to briefly review the response of a thick ring under diametral loading. Results should be applicable not only to chain links but also to such machine components as bearing rings and rims of heavy gears. We will base our analysis upon elastic strain energy [1–4]. Timoshenko [5,6] provides more rigorous analyses.

38.2 THICK-RING THEORY

Figure 38.1 depicts a thick ring being compressed along its vertical diameter by a load P. In the figure, H is a fictitious force introduced to enable evaluation of the change in the horizontal diameter due to load P. As before, M, Q, and N are the bending moment, shear force, and normal force at a cross section located at the angular coordinate θ. M_f is the restraining moment at $\theta = 0$. R is the midring radius as shown.

Due to the symmetry, we can focus upon a quarter of the ring. Equilibrium considerations produce the following relations:

$$M = \frac{PR(1 - \cos\theta)}{2} + \frac{HR\sin\theta}{2} - M_f \tag{38.1}$$

$$N = \frac{H\sin\theta}{2} - \frac{P\cos\theta}{2} \tag{38.2}$$

and

$$Q = \frac{P\sin\theta}{2} + \frac{H\cos\theta}{2} \tag{38.3}$$

From Castigliano's theorem (see Section 37.2), we can determine the statically indeterminant fixing moment M_f from the expression:

$$\int_0^{\pi/2} (M - N\delta)\frac{\partial M}{\partial M_f}\,d\theta = 0 \tag{38.4}$$

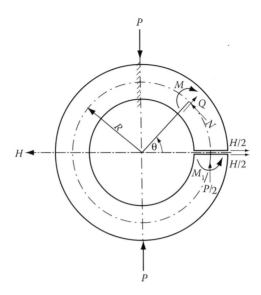

FIGURE 38.1   Thick ring in diametral compression.

where δ is the displacement of the neutral axis. By substituting from Equations 38.1 and 38.2 into Equation 38.4 and integrating, we can determine M_f as

$$M_f = \frac{P(\pi R - 2R + 2\delta)}{2\pi} \tag{38.5}$$

Then from Equation 38.1 with $H = 0$, M becomes

$$M = \frac{P(2R - 2\delta - \pi R \cos \theta)}{2\pi} \tag{38.6}$$

In this manner, using Castigliano's theorem, the vertical and horizontal displacements are found to be

$$Y = \frac{P}{4AE} \left[\frac{\pi^2 R(R - 2\delta) - 8(R - \delta)^2}{\pi \delta} + \pi R(1 + 2.5\xi) \right] \tag{38.7}$$

and

$$X = \frac{P}{2AE} \left[\frac{(4R - \pi R - 4\delta)(R - \delta)}{\pi \delta} + 2.5R\xi \right] \tag{38.8}$$

where ξ is a shear distribution factor (multiple of maximum shear stress to mean normal stress on a cross section).

38.3   THEORY OF CHAIN LINKS

Figure 38.2 presents a theoretical model of a chain link where a central support stud may or may not be present. Since we are usually interested in knowing the maximum stress under a given load W, it is necessary to calculate the bending moment M_O under load W. Since the link is statically

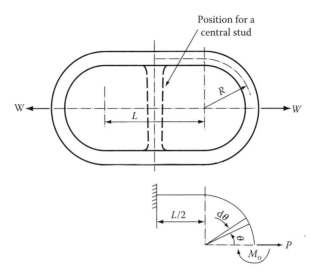

FIGURE 38.2 Model of a chain link.

indeterminant we can use Castigliano's theorem to determine M_O, assuming that due to symmetry there is no rotation at $\theta = 0$. This procedure leads to the expression:

$$\int_0^{\pi/2} M_1 \frac{\partial M_1}{\partial M_0} R \, d\theta + \int_0^{L/2} M_2 \frac{\partial M_2}{\partial M_0} \, dx = 0 \tag{38.9}$$

where M_1 and M_2 are the bending moments in the curved and straight portions of upper right quadrant of the link (see Figure 38.2).

For the model shown, M_1 and M_2 are

$$M_1 = M_0 - PR \sin \theta \quad \text{and} \quad M_2 = M_0 - PR \tag{38.10}$$

Observe that $\partial M_1/\partial M_0 = \partial M_2/\partial M_0 = 1$. Then by substituting from Equation 38.10 into 38.9 and integrating, we obtain

$$\left(\frac{M_0 \pi}{2} - PR \right) R + (M_0 - PR) \frac{L}{2} = 0 \tag{38.11}$$

By solving for M_0 we have

$$M_0 = \frac{WR(2R + L)}{2(\pi R + L)} \tag{38.12}$$

Knowing the bending moment M_0 together with the link geometry, we can calculate the bending stress. In addition, for design purposes we should evaluate the contact stresses as chains often fail due to wear at the contact between two links.

Finally, for the model of Figure 38.2 if r is the radius of a circular cross section, then for most chains the ratio r/R is between 0.2 and 0.5. The ratio L/R may vary between 0 and 4.

38.4 LINK REINFORCEMENT

When the strength of a typical open link such as that shown in Figure 38.2, is found to be insufficient, the insertion of a central stud shown by dashed lines, is known to increase the strength significantly. Figure 38.3 provides a free-body (equilibrium) diagram for studying this special case.

As before, this case is also statically indeterminant. Here there are two redundant quantities: M_0 and H. From Figure 38.3, the moments M_1 and M_2 in the curved and straight portions are

$$M_1 = M_0 - \frac{WR}{2} \sin\theta + HR(1 - \cos\theta) \tag{38.13}$$

and

$$M_2 = M_0 - \frac{WR}{2} + H(R + x) \tag{38.14}$$

Due to symmetry, there is no change in slope at the ends of the quadrant. Also, we can reasonably assume that there is virtually no deflection in the direction of the stud. Using Castigliano's theorem, the first of these conditions is satisfied by Equation 38.9 as

$$\int_0^{\pi/2} M_1 \frac{\partial M_1}{\partial M_0} R \, d\theta + \int_0^{L/2} M_2 \frac{\partial M_2}{\partial M_0} dx = 0 \tag{38.15}$$

Similarly, the condition of zero deflection leads to the expression:

$$\int_0^{\pi/2} M_1 \frac{\partial M_1}{\partial H} R \, d\theta + \int_0^{L/2} M_2 \frac{\partial M_2}{\partial H} dx = 0 \tag{38.16}$$

From Equations 38.13 and 38.14 we have

$$\frac{\partial M_1}{\partial M_0} = \frac{\partial M_2}{\partial M_0} = 1 \tag{38.17}$$

$$\frac{\partial M_1}{\partial H} = R(1 - \cos\theta) \tag{38.18}$$

and

$$\frac{\partial M_2}{\partial H} = R + x \tag{38.19}$$

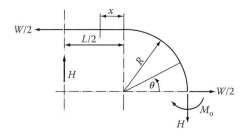

FIGURE 38.3   Equilibrium model for a quarter of a studded link.

By substituting from Equations 38.13, 38.14, 38.17, 38.18, and 38.19 into Equations 38.15 and 38.16 yields

$$M_0 = \frac{WRC_1}{2} \quad \text{and} \quad H = \frac{WC_2}{2} \tag{38.20}$$

where C_1 and C_2 are

$$C_1 = \frac{(k+2)[k^3 + 6k^2 + 12k(4-\pi) + 48(\pi-3)]}{k^4 + 4\pi k^3 + 48K^2 + 24\pi k + 24(\pi^2-8)} \tag{38.21}$$

and

$$C_2 = \frac{12(k+2)[(\pi-2)k + 2(4-\pi)]}{k^4 + 4\pi k^3 + 48K^2 + 24\pi k + 24(\pi^2-8)} \tag{38.22}$$

where k is L/R.

Previous investigations indicate [7,8] that the provision of a link study could decrease the maximum tensile stress by about 20%. The relevant maximum compressive stress can be reduced by as much as 50%, although this type of stress is generally less important in link design.

38.5   PROOF RING FORMULAS

In the past, whenever steel rings with circular cross section were used as lifting components, it was customary to use the following simplified expression for determining the maximum safe or "proof" load W as [8]:

$$W = \frac{33,000d^3}{D_i + 0.3d} \tag{38.23}$$

where as before

D_i is the inner diameter of the ring (in.)

d is the cross section diameter (in.)

In the development of this formula, the maximum material strength was taken to be of the order of 54,000 psi.

The analysis of circular links provides an introduction to the study of eyebars and similar mechanical joints [9,10], which despite their apparent simplicity, often become a point of contention as to the design criteria and the potential modes of failure. There is surprisingly little information on the effect of basic variables on the critical stresses in eyebars. In particular, local areas of the maximum tensile stresses may be of concern because of the modern requirements of fracture-safe design.

38.6   KNUCKLE JOINT

Figure 38.4 shows the typical knuckle joint geometry. In this design, the pin strength is of primary importance. The remaining two components, however, behave essentially as eyebars which theoretically can fail in various ways. Figure 38.5 illustrates three common failure modes: (1) local compression due to pin contact; (2) primary tension; and (3) tear-out shear.

The average compressive stress for the eyebar given in stress calculations can be obtained from the projected area. This procedure is probably satisfactory provided the pin fits the eyebar with zero

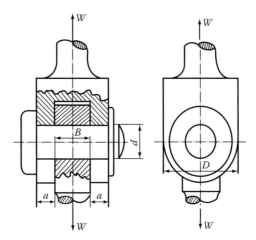

FIGURE 38.4 Typical knuckle joint geometry.

clearance. However, even under these assembly conditions, the pressure around the pin is expected to vary according to a definite pattern.

A reasonable approach for calculating compressive stresses is based upon the "cosine load distribution." Figure 38.6 illustrates the concept, where q is the load per unit pin length. Due to symmetry, the horizontal components of q balance out. The vertical components are then related to the external load W by integrating around a quarter of the pin as

$$W = 2r \int_0^{\pi/2} q \cos \theta \, d\theta \tag{38.24}$$

From Figure 38.6, we see that q may be expressed in terms of q_{max} as

$$q = q_{max} \cos \theta \tag{38.25}$$

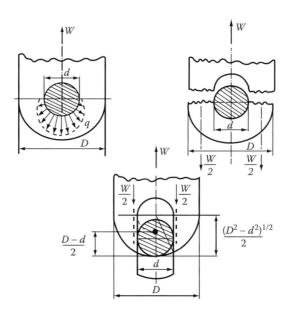

FIGURE 38.5 Typical eyebar failure. Local compression, primary tension, and tear-out shear.

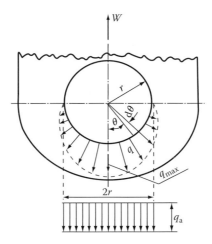

FIGURE 38.6 Cosine loading on an eyebar.

By substituting from Equation 38.25 into Equation 38.24, we obtain

$$q_{max} = 2W/\pi r \qquad (38.26)$$

By dividing q_{max} by the width of the eyebar cross section, we have the compressive stress.

38.7 EYEBAR WITH ZERO CLEARANCE

The approach described so far provides a quick answer to the question of contact stresses if the clearance between the pin and the eyebar is not excessive. However, the magnitude of the acceptable clearance in a pin joint appears to be rather poorly defined in engineering literature because of the theoretical complexity of the problem where the elastic theory breaks down. All that can be stated for certain is that the cosine load distribution is probably a reasonable approximation to the manner of loading under zero clearance, while a concentrated load model should correspond to a relatively large clearance between the pin and the eyebar.

When the clearance is found to be relatively small and the pin can be assumed to be rigid, the analysis of the maximum stresses in the eyebar portion of the joint can be performed with the aid of the following simple expression [11]:

$$S = \frac{W\varphi}{BR} \qquad (38.27)$$

where φ is a design factor depending upon the angular position θ and the nominal radii R and r, as in Figure 38.7. Figure 38.8 provides numerical values of φ as a function of θ and R/r.

According to this model, the maximum tensile stress is found at the inner surface of the eyebar, where θ is $\pi/2$. The graph of Figure 38.8 is intended for a typical range of R/r and it shows that the effect of this ratio on the maximum stress in the eye is relatively small in the critical regions of $\theta = 0$ and $\theta = \pi/2$.

As far as the theory of fracture-safe criteria is concerned, only the maximum tensile stress is of an immediate interest to the designer. Accordingly, in line with the eyebar theory discussed above, the tensile stress S becomes

$$S = \frac{3.52W}{BR} \qquad (38.28)$$

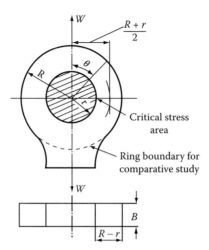

FIGURE 38.7   Eyebar geometry by Faupel. (From Faupel, J.H., *Engineering Design*, Wiley, New York, 1964.)

where Figure 38.7 illustrates W, B, and R. Equation 38.28 should be applicable for typical eyebar geometry for all ratios of R/r between 2 and 4.

The primary tensile stress S_t corresponding to the simplified mode of failure depicted in Figure 38.5 is

$$S_t = \frac{W}{2B(R - r)} \tag{38.29}$$

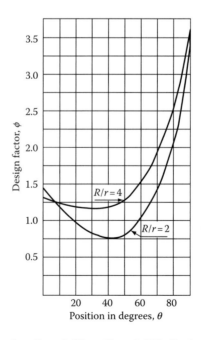

FIGURE 38.8   Design factor based on Faupel. (From Faupel, J.H., *Engineering Design*, Wiley, New York, 1964.)

In comparing the two models based on Equations 38.28 and 38.29, we can conveniently introduce a dimensionless ratio λ simply defined as

$$\lambda = R/r \qquad (38.30)$$

This in turn leads to the expression:

$$\frac{S}{S_t} = \frac{7.04(\lambda - 1)}{\lambda} \qquad (38.31)$$

In conventional machine design practice, λ typically has a value of 2. Equation 38.31 then produces a stress ratio value of 3.52. The nominal tensile stress of the type given by Equation 38.29 is the most damaging from the point of view of fracture propagation because of the relatively large amount of elastic energy stored and available to develop the crack. Furthermore, as the crack develops, the net area is being progressively lost, thereby increasing the nominal stress. Since the elastic energy stored is proportional to the square of the stress per unit volume of the stressed material, it is easy to see the role of the nominal stress. Hence, the local tensile stress of the type described by Equation 38.28 is most likely to be responsible for crack initiation. By the application of relatively high factors of safety, the nominal stresses can be kept at a low level and the corresponding calculations should not present undue difficulties. Unfortunately, the more rigorous analysis of a local stress concentration can be very complicated and requires advanced knowledge of material behavior supported by a well-conceived experimental program.

The characteristics of eyebar geometry reported by Faupel [11] can also be investigated on the basis of a thick-ring model. The hypothetical boundary for such a ring is illustrated in Figure 38.7.

38.8 THICK-RING METHOD OF EYEBAR DESIGN

In general, when performing the analysis of a curved member of a relatively sharp curvature, it is customary to assume that plane sections remain plane during bending while the neutral axis is displaced toward the center of curvature, which in this particular case coincides with the center of the eyebar. For a circular ring of a rectangular cross section, such a displacement can be obtained in a closed-form solution and then approximated using the theorem of Maclaurin. Denoting the relevant displacement of the neutral axis by δ, the following simplified relation may be obtained

$$\delta = rF(\lambda) \qquad (38.32)$$

where $F(\lambda)$ is a geometric factor given by

$$F(\lambda) = \frac{(\lambda - 1)^2(\lambda + 1)}{8(\lambda^2 + \lambda + 1)} \qquad (38.33)$$

where λ is the eyebar ratio defined as

$$\lambda = R/r \qquad (38.34)$$

where
  R is the outer radius
  r is the inner radius

When we compare the geometry factor $F(\lambda)$ of Equation 38.33 with more exact factors [1], we find the difference is 15%–20% for $\lambda = 2$ and $\lambda = 4$, respectively. But since the neutral axis shift δ

has only a limited effect on the magnitude of the bending moment, the use of Equation 38.33 may be justified within the range of λ considered.

From Figure 38.7, we see that the bending moment M may be expressed as

$$M = \frac{W}{2\pi}\left[R + r - 2\delta - \frac{\pi(R+r)}{2}\sin\theta\right] \qquad (38.35)$$

Since the maximum tension develops at the inner radius where $\theta = \pi/2$, putting $\lambda = R/r$ and $\delta = (R-r)^2\,(R+r)/8\,(R^2+Rr+r^2)$, Equation 38.35 gives

$$M = WrG(\lambda) \qquad (38.36)$$

where $G(\lambda)$ is a bending moment factor defined as

$$G(\lambda) = \frac{\lambda+1}{4\pi}\left[\pi - 2 + \frac{(\lambda+1)^2}{2(\lambda^2+\lambda+1)}\right] \qquad (38.37)$$

Figure 38.9 provides graphical values of the geometry and bending moment factors $F(\lambda)$ and $G(\lambda)$.

From Figure 38.7, we see that the section modulus Z for the eyebar is

$$Z = Br^2(\lambda-1)^2/6 \qquad (38.38)$$

Thus the bending stress S_b becomes

$$S_b = \frac{6WG(\lambda)}{Br(\lambda-1)^2} \qquad (38.39)$$

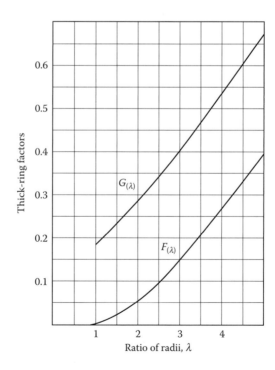

FIGURE 38.9   Moment $[G(\lambda)]$ and geometry $[F(\lambda)]$ factors in thick-ring theory.

The total stress at the extreme fiber on the inner eyebar diameter is the sum of bending and tension. According to the original Winkler-Bach theory, the general expression for the maximum stress S_{max} may be expressed as

$$S_{max} = \varphi_0(S_t + S_b) \tag{38.40}$$

where φ_0 is a correction factor allowing for a hyperbolic distribution instead of linear distribution of the normal stress over the depth of the cross section (see Figure 14.5). Consequently, the stress developed at the inner face of the curved beam may be assumed to be substantially higher than that predicted by the theory of straight members.

So far the calculations of maximum stresses using the curved-beam model were made for a single concentrated load. This discussion has been included here because many engineering estimates are based on a single-load assumption, which often proves to be highly conservative.

Alternatively, the expression reported by Faupel [11] in Equation 38.27 is for a rigid pin with zero clearance, with a loading distribution similar but not necessarily identical to that shown in Figure 38.6.

38.9   EYEBAR WITH FINITE PIN CLEARANCE

When an eyebar has a finite clearance with its pin, the load distribution is affected and this in turn will affect the stress levels. With finite clearance, the geometry is more complex and thus exact analyses are elusive. From a design perspective, however, with typical eyebar geometry (say, R/r approximately equal to 2), we can interpolate between the results for a rigid pin and those for a thick ring, under the assumption that those two cases represent extreme loading conditions for most practical purposes.

To provide some guidelines, recall from Equation 38.27 that the maximum stress S in an eyebar with very small clearance is approximately

$$S = W\varphi/BR \tag{38.41}$$

where φ is a stress (or design) factor given by Figure 38.8.

For an eyebar with a finite clearance, we can estimate the maximum tensile stress S_{tmax} in the eyebar by an expression analogous to Equation 38.41 as

$$S_{t\,max} = W\psi/BR = W\psi/\lambda rB \tag{38.42}$$

where ψ is a stress factor depending upon the clearance ratio e/r, where e is the clearance. Figure 38.10 provides a graphical representation for ψ in terms of the clearance ratio.

It may also be convenient to express the maximum stress S_{tmax} of Equation 38.42 in terms of the nominal tension S_t of Equation 38.29 corresponding to the failure mode of Figure 38.6. Specifically, from Equation 38.29 S_t is

$$S_t = \frac{W}{2B(R - r)} \tag{38.43}$$

Then the ratio S_{tmax}/S_t is

$$\frac{S_{t\,max}}{S_t} = \frac{2\psi(\lambda - 1)}{\lambda} \tag{38.44}$$

where, as before, λ is the radius ratio R/r.

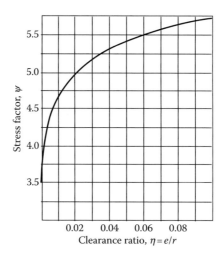

FIGURE 38.10 Stress factor for typical eyebar design.

38.10 MODES OF EYEBAR FAILURE, FACTORS OF SAFETY

Figure 38.6 shows various modes of eyebar failure. The question of a tear-out mode can be resolved rather simply. Since this failure requires a double-shear equilibrium, and as the shear areas are assumed to be rectangular, the theoretical values of the maximum shear stress S_{smax} becomes

$$S_{smax} = \frac{3W}{4Br(\lambda^2 - 1)^{1/2}} \tag{38.45}$$

For the purpose of the calculations involving shear, the ratio of the allowable shear strength to yield strength in tension should be taken as $1/\sqrt{3}$. Hence, the factor of safety based on shear is

$$F_{ss} = \frac{S_y}{\sqrt{3}S_{smax}} \tag{38.46}$$

To correlate this with tension, using Equation 38.42 the corresponding factor of safety may be defined as

$$F_{st} = \frac{S_y}{S_{tmax}} \tag{38.47}$$

Then by utilizing Equation 38.42 and Equations 38.45 through 38.47 yields

$$\frac{F_{st}}{F_{ss}} = \frac{3\sqrt{3}\lambda}{4\psi(\lambda^2 - 1)^{1/2}} \tag{38.48}$$

For the case of zero clearance and the practical range of values of λ equal between 2 and 4, Equation 38.48 gives safety factor ratios of 0.426 and 0.331, respectively. Therefore, one of the primary considerations in eyebar joint design is to evaluate the factor of safety related to the critical tensile stress criteria rather than shear. This is particularly important in fracture-safe design.

SYMBOLS

A	Area of bar cross section
A_c	Core area of shank
A	Width of knuckle joint
B	Width of eyebar
B'	Width of contact area
C	Correction factor for eyebars
C_1, C_2	Force factors in studded link
D	Maximum eyebar diameter
D_i	Inner diameter of link
D_s	Nominal diameter of shank
d	Bar diameter
d_s	Core diameter of shank
E	Modulus of elasticity
e	Clearance
$F(\lambda)$	Auxiliary factor
F_{ss}	Factor of safety for shear
F_{st}	Factor of safety for tension
G	Shank diameter factor
$G(\lambda)$	Auxiliary factor
H	Fictitious ring load
K_y, K_x	Ring deflection factors
$K = L/R$	Length-to-radius ratio
L	Length of straight portion
M, M_1, M_2	Total bending moments
M_f	Redundant moment in thick ring
M_0	Redundant moment in link
N	Direct force
n	Proportionality constant
P	Concentrated load
Q	Shear load
q	Unit load
q_a	Average unit load
q_{max}	Maximum unit load
R	Mean radius; major radius of eyebar
r	Radius of bar; radius of eye
r_0	Radius of pin
S	General symbol for stress
S_b	Bending stress
S_{max}	Maximum stress
S_s	Shear stress
S_{smax}	Maximum shear stress
S_t	Tensile stress
S_{tmax}	Maximum tensile stress
S_y	Yield stress
W	Link or eyebar load
X	Horizontal deflection
x	Length along straight portion
Y	Vertical deflection
Z	Section modulus

α	Angle of inclined load
β	Angle subtending contact arc
δ	Displacement of neutral axis
$\eta = e/r$	Clearance ratio
θ	Angle at which forces are considered
$\lambda = R/r$	Eyebar ratio
ξ	Shear distribution factor
φ	Eyebar design factor
φ_0	Curved-beam stress factor
ψ	Pin clearance correction factor

REFERENCES

1. A. Blake, *Design of Curved Members for Machines*, Robert E. Krieger, Huntington, NY, 1979.
2. E. R. Leeman, *Stresses in a Circular Ring*, Engineering Ltd., London, 1956.
3. A. Blake, Deflection of a thick ring in diametral compression by test and strength of materials theory, *Journal of Applied Mechanics*, 1959.
4. A. J. S. Pippard and C. V. Miller, The stresses in links and their alternation in length under load, Proceedings, Institute of Mechanical Engineers (London), 1923.
5. S. Timoshenko and J. N. Goodier, *Theory of Elasticity*, McGraw Hill, Englewood Cliffs, NJ, 1957.
6. S. Timoshenko, On the distribution of stresses in a circular ring compressed by two forces acting along a diameter, *Philosophical Magazine*, 44 (263), 1922.
7. G. A. Goodenough and L. E. Moore, Strength of chain links, Bulletin 18, Engineering Experiment Station, University of Illinois, Urbana, IL, 1907.
8. H. J. Gough, H. L. Cox, and D. C. Sopwith, Design of crane hooks and other components of lifting gear, Proceedings, Institute of Mechanical Engineers (London), 1934.
9. A. S. Hall, A. R. Holowenko, and H. G. Laughlin, *Theory and Problems of Machine Design*, McGraw Hill, New York, 1961.
10. A. Blake, Stresses in eye bars, *Design News*, 1974.
11. J. H. Faupel, *Engineering Design*, Wiley, New York, 1964.

39 Springs

39.1 INTRODUCTION

The technology of conventional mechanical springs has been developed over the past 100 years. Sufficient data are currently available on materials, design, and manufacture so that a designer can size a spring for a particular application. An excellent summary of the mechanical properties of spring materials, working formulas, and fabrication variables has been compiled by Carlson [1]. In addition, a number of excellent textbooks on mechanical design provide useful discussions on spring design [2–4].

Basic derivation of spring formulas and some of the more complicated equations can be found in a classical text on mechanical springs by Wahl [6]. This chapter is limited to conventional formulas and the elementary principles of stress analysis.

Principal variables of importance to spring design are force, deflection, and stress. Once the preliminary design for a given material and environment is accomplished, the appropriate production techniques are best selected with the help of the established companies specializing in spring manufacture.

The methodology of spring calculation has evolved through the use of special slide rules, charts, tables, and finally electronic computers, which have considerably reduced the laborious process of repetitive trial and error computations. The principal design relationships, however, remained the same since they evolved from the theoretical concepts of stress and strain.

Design, manufacture, and application of mechanical springs result in a customary subdivision of this topic into compression, extension, and torsion categories. The basic stress and deflection formulas are essentially the same for the compression and extension springs. The only difference between these two categories lies in the fact that in most applications extension springs have initial tension wound into them. In the case of extension springs, the errors in both stress and deflection calculations result when the helix angle exceeds 12.5°. Furthermore, if the extension spring was pulled out far enough, the tensile and not the torsional component of stress would become more significant.

39.2 COMPRESSION SPRINGS

Compression springs in general are open-coil, helically wound springs which may be cylindrical, conical, barrel-shaped, or concave in form. Since the solid height and end conditions are important factors in design, it is necessary to know whether a compression spring under consideration is expected to have plain or squared ends with or without ground end coils. These conditions determine the appropriate corrections developed by the industry for the total number of coils, load tolerances, operational characteristics, and other features affecting the performance and cost. The compression springs are the most popular and represent 80%–90% of all springs produced by the industry.

Basic design parameters in the case of a round wire cylindrical spring include the number of active coils N, wire diameter d, mean coil diameter D, and the applied force P, which can be either compressive or tensile. Figure 39.1 shows the various symbols used for helical springs.

From the point of view of stress analysis, helical "compression" spring is somewhat of a misnomer because the compressive force produces torsional stress. When using a round wire, the basic design formula for a helical compression spring is

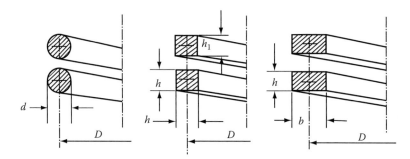

FIGURE 39.1 Typical symbols for helical springs.

$$\tau = \frac{2.55\,PD}{d^3} \tag{39.1}$$

where

   τ is the torsion (shear) stress (psi)
   P is the applied (axial) load

This formula is based on the assumption that the spring wire responds as a straight bar of circular cross section when acted upon by a torsional moment. The simplified model cannot yield the correct value for the fiber stress because the coiled wire behaves as a curved member loaded out of the plane of curvature. The actual stresses should vary in a hyperbolic fashion, with the higher stress level existing on the inner surface of the spring.

Equation 39.1 is an approximate expression which is sufficiently accurate for most practical purposes where static or slowly applied loads are involved. When helical springs operate under severe dynamical conditions, however, additional stresses caused by the curvature of the spring and shear loads need to be included.

A stress correction factor K_0 for dynamic loading, proposed by Wahl [6], varies with the ratio of mean to wire diameter. Specifically, Wahl's formula is

$$\tau_{\max} = PDK_0/d^3 \tag{39.2}$$

where Figure 39.2 shows the variation of K_0 with D/d.

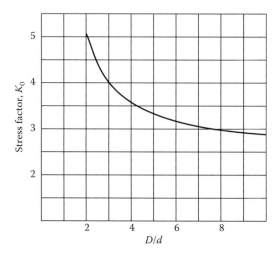

FIGURE 39.2 Stress correction factor. (From Wahl, A.M., *Mechanical Springs*, McGraw-Hill, New York, 1963.)

For square wire cross sections, Wahl's formula is

$$\tau_{max} = \frac{0.94 \, PDK_0}{h^3} \tag{39.3}$$

where h is the side of the square cross section (see Figure 39.1) and where d is replaced by h in Figure 39.2.

The initial step in spring sizing is to estimate the proper wire diameter for a given external load P, the approximate mean diameter D, and the allowable torsional stress τ. The convenient formula for starting the calculation procedure is given by Equation 39.1. The number of coils can be obtained from the standard spring formula

$$Y = \frac{8PD^3 N}{Gd^4} \tag{39.4}$$

where

Y is the spring compression
P, D, and d have the same meaning as before
N is the number of active coils
G is the shear modulus

Depending upon the geometry of the ends of the spring, there may be one or two "dead" or noncontributing coils. The solid height of the spring is taken as the total number of coils times the wire diameter. A compression spring, however, should never be designed to deflect in its working travel until the coils actually contact each other. Therefore in selecting the working range Y, certain allowance should be made so that the spring can carry the load above the solid height.

For a spring with a square wire cross section (rarely used), the expression analogous to Equation 39.4 is

$$Y = \frac{5.58PD^3 N}{Gh^4} \tag{39.5}$$

39.3 EXTENSION SPRINGS

Extension springs are closely coiled helixes that offer a finite resistance to a pulling force. The load buildup obtained by the coiling process is defined as the initial tension. Extension springs are normally made out of round wire, although in special cases, square or shaped wires can also be found. From the design point of view, initial tension means the presence of a definite stress in the spring wire in the original undeflected condition. The initial tensile stress should be added to the subsequent working stress of the spring. When there is no initial tension, the compression spring formulas, quoted in this chapter, are directly applicable to extension spring design.

Studies show that extension springs represent about 10% of all springs produced by the various manufacturers. The design lead time for these springs is longer because of the additional requirements for the control of initial tension, special coiling methodology, and stress analysis of end hooks. There is quite a variety of standard and special extension spring ends, designed according to the military and industrial applications, to which the designer is referred during the process of selecting the appropriate configurations. This is not a mundane problem because of the end curvature and the stress concentration phenomena involved.

Figure 39.3 depicts a typical end hook geometry of an extension spring. Due to the transition from the plane of the end hook to the coil geometry, there are four different radii of curvature between the hook and the coil. Figure 39.3 shows these radii.

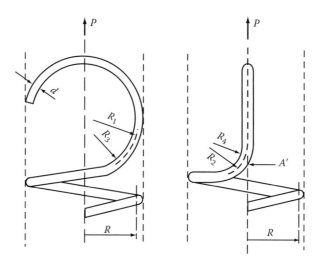

FIGURE 39.3   Typical end hook geometry for an extension spring.

It appears that, in general, the maximum bending and torsional moments can be taken equal to *PR*. The exact solution of the maximum bending and torsional stresses may require certain design corrections due to the sharp curvature. However, the relevant industrial practice suggests that in a well-proportioned spring end, the probability of a structural failure due to the torsional moment is rather low. Hence the following simplified formula for bending is all that might be needed [1]:

$$S_b = \frac{5\,PD^2}{(D-d)d^3} \tag{39.6}$$

where
   S_b is the bending stress
   P is the axial load
   D is the mean coil diameter
   d is the wire diameter (see Figure 39.3)

Another approximate equation for the bending stress uses the radii of curvature R_1 and R_3 [7]:

$$S_b = \frac{10.19\,PRR_1}{R_3 d^3} \tag{39.7}$$

where, as before, R is the mean coil radius as in Figure 39.3.

   Equations 39.6 and 39.7 provide useful approximations for the wire bending stress applicable for the majority of spring configurations including regular and hooks over center, crossover center hooks, as well as machine or double hooks. Exceptions may involve half and side hooks subjected to additional stresses.

   When R_1 is equal to R_2, and the wire diameter d is small compared with the mean coil diameter D (see Figure 39.3), Equations 39.6 and 39.7 give approximately the same numerical results.

   In the region of A' shown in Figure 39.3, where the bend joins the helical portion of the spring wire, the stress is caused by twisting. The corresponding torsional shear stress can be obtained by multiplying the result from Equation 39.1 by the ratio R_2/R_4. This approximation should be sufficiently conservative for most practical needs.

   The amount of stretch in an extension spring has no well-defined limit similar to the solid-height limit found in a compression spring. Hence the amount of extension has to be governed by the maximum fiber stress dependent on the wire size and the diameter ratio. From the point of view of production, the extension springs can be manufactured with a definite space between coils, with zero space and zero initial tension, or close-wound with a given initial tension. In the latter case, after the initial tension has been broken to create a space between the coils, the extension has the same rate as that which can be calculated from the deflection formula for the compression spring. The specific extension spring which requires a uniform rate of load from zero to the maximum allowable deflection, must be wound with zero initial tension.

39.4 TORSION SPRINGS

Torsion springs used in the industry relates to a helical coil spring that exerts a torque or moment, hence the term "torsion." From another perspective, however, the term "torsion" may be a misnomer in that torsion springs operate via bending stresses, whereas compression and extension springs operate via torsion stresses.

Torsion springs have commonly been used in clocks, doors, toy motors, and measuring devices. In their design, the spring usually winds up from the free position, causing reduction in the coil diameter. Although many torsion springs are made from round wire, a more efficient cross-section shape is rectangular, the so-called "flat coil."

Figure 39.4 shows the basic principle of a torsion spring. The torque produced depends upon the winding geometry. That is, in referring to the figure, the torque is not $P \times L_r$ but instead $P \times L_0$.

There are numerous working tables, formulas, and recommended design steps for the analysis of torsion-type springs [1–7]. The underlying principles follow the conventional beam theory on the premise that a constant moment acts on the entire wire cross section.

The bending stress S_b for the round wire spring is

$$S_b = \frac{10.2\,M}{d^3} \tag{39.8}$$

where
M is the applied moment (in.-lb)
d is the wire diameter (in.)

For a rectangular cross section with width b and depth h, the bending stress S_b is

$$S_b = \frac{6\,M}{bh^2} \tag{39.9}$$

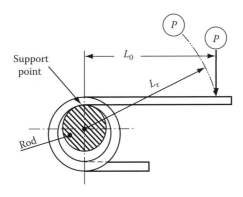

FIGURE 39.4   A torsion spring.

The relationship for a round wire, between the applied torsional moment M and the number of turns T is

$$M = \frac{0.093\ ETd^4}{ND} \tag{39.10}$$

where
  N is the number of active coils
  D is the mean coil diameter
  E is the modulus of elasticity

Similarly the relation between the applied torsional moment M and the number of turns for a rectangular wire is

$$M = \frac{0.152\ EbTh^3}{ND} \tag{39.11}$$

where
  h is the cross-section thickness (measured in the radial direction)
  b is the cross-section dimension parallel to the axis of the coil

From the point of view of practical stress analysis, the coiling procedure and the degree of springback of a torsion spring depend on the D/d ratio. In the case of smaller D/d ratios, the tensile stresses induced by the coiling process can be in excess of the proportional limit of the material so that the calculations can only indicate the apparent rather than real stresses.

39.5 BUCKLING COLUMN SPRING

When a flat and initially straight thin strip (or band) is subjected to a compressive end load, large deflections can develop. The spring formed in this way is essentially a special curved member characterized by what is known as a "zero rate response."

Figure 39.5 illustrates the concept. The desired spring effect is provided by the large displacement of the flat steel band.

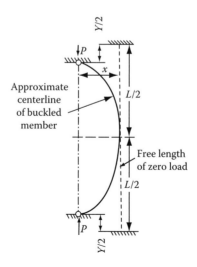

FIGURE 39.5 Buckling column spring.

Analysis of the band structure leads to the following nonlinear differential equation:

$$EI\left(\frac{d\psi}{ds}\right) = -Px \tag{39.12}$$

where
  x is the horizontal displacement
  ψ is the slope of the band
  s is the distance along the band
  E, I, and P are the elastic modulus, second moment of area, and axial load

The nonlinearity in Equation 39.12 arises upon differentiation with respect to s, yielding

$$EI\frac{d^2\psi}{ds^2} = -P\sin\psi \tag{39.13}$$

where dx/ds is identified as $\sin\psi$ [8].

Solution of Equation 39.13 involves elliptic integrals which can be simplified to provide design formulas and charts suitable for practical use [9].

Relevant working formulas are

$$P = \frac{h^2\,AEF_1}{L^2} \tag{39.14}$$

$$S_b = \frac{hEF_2}{L} \tag{39.15}$$

and

$$A = \frac{PEF_2^2}{F_1 S_b^2} \tag{39.16}$$

where
  S_b is the bending stress
  h is the thickness of the band
  L is the length of the band
  A is the band cross-section area
  F_1 and F_2 are design factors, provided graphically in Figure 39.6

From Equation 39.14 and Figure 39.6, we can determine the vertical displacement Y and then also the lateral displacement x. By iteratively using Equations 39.15 and 39.16, we can adjust the cross-section area A so as not to exceed a given bending stress S_b.

Since relatively thin strips in axial compression cannot support substantial loads, parallel arrangement of a number of strips may be required for a particular design. The controlling factor is the stress given by Equation 39.14 and the type of load–deflection characteristics required.

For best approximation to zero rate spring response, the Y/L ratio should be kept between 0.1 and 0.3, while L/h should not be appreciably smaller than about 200.

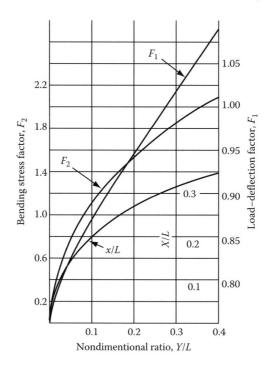

FIGURE 39.6   Design factors for a buckling column spring.

SYMBOLS

A	Area of cross section
a	Length of vertical drop
D	Mean coil diameter
d	Wire diameter
E	Modulus of elasticity
e	Spring pitch
F_1, F_2	Design factors
G	Shear modulus
g	Acceleration of gravity
h, h_1	Depths of sections
I	Second moment of area
K_0	Stress correction factor
k	Spring constant
L	Length of spring or flat stock
L_0	Arbitrary length
L_r	Radial distance
M	Applied torque
M_a	Moment per one turn of torsion spring
N	Number of active coils
n	Frequency (cycles/s)
P	Spring force
R	Mean radius of coil
$R_1, \ldots, R_4$	Radii of curvature
S_b	Bending stress

s	Length of deflection curve
T	Number of turns
V	Impact velocity
W_s	Weight of spring
x	Horizontal deflection
Y	Deflection
γ	Weight density
Δ	Diameter increase
τ	Torsional stress
τ_{max}	Maximum torsional stress
ψ	Slope

REFERENCES

1. H. Carlson, *Spring Designer's Handbook*, Marcel Dekker, New York, 1978.
2. B. J. Hamrock, B. Jacobson, and S. R. Schmid, *Fundamentals of Machine Elements*, WBC/McGraw Hill, New York, 1999.
3. J. E. Shigley, C. R. Mischke, and R. G. Budynas, *Mechanical Engineering Design*, 7th ed., McGraw Hill, New York, 2004.
4. M. F. Spotts and T. E. Shoup, *Design of Machine Elements*, 7th ed., Prentice Hall, Upper Saddle River, NJ, 1998.
5. R. C. Juvinall and K. M. Marshek, *Fundamentals of Machine Component Design*, 2nd ed., John Wiley & Sons, New York, 1991.
6. A. M. Wahl, *Mechanical Springs*, McGraw-Hill, New York, 1963.
7. Associated Spring Corporation, *Handbook of Mechanical Spring Design*, Associated Spring Corporation, Bristol, CT, 1964.
8. S. Timoshenko, *Theory of Elastic Stability*, McGraw Hill, New York, 1936, p. 69ff.
9. A. Blake, Analysis of buckling column spring with pivoted ends and uniform rectangular cross-section, *Transactions*, American Society of Mechanical Engineers, 1960, Paper 60-SA-10.

40 Irregular Shape Springs

40.1 INTRODUCTION

By varying the geometry of long flexible members, we can construct springs with shapes appropriate for special application. Since there is theoretically no limit to the number of shapes we can envision, there is no limit to the different shaped springs we can design. An exhaustive discussion of this is therefore beyond our scope. Instead, we limit our discussion to a few typical shapes useful in design and also helpful for development of similar components [1].

Predicting deflections and stresses of complex shaped springs has obvious advantages, as with the help of the calculations fewer test samples need be built in the shop before any final design is firmed up and mass production established.

In developing various analytical expressions, we assume that the external forces applied to a spring are delivered without shock. The cross-section areas are constant and only small deflections will be considered so that the principle of superposition remains valid.

40.2 SNAP RING

Figure 40.1 shows a snap-ring spring. It consists of flat strip of thickness h and width b shaped into a portion of a circle with radius R.

Due to symmetry, it is sufficient to analyze one half of the spring. At a typical point B, at an angle θ of the ring as in Figure 40.1, the bending moment M is

$$M = PR(\cos \alpha - \cos \theta) \tag{40.1}$$

where

P is the applied load
α is the half-angle of the spring opening as shown

From Equation 40.1, and also from Figure 40.1, we see that the maximum bending moment M_{max} occurs at C ($\theta = \pi$). That is,

$$M_{max} = PR(1 + \cos \alpha) \tag{40.2}$$

The maximum bending stress S_{bmax} is then

$$S_{bmax} = M_{max}(h/2)/I = \frac{6PR(1 + \cos \alpha)}{bh^2} \tag{40.3}$$

where, as before, I is the second moment of area $bh^3/12$.

When α increases, arc AC decreases until $\alpha = \pi$ and $\cos \alpha = -1$. The bending effect then disappears. Alternately, when α approaches 0, the maximum bending stress S_{bmax} approaches that of a split ring:

$$S_{bmax} = \frac{12PR}{bh^2} \tag{40.4}$$

Using Castigliano's theorem (see Section 3.7), the total deflection Y between A and D is

611

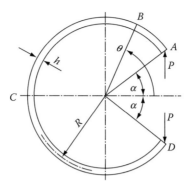

FIGURE 40.1 Snap-ring spring.

$$Y = \frac{2}{EI} \int_{\alpha}^{\pi} M \frac{\partial M}{\partial P} R \, d\theta$$

By substituting from Equation 40.1, we obtain

$$Y = \frac{PR^3}{EI} [(\pi - \alpha)(1 + 2\cos^2 \alpha) + 1.5 \sin 2\alpha] \tag{40.5}$$

Note that the foregoing expressions are also valid when the direction of the loading is reversed.

When α is zero, that is, for a snap-ring spring with a minute gap between points A and D, the deflection becomes

$$Y = \frac{3\pi PR^3}{EI} \tag{40.6}$$

where, as before, E is the elastic modulus.

Figure 40.2 shows a split ring subjected to pull along the vertical diameter AB. This configuration applies to the design of piston rings, and the gap at the split end is regarded to be rather small compared with the radius of curvature R. The design equations can be developed on the assumption

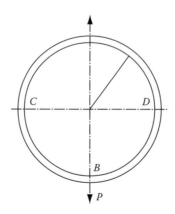

FIGURE 40.2 Snap-ring spring under diametral load.

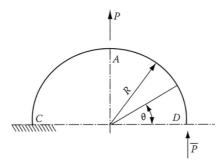

FIGURE 40.3 Arched cantilever model for split-ring analysis.

that the stresses are essentially those for a conventional straight member and the deflection is caused by the bending stresses alone.

The increase in opening at D is equal to twice the displacement of the free end D relative to C, as shown by the arched cantilever model in Figure 40.3. Since there is no force at point D, a fictitious force $\bar{P}$ is applied at the point and in the direction of the required deflection. The bending moment for the portion AD is then

$$M_1 = \bar{P}R(1 - \cos\theta) \tag{40.7}$$

Similarly, for the second quadrant we have

$$M^2 = PR\cos(\pi - \theta) + \bar{P}R[1 + \cos(\pi - \theta)] \tag{40.8}$$

Using Castigliano's theorem, the general expression for the deflection Y is

$$Y = \frac{2}{EI}\int_0^{\pi/2} M^1 \frac{\partial M_1}{\partial \bar{P}} R\,d\theta + \frac{2}{EI}\int_{\pi/2}^{\pi} M_2 \frac{\partial M_2}{\partial \bar{P}} R\,d\theta \tag{40.9}$$

where $\partial M_1/\partial \bar{P}$ is

$$\frac{\partial M_1}{\partial \bar{P}} = R(1 - \cos\theta) \tag{40.10}$$

and $\partial M_2/\partial \bar{P}$ is

$$\frac{\partial M_2}{\partial \bar{P}} = R(1 + \cos(\pi - \theta)) \tag{40.11}$$

Finally, by substituting into Equation 40.9 and setting $\bar{P}$ equal to zero, we find the displacement to be

$$Y = \frac{PR^3(4 + \pi)}{2EI} \tag{40.12}$$

40.3 SPRING AS A CURVED CANTILEVER

In some applications, we can use a curved cantilever spring as a loading or a support member. Figure 40.4 shows such a spring. One end of the spring can be considered fixed while the loaded end

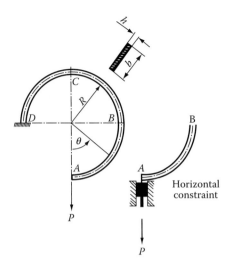

FIGURE 40.4 Three-quarter circular cantilever spring.

may be either free or constrained. For the case of the constrained end A, the bending moment M at any section is

$$M = PR \sin \theta \qquad (40.13)$$

The maximum bending stress S_{bmax} for this spring is then

$$S_{\text{bmax}} = \frac{M_{\text{max}}(h/2)}{I} \qquad (40.14)$$

where
  M_{max} is the maximum bending moment
  h is the spring thickness (see Figure 40.4)
  I is the second moment of area $bh^2/12$

From Equation 40.13 we see that M_{max} is PR. Therefore, S_{bmax} is

$$S_{\text{bmax}} = \frac{6PR}{bh^2} \qquad (40.15)$$

Using Castigliano's theorem (see Section 3.7), the displacement Y at the free end is determined as

$$Y = \int_0^{3\pi/2} \frac{M}{EI} \frac{\partial M}{\partial P} R \, d\theta = \frac{3\pi PR^3}{4EI} \qquad (40.16)$$

If the unsupported end AB of the spring is constrained by guides, to move in a vertical direction only as shown in Figure 40.4, the bending moment equation, Equation 40.13 must be modified by introducing a constraining moment M_f as a redundant and unknown quantity:

$$M = PR \sin \theta - M_f \qquad (40.17)$$

Due to the constraint, the rotation at end A is zero. Then by again using Castigliano's principle, we can obtain the constraining moment M_f from the expression

$$\int_0^{3\pi/2} M \frac{\partial M}{\partial M_f} d\theta = 0 \tag{40.18}$$

By substituting from Equation 40.17 into Equation 40.18 and integrating, we obtain M_f as

$$M_f = 2PR/3\pi \tag{40.19}$$

Hence, for the end constrained spring the bending moment of Equation 40.17 becomes

$$M = PR\left(\sin\theta - \frac{2}{3\pi}\right) \tag{40.20}$$

By again using Castigliano's theorem as in Equation 40.16, we obtain

$$Y = \frac{PR^3}{EI}\left(\frac{9\pi^2 - 8}{12\pi}\right) \tag{40.21}$$

The constraint reduces downward deflection by about 9%. However, this effect is not constant and depends on the angle subtending the spring arc. It may be of interest to note that the maximum bending stress in the spring is still given by Equation 40.15. The deflection can also be expressed as a function of stress for the free-end and the guided-end design conditions, giving

$$Y = \frac{3\pi R^2 S_{bmax}}{2Eh} \quad \text{(free end)} \tag{40.22}$$

Similarly, by eliminating P between Equations 40.15 and 40.21, we have

$$Y = \frac{(9\pi^2 - 8)R^2 S_{bmax}}{6\pi EH} \quad \text{(guided end)} \tag{40.23}$$

40.4 HALF-CIRCLE S-SPRING

Figure 40.5 shows a half-circle S-spring supporting a horizontal load P. Let the device have a rectangular cross section with width b and depth h.

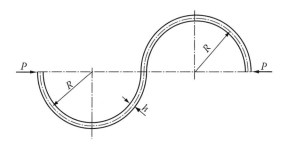

FIGURE 40.5 Half-circle S-spring.

Due to the symmetry, we can consider the spring to be an assemblage of four arched cantilevers. Therefore, the analyses of the foregoing section are directly applicable. Using these procedures, we find that the spring deflection Y in the direction of the applied load P is

$$Y = \frac{12\pi PR^3}{bEh^3}$$
(40.24)

where, as before,
   E is the elastic modulus
   R is the half-circle radius, as in Figure 40.5

Similarly, the maximum bending stress S_{bmax} is

$$S_{bmax} = \frac{6PR}{bh^2}$$
(40.25)

By eliminating P between Equations 40.24 and 40.25, we have

$$Y = \frac{2\pi R^2 S_{bmax}}{Eh}$$
(40.26)

40.5 THREE-QUARTER WAVE SPRING

Consider a three-quarter circular wave spring as shown by Figure 40.6. We can analyze portions AC and CF separately by writing the bending moment equations for points B and D, respectively. For point B, we see from Equation 40.24 that

$$M = PR\sin\theta$$
(40.27)

For point D, however, the bending moment equations should be modified to

$$M = PR(2 - \cos\theta)$$
(40.28)

Observe that at $\theta = 0$ for the CDF portion of the spring, $M = PR$; and at $\theta = \pi$, $M = 3PR$.
   The maximum bending stress S_{bmax} is

$$S_{bmax} = \frac{18PR}{bh^2}$$
(40.29)

Using Castigliano's theorem (see Section 37.2), we see that the displacement Y under load P can be obtained from the sum of the two component values for portions ABC and CDF as

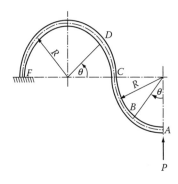

FIGURE 40.6   Three-quarter circular wave spring.

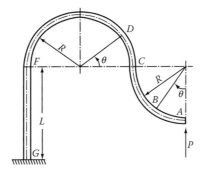

FIGURE 40.7 Three-quarter circular wave spring with extension.

$$Y = \frac{PR^3}{EI} \int\limits_0^{\pi/2} \sin^2\theta \, d\theta + \frac{PR^3}{EI} \int\limits_0^{\pi} (2 - \cos\theta)^2 d\theta \qquad (40.30)$$

where, as before,
  E is the elastic modulus
  I is the second moment of area $bh^3/12$

By carrying out the indicated integrations in Equation 40.30, we find Y to be

$$Y = \frac{19\pi PR^3}{4EI} \qquad (40.31)$$

If a straight portion of length L is added as shown in Figure 40.7, the maximum bending stress remains unchanged because the extension FG acts as a cantilever beam, built-in at G and loaded by a couple (ePR) at F provided that the direct stresses are ignored. The total deflection (or resilience) of the spring, however, is increased by the effect of the end couple on the cantilever so that the modified deflection formula becomes

$$Y = \frac{PR^3}{EI}\left(9k + \frac{19\pi}{r}\right) \qquad (40.32)$$

where $k = L/R$, as before, so that for $k = 0$, Equation 40.32 reduces to Equation 40.31. The direct extension of the portion FG under load P is ignored as being small relative to the deflection at point A.

The term "resilience" implies the elastic strain energy stored in the spring, which can be totally recovered upon the release of load P.

40.6 CLIP SPRING

Figure 40.8 shows the geometry and loading of a typical clip ring. We can use the result of Equation 40.32 to calculate the spread of a clip ring under loads P as in Figure 40.8. Specifically,

$$Y = \frac{2PR^3}{EI}\left(9k + \frac{19\pi}{4}\right) \qquad (40.33)$$

where the notation is the same as that of Section 40.5.
Recall from Equation 40.29 that the maximum bending stress S_{bmax} is

$$S_{\text{bmax}} = \frac{18PR}{bh^2} \qquad (40.34)$$

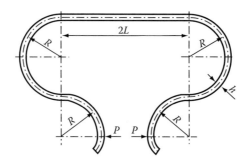

FIGURE 40.8   A clip ring.

By combining Equations 40.33 and 40.34, we can express the maximum bending stress S_{bmax} in terms of the spring deflection Y and the spring dimensions as

$$S_{bmax} = \frac{hYE}{R^2(12K + 19.9)} \tag{40.35}$$

where, as before,
  E is the elastic constant
  k is the dimension ratio L/R

40.7   GENERAL U-SPRING

Figure 40.9 shows the geometry and loading of a general U-spring. We can develop design formulas for this component by treating it as a joined, curved cantilever beams as discussed extensively by Blake [2].
  Specifically, in Figure 40.9 the component force P_1 acting perpendicular to the straight leg is

$$P_1 = P \cos \nu \tag{40.36}$$

where ν is the angle shown in Figure 40.9 (not Poisson's ratio, as often designated by ν).
  If Y_1 is the displacement of the loaded end A in the direction of P_1 and Y, the total change in the distance AB due to P, then Y_1 and Y are related as

$$Y_1 = 0.5Y \cos \nu \tag{40.37}$$

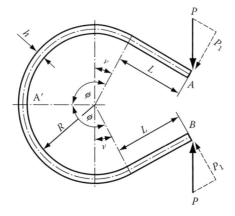

FIGURE 40.9   A general U-spring.

From previous results documented by Blake [2], we have

$$Y_1 = \frac{P_1 R^3}{EI} F(k, \phi) \tag{40.38}$$

where, as before, k is the ratio L/R; ϕ is the angle shown in Figure 40.9; and from Blake [2] $F(k, \phi)$ is

$$F(k, \phi) = 0.33k^3 + k^2\phi + 2k(1 - \cos\theta) + 0.5\phi - 0.25\sin 2\phi \tag{40.39}$$

Thus from Equations 40.37, 40.38, and 40.39, Y is

$$Y = \frac{2PR^3}{EI} F(k, \phi)/\cos\nu \tag{40.40}$$

The maximum bending stress S_{bmax} at the arch point A' is found to be

$$S_{bmax} = \frac{6PR}{bh^2}\left(1 + \sin\nu + \frac{k}{\cos\nu}\right) \quad (\phi \geq \pi/2) \tag{40.41}$$

and

$$S_{bmax} = \frac{6PR}{bh^2}(1 - \cos\phi + k\cos\nu) \quad (\phi < \pi/2) \tag{40.42}$$

For the special case of $\nu = 0$ (and thus $\phi = \pi/2$) Equations 40.41 and 40.42 both reduce to

$$S_{bmax} = \frac{6PR(1 + k)}{bh^2} \tag{40.43}$$

40.8 INSTRUMENT TYPE U-SPRING

Designers use U-springs in many applications and especially, in instruments and precision equipment. We can manufacture U-springs with relatively simple machinery, and assembly is easy. U-springs are generally loaded in the plane. They can be used as both tension and compression devices.

Figures 40.10 through 40.12 depict the three most common types of U-springs.

Table 40.1 provides formulas for the maximum moment and deflection of these three spring types.

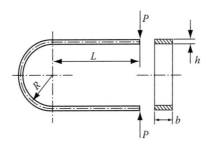

FIGURE 40.10 Flat U-beam/spring.

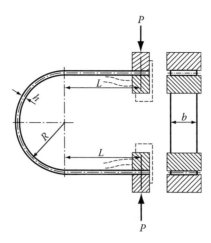

FIGURE 40.11   U-beam/spring with ends fixed as to slope.

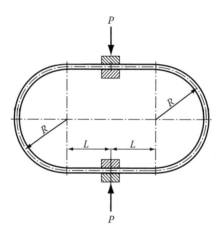

FIGURE 40.12   Double U-beam/spring.

TABLE 40.1

Moment and Deflection Formulas for Three Common U-Springs

Type of Spring	Moment	Deflection
Single U-spring without constraint	$M = PRC_3$	$Y = \dfrac{PR^3}{EI}C_1$
Single U-beam with constraint	$M = PRC_4$	$Y = \dfrac{PR^3}{EI}C_2$
Double U-beam	$M = 0.5\,PRC_4$	$Y = \dfrac{PR^3}{2EI}C_2$

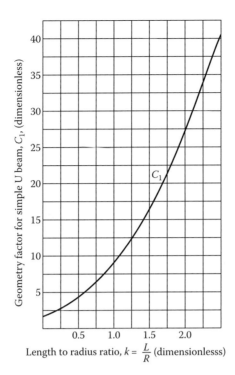

Geometry factor for simple U beam, C_1 (dimensionless)

Length to radius ratio, $k = \dfrac{L}{R}$ (dimensionlesss)

FIGURE 40.13 Deflection factor for U-springs without constraint.

These formulas contain design factors $C_1, \ldots, C_4$, which in turn are provided by Figures 40.13 through 40.16.

40.9 SYMMETRICAL WAVE SPRING

We conclude with a brief discussion about the symmetrical wave spring which has a variety of applications in machine design. Figure 40.17 depicts the device and its loading. Due to the symmetry, we can analyze the structure by considering the cantilever portion ABC as fixed at C and loaded by $P/2$ at end A.

Assuming that contact friction between the supporting surface and the end of the spring at A is negligible, the bending moment equations can be stated in customary terms:

$$M_1 = \frac{PR\sin\theta}{2} \quad \text{and} \quad M_2 = \frac{PR}{2}(2\sin\phi - \sin\theta) \tag{40.44}$$

where Figure 40.17 shows the notation. In the figure, M_1 is the bending moment from the load $(P/2)$ at A in the spring section from A to B. Correspondingly, M_2 is the bending moment from the load at A in the section from B to C.

By again using Castigliano's theorem (see Section 37.2), the displacement Y at A relative to C may be computed from the expression:

$$Y = \int_0^\phi M_1 \frac{\partial M_1}{\partial P} R \, d\theta + \int_0^\phi M_2 \frac{\partial M_2}{\partial P} R \, d\theta \tag{40.45}$$

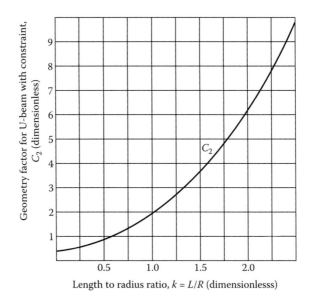

FIGURE 40.14   Deflection factor for U-springs with constraint.

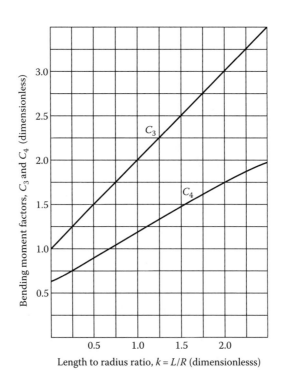

FIGURE 40.15   Bending factors for U-springs.

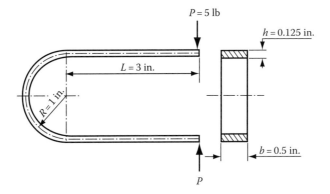

FIGURE 40.16 Flat U-spring (Design problem 26.2).

where from Equation 40.44, $\partial M_1/\partial P$ and $\partial M_2/\partial P$ are

$$\frac{\partial M_1}{\partial P} = \frac{R\sin\theta}{2} \quad \text{and} \quad \frac{\partial M_2}{\partial P} = \frac{2(\sin\phi - \sin\theta)}{2} \tag{40.46}$$

By introducing Equations 40.44 and 40.46 into Equation 40.45 and integrating, we find the displacement Y to be

$$Y = \frac{PR^3}{EI}(0.25\phi + \phi\sin^2\phi - \sin\phi + 0.375\sin 2\phi) \tag{40.47}$$

By varying ϕ, we can obtain a number of interesting spring configurations. Equation 40.47 then provides the spring constants.

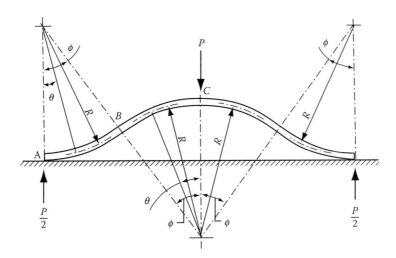

FIGURE 40.17 Symmetrical wave spring.

SYMBOLS

a	Length
b	Width of rectangular section
c	Length
C_1, C_2, C_3, C_4	Moment and deflection factors for U-springs
d	Length
E	Modulus of elasticity
$F(k, \phi)$	Function of K and ϕ (see Equation 40.39)
H	Horizontal load
h	Depth of cross section
I	Second moment of area
K_1	Deflection factor for arched cantilever
$k = L/R$	Straight length-to-radius ratio
L	Straight portion of spring
M, M_1, M_2	Bending moments
M_f	Fixing moment
M_{max}	Maximum bending moment
P	Vertical load
$\bar{P}$	Fictitious vertical load
P_1	Component of vertical load
R	Mean radius of curvature
S_b	Bending stress
S_{bmax}	Maximum bending stress
x	Arbitrary distance
X	Horizontal deflection
Y	Vertical deflection
Y_1	Component of vertical deflection
α	Half-angle of snap spring
β	Angle at which load is applied
θ	Angle at which forces are considered
ν	Auxiliary angle in U-springs
ϕ	Angle subtended by curved portion

REFERENCES

1. A. Blake, Complete flat springs, *Product Engineering*, 1961.
2. A. Blake, *Practical Stress Analysis in Engineering Design*, 2nd ed., Marcel Dekker, New York, 1990 (Chap. 25).

Index